AF826940

Amazon Pollen
Manual and Atlas

Manual e Atlas
Palinológico da Amazônia

Amazon Pollen Manual and Atlas

Manual e Atlas Palinológico da Amazônia

Paul Colinvaux

Marine Biological Laboratory
Woods Hole, Massachusetts, USA
formerly at
Smithsonian Tropical Research Institute
Balboa, Republic of Panama

Paulo Eduardo De Oliveira

Department of Botany
Field Museum of Natural History
Chicago, Illinois, USA
and
Departamento de Paleontologia e Estratigrafia
Instituto de Geociências
Universidade de São Paulo
São Paulo, Brasil

and

Jorge Enrique Moreno Patiño

Smithsonian Tropical Research Institute
Balboa, Republic of Panama

CRC Press
Taylor & Francis Group
Boca Raton London New York

CRC Press is an imprint of the
Taylor & Francis Group, an informa business

CRC Press
Taylor & Francis Group
6000 Broken Sound Parkway NW, Suite 300
Boca Raton, FL 33487-2742

© 1999 by Taylor & Francis Group, LLC
CRC Press is an imprint of Taylor & Francis Group, an Informa business

No claim to original U.S. Government works

This book contains information obtained from authentic and highly regarded sources. Reasonable efforts have been made to publish reliable data and information, but the author and publisher cannot assume responsibility for the validity of all materials or the consequences of their use. The authors and publishers have attempted to trace the copyright holders of all material reproduced in this publication and apologize to copyright holders if permission to publish in this form has not been obtained. If any copyright material has not been acknowledged please write and let us know so we may rectify in any future reprint.

Except as permitted under U.S. Copyright Law, no part of this book may be reprinted, reproduced, transmitted, or utilized in any form by any electronic mechanical, or other means, now known or hereafter invented, including photocopying, microfilming, and recording, or in any information storage or retrieval system, without written permission from the publishers.

For permission to photocopy or use material electronically from this work, please access www.copyright.com (http://www.copyright.com/) or contact the Copyright Clearance Center, Inc. (CCC), 222 Rosewood Drive, Danvers, MA 01923, 978-750-8400. CCC is a not-for-profit organization that provides licenses and registration for a variety of users. For organizations that have been granted a photocopy license by the CCC, a separate system of payment has been arranged.

Trademark Notice: Product or corporate names may be trademarks or registered trademarks, and are used only for identification and explanation without intent to infringe.

Visit the Taylor & Francis Web site at
http://www.taylorandfrancis.com

and the CRC Press Web site at
http://www.crcpress.com

CONTENTS

PREFACE

This is both a pollen manual and an atlas. The Manual, Part I, is a practical guide to researching the Amazon past by pollen analysis, with discussions of the theory of tropical pollen analysis followed by "how to" chapters. The Atlas, Part II, is an illustrated guide to pollen found in Amazon sediments.

The Manual begins with our assessment of pollen analysis as applied to the Amazon. We find the Amazon pollen record both rich and informative, although its interpretation requires knowledge of the ways in which tropical pollen signals differ from those in temperate latitudes where pollen analysis was invented. We discuss both the interpretation of Amazon pollen data and the use of surface sampling. Chapter 2 reviews the present state of knowledge of the history of Amazon climate and diversity, although much of Amazon history is still controversial or completely unknown.

Chapter 3 describes how to core a lake and chapter 4 describes standard laboratory procedures. Coring sediments under the open water of lakes seems forbidding to some who have not yet tried it, but it is in fact a simple exercise in drilling that is easy to learn. We show how the coring is done, what equipment is needed to core, and how to procure the equipment.

Likewise, we provide basic laboratory protocols: We begin with subsampling of sediments, then cover pollen extraction, pollen counting, and the presentation of data. For both field and laboratory work, we provide an appendix listing sources of supply, from rubber boats through coring equipment and radiocarbon dating, to exotic pollen markers and software. The final chapter of Part I is a condensed account of pollen and spore morphology, written as a practical guide for the working paleoecologist.

The Atlas, which has its own introduction, provides keys, descriptions, and photomicrographs of the pollen taxa most likely encountered in the Amazon rain forest. Our list of 421 pollen taxa seems humble when compared with the more than 80,000 species of vascular plants so far described for the Amazon basin. Yet this list, mostly identified only to family or genus, is sufficient to carry a profound signal of forest communities and climate. The number is far larger than is normally identified in temperate pollen diagrams. Our reduction of the vast Amazon flora to a few hundred identifiable taxa should make Amazonian palynology accessible to the next generation of researchers.

A special feature of the Atlas is that the photomicrographs are printed to the scale of one millimeter to one micron, which means that dimensions can be read off directly with an ordinary metric ruler.

This Manual and Atlas are the result of many years' work. De Oliveira and Colinvaux first went coring together in Brazil in 1988, taking advantage of the many years of prior experience of Colinvaux's Ohio laboratory on lake coring and pollen analysis in Ecuador. Meanwhile, Moreno had compiled a pollen atlas for the tropical forest of Barro Colorado Island, the Smithsonian research site in Panama. When we three were brought together at the Smithsonian Tropical Research Institute for our studies of the environmental history of the Amazon rain forest, the writing of this book was a natural outcome.

Many people and organizations contributed to our efforts. The Amazon fieldwork of our laboratories began in Ecuador in association with Drs. Miriam Steinitz-Kannan, Michael C. Miller, Kam-biu Liu, Mark B. Bush, Melanie Riedinger and Mark Schutzow, and with the encouragement of many institutions and people in Ecuador. For the Brazilian fieldwork we thank the Brazilian Research Council, CNPq and the Brazilian Institute for Natural Resources (IBAMA) for permission to collect the Lake Pata sediments, in collaboration with Dr. Elena Franzinelli of the University of Amazonas in Manaus at the Parque National do Pico da Neblina.

We thank Dr. Peter Crane of the Field Museum of Natural History in Chicago for his enthusiastic support and provision of a laboratory for the Lake Pata pollen analysis. Our pollen reference collection was built using the herbaria of the Field Museum, the New York Botanical Gardens (with particular help from Dr. Douglas Daly), and the U.S. National Herbarium. Without these great collections, a work of this kind could not be attempted. We thank particularly Dr. Maria Léa Salgado-Labouriau, University of Brasilia, and Dr. Kenitiro Suguio, Institute of Geosciences of the University of São Paulo, for their continued encouragement of our paleoecological studies in Brazil.

Mark B. Bush initiated and developed the program of pollen trapping in the Amazon that made possible the naming of many pollen taxa that were up to then unknown. Eduardo Asanza monitored pollen trapping at the Cuyabeno Field Station in Ecuador.

Editing and revision of the Portuguese text was by Maria Cleonice L. Sanches. Maria Josefa Villar was invaluable as the trilingual administrative assistant for the project. Salomon Vergara and Marcos Guerra assisted with the production of pollen plates. Eduardo Montenegro worked with artist Alberto Castillo to produce the illustrations for Part I.

Our Amazon research has been possible only because of generous support from the U.S. National Science Foundation (NSF), the National Geographic Society, the U.S. National Oceanographic and Atmospheric Administration (NOAA), and the Andrew Mellon Foundation.

PREFÁCIO

Este livro é, ao mesmo tempo, um manual e um atlas palinológico. O Manual — Parte I — é um guia às técnicas palinológicas adequadas ao estudo do passado da Amazônia e contém discussões sobre a teoria da análise palinológica, seguido de capítulos que ensinam ao leitor os procedimentos básicos envolvidos nessa pesquisa. O Atlas — Parte II — é um guia ilustrado dos tipos polínicos encontrados em sedimentos quaternários da Amazônia.

Começamos o Manual com a nossa interpretação da análise palinológica aplicada à Amazônia. O registro palinológico da Região Amazônica é rico e informativo. Entretanto, sua interpretação requer conhecimentos para diferenciar seus sinais polínicos, daqueles encontrados nas regiões de clima temperado, onde foi criada a análise palinológica. Discutimos não só a interpretação dos dados palinológicos da Amazônia, como também o uso de amostragem de sedimentos superficiais. No Capítulo 2 apresentamos uma revisão do estado atual dos conhecimentos da história climática do Quaternário Tardio e da diversidade da Região Amazônica, embora, ainda, muito controverso e desconhecido.

No Capítulo 3 descrevemos o procedimento de coleta de sedimentos lacustres e no Capítulo 4, a metodologia palinológica padrão. Embora a coleta de sedimentos em lagos possa parecer difícil para aqueles que ainda não a experimentaram, sua realização é um simples exercício que pode ser facilmente aprendido. Mostramos o equipamento básico necessário para a remoção de sedimentos e como ela é feita, onde conseguir o equipamento e como usá-lo.

Da mesma forma, detalhamos a metodologia básica para o trabalho de laboratório. Começamos com a sub-amostragem dos sedimentos, seguida da extração do pólen, sua contagem e apresentação dos dados. Para ambos os trabalhos de campo e laboratório, fornecemos, no Apêndice, uma lista de fontes de materiais que inclui: barcos infláveis, equipamento coletor de sedimentos, pólen e esporos exóticos marcadores e programas computacionais empregados nesse estudo, bem como informações sobre laboratórios para datação radiocarbônica. O capítulo final da Parte I é um resumo simplificado da morfologia de pólen e esporos, escrito com o objetivo de ser, apenas, um guia prático para o palinólogo.

O Atlas, que possui sua própria introdução, contém chaves, descrições e fotomicrografias dos táxons polínicos mais prováveis de serem encontrados na floresta amazônica. Nossa lista de 421 táxons é humilde quando comparada às 80.000 ou mais espécies de plantas vasculares descritas, até o momento, para a Bacia Amazônica. Todavia, essa lista, cujos táxons são identificados apenas a nível de família e gênero, é suficiente para identificar nos sedimentos, sinais climáticos e de comunidades florestais. O número desses táxons é amplamente superior àquele identificado em diagramas polínicos das regiões de clima temperado. Nossa redução da vasta flora amazônica a algumas centenas de táxons identificáveis, deve tornar a palinologia da Amazônia mais acessível à nova geração de pesquisadores.

Um aspecto especial do Atlas refere-se ao fato de que as fotomicrografias estão impressas em uma escala de 1 milímetro para 1 mícron, o que permite a leitura direta das dimensões com uma simples régua métrica.

O Manual e o Atlas são o resultado de muitos anos de trabalho. De Oliveira e Colinvaux coletaram sedimentos no Brasil, já em 1988, aproveitando-se dos muitos anos de experiência do laboratório de palinologia de Colinvaux, em Ohio, EUA e da coleta de sedimentos e de análises palinológicas no Equador. Enquanto isso, Moreno produzia um atlas palinológico para a floresta tropical da ilha de Barro Colorado, Panamá, local da base das pesquisas tropicais da Instituição Smithsonian. Escrever este livro foi um resultado natural da nossa união para estudar a história ambiental da floresta amazônica no Instituto de Pesquisas Tropicais Smithsonian, pertencente à Instituição Smithsonian.

Muitas pessoas e organizações contribuíram para este projeto. O nosso laboratório iniciou o trabalho de campo na Amazônia em associação com os doutores Miriam Steinitz-Kannan, Michael C. Miller, Kam-biu Liu, Mark B. Bush, Melanie Riedinger e Mark Schutzow e com o apoio de muitas instituições e do povo equatoriano. Pelo trabalho de campo no Brasil, agradecemos o Conselho Nacional de Desenvolvimento Científico e Tecnológico-CNPq e o Instituto Brasileiro do Meio Ambiente e dos Recursos Renováveis-IBAMA, pela permissão de coleta dos sedimentos do Lago da Pata, no Parque Nacional do Pico da Neblina, em colaboração com a Dra. Elena Franzinelli da Universidade do Amazonas, em Manaus.

Dr. Peter Crane, do Museu Field de História Natural de Chicago, EUA pelo seu apoio entusiástico e pelo fornecimento do laboratório para a análise palinológica dos sedimentos do Lago da Pata. Nossa coleção de referências palinológicas foi montada usando plantas dos herbários do Museu Field em Chicago, do Herbário Nacional dos Estados Unidos em Washington, DC e do Jardim Botânico de Nova York, com ajuda especial do Dr. Douglas Daly. Sem essas grandes coleções, o trabalho presente não poderia ter sido realizado. Agradecemos à Dra. Maria Léa Salgado-Labouriau, da Universidade de Brasília e ao Dr. Kenitiro Suguio, do Instituto de Geociências da Universidade de São Paulo o apoio e estímulo aos nossos estudos no Brasil.

Mark B. Bush iniciou e desenvolveu o programa de amostragem da chuva polínica na Amazônia, o que possibilitou a identificação de muitos tipos polínicos, que até então permaneciam desconhecidos. Eduardo Asanza fez o monitoramento dos coletores de chuva polínica na Estação de Cuyabeno, Equador. O trabalho de ambos enriqueceu nossa pesquisa.

A revisão do texto em Português ficou a cargo de Maria Cleonice L. Sanches. O trabalho de Maria Josefa Villar, como assistente-administrativa trilingüe do projeto, foi de inestimável valor. Salomon Vergara e Marcos Guerra participaram da produção das pranchas polínicas. Eduardo Montenegro trabalhou com o artista Alberto Castillo nas ilustrações da Parte I.

Nossa pesquisa na Amazônia tem sido possível graças ao generoso apoio da Fundação de Ciências dos Estados Unidos (NSF), da Sociedade Geográfica Nacional (NGS), da Administração Nacional Oceanográfica e Atmosférica dos Estados Unidos (NOAA), e da Fundação Andrew Mellon, EUA.

Part I

Amazon Pollen Manual

CHAPTER

1

NEOTROPICAL POLLEN ANALYSIS

The pollen record of the neotropical flora is potentially rich, able to yield an history not only of vegetation change (and hence of climate change) but also of community dynamics. Yet for a long time pollen analysts neglected the region — the Amazon in particular. This was due partly to the forbidding size of the pollen flora, or practical difficulties of working in tropical forests. More important, however, was the fact that pollen analysis was a creature of the north temperate zone. The first palynologists, like most ecologists, worked in research universities placed by historical accident in a thin band round the northern hemisphere between latitudes 40° and 55° north. Pollen analysis was invented to study the late Pleistocene forest history in these northern latitudes.

The northern forests have some peculiar properties that simplified the work of palynologists. Species diversity is low — so low that a whole country may be home to at most only a few tens of tree species (in the island of Britain, for instance, there are only a dozen native trees). Nearly all these northern trees are pollinated by wind, and their pollen production is visibly tremendous, such that a parked car can turn yellow with pine (*Pinus*) pollen in a few hours. Abundant herbs like grasses (Gramineae/Poaceae) or sedges (Cyperaceae) are also wind-pollinated, and their pollen production is so large that many people in the north temperate belt suffer the allergy disease known as 'hay fever' from breathing the air in which the pollen grains float.

Wind blows this pollen over a wide area so that in the course of a year, a local 'pollen rain' represents the vegetation of a considerable stretch of country. Concentration of the dead husks of this pollen rain in the sediments of lakes and bogs is high. In these circumstances, a signal both for climatic change and community composition is given by relatively simple percentage counts of the few species of trees and herbs that make up the bulk of the annual pollen rains preserved in the sediments.

In the Amazon forest, however, pollen production and dispersal are very different. Species diversity is vastly greater — so great, in fact, that three hundred species of tree have been counted in a single hectare (Gentry, 1988). Moreover, the vast majority of Amazonian trees do not loose pollen into the air for the wind to carry, but rather rely on animals — insects, birds, or bats — to collect and distribute their pollen. These facts were for a long time daunting to pollen analysts who were used to the assumptions of wind dispersed pollen in simple forests and questioned the viability of pollen analysis in tropical forests (Faegri and van der Pilj, 1979).

Insect-pollinated trees of the Amazon forest should, according to this pessimistic view, produce few pollen grains, just enough to load their insect vectors. Of this sparse pollen production, surely none would float free in the wind to make its way to lake or bog? Thus, to a pollen analyst trained in temperate latitudes as they all were, the Amazon forest should be filled with 'silent trees' that would leave no record of pollen in sediments. And the sediments themselves were expected to be pollen-poor, because they were denied vast influxes of windblown pollen. The logical result of these fears should be a dearth of pollen to be analyzed, and what there was might give so distorted a view of the parent vegetation as to be nearly useless. Yet both these fears have proven to be unfounded — the truth being that Amazon lake sediments have ample pollen, and that this can give a remarkably clear picture of local forest or other communities.

AMAZON POLLEN PRODUCTION

Amazon lake sediments consistently hold large concentrations of pollen. In ancient Lake Pata in the lowland rainforest of northwest Brazil, pollen concentrations vary about a mean of 247,600 grains/ml^{-3}; in Lake Kumpak[a] in the Oriente Province of Ecuador, which receives sediment from a larger watershed, median pollen concentrations are about 100,000 grains/ml^{-3}.[1] These concentrations compare favorably with pollen concentrations in lake sediments from Europe or the midwest of the United States (e.g., *ca.* 100,000 grains/ml^{-3} in Mirror Lake in New Hampshire (Davis *et al.*, 1984).

That pollen is produced copiously in the Amazon forest is also shown by our program of trapping airborne pollen from within the forest itself. We set out simple traps (described in Chapter 3) in batteries of 10–100 in forest plots of known species composition: near Manaus in the Central Amazon,[2] in the western Amazon of Ecuador, and in other neotropical forests such as Barro Colorado Island in Panama. We collect the traps at 6-month or yearly intervals which lets us calculate pollen influx as grains/cm^{-2}/year: typically, our traps collect pollen at between 7,000 and 30,000 grains/cm^{-2}/year (Bush, 1992). These results are comparable to pollen influx measures into lake sediments in temperate forests of the U.S. (10–15,000 grains/cm^{-2}/year at Mirror Lake (Davis *et al.*, 1984).

Yet the Amazonian pollen assemblagies are qualitatively different from temperate pollen. They are, reasonably enough, far more diverse, with total taxon-richness in one trap or sediment sample many times what would be found in their north temperate equivalents. But this high taxon-richness is comprised of taxa that appear in low numbers — often just one grain identified in a pollen count. When added together, however, these rare pollen types can make up a significant percentage of the total pollen. What this suggests is that the extremely high species richness of animal-pollinated trees in the Amazon forest is represented in the pollen rain by a comparable richness of pollen types, each at low concentration.

Results from our pollen traps do confirm one of the presumptions of temperate zone pollen analysts, that animal dispersed pollen does not usually get carried in the wind. We found such pollen up to 50 m from the nearest parent tree, but no further. Most fell into traps that were directly under the parent trees. And yet many of these pollen types are sometimes found in lake sediments. The conclusion seems inescapable that the pollen is carried from tree to lake in the flow of surface water. Rainstorms knock pollen or flowers to the ground, and surface runoff then moves the pollen across the catchment to the sedimentary sink within the lake.

Therefore, the pollen signal of these animal-pollinated plants records only the community actually growing within the watershed or catchment of the local drainage. To this extent, the fears of classical pollen analysts are justified: the pollen of animal-pollinated trees does not give a regional signal because it records only local forest communities. For studies of forest communities, however, this washing of pollen only from the immediate watershed has the powerful advantage of potentially allowing detailed historical studies of forest associations within the watershed itself.

Thus the great majority of Amazon trees that are pollinated by animals are not 'silent' in the pollen record but instead have the potential to provide detailed local records. Their total pollen production is surprisingly high. This could possibly reflect competition for insect vectors in the forest in which the pollen itself is part of the 'bait' or 'reward' to the transporting insect. This overproduction of pollen, however, might just as well be a simple function of flowering structures that expose both flowers and pollen bearing anthers to the environmental vicissitudes of storm winds and falling rain. Anecdotal evidence in support of this idea is that we have never found orchid pollen in sediments or traps. Perhaps the refined pollinating mechanisms of orchids avoid the pollen losses suffered by other plants, thus denying to pollen analysts the signal given by multitudes of redundant and lost pollen grains.

Amazon ecosystems also sometimes produce large quantities of pollen that can be dispersed by wind, even in the forests. The families Moraceae, Urticaceae, Ulmaceae and Myrtaceae are represented in the lowland forests by many such species and their pollen production can be both large and widely dispersed. The tiny pollen of the moraceous weed tree *Cecropia*, for instance, can form 40% of the pollen rain in some Amazonian lakes (Bush and Colinvaux, 1988).

Other Amazon tree genera appearing in different forest or woodland types are also wind-pollinated, particularly the tree genera *Ilex*, *Podocarpus*, *Bursera*, and shrubs of the family Euphorbiaceae. And most important, grasses (Gramineae/Poaceae), together with other herbs such as Cyperaceae, are producers of copious quantities of windblown pollen in the Amazon basin as elsewhere.

Pollen production in Amazon ecosystems, therefore, is ample for the purposes of pollen analysis, but Amazonian pollen signals must be read with awareness that they are not like those of North Temperate forests where pollen methods were developed. The Amazon yields an highly diverse record of forest communities within watersheds, but regional records must be inferred by combining data for strictly local forest communities with regional records provided by the small subset of genera whose pollen is blown about in the wind.

INTERPRETING AMAZON POLLEN SPECTRA

We start with the pollen percentage diagram from Lake Pata in the lowland rain forest of the Brazilian Amazon given as Figure 1.1. Lake Pata occupies a closed basin without significant inlet or outlet streams, probably a pseudokarst basin formed by the slow solution of silicates from the parent plutonic rocks. The lake lies at about 300 m elevation on a low inselberg rising above an immense stretch of lowland plains, at roughly 100 m elevation and covered with lowland tropical rain forest. From a vantage point near Lake Pata, one can look out over unbroken forest as far as the eye can see. The inselberg itself is also covered with tropical forest, though somewhat stunted and modified in places because of shallow soils on the hard rocks that have resisted weathering, forming an edaphically constrained forest. The small watershed of Lake Pata is of course vegetated with the local facies of tropical lowland forest, though an apparent absence of the rock outcrops seen elsewhere on the inselberg suggests that the forest community in the small Lake Pata watershed itself has minimal edaphic constraint. A Livingstone piston core shows the sediments to be nearly 7 m thick, with the top 1.7 m spanning the last 40,000 radiocarbon years. Pollen analysis of this top section provides the first history of the lowland Amazon forest through the last glacial cycle (Colinvaux *et al.*, 1996a).

The upper pollen spectra of Figure 1.1 illustrate the pollen signal for lowland Amazon forest of the kind now present in the area. Four things are apparent about this part of the diagram:

1. The prominence of wind-dispersed taxa (Moraceae, Urticaceae and Euphorbiaceae (*Alchornea*).
2. The presence of significant amounts of one or two taxa of animal-pollinated trees (*Mauritia* [Palmae/Arecaceae]; *Copaifera* [Leguminosae/Caesalpinioideae]).
3. When all the tree pollen, including the low counts of grains found as a trace (listed as 'other arboreal') are added together, the total tree (arboreal) pollen is nearly 90%.
4. The very low count of herb pollen, particularly the almost complete absence of Gramineae.

Thus the essential pollen signal from deep in a large expanse of lowland tropical rain forest in Amazonian Brazil is given by tree pollen, with negligible grass or other herbs.

The essential properties of the Lake Pata pollen percentage diagram characterize the older parts of the pollen diagram also, showing that lowland tropical forest has occupied the site for at least the last 40,000 years, through both glacial and postglacial time. This important conclusion is, of course, derived from only a tiny portion of the rich diversity in the forest, but the expected watershed record of pollen from animal-pollinated trees delivered to the lake in runoff is also present. Their presence is recorded in the taxon listed as 'other arboreal,' which includes such tropical forest genera as *Apeiba, Cedrela, Cordia, Didymopanax, Guarea, Iriartea, Luehea,* etc.

On the Lake Pata Animal-Pollinated Trees

The taxon *Mauritia* is prominent in the Holocene portion of the section, but not earlier. *Mauritia* is a palm of wet or marshy ground. It appears to be animal-pollinated, but stands of the trees are very common in marshy areas at the edges of lakes so that their pollen tends to be present in most Amazonian lacustrine pollen diagrams. There is an extensive stand of *Mauritia* beside modern Lake Pata and the pollen diagram suggests that this has been present for all the 10,000 years of Holocene time. Before 10,000 years ago, however, the *Mauritia* stand was not present. This is a purely local record of a patch of wetland within the watershed that might or might not record changed climate — increased precipitation could create a swamp for *Mauritia*, but so could impoundment of drainage or rising of the water table. The coincidence of the *Mauritia* rise at Lake Pata with a stratigraphic change in the sediments and the onset of postglacial time gives strength to the hypothesis that this particular swamp was associated with the end of the last glaciation as increased precipitation raised the water level creating the swamp beside the lake. An alternative consequence of the end of glacial times, however, is simply that the local water tables were raised consequent on the rise of sea level to its present height and that this raised water table flooded the ground that was to become a *Mauritia* swamp.[3]

Persistence of tropical forest both before and after the establishment of the *Mauritia* swamp shows that any change in the rainfall regime cannot have been major. Thus the record of *Mauritia* pollen strongly suggests that mean lake level rose at the end of glacial time to flood the shallow portion of the basin now occupied by the *Mauritia* swamp, but in and of itself it gives no indication of the cause of lake level rise.

In other Amazon deposits, a rise of *Mauritia* pollen is clearly more due to ecological process than climatic change. This is so in those Amazon pollen histories that are essentially records of infilling of old basins or river channels. In these deposits, the pollen data record the establishment of *Mauritia* stands on the wetlands after these have been sufficiently drained by infilling or stream down-cutting.

The taxon *Copaifera* accounts for 10% of total pollen in the mid-Holocene. *Copaifera* is a genus of large forest trees in the family Caesalpinioideae (Leguminosae). Twenty-five species are known from the neotropics, any one of which might be represented in the Lake Pata pollen diagram. Our pollen data suggest a local *Copaifera* population event in the watershed lasting many tree lifetimes and starting about 6,000 years ago. This record illustrates the kind of detail inherent in Amazonian pollen diagrams.

Other animal-pollinated tree taxa appear regularly, though in the low percentages expected for most animal-pollinated trees: Caesalpinioideae, Lecythidaceae, *Tapirira* (Anacardiaceae). Of these, *Tapirira* is a genus only of lowland forest trees. With the exception of *Copaifera,* we have not yet been able to identify the pollen from most Caesalpinioideae and Lecythidaceae below family rank, but both families are well represented among trees of the lowland rain forest, and Caesalpinioideae includes forest lianas. Together with *Tapirira,* the presence of these familial taxa demonstrates the presence of lowland rainforest trees actually growing within the small Lake Pata watershed from whence their pollen could be washed into the lake.

Cecropia and Melastomataceae: The Generalist Problem

Cecropia is a genus of weed trees in the generally wind-pollinated family Moraceae. The trees could be called 'early colonizers,' 'opportunist species,' 'fugitive species' or even 'r-strategists' in the various English usages of theoretical ecology. They are present across the whole range of reasonably moist vegetation types in the neotropics, from the lowlands to at least 2000 m elevation. Pollen production is very heavy and the characteristic, tiny (6–14 micron, biporate) pollen grains appear in virtually all neotropical pollen diagrams, even those from the high Andes (4000 m) of Peru where they have apparently been blown up the mountains from below (Hansen *et al.*, 1984). The presence of a modest percentage of *Cecropia* pollen against a background signal that is clearly that of lowland tropical forest as in the lake Pata diagram is to be expected as the normal presence of gap-colonizers in the forest. Similar and relatively low *Cecropia* percentages in other vegetation types, however, should be treated with caution as possibly being due to long distance transport (as in the Peruvian Andes and in dry regions). Because lake margins are by nature open and disturbed, it is likely that *Cecropia* will sometimes be overrepresented in lake sediments. In crater Lake Ayauch[i] in the lowland rainforest of Ecuador, *Cecropia* pollen was present at more than 70% of total pollen, possibly from the lake margin effect (Bush and Colinvaux, 1988). Large percentages of *Cecropia* can also be expected in the early stages of infilling of hydrarch successions, as in abandoned river channels when dense stands of *Cecropia* develop as early colonists. Some mudflat or sand-flat islands in the Amazon system support nearly monospecific stands of *Cecropia,* which should surely result in heavy overrepresentation of *Cecropia* in the local pollen rain.

The ubiquity of *Cecropia* pollen in many neotropical vegetation or community types is but a special instance of a widespread problem. The broad taxa — often at family rank — into which we classify neotropical pollen are inherently likely to include species adapted to many different habitats, environments, or communities. A familiar example is the family Melastomataceae. Melastomes are ubiquitous as shrubs and small trees throughout the moister parts of the neotropics, from the lowland tropical rain forest to cloud forests high in the Andes or even the Galapagos islands, where the endemic *Miconia robinsoniana* forms a monospecific belt of low bushes beyond the tree-line on the south flank of Isla Santa Cruz (Lawesson *et al.*, 1990).

Melastomes are pollinated by insects or hummingbirds (Cronquist, 1988). The small bell-like flowers of many of them, and the large showy flowers of others, do not suggest loosing pollen to the wind, yet it is common to find respectable percentages of their distinctive pollen in sediments. Over much of the Lake Pata pollen diagram the taxon Melastomataceae fluctuates between 10% and 30%, and even in its lowest zone in the Holocene part of the section it maintains a presence at 5% to 10%. We have found comparable percentages in other rain forest lake sediments and in Andean lakes below the tree-line. These are large numbers for animal-pollinated plants. The best proximal explanation of these high percentages seems to be that Melastomes produce much redundant pollen that is dropped to the ground before being transported to the lake in surface water. For the significance of Melastomataceae pollen percentages, however, it is necessary to evaluate the pollen as a function of its communal associates, usually by multivariate statistical methods. For the reduction in Melastomataceae in the Holocene at Lake Pata, a possible explanation is that the local melastomes were inhibited by postglacial warming.

REASSORTMENT OF FOREST TAXA IN RESPONSE TO TEMPERATURE CHANGES

Figure 1.2 plots surface pollen spectra along an elevational gradient on the eastern flank of the Andes mountains at roughly zero latitude in Ecuador. All the samples except Puyo Bog and Mera (which are peat and moss polster samples) are from mud–water interfaces of lakes or ponds. The parent vegetation of these pollen spectra ranges from lowland tropical rain forest at 280 m elevation to the edge of the treeless paramo at 3400 m. The counts were made early in our neotropical investigations when our reference collections and knowledge were still rather meager, so that the sum of unnamed taxa (listed as 'other') is rather large (20–40%) at all elevations. However, the 60% to 80% of taxa identified are easily correlated with the main features of modern vegetation along the transect (Bush, 1991).

The altitudinal distribution of *Cecropia* is roughly coincident with closed tropical forest in Amazonia. In the eastern coastal forest of Brazil (Hueck, 1966) and in the northwestern Brazilian Amazon at Pico da Neblina, the upper limit is close to 2000 m (Radambrasil, 1976), but in the western Amazon on the flank of the Andes (where our transect was taken), the upper limit of rain forest is conventionally taken to be about 1200 m. In Figure 1.2, large percentages of *Cecropia* pollen appear to be crude signals of the presence of forest.

The wind-dispersed pollen of the family Moraceae, to which *Cecropia* also belongs, is likewise prominent in the tropical rain forest below 1200 m in the Ecuadorian section, with low percentages at higher elevations plausibly attributed to dispersal upslope by wind. The percentage spike of Moraceae at 2600 m at Lake Yambo is consistent with this hypothesis of wind distribution, because this is a desert region of the inter-Andean plateau with very little local pollen production so that pollen brought by long distance transport appears as a high percentage of the total pollen rain (Colinvaux *et al.*, 1988). The roughly constant percentage of Melastomataceae pollen at all elevations certainly reflects the presence of different melastome species in virtually all neotropical woodlands.

On this transect, the higher elevations are easily characterized by high percentages of grass (Gramineae), sedge (Cyperaceae) and Compositae (Asteraceae) pollen. Of course, none of these families, are restricted to high elevations and plants of all three can be found abundantly in various parts of the Amazon lowlands. Their prominence in pollen percentages at elevations above 2000 m along this transect simply reflects the progressive replacement of forest with open vegetation, notably the high altitude 'grassland' known in the equatorial Andes as 'paramo.' The appearance of *Podocarpus*, *Alnus* and *Hedyosmum* above 3000 m, however, suggests woody plants restricted to high elevations, presumably by temperature.

The genus *Podocarpus* has many species in the neotropics, one source suggesting a total of 29 species (Bucholtz and Gray, 1948). Most modern populations are confined to high elevations as suggested in Figure 1.2 but in Central Brazil two species, *P. sellowii* and *P. lambertii*, grow down to 750 m in the more humid gallery forests (De Oliveira, 1992). Widely scattered individuals have been recorded at near sea level in special sites, in the Amazon particularly on white sands (Gentry, 1993). Like the family Melastomataceae, therefore, the genus *Podocarpus* has species individuals of which can be found at any elevation in the neotropics at which tree-growth is possible. Thus, *Podocarpus* cannot be used as an index fossil for temperature. But unlike Melastomataceae, its pollen is not routinely found in significant amounts at all elevations. To find more than the few grains that can be attributed to long distance transport in surface samples or traps it is necessary to sample well above 1000 m elevation, suggesting that significant recent populations were confined to high elevations.

Finds are now being made of significant percentages of *Podocarpus* at low elevations in rain forest sites in sediments dated to glacial times, of which the near 10% *Podocarpus* contents of pollen spectra dated to between 18,000 and 14,000 years ago at Lake Pata (Figure 1.1) are an example. At Lake Pata (and at some low-elevation Ecuadorian sites of glacial age), the high *Podocarpus* percentages are accompanied by other pollen taxa suggestive of the environments of higher elevations such as *Hedyosmum* or even *Alnus* and the exclusively montane genus *Drimys* (Winteraceae). None of these taxa (with the possible exception of *Drimys*) are satisfactory index fossils for temperature, yet a glacial cooling hypothesis seems the most parsimonious to explain their co-occurrence in low elevation tropical forests.[4]

Thus, although the Amazonian pollen data offer few, if any, clear index fossils for temperature, they probably yield a signal for temperature in combination. This conclusion mirrors the experience of pollen analysts in the north temperate zone who have used transfer functions of percent composition of prominent tree genera to construct maps of mean annual temperature in past millennia. The technique is to use pollen percentages as importance values that can be correlated with temperature in surface samples.[5] Pollen percentages in past (often very different) communities are then transferred as temperatures.

The appearance of 10% *Podocarpus* pollen and the rest in the ice-age Amazon forests suggests that a similar approach is worth pursuing in the neotropics. But first we must have a database of several hundred surface samples across the Amazon basin, all keyed to local plant communities and climatic parameters. Until we have these data, inferring past temperature from neotropical fossil pollen will remain controversial.

POLLEN SIGNALS FROM CERRADOS AND SAVANNA

All palynological investigations of the late-Pleistocene Amazon will be alert to the possibility of evidence for reduced precipitation in the past. This is because strong evidence exists for ice-age aridity in other parts of the tropics (particularly Africa) and because of the biogeographical 'refuge theory' of Jurgen Haffer and others (Prance, 1982; Whitmore and Prance, 1987). A drier ice-age is made plausible by the increasing evidence for cooling (cooler tropics would necessarily be drier tropics). In its more extreme forms, the arid ice-age hypothesis has much or all of the lowland Amazon forest replaced by cerrados or savanna vegetation. Fortunately, the pollen signal of the more arid vegetation — dry woodland (cerradão), cerrados, and savanna — can be distinguished unequivocally from that of tropical forest.

Figure 1.3 is a pollen percentage diagram of glacial and postglacial time from in the cerrado (sensu stricto) phytogeographical domain in the relatively dry central Brazilian highlands, south of the Amazon at Crominia, State of Goiás (Ferraz-Vicentini and Salgado-Labouriau, 1996). The work is from the laboratory of the pioneer Brazilian palynologist M.L. Salgado-Labouriau and is from a core of deposits from a *Mauritia* swamp, representing a region between 500 m and 1000 m elevation with a dry season so long and severe that it prevents the establishment of closed canopy forest except for gallery forests along streams. The present cerrado vegetation is described as "savanna-like vegetation in which the low stratum is dominated by grasses and has scattered shrubs and low twisted trees, mainly Leguminosae." In the even drier vegetation of savanna, the proportion of shrubs and trees would be reduced.

This cerrado pollen diagram is characterized by 50% of grass (Gramineae) pollen in the Holocene samples. Surface samples of soil from the gallery forest and from lake and the swamp surface also had about 50% grass pollen. In the samples of glacial age, grass pollen reached 80%. The main contributors to the pollen sum other than grass are herbs of the families Compositae and Cyperaceae. Apart from a few percent of the ubiquitous Melastomataceae, the input of tree pollen (even *Cecropia*) is little more than a trace, despite the nearby presence of gallery forests. The rise to 80% of Gramineae pollen in glacial times possibly suggests local vegetation tending towards actual savanna, although it is equally plausible that this extra 30% of grass is a statistical artifact of the withdrawal of Compositae, *Mauritia* and Melastomaceae from the pollen rain without their being fully compensated by rises of taxa other than Gramineae.

Almost entirely lacking from the Crominia pollen diagram are the characteristic pollen assemblies of trees of diverse tropical forest, both wind- and animal-pollinated. This pollen profile of dry regions of tropical Brazil, therefore, is utterly different from the pollen signature of lowland tropical forest illustrated by the Lake Pata record. If such drier climate vegetation ever replaced lowland tropical forest in the Amazon lowlands, the signal would be overwhelming in a local pollen diagram. Obviously, there was no such replacement in the west central Amazon where Lake Pata is sited (Figure 1.1).

The Cyperaceae record in this core probably emphasizes plants actually growing on the bog, as of course does the *Mauritia* pollen. Here *Mauritia* pollen gives a local habitat (presence of a swamp) signal as usual, leading to the suggestion that in the glacial age deposits (in which *Mauritia* pollen is largely absent), the site was too dry for *Mauritia*. An alternative explanation, however, is that a swamp was present, but that it was too cold for *Mauritia*. This argument is plausible because *Mauritia* populations do not extend further from the equator than 20° S (Lorenzi *et al.*, 1996). Temperature sensitivity of *Mauritia* would also explain its appearance in the Holocene, but not glacial times at the vereda of Aguas Emendadas, central Brazil (Figure 1.4) as the author suggests (Barberi-Ribeiro, 1994). In this example, an aridity explanation for the absence of *Mauritia* in glacial times seems improbable because the presence of aquatic plants (Pontederiaceae) demonstrate the continued presence of water.

Interpreting Grass (Gramineae) and Herb Pollen in Amazon Pollen Spectra

The dry facies of neotropical vegetation (principally savanna, but including the more wooded forms locally known as cerradão, caatinga, etc.) all have thick ground covers of grass and all yield pollen spectra in which Gramineae pollen is predominant, possibly at 50% or more. Thus, any Amazon pollen spectra without large Gramineae percentages cannot represent dry habitats or climates. The converse proposition, however, that high Gramineae always means drying, is unfortunately not true.

A common habitat for grasses in the humid Amazon lowlands is the floating mats that form in shallow lakes or slow-moving streams, or in *varzea*. The grass species that form much of the matrix of these mats are, of course, not the same species that grow in dry savannas, but their pollen cannot be distinguished with a light microscope — both sets of grasses appear simply as 'Gramineae.' Other common habitats for grasses along Amazon rivers are low islands and mudbanks exposed by movements of river channels, which are often colonized by dense stands of savanna-like grasses. Both floating mats and mudbank islands inject grass pollen into what might well be a primary pollen signal for lowland tropical forest. When interpreting a blip of Gramineae in a pollen diagram, therefore, the guiding principal should be to look for evidence of change elsewhere in the diagram. If forest taxa continue alongside the expanded Gramineae spectra, the grass pollen is not *prima facie* evidence for drying.

Another more subtly wrong interpretation of Gramineae maxima is inherently likely in pollen diagrams of bog or marsh deposits. Unlike sediments in deeper lakes, bog deposits are subject to surface drying from small changes in precipitation or from changes in drainage patterns in the local landscape. A result is that grasses and other weeds grow densely on the dry bog surface, resulting in local overrepresentation of grass and other herb pollen. Percentage statistics then require that the percentages of all other taxa in the pollen sum are reduced to accommodate the increased influx of grass and herb pollen derived from plants growing on the dried surface of the bog. Pollen analysts have sometimes fallen into the trap of using summary diagrams from such bogs to suggest that regional grasslands were expanding at the expense of forest, when the only real change was the drying of the sampled bog — possibly from climatic change, but possibly not.

Gramineae pollen (or sum herb pollen), therefore, cannot be used as an index fossil for savanna or other dry vegetation type. Nor are other herb pollen taxa useful as index fossils for savanna. One candidate pollen index fossil for savanna that has been proposed is the genus *Cuphea* (Lythraceae), which has the nice advantage of being an elegant and easily identified pollen type. It is true that *Cuphea* species are prominent in savannas and *Cuphea* pollen has recently been shown to reach 40% in some caatinga pollen spectra from northeastern Brazil (De Oliveira *et al.*, 1999, in press). But *Cuphea* shares with *Podocarpus*, Melastomataceae, and many other pollen taxa the property of having species adapted to many habitats. Some *Cuphea* species are understory shrubs of tropical forest. Some species are weeds of moist sites like stream banks, plants of which probably account for the few percent of *Cuphea* in the Crominia pollen diagram of Figure 1.3. More importantly, other *Cuphea* species are weeds of drying bogs. Thus a bog with a dry surface, for whatever reason, is likely to yield pollen spectra rich in Gramineae, supported by a few percent of *Cuphea* pollen, and probably with other common tropical weeds like *Borreria* as well. These taxa, severally or individually, are not index fossils for savanna.[6]

Thus, although a high percentage of grass pollen, probably with associated herbs, is always present in savanna or cerrados pollen profiles, high percentages of these herbs can be present for other reasons and do not stipulate regional aridity. When these pollen types are produced in floating mats or on river banks, they may not imply drying at all.

Arboreal Plant Signals for Savanna

The pollen of three woody plant genera are useful as indicators of savanna, when considered in the context of the rest of the pollen spectra in which they appear. They are *Curatella* (Dilleniaceae), *Stryphnodendron* (Mimosoideae), and *Caryocar* (Caryocaraceae). The genus *Curatella* has only two species in the neotropics, both of which are plants of dry habitats, and *Curatella americana* being characteristic of savannas. *Stryphnodendron* pollen is in polyads like most of the large subfamily Mimosoideae, but distinctive enough to identify at the genus level. Both of these genera are strongly suggestive of savanna if the rest of the pollen spectra are also consistent with savanna. But *Caryocar* has to be used, with more caution. *Caryocar* trees are an important constituent of savanna such that their pollen is expected to show in savanna pollen diagrams. Yet the genus also provides some of the largest trees in lowland tropical rain forest, so that the presence of the genus in a pollen diagram is not, by itself, an indicator of savanna. It is necessary to interpret the presence of all supposed index taxa in the context of the whole pollen spectrum.

MULTIVARIATE STATISTICS OR ECOLOGICAL GROUPINGS?

It is clear that a fundamental difficulty for Amazonian pollen analysis is the ubiquity of the most abundant taxa across most vegetation types. Gramineae, Melastomataceae, Moraceae, Urticaceae, *Cecropia*, Compositae and others contribute to pollen rains from lowlands to highlands and from wet to dry. These are the heavy pollen producers, many of them wind-pollinated. They are taxa whose pollen are only poorly differentiated, sometimes as with Moraceae and Urticaceae not even distinguished at the family level. Even allowing for the many species in each of the groups of this crude taxonomy, they include but a tiny minority of the total Amazonian species pool.

As we have shown, it is yet possible to distinguish the most major vegetation types from pollen spectra dominated by these taxa by purely subjective criteria — the tropical rain forest of Figure 1.1 cannot be confused with the savanna-plus-gallery forests of Figure 1.3, despite the overlap of many taxa. Our difficulties begin when we need to make more subtle distinctions — to separate seasonal forest from rain forest, *varzea* from *igapo*, all flooded forests from *terra firme*, savanna from flooded grassland, etc. Pollen experience in temperate latitudes of low diversity and anemophilous trees suggests that we get around this difficulty by assigning the different pollen taxa to ecological groupings, so that we plot summed taxa from trees, shrubs, wet habitats, dry habitats, etc. In some published neotropical pollen studies, this technique has been used to identify such ecological groupings as 'savanna group,' 'forest group,' 'aquatics,' etc. We suggest that this is a dangerous practice likely to lead to serious error when applied to Amazonian vegetation. The practice requires that the pollen analyst subjectively assign our broad taxa, like Gramineae, Melastomataceae or Moraceae to an ecological group when our problem is that these broad taxa span most of the ecological groups we can imagine. It is better to appeal to multivariate statistics to do the ecological grouping.

M.B. Bush was the first seriously to investigate the possibility of discriminating ecosystems in neotropical pollen spectra by multivariate analysis in this way (Bush, 1991). He considered only pollen types that appeared as >2% total pollen, thus removing from the analysis the diverse assembly of animal-pollinated plants such as were used in our Lake Pata analysis to confirm that this was an history of lowland tropical forest. Bush let himself work only with the heavy pollen producers and their crude taxonomy, seeking to discriminate between surface samples from as many neotropical habitats as were available (Figure 1.5).

Bush's analysis began by excluding *Cecropia* from the pollen sums on the grounds that its overabundance would distort the analyses (in many samples *Cecropia* was 40% of total pollen). Then he applied a TWINSPAN analysis to the remainder. TWINSPAN is the acronym for 'two-way indicator species analysis,' a program that classifies 'species' (in this case pollen spectra) by comparing their characters (pollen percentages). Figure 1.6 shows the resulting hierarchical classification. Some of the surface spectra are grouped unequivocally and satisfactorily: terra firme forest, semi-deciduous forest, *igapo* and *varzea* being well delineated. Four surface samples (cerrado, and lakes Geral, Comprida and La Yeguada) are not separated. All four, however, have in common evidence of a dry season. Cerrado is inherently dry. Comprida and Geral are two lakes between Monte Alegre and Santarem in central Brazilian Amazonia. These two lakes are in a region with a strong and prolonged dry season; moreover, agriculture was established in the region for some ten years at the time of sampling. La Yeguada is a lake in the pacific coastal forest of Panama where the landscape has been influenced by human activities for most of the Holocene and where the archaeologist D. Piperno describes the resulting local vegetation as 'anthropogenic cerrado,' Thus, TWINSPAN identifies common denominators in these four surface samples that we can readily understand.

More subtle distinctions between pollen groupings can be made by Detrended Correspondence Analysis (DCA). DCA ordinates data assemblagies along hypothetical environmental gradients (the axes) that explain major variance in the data set. The axes are driven by taxa that are relatively abundant in a sample or group of samples, but absent or rare in others. Minor variations in the major taxa are unimportant. Figure 1.7 is a DCA interpretation of the surface spectra of Figure 1.6. As with the TWINSPAN classification, the pollen signals of the principal vegetation types separate clearly, particularly on the first axis where *varzea* and terra firme forest are at opposite ends of the ordination. It is interesting to note that on this axis *varzea* is closer to sites with a dry season, or to actual cerrados, than it is to moist terra firme forest, a result possibly due to the conflicting signals given by Gramineae and other herb pollen.

The successful separation of these surface pollen spectra serves to refute the old fear that the wind dispersed pollen of the Amazon, being all from ubiquitous and invasive species, would give little information about forest composition. Bush's analyses are restricted to pollen taxa present at >2%, and are thus predominately these anemophilous invasive species, but the principal community types are clearly differentiated. Thus Amazon sediments can be expected to yield regional signals from their windblown pollen in addition to the histories of communities within the watershed of a lake that is provided by the richly diverse array of the lesser fluxes of pollen from animal-pollinated plants. Amazon pollen rains potentially yield two sets of climate signals: the regional signal brought in the wind and the local signal brought by runoff water from the catchment.

The finding of a regional signal gives strong encouragement to the view that calculating transfer functions between neotropical pollen percentages and climate variables will work in the neotropics just as it did in temperate latitudes. What is wanted is a large set of surface samples calibrated both to vegetation and to climate variables. It will then be possible to use pollen from a grid of Amazon lake cores to map the climate of the Amazon basin at chosen time intervals of the late Pleistocene.

NOTES

1. Lake Pata has a very small catchment so that runoff into the lake is minimal (Colinvaux *et al.*, 1996a). In Lake Dragão, a similar lake close to L. Pata we found the mean concentration to be 76,800 grains/ml^{-3}. The mean figure of 100,000 grains/ml^{-3} found at L. Kumpak[a] probably records the concentration of grains from a large and well-washed watershed (Liu and Colinvaux, 1988).
2. Pollen trapping at Manaus was by courtesy of INPA and in association with the Minimum Critical Size of Ecosystem Project. Pollen trapping in the Oriente Province of Ecuador was at the Cuyabeno Research Station under the direction of E. Asanza.
3. Lake Pata lies 300 m above modern sea level. With the eustatic lowering of sea level in glacial times (i.e., before the *Mauritia* rise), the lake would have been at least 400 m above mean sea level possibly effectively raising the basin relative to the water table.
4. The evidence of these Ecuadorian pollen spectra for temperature depression in glacial times is discussed in Chapter 2. Principal references for tree population descents in Ecuador in glacial times are: Liu and Colinvaux (1985), Bush *et al.* (1990), Colinvaux, *et al.* (1996b, 1997).
5. See for instance: Huntley and Webb (1988) or Berglund (1986).
6. We suggest that some interpretations of the pollen data from a marsh on the Carajas Plateau of east central Brazil have been mistaken for these reasons. That the marsh surface dried is elegantly demonstrated by mineralogical analyses of the cores but the claim that the pollen show synchronous replacement of the lowland forest by savanna cannot be substantiated (Absy *et al.*, 1991; Van der Hammen and Absy, 1994).

CHAPTER

2

Review of Amazon Paleoecology: The 1997 NWO/Huygens Lecture

On November 12, 1997, at the invitation of Professor Henry Hooghiemstra, Colinvaux gave the annual Christian Huygens lecture of the Netherlands National science foundation (NWO) at the Nieuwe Kerk in The Hague. The theme of the lecture was *The ice-age Amazon and the problem of diversity*. The lecture was followed by a commentary from Professor Hooghiemstra with the title *Tropical Rain Forest versus savanna: two sides of a precious medal?* This chapter is a shortened version of the printed Colinvaux lecture that was published as a brochure by NWO. Some, but not all, of the first-person language of a lecture has been removed from this version.

Some parts of the lecture — particularly the conclusion that the lowland tropical forest of the Amazon basin was never fragmented by aridity at any time in the last glacial cycle — are controversial. In his commentary, Professor Hooghiemstra drew particularly on the inference of Van der Hammen and Absy (1994) from the Carajás record to suggest that parts of the east-central Amazon basin did indeed support savanna in glacial times, contrary to Colinvaux's conclusion. But controversy is the life-blood of science and today's controversy is tomorrow's hypothesis for testing. It is possible to hold different opinions about something so important as the stability and antiquity of Amazon rain forests only because we have so few data. An adequate series of pollen histories from all parts of Amazonia should quickly settle this controversy. To encourage the acquisition of these pollen profiles is our purpose in writing this manual.

Introduction: The Diversity Problem

Our favorite statistic for the Amazon comes from the work of Al Gentry in the Amazon rain forest of Peru. Gentry counted and identified every tree in a single hectare of lowland rain forest. There were 600 trees in his chosen hectare, but they were spread among 300 species. This is about as many tree species as are known for the complete floras of Europe and North America combined (Gentry, 1988).

We do not have many such complete counts from the Amazon, for the work of census requires not only awesome knowledge of taxonomy but also extraordinary perseverance. Gentry had to climb many of those 600 trees to identify them. Later, he was killed when the light plane he was using for a Rapid Assessment Program (RAP) to report on the value of a Peruvian Amazon forest for conservation crashed into the trees he was surveying.

Counts of both animal and plant species across the Amazon, though less splendid in execution than Gentry's, are sufficient to show that the grand total is very large: 80,000 plant species, insects perhaps in the millions, and other animals somewhere in between. The Amazon forest is the richest ecosystem on earth.

So why are there so many different kinds of plants and animals in the Amazon? Why so many more than in the fertile high latitudes of Europe and North America? These are among the most contentious questions of biological theory.

There are two families of explanation for high Amazon diversity: 'Amazon as museum' and 'Amazon as engine.'

In 'museum' models, the Amazon is a place of safety where extinction is rare so that species accumulate over geological time: no shocking winters, no incursions of the North Sea, above all no grinding glaciers to wipe out forests with the frequency of ice-ages. 'Amazon as Museum' thus holds the species of long ago as well as those made more recently, the whole adding up to the colossal numbers we see. 'Amazon as museum' is a nice, warm, tropical paradise where very little happens.

'Amazon as engine' models are almost the opposite, saying that the Amazon is in fact an eventful place of continual, but less than catastrophic, happenings. Trees fall in violent tropical storms, rains fail to come on time or rivers flood, above all predators and seed-eaters rampage unchecked by winter. Populations are continually culled by these stresses of tropical life, but seldom to extinction. The practical effect is that populations seldom get large enough to dominate their communities in ways that keep other kinds out. Coexistence of the many is possible. Life is lived, as it were, on the run, with resources always available for a new species to find a niche. Innovative ways of life are at a premium so that new species arise like ideas in a competitive market place. Where 'museum' models see the Amazon as safe and stable, 'engine' models see violence that checks the bullies and land grabbers, leaving room for newcomers.

Both museum and engine models rely on a special property of low latitudes: temperature varies little over huge distances. A degree or two of latitude sees little difference in ambient temperature near the equator, whereas that same change in latitude in Northern Europe can take you from one climate zone to another. The Amazon is a huge piece of real estate to start with, larger for instance than the European Union, and the temperature constancy of low latitudes makes it all one (Figure 2.1). Thus every Amazon habitat is backed up, as it were, with a copious collection of similar habitats to which the hard-pressed can escape.

But whether museum or engine, Amazon diversity is contingent on history. We live in an epoch of ice ages when all the environments of the earth suffer periodic climatic shocks. Glaciers advance to occupy northern lands for fifty or more thousand years at a time. Then the glaciers melt and an interval of warmth like that in which we live engulfs the earth. Then comes another time of glaciation. This cycle has been repeated again and again for more than the last million years.

The existence of glacial cycles is common knowledge — less familiar is the idea that the glacial condition is the norm. Glacial times typically last longer than warm interglacial times like the present. Thus the ice-age condition has been the normal condition these last two million years. The warmth of an interglacial like ours is exceptional (Imbrie *et al.*, 1984).

Species last longer than glaciations — generally several million years, depending on the kind of organism. Thus the life-span of a species is much longer than a glacial cycle. Nearly all contemporary species, therefore, have passed most of their existence in ice-age time. Modern diversity is ice-age diversity, accommodated for the moment to the unusual climate of our comparatively brief interglacial.

Thus, to understand Amazon diversity, we must understand the ice-age Amazon. The last ice-age ended a scant ten thousand years ago, a time so recent that only forty lifetimes of a big tree have passed since the last retreat of the ice.[1] For comparison note that forty human lifetimes (giving us 50 years each) only separate us from the time when Julius Caesar crossed the River Rubicon with his army, signaling the final destruction of the Roman Republic. The last ice age was very, very recent.

AIR STARVED OF CARBON DIOXIDE

An ice-age climate might have differed from ours in temperature or precipitation. Northern latitudes were certainly colder than now, and some parts of the tropics were certainly drier. But the universal change of which we are most certain is that the carbon dioxide (CO_2) of the air was reduced.

The global CO_2 concentration of ice-age time has been measured in bubbles of ancient air trapped in the ice of Antarctica and Greenland. It was only two-thirds of the CO_2 concentration in our interglacial at the start of the industrial revolution: 0.02% by volume as opposed to our 0.03%. This is a profound difference. CO_2 is the prime raw material of growth for all green plants. The low concentration of this gas alone made certain that the relative abundance of many species of plants were different in the ice-age Amazon.

Plants called 'C4 plants,' and others called 'CAM' (for Crassulacean Acid Metabolism) are equipped with special CO_2 collecting enzymes that can sweep air clean of CO_2, even at low concentrations. The C4 and CAM photosynthetic mechanisms are adaptations to life in stressful conditions, particularly to aridity or heat (Colinvaux, 1993a). Typically the system is used by a cactus or a bromeliad epiphyte, in both of which it is an adaptation to relative aridity: in a cactus the adaptation is to life in a desert, in an epiphyte to a life without roots in the soil. C4 photosynthesis in particular is used by savanna grasses, where it adapts them to the hot and dry conditions in which they are typically found. Other grasses are C3, not needing this special adaptation to drought. But the chemistry of C4 and CAM plants has the special peculiarity that the plants can absorb CO_2 from CO_2 deficient air more rapidly than can the more usual C3 plants. Thus cacti, bromeliads, and most importantly savanna grasses, might be expected to have had an advantage in the CO_2-starved air of the ice age.

We see the ice-age forests as even more heavily draped with bromeliad epiphytes than contemporary forests, perhaps with more bromeliads on the ground cover too. Grassy areas on seasonally flooded river flats would have had more of the C4 savanna grasses, and perhaps there was a similar switch from C3 to C4 grass species high on mountain sides, where the effects of altitude would exacerbate the shortage of CO_2.

These changes arising from impoverished CO_2 supplies should have been enough to ensure that the plant associations of the ice-age Amazon were not quite what we see today, yet the changed CO_2 surely had less effect than would changes in either temperature or precipitation.

THE COOLED EQUATOR

Common sense might seem to say that times were cooler in an ice-age but, surprisingly and until very recently, prevailing opinion was that temperatures at the equator had changed very little from the present day. This 'minimal cooling' hypothesis was a construct of paleooceanography, particularly of a research group known by the acronym 'CLIMAP.'

CLIMAP set out to map sea surface temperatures of all the oceans at the last glacial maximum, defined by them as at 18,000 years ago, using data from cores of ocean sediment. The CLIMAP thermometers for the ocean surface were arrays of Foraminifera fossils in the cores, calibrated by fitting living foraminiferal species to the temperature of the water in which they were found. The CLIMAP conclusion was that, although the seas of middle and northern latitudes were six or more degrees colder in the ice-age, tropical oceans were scarcely cooler (perhaps 2°C) than now (CLIMAP, 1976).

If tropical seas did not cool in the ice age, it followed that the adjacent land could not cool either. Such was the prevailing consensus, even paradigm, of science for the two decades that followed the first publication of the CLIMAP results in 1976.

Yet geologists had long had evidence that glaciers on high mountains of the tropics had descended in synchrony with the continental glaciations. That this was so was demonstrated conclusively in 1962 for Africa when Daniel Livingstone (1962) dated to the last ice-age by radiocarbon the oldest sediments in a glacial lake high in the Ruwenzori Range. Radiocarbon dates for end moraines below montane glaciers of the Andes also demonstrated expanded glaciers on tropical mountains during the last ice-age. This looked like cooling on tropical mountains when CLIMAP said there was no cooling on the lands below.

Soon it was clear that glaciers had expanded on tropical mountains in all continents. The simplest explanation for these descents of montane glaciers in the tropics was cooling of between 6 and 9° Celsius. But the CLIMAP paradigm of tropics with minimal cooling stood firm and our profession faced the apparent paradox of colder highlands over lowlands that remained warm.

The hold that the 'minimal cooling hypothesis' had on our science can be judged by the efforts made to explain how glaciers advanced without cooling. Mathematical models appeared showing that tropical glaciers would grow without additional cooling if precipitation was increased: supply enough snow at the top and you can overwhelm the melting at the bottom.

Meanwhile came the remarkable discoveries from Colombia by T. Van der Hammen, H. Hooghiemstra and their colleagues. The city of Bogota lies near tree-line in the Andes and is built on the sediments of a caldera lake that drained some 20,000 years ago. 300 m of lacustrine deposits span about 3,000,000 years. Pollen analysis shows that total tree pollen percent fluctuates in synchrony with northern hemisphere glaciations and with montane glaciers on the higher peaks of the Andes themselves, the vertical range of tree-line being about 1500 m (Hooghiemstra, 1989).

The pollen data provided evidence of, shall we say 'frigid phenomena' not only near the ice caps on the highest mountains but also in the vegetation one or two thousand meters below. Special pleading for glaciers that advanced without additional cooling, in particular, seem scarcely convincing when the vegetation far below responds as if to cooling.[2]

Instead the CLIMAP conclusion remained apparently inviolate. Biogeographers drew diagrams of tropical mountainsides with all the modern vegetation belts present but compressed into a narrow band on the lower flanks of the mountains, with astonishingly steep temperature gradients from sea level to the lowered tree-line. The physics of these temperature gradients was a little tricky but by postulating dry air at the base and wet air at the top they could be made to work: just.

This paradox of colder highlands poised over still warm lowlands was profoundly unsatisfactory. In the face of fashion we began to suspect that tropical lowlands cooled in synchrony with the highlands. The test should be actual temperature records from the lowlands. We started looking for climate records in the equatorial lowlands, first in the Galapagos islands, then in the Amazon basin itself. Pure chance gave us an Amazon pollen record at the foot of the Andes, at a place called Mera in Ecuador.

Kam-biu Liu found the Mera section. We were using a nearby airstrip as a base from which to fly to remote Amazonian lakes but were grounded by bad weather. So we walked about to kill time, and Dr. Liu sampled wood, twigs, and old lacustrine sediments exposed in a road cut at about 1100 m elevation, still in the upper reaches of the tropical rain forest. Radiocarbon ages of the wood samples turned out to be close to 30,000 years. This was a glacial age. Liu's pollen analysis showed pollen of lowland forest, but in addition pollen of some trees now largely confined to the high Andes 1000 m higher up, trees like *Podocarpus* and *Alnus*. We published the result as a note in *Nature* in 1985 as showing that cooling of more than 4°C prevailed at all elevations, from mountain top to the bottom lands of the Amazon (Liu and Colinvaux, 1985).

Mera was a break, but not definitive. If you did not like the result, reasons for rejection could be found fairly easily. Radiocarbon dates in the 30,000 year range can always be questioned, "Obviously contamination, my dear chap, the samples must be much older!" etc. The best answer to skepticism would be more sections. We wrote a grant proposal to the National Science Foundation asking for funds to search the foothills of the Ecuadorian Andes for another section: incredibly they funded the proposal and we found the section.

The second section was found in the forest at San Juan Bosco, 160 kilometers away from Mera. Mark Bush and Paulo De Oliveira found it, after Colinvaux had been left in Quito, comatose from a sickness he had contracted in the forest. The dates overlapped with those of Mera, and Bush's pollen analysis gave the same result, wet lowland forest invaded by *Podocarpus, Alnus* and others from on high. Putting the Ecuadorian and Colombian data together we now had that glaciers descended 1500 m, below them the tree line also descended 1500 m, and below that some trees of the Andean forest descended 1500 meters to enter the lowland forest. We calculated the extreme of the cooling episode as requiring a reduction of 7.5°C over the present day (Bush *et al.*, 1990).[3]

These two Ecuadorian sites were the first within the tropical rain forests of the Amazon to give evidence of cooling. The descents of temperature sensitive trees that they revealed demonstrated that the synchronous movements of tree-lines and glaciers above them were associated with declines of temperature and should not be ascribed to such factors as reduced CO_2.[4]

We found a similar history in lake deposits from the Pacific coastal forests of Panama. The small Panama town of El Valle is built on the naturally drained sediments of an ancient caldera lake, a miniature version of Bogota but in the coastal forest at only 500 m elevation. The history of ice-age time revealed by pollen analysis was directly parallel to those of San Juan Bosco and Mera, with forests of glacial times being invaded by plants descending a thousand meters from their present elevations. As at the Ecuadorian sites, this was an history of more cold adapted species invading the lowland forest. It was not an history of vegetation moving in belts, rather the pollen showed that tropical forest composition kept changing though still recognizable as tropical forest (Bush and Colinvaux, 1990).[5] But this was a continuous record with good radiocarbon control throughout. So we had shown that the forested lowlands of both Panama and Ecuador cooled in an ice-age, contrary to the implications of CLIMAP.

The record of cooling was then spread across the whole width of the Amazon basin to eastern Brazil with pollen evidence that the *Araucaria* forests of the State of Paraná had spread northwards some 700 kilometers, crossing the Tropic of Capricorn in the process. The signal for cooling of at least 6°C is found in the *Araucaria* pollen record from Serra Negra (De Oliveira, 1992) and Salitre (Ledru, 1993).

At this point we pollen analysts were still all alone with our cooling records: a few cores and biological inference to set against the massed data of sea surface temperature from CLIMAP. But help was at hand. Martin Stute found a record of ice-age temperatures for northeastern Brazil in fossil groundwater. His thermometer was the ratio of noble gasses in solution. In Brazilian groundwater dated by radiocarbon to the glacial maximum Stute's thermometer read a temperature depression of 5 degrees Celsius (Stute *et al.*, 1995). Meanwhile some of those seeking paleotemperature records in chemical and isotopic ratios in the matrix of coral reefs found evidence that sea surface temperatures themselves had been lowered, despite the calculations of CLIMAP (Guilderson *et al.*, 1994). At last, in February 1997, a climate modeling team published in *Nature* a global climate model that allowed tropical oceans and lands to cool by our suggested 6 degrees (Webb *et al.*, 1997). An e-mail from one of the authors to Colinvaux said, "In my view, your vindication came from (the) coral and groundwater work, whereas our article provided a mechanism.... a lot of people now are redirecting the grief that you have received over the years to me."[6]

So the paradigm is shifting towards acceptance of cooler tropics in glacial times. This feels sweet after so many years of having anonymous reviewers trounce manuscripts and grant proposals for the suggestion that the lowland tropics cooled. Even scientists have human pleasures.

WAS THE AMAZON EVER ARID?

But what of the third likely parameter of climatic change, precipitation? The prevailing paradigm is that the tropics were arid in ice-age time. In Africa, for instance, lake levels were undoubtedly lower and great Lake Victoria possibly dried up completely. Closer to the Amazon, in parts of central America and the Caribbean, incontrovertible evidence shows stark and regional drought in the last glaciation, the best data being from Guatemala where pollen in ice-age lake sediments shows an arid landscape of savanna in place of the modern tropical rain forests (Deevey *et al.*, 1983).

Our demonstration of tropical cooling in part strengthens the aridity hypothesis, since cooling should reduce evaporation. That a cooler Amazon should be a drier Amazon seems inescapable.

But how much drier? One recent geological review takes the argument for tropical aridity so far as to map the entire extent of the ice-age Amazon lowlands as savanna (Clapperton, 1993). Biologists mindful of the necessity for maintaining diversity through the ice-age are more modest, allowing islands of forest, which they call 'refuges,' to persist in a grand savanna sea. These maps owe as much to inductive reasoning as to actual data (Figure 2.2).

But maps showing the Amazon to be arid do serve as statements of an hypothesis that is open to testing. They state that the Amazon lowlands held savanna in ice-age time, a proposition easily tested by core drilling and pollen analysis of ancient Amazonian lakes. Lake Pata is well suited for this purpose in that it lies in tropical lowland forest typical of the west-central Amazon lowlands where both geological and biogeographical maps based on the arid Amazon hypothesis show the local vegetation in ice-age times as savanna.

Lake Pata occupies a closed basin with neither inlet nor outlet streams, though water is apparently lost by seepage as well as by evaporation. Sediment accumulation was remarkably slow, with the CLIMAP 18 k datum demonstrated by accelerator mass spectrometry (AMS) radiocarbon dating to fall at 0.7 m. Sediments deeper than one meter are more than 30,000 years old.

Pollen analysis shows unequivocally that the region was forested throughout the time the sediments collected (see Figure 1.1 and discussion in Chapter 1). Some 80% of total pollen grains can be assigned to trees of the tropical lowland forest throughout. Even more decisive is that the percentage of grass pollen is minimal at all levels, both glacial and postglacial. Sediments in savanna regions of the central Brazilian highlands have grass pollen as 40–80% of the total (see Figure 1.3 of Chapter 1). The fact that grass pollen is negligible in the ice-age deposits of L. Pata is an unequivocal demonstration that savanna was not present in the region in ice-age time. This part of the Amazon lowlands was always forested (Colinvaux *et al.*, 1996a).[7]

Yet the L. Pata core does provide one hint of reduced precipitation consistent with cooling. From roughly 30,000 to 14,000 years ago the color and structure of the deposit suggests intermittent exposure to the air, hence lower lake levels. The pollen assemblies show, however, that sufficient precipitation remained to support forest. The annual local precipitation is now in the 3000–4000 mm range. The most prudent interpretation of the Lake Pata result is that cooling did reduce this copious precipitation but never enough to disrupt the forest.

EVIDENCE FROM THE AMAZON FAN

Pata, with its sister lakes Verde and Dragão, are valuable because they have yielded a continuous pollen record since glacial times from within the lowland rainforest itself. The long pollen record from Carajás also spans ice age time, but is from a high plateau covered with edaphically controlled, dry facies vegetation, not tropical lowland forest (see note 10 below), and the sections at Mera and San Juan Bosco in Ecuador span only a short interval in isotope stage 3 time of the last glaciation. Thus Lake Pata and its sister lakes are still the only lakes actually within the lowland Amazon forest yet to have yielded a continuous record back to ice-age times. We thus have a long and complete record from only the one site in the immensity of the forested Amazon lowlands. The hazards of extrapolating from this one site to the whole Amazon basin are obvious. But a more general test of the savanna hypothesis is available from pollen analysis of sediments in the Amazon fan.

The Amazon fan is a vast accumulation of river sediments in the deep sea off the mouth of the Amazon river. Sediment reaches the fan only during continental glaciations because sea levels are then more than 100 m lower than now, so that sediment laden water pours through a gorge cut into the shallow continental shelf and so out into the deep sea. The sediment is finally dropped beyond the shelf into water 4000 m deep to form the fan.

Now, with the high sea level of an interglacial, the Amazon drops its sediment load on the continental shelf. Thus there exist off the mouth of the Amazon river separated samples of the sediment for glacials and interglacials.

The International Ocean Drilling Program (ODP) drilled into the ice-age sediments of the fan, at the same time coring the interglacial deposits on the shelf. Our colleague, Simon Haberle, was aboard the drilling ship and sampled the cores to patch together a continuous sedimentary record for all of the last glacial cycle (Haberle, 1997; Haberle and Maslin, 1998).[8] He then made a pollen analysis of this long record to determine changes in the vegetation of the entire Amazon basin as the climate changed from glacial to interglacial and back.

While waiting for Haberle's pollen analysis to be finished we were convinced that the results would show some evidence of extended savannas, despite our rejection of the idea of replacement of the lowland forests. The Amazon system drains the entire basin so that an increase in grass-rich vegetation (savanna, cerrado, caatinga) in any part of the basin should send its signal of grass pollen down to the sea. Parts of the basin, such as its northern and southern edges, and some elevated plateaus have savannas or other dry facies vegetation today. Cooling, and the sort of reduced precipitation suggested by the physical sediment at L.

Pata, seemed likely to increase the area of these drier types of vegetation at the periphery, even if most of the forest was left intact. Thus we expected the ice-age pollen to include considerably more grass, the signal of savanna. We were wrong. There was no excess grass pollen in the Amazon fan (Figure 2.3). Haberle's diagram of pollen deposited by the entire Amazon system is like that from Lake Pata in showing minimal changes between glacial and interglacial times. In particular no more grass pollen came down the rivers in glacial times than during our interglacial.

It is still possible, even likely, that pollen from the remoter parts of the basin are underrepresented in the fan deposits, having been destroyed, filtered, or swamped by other pollen in the long journey to the sea, so that savannas could have expanded in the distant drier parts of the basin without being signaled in Haberle's pollen diagrams. What does seem certain, however, is that savannas cannot have occupied the broad lowlands through which the rivers pass on the last part of their journey and which are now the domain of tropical rain forest.[9] Haberle had tested the hypothesis that Amazon lowland forest was replaced by savanna in ice-age time and found it to be false.

CONTROVERSY ON AMAZONIAN ARIDITY

It is our conclusion that the Amazon lowlands remained forested throughout glaciations. The forests were never replaced by savanna, and the Amazon lowlands were never 'arid,' as the word is normally understood (Colinvaux *et al.*, in press). But the likelihood of significantly less precipitation in glacial times remains strong.

The best direct evidence for reduced precipitation is in the Carajás core of well-dated marsh sediments taken from a plateau rising out of the east-central part of the Amazon lowlands in Brazil (Absy *et al.*, 1991). The plateau lies between 700 and 900 m in a relatively dry part of the Amazon covered now by various of the drier facies of Amazon vegetation, including savanna. Excellent data on the mineral sediments show that water levels in the marsh-lake system were significantly lower in glacial times. Where our observations at L. Pata merely suggest lowered water levels at 18 k on the basis of gross appearance of the sediments, the work at Carajás demonstrated convincingly that the plateau was drier. These are the only two well-dated records of drying from within the Amazon basin in this time interval. Although the records are nearly 2,000 km apart it seems reasonable to correlate them as showing a widespread reduction in rains (Absy *et al.*, 1991).[10]

Observations of Amazonian soils, and such features as 'stone lines' within them have long been used to argue that the Amazon was not just arid but actually denuded of vegetation in places. This, we suggest, cannot be true. Such phenomena as 'stone lines' need reevaluation in light of the strong evidence that the basin remained vegetated. Dust in ice cores from the Peruvian Andes likewise cannot be attributed to the Amazon lowlands as suggested by Thompson *et al.* (1995).

Fossil sand dunes in the Orinoco, the mid Rio São Francisco and in various parts of northeastern Brazil are all outside the Amazon basin. Possibly the limited drying suggested by the Lake Pata and Carajás records was associated with dune formation and mobility in these drier regions outside the basin, but current studies on the São Francisco dune system under caatinga vegetation show that dune activation has been more intense in the last 4000 years than it was in the Pleistocene/Holocene transition between 11,000 and 9,000 years BP (De Oliveira *et al.*, 1999, in press).

Thus the preponderance of the evidence is that most of the Amazon lowlands remained forested in the ice-ages although precipitation was reduced. We should expect subtle changes in forest composition in the drier times.

This, meanwhile, is our working model of the environment of the ice-age Amazon. Carbon dioxide and temperature were both down. The Amazon lowlands were never arid, but remained covered with the various facies of closed tropical forest appropriate to cooling and modest drying.

MUSEUM OR EVOLUTIONARY ENGINE?

The 'museum' model of Amazon diversity is contingent on safe storage for tropical forest species. This 'safe storage' is provided by the ice-age Amazon as we reconstruct it. We deny that aridity eliminated much forest in glacial times. Instead, vast lowland tropical forests offered living space for Amazon forest species at all stages of the last glacial cycle, making reasonable and parsimonious the conclusion that this was true for preceding glacial cycles also. A safe warehouse has existed for the museum since at least the start of the Pleistocene epoch, two million years ago.

But the 'engine' model receives strong support also, because of the way that the long periods of ice-age normality are interrupted by brief interglacials, characterized by warmer temperatures and enriched CO_2. It should be emphasized that it is the warm interglacials that are unusual, not the ice-ages. Tropical biologists should learn to think of an interglacial like the present as the bad time, the time of troubles when long periods of placid ice-age normality are afflicted by dreadful heat.

The pollen data show that tropical plant species can live through these cycles, both in the Amazon and elsewhere in the tropics. But there can be no doubt that different plant species tolerate these fluctuations differently. Each time that ice-age normality was interrupted by the warming and enriched CO_2 of an interglacial there should have been a general reshuffling of plant species in the Amazonian communities.

Our pollen data give us a glimpse of this reshuffling in progress. The Lake Pata pollen record of ice-age forests shows that such genera as *Podocarpus* and *Hedyosmum* were included in what was otherwise lowland tropical forest. These are plants whose modern comparable populations are found a thousand meters higher, the nearest to L. Pata being on mountains a hundred kilometers to the north. We take their presence in the lowland forest to be an indication of cooling, a datum that brings our cooling record into the lowland Amazon itself. More importantly, these pollen spectra of ice-age times in which the presence of heat-intolerant plants is superimposed on the modern pollen signal for rain forest is direct evidence for the reshuffling of plant communities.

The greatest triumph of pollen analysis in recent decades has been the demonstration by Margaret Davis, Tom Webb and others working in eastern North America that communities and associations of forest plants are never permanent (Davis, 1986; Huntley and Webb, 1988). It used to be imagined that plants lived together in 'associations' that endured through the ages, that in North America or Europe whole plant communities retreated south when glaciers advanced, then went back up north again with glacial retreat. The same idea was applied to mountains where vegetation was supposed to exist in belts that marched up and down the mountains as intact plant communities. The pollen triumph was to show that this did not happen. Instead forest communities dissolved before advancing ice as every species found its own route to survival. The comparatively rich forests of eastern North American were shown, by a grid of about 300 pollen diagrams, to have been built slowly by sequential immigrations over the whole of the last ten thousand years.

Our pollen data from Lake Pata show that the trees of the Amazon forest communities obey the same rules. A tropical rain forest can include more podocarps if cooled just a few degrees, without ceasing to be a tropical rain forest. The forest communities change as the environments change. Species last, somewhere in that huge piece of real estate, but the communities of species are ephemera that are constantly reforming.

'Engine' models require disturbance that is less than catastrophic; disturbance that prevents aggressive competitors or predators from becoming so numerous that they eliminate other species. The historical record of Amazonia now shows that this level of disturbance operates on glacial and millennial time-scales as well as within the lifetimes of individual plants. Games of survival are played in the Amazon under rules which change subtly with the rhythm of ice-ages: this stokes the engine. But the gaming board is large and permanent: this provides escape, lowering the penalty for losing and letting the museum be built.

THE PROBLEM OF AMAZONIAN ENDEMICS

The answer to another diversity problem also lies in the ice-age Amazon, that of the local endemics. Forest birds and butterflies of some genera exist as distinct species in different parts of the Amazon basin such that their mapped distributions look like separate islands in a sea of green forest. Moreover some of the butterfly and bird isolates have roughly concentric distributions, occupying as it were the same 'isolate' in the unbroken sea of forest. Maps of these 'endemic isolates' look remarkably like maps of an archipelago in the oceans: only there is no ocean (Figure 2.2).

The birds and butterflies of these isolates are close relatives and are thus species that have certainly evolved from common ancestors within the forest. But such evolution should be possible only when populations are effectively prevented from cross breeding for a very long time. We say that there must be 'vicariance,' by which we mean populations separated by a physical barrier to 'take each other's place.' In the continuous Amazon forest geographical boundaries to separate populations for the necessary vicariance are not immediately obvious.

In a classic paper published in 1969, Jurgen Haffer noted that the endemic isolates cluster at the edges of the basin on the higher-lying bits of the forest, which led him, geologist as he is, to suggest that the forests of these elevated bits of land might well have been cut off from each other in the recent geological past although they are now an integral part of the main forest. Ice-ages were the only geological happening that seemed both drastic and recent enough. None of the recent data for the ice-age Amazon were available in 1969, of course, but the hypothesis of tropical aridity was already in vogue. Haffer suggested that in ice-age time forest patches lingered on high ground, and he further suggested that the intervening low areas were dry savanna so that the forest patches were effectively isolated. So here was the vicariance: new species evolved in forest patches separated for thousands of years in ice-age time by arid savannas, a beautiful hypothesis.[11]

If we are correct in concluding that the Amazon lowlands were always forested, the Haffer hypothesis has now been tested and falsified: there was no geographical isolation by aridity to promote the necessary vicariance. But the main tenet of the Haffer hypothesis, "vicariance in ice age time," can still hold. We just need to substitute cooling and reduced carbon dioxide for Haffer's aridity.

Cooling and CO_2 concentration should always be most important in elevated areas where their effects are compounded by altitude. Thus their effects are not to be seen in the separating areas of Haffer's sea of savanna but rather in the isolates themselves. The ice-age did not make the 'sea' as Haffer thought, instead it made the 'islands.' Vicariance is preserved by the discoveries of ice-age cooling and reduced carbon dioxide, which were not known when Haffer first put forward the hypothesis.[12]

NOTES

1. The calculation of forty tree lifetimes since the end of the last ice age makes the arbitrary assumption that a forest tree lives for 250 years, if it does not die young. 250 years, though possibly a common age for a tree, has now been shown to be very conservative. We cannot date rain forest trees as we date northern trees by counting tree-rings because most Amazon trees, experiencing no winters, do not have tree rings. Recently, however, large Amazon trees have been aged by radiocarbon dating of their heart wood, when some have been found to be up to 1500 years old (Chambers, 1998). There have only been seven life times of a 1500 year old tree since the end of the ice age.
2. In some circumstances it is possible that tree-line should also be sensitive to CO_2 concentration so that pollen data for depressed tree-line could reflect only the reduced CO_2 of ice-age time. This has recently been argued for the tree-lines of East Africa and defended by using a biome simulation model with pronounced aridity as one of the boundary conditions (Jolly and Haxeltine, 1997). The model is for a forest scrub biome in Africa, not the moister tree-lines of the Andes. Our data (described below) for Ecuador show that the phenomenon recorded on mountain sides is not in fact the movement of 'biomes' but of individual species populations, many at lower elevations than the tree line. These are C3 forest plants living in forests for which the biome model does not apply. Another model by Street-Perrott *et al.* (1997) argues that, because trees are C3 plants the depression of tree-line in glacial times reflected their inability to maintain themselves at high altitudes in conditions of reduced CO_2. This begs the whole question of the physiological limits of trees (Tranquillini, 1979) and can be argued in detail. However, the point is moot for the Amazon flank of the Andes where tree populations well below the tree-line have been shown by the Mera and San Juan Bosco data to have descended in synchrony with glacial advance.
3. The Mera record, and by implication, the San Juan Bosco record also were directly challenged in the literature by K. Heine, essentially on the grounds that the radiocarbon dates of 26,000 to 30,000 years could conceivably be the result of contamination of ancient wood by a little modern carbon. Heine falsely described the Mera lacustrine deposits as 'lignite' and declared that the weathering of the deposit appeared such as to suggest great age. Heine attempted to date the Mera lahar by the uranium–thorium method getting 'ages' that ranged from 8,000 to 86,000 years, these varied dates presumably depending on the histories of the samples before they were emplaced by the lahar (flow of volcanic debris, mud, rock and water set in motion by the eruption). Thus Heine was not able to impeach the radiocarbon dates except by assertion (Heine, 1994). Subsequent demonstration of similar vegetation descents at San Juan Bosco, in Pacific Panama, and at Lake Pata, together with other evidence of tropical cooling, support the original interpretation of the Mera record as an history of glacial cooling 26,000–30,000 years ago in isotope stage 3 time.
4. We have since recovered a sedimentary record from near tree line a thousand meters directly above the San Juan Bosco section that provides pollen evidence for the descent of tree-line in Ecuador (Colinvaux *et al.*, 1997).
5. The El Valle crater had drained in the early Holocene but we were fortunate in finding a complete late glacial and Holocene portion of the record in nearby Lake La Yeguada, the pollen percentage diagram from which is given as Figure 2.4. Notice that the Urticaceae/Moraceae signal of tropical forest is present throughout, despite the cooling signals of *Quercus* and other pollen in the late glacial. The high Gramineae percentages of Holocene time describe the anthropogenic savanna (see surface samples in Chapter 1); Gramineae pollen is low in the cooled tropical forests of the late glacial (Bush *et al.*, 1992).
6. More climate models are appearing that provide for cooler tropics in glacial times, as for instance (Ganopolski *et al.*, 1998).
7. In this paper we publish the pollen data from only the top 1.7 m of the sediment column, spanning the last 42,000 years. Extrapolation suggests that the complete 7 m core might span the entire glacial cycle. A survey pollen analysis shows that

forests can never have been replaced by grassland at any time in this long period.

8. Haberle sampled sediments from the Holocene through oxygen isotope stage 6 (the glaciation before last), the sediments being assigned ages by standard oceanographic criteria by the ODP team. Haberle also sampled the sediment load of the modern river on a west to east transect by river boat in order to calibrate the fossil pollen assemblies against pollen being delivered from the Amazon vegetation of the present day.
9. Objections to this conclusion, although not yet published, have been offered privately. Essentially they claim that pollen in the fan sediments represents 'only' the river banks, so that in glacial time ribbons of forest existed along the rivers but not on the interfluves. This is a most unreasonable bit of special pleading: the idea that the grass pollen clouds from huge areas of savanna could have been masked by narrow strips of riverine vegetation along the rivers beggars the imagination: compare the pollen diagrams from forest at L. Pata with the savanna pollen from Central Brazil (Figures 1.1 and 1.3).
10. We differ with these authors, and with Van der Hammen and Absy (1994), in our interpretations of the pollen records. The Carajás pollen diagram throughout has high grass and herb components and is without virtually all the pollen types we associate with tropical forest, and which make up 80% or more of the pollen counts at L. Pata, as well as the counts from our extensive pollen trapping in the modern rain forest. The arboreal pollen found at Carajás appears to us to be entirely of trees typically associated with the various forms of open and grassy woodlands of the kind reported from the modern Carajás Plateau. The herb to arboreal ratio of pollen types certainly increased in the drier ice-age climate, but in our opinion this is a result of edaphic constraints at the site itself and conveys no information about the lowland forest 600 m below the plateau (see note 6 in Chapter 1). Professor Hooghiemstra differs with us on this, accepting that the lowered percent of tree pollen in glacial times at Carajás requires disruption of the forest below the plateau. His reconstruction of the ice-age Amazon accordingly shows a wide band of possible savanna running north and south across the eastern part of the Amazon basin. This reconstruction, however, like ours preserves an immense tract of forest in the western Amazon basin.
11. Haffer's refugial hypothesis now has a large literature. The key reference is Haffer (1969). Elaborations of the hypothesis are provided by the papers in Prance (1982) and in Whitmore and Prance (1987).
12. The obvious alternative to a Pleistocene vicariance hypothesis has always been that the endemics, at least the birds, are older than the Pleistocene. This possibility has moved closer to the probable as the result of recent molecular studies of relationships between North American birds. It has long been supposed that the pattern of different bird species between the eastern and western United States developed in the Pleistocene when ancestral populations were fragmented by glaciers, but the new molecular studies most strongly suggest that the separations were much older than the Pleistocene. If the Amazon endemics are also of Tertiary age (> three million or more years old) then it is only necessary to explain how they have remained separated throughout the Pleistocene. They are at the periphery of a vast forest which, on our reconstruction, persisted throughout the Pleistocene, suggesting that local environments (Haffer's elevated areas) are sufficiently different for the local populations to persist.

CHAPTER

3

A Guide to Piston Coring of Lake Sediments

A pollen analyst seeking long or ancient records from the Amazon basin will be glad to accept any polliniferous deposit that fortune and geography provide: bogs, road cuts, valley fill, old river beds, fossil soil profiles, even archaeological digs. But lakes, particularly those between 100 m and 1000 m across, have advantages that make them the targets of choice. In lakes of this size, the regional pollen rain (the windblown part of the pollen signal) is most reliably represented, with minimal chances of the local overrepresentation that happens in bog or marsh deposits. Pollen preservation is usually excellent in lake sediments. Sedimentation rates are often constant over long periods, thus allowing sedimentation rates to be accurately measured with an adequate array of AMS radiocarbon dates, which in turn allows calculation of pollen influx. Lake sediments are less susceptible to postdepositional disturbance than other deposits. Other paleoecological signals (such as diatoms, sediment chemistry, plant pigments, magnetic intensities, isotope ratios, etc.) are more likely to be valuable in lake deposits than in bogs or soils. And, in the Amazon, lake sediments are the only likely sink for animal dispersed pollen moved from the local watershed in runoff water.

Collecting the best samples for pollen analysis, therefore, becomes a matter of choosing a lake and raising complete cores of its sediments. This is a straightforward undertaking. A raft, usually made of rubber boats, is set in the middle of the lake to at least three anchors on long ropes so that its position remains fixed. A sectional pipe (the casing) is lowered into the water so that it connects the raft to a fixed point on the bottom. The core is then taken in sections with a hand-operated drill (a piston sampler) operated inside the casing. Each successive sample tube is able to find and reenter the old hole because guided to it by the casing (Figure 3.1).

PISTON SAMPLERS

The more organic lake muds (*sapropels* or *gyttja*) are, when underwater, soft and easily penetrated with a tube. Yet just pushing a simple tube into the mud does not result in a complete core because the mud tends to block the end of the tube, after which continued pushing on the blocked tube merely pushes the sediment aside rather than letting the sediment enter. This happens because friction between the inside of the tube and the sediment quickly becomes greater than the force needed to push the sediment aside. In lake coring, we overcome this inner friction by driving the sample tube past a captive piston so that sediment enters the sample tube for the whole of its length.

Piston samplers act like bicycle pumps in reverse. When pumping air into a bicycle tire, one pushes a piston down a tube to expel air from the bottom. In a sediment sampler, the piston starts at the bottom of the tube which is then pushed down past the piston. Because the piston has a good seal to the inner walls of the descending tube, pressure within the tube can only be equalized if sediment enters at the bottom as fast as the tube is forced down. The weight of water on top of the mud (hydrostatic head) pressing on the mud surface equals the weight of water on top of the piston as long as the mud fills the tube. In this way hydrostatic pressure is used to overcome the friction of the mud inside the tube. The process can be visualized by saying that sediment is sucked into the tube — the bicycle pump in reverse.

Thus, the essentials of a piston sampler are the sample tube, a system of rods and connector to lower it through the water, and a means of holding the piston in place as the tube is pushed down past it. Holding the piston in place when at sampling depth is easy. It is simply attached to a flexible cable (wire rope), the end of which can be clamped to the raft. Further pushing on the extension rods then drives the tube down past the stationary piston in the desired manner (Figure 3.1).

For a piston sampler to work properly, the piston must not be able to move before its time. The sampler must be lowered to the bottom of the lake and down the old hole with the piston safely positioned at the bottom of the sample tube. Then and only then can it be set free, the piston cable clamped to the raft, and the tube pushed down past it. But there can be strong pressures tending to move the piston from its correct place at the bottom of the sample tube, particularly when pushing the sampler down the old hole. It is absolutely essential that the piston not slip from its position closing the mouth of the sample tube until it is precisely at the depth within the sediment column at which the next drive is to begin.

Two alternative methods are used to secure the piston in place until deliberately freed. One method is to have a rod inside the sample tube that physically prevents the piston from being forced upward by water pressure. At sampling depth this 'sliding rod' is withdrawn and its end locked in place as the bottom of the drill string. The alternative is to design a piston that locks inside the mouth of the sample tube until released by tension on the piston cable.

The widely used 'square rod' sampler is of the first or 'sliding rod' type. The push rods of the drill string are connected to the sampler by a special last rod that is square in cross section. The square rod slides through a square aperture in the sample tube holder until its bottom rests against the piston. The sample tube and piston are then pushed down together until the sampling depth is reached, at which point pulling on the drill string withdraws the sliding square rod until the end is clear of the inside of the sample tube. The end of the square rod is now clear of the square aperture, although still within the tube holder. Rotating the drill string (and hence the square rod) through 45° then lets its corners bear down on the sides of the square hole through which it had been withdrawn. The piston cable is then clamped to the raft so that the piston can descend no further. Pushing on the drill string now drives the sample tube down past the captive piston.[1]

This 'square-rod' system works well in many lakes but it has some characteristics that we have found unsatisfactory. For packing into remote jungle sites, the cumbersome square rod assembly is both heavy (the rod is solid metal and about 1.2 m long) and sensitive to damage (the slightest bend will cause it to bind in the square hole). More serious in our view is the fact that it is difficult to hammer on the drill string in order to penetrate stiff clays or volcanic ash layers because there is no rigid connection between the end of the drill string (the square rod) and the sample tube holder. Accordingly, we have obviated the need for a sliding rod altogether by designing pistons that lock into the mouth of the sample tube until released by tension on the piston cable.

The locking piston designed by the engineer Vince Vohnout is shown in Figures 3.2 and 3.3. The principle of this design is that the lower rubber piston seal can be squeezed between a rotatable cam and a terminal hex nut reachable from outside the tube (Figure 3.4a). The rubber seal then presses against the inside of the sample tube (arrows in Figure 3.4a). Quite moderate tightening of the nut locks the piston so firmly in place that even blows with a hammer will not move it. However, the pressure on the rubber is easily released by rotating the cam so that its flat surface is horizontal (Figure 3.4b). The rubber seal resumes its normal shape and the piston is free to move. To rotate the cam it is only necessary to pull on the piston cable.

To use this sampler, the piston is inserted into the sample tube in the 'cocked' position with the cam rotated so that the rounded face presses against the metal washer on the rubber piston seal. The cocked piston is then pushed down the sample tube with a rod until flush with the bottom of the tube, when the nut holding the bottom piston seal in place can be tightened with a wrench until the piston is firmly in place (Figure 3.4a). The assembled sampler is then pushed down into the lake or old hole with connecting rods until the required depth is reached (one can even bulldoze through overlying sediment if desired, for the locked piston will not slip when pushed through lake muds). The piston cable is then clamped to the raft. A sharp jerk on the rods tightens the piston cable, the cam rotates, the piston no longer grips the tube so firmly, and one can drive the tube down past the stationary piston by pressure on the rods (Figure 3.4b).

An obvious advantage of this system over a square rod is that the sample tube is rigidly fixed to the drill string so that hammer blows can be transferred from rods to sample tube with minimal loss of impact. We use a ten-pound (5 kg) drop hammer that slides on the drill string and impacts an anvil fitted between the top two drill rods. By threading the drop hammer on below the anvil this hammering system can even be used for hammering upwards to free a stuck sampler in an emergency (Figure 3.5).

This piston design has the added advantage of being able to lock into the sample holder at the end of a drive. At the end of a drive the tube is filled with sediment and the piston is at the top of the sample tube, pressed against the tube holder. But now one must pull the tube out, creating a powerful suction that will tend to pull both piston and sediment out of the tube again. To prevent this, the piston must be pulled up at the same rate as the sample tube. We can do this by pulling on the piston cable — one person pulling up rods while a second pulls up the cable at the same rate. This is what must be done with square rod samplers. But with our design there is no impediment to fitting a locking device that will let the sample holder grip the piston so that both sample tube and piston, locked together, can be pulled up by the rods. We use standard quick disconnect hose couplings, with the male fixture in the top of the piston and the female part of the coupling in the holder (Figure 3.6). The piston at the end of the drive is now safely locked to the holder and cannot be pushed back down by water pressure. The danger of losing the core is significantly reduced. The relief to the raft crew who do not have to hold the cable tight during the recovery is also considerable.

Cable Clamps and Recordkeeping

With the sampler lowered to the correct depth and the piston in place blocking the mouth of the tube, the piston cable must be fastened to the raft. Two ways of doing this are to wind the slack of the wire rope round an ordinary cleat or to employ a clamp of some kind. We recommend using a clamp and slot technique as shown in Figures 3.7 and 3.8. The clamp is attached to the cable, which is constrained because the clamp will not pass through a slot in a bridge over the top of the casing (Figure 3.8). This fastens the cable directly over the hole and vertically above the sampler. More importantly, the position of the clamp on the cable yields a precise record of the depth at which sampling began, an immense advantage when preparing to take the next section of core.

When working in 10, 20, or even 30 m of water, it is not possible to feel through the rods (drill string) when the sampler is entering the mud or when it reaches the bottom of the hole, so one must keep track of the depth reached by measuring from the raft or water surface. Without a clamp on the piston cable, this measuring becomes a tiresome business of keeping count of the number of rods down the hole and of actually marking the top rod each time at the level that it passed through the hole in the raft. Having a clamp firmly fixed to the cable at the point where the preceding drive started avoids all this.

The clamp marks the depth at which the piston had been locked into place at the start of the preceding drive. During the drive, the tube was thrust past the piston for one meter. The piston of course had not moved during the drive, but nevertheless ended at the top of the tube above a 1 m section of sediment core. Therefore, to start the next section, the new tube carrying the piston in its mouth must be pushed exactly one meter further than the last. This is easily achieved by moving the cable clamp exactly one meter farther back along the cable, then pushing the sampler down until the one meter of slack in the cable is taken up. The locked piston is now at the bottom of the old hole and the new drive can begin. This is a much simpler (and more accurate) procedure than keeping count of the number of rod lengths and partial rod lengths that have gone down the hole.

There remains the problem of finding the depth of the mud–water interface where the first section of core should begin. To sample the top few centimeters of unpacked sediment, it is necessary to use a special sampler (described in a later section). But the approximate sediment surface must be found to start the main core. This cannot be felt through the rods, but its approximate position can be found by lowering a 12-inch (25 cm) plexiglas (or similar material) disk on a line. When we have had a recording sonar with us to make bathymetric maps, we have used this to watch the descent of the sample tube towards the mud, stopping when we see contact on the sonar (for this first meter drive the casing is kept deliberately short so that the sampler can be seen as it projects below the casing).

Once the piston cable is clamped for the first drive, however, an accurate tally of relative depth is easily kept by simply moving the clamp successively along the cable by the amount of sediment cored in the preceding drive. Usually, this distance is one full meter, but if for some reason a drive was only partial, then the length of sediment in the recovered tube is measured and the clamp moved along the cable for this distance instead of a full meter.

Fitting a Shoe for Stiff Sediments

Soft organic sediments (*sapropel* or *gyttja*) are easily penetrated by simple pushing — the weight of even a small 110-pound (50 kg) human is ample for the job. Sticky silts and clays are another matter and are sure to be encountered in any prolonged coring operation, if not throughout the column, then very likely in the basal sediments near bedrock. When one can push no further, hammering is necessary. But the task can be made easier by reducing friction with a shoe.

The serious resistance offered by stiff sediment is not caused by the need to deform the sediment to make room for the sampler, rather the resistance is from friction against the sides of the sample tube. Sticky sediment clings to the sampler like the mud that seizes rubber boots when wading across a muddy pond. To penetrate stiff sediments, therefore, we must do what we can to reduce friction. The most practical way is to arrange for a short length of the bottom of the tube to be wider than the rest, thus punching a hole through which the tube follows easily. A shoe achieves this (Figure 3.9). Our shoes are essentially short (5 cm) tubes of mild steel, tapered at the leading edge for easy entry into the mud (not too sharp, as knife edges are not wanted in a rubber boat). The shoe slips over the end of the tube, which rests against a shoulder inside the shoe. The point of a screw in the shoe projects into a slot in the sample tube to hold the shoe in place during recovery (Figure 3.9).

It is important to realize that the purpose of the steel shoe is NOT to cut into the sediment better than the bare tube does. The purpose is to reduce friction by punching out a hole of wider diameter than the core tube that will follow the shoe down the hole. In theory, the only friction encountered should then be over the length of the shoe, with the sample tube that follows the shoe through the enlarged hole being loose in the boring and in contact with water rather than mud. The practice is not so neat, some mud managing to remain in contact with the tube, but penetration while hammering is much improved when a shoe is used all the same.

The practical limits to coring in stiff sediments are set not so much by difficulty in penetration as by the difficulty of recovery. You can hammer your way through stiff sediments, but you may not be able to pull the sampler out again. In his 37 years of lake coring, Colinvaux has lost two piston samplers in this way, one in a large lake in Arctic Alaska, the other in the central Amazon of Brazil. They remain as underwater relics, potential trophies for an industrial archaeologist of the future.

The force we can use to extract stuck samplers is limited to the flotation of the raft. It is comparatively simple to mount a winch or jack to the wooden raft platform, but applying this kind of force to the rods of a well-stuck sampler merely pushes the raft underwater. Indeed, experience shows that the pulling power of two or three strong people is sufficient to swamp our standard raft of two rubber boats. We have no way of applying more extractive force from lightweight rafts, except perhaps for hammering up.

Hammering up will free a stuck sampler, given our rigid system of attachment of rods to sampling tube (i.e., no square rod) but the method is narrowly constrained by the fact that the couplings on the rods are likely to break before the sampler is freed. Hammering down works well because the rods all push shoulder to shoulder. But for pulling out, the rods are joined only by the screw couplings. Each coupling is held into its parent rod by adhesives and one or more through pins. The adhesives can yield and the pins can shear.

The best advice to offer is to hammer down cautiously. If each hammer stroke drives the sampler down perceptibly, you can probably continue until the full meter has been sampled. But be prepared to settle for less than a full sample tube if you are making no progress. When the research goal suggests that you must do your best to sample very stiff sediments, it is a good idea to use half-meter core tubes instead of meter tubes (cut them in half with a hack saw if you must). Always use shoes; the resulting reduction in friction can be even more important when pulling out than when hammering in. When you do get stuck, so that no amount of strong-back pulling works, have your crew take the strain as hard as they can and, while they are pulling, try hammering up with a few strokes. Usually this will move the sampler those first few critical centimeters, after which a sustained pull will retrieve it. If this does not work, try filling the hulls of the rubber boats with water to sink them, then take the strain as before and, at maximum strain, clamp the piston cable to the raft. Then bail out the hulls and leave the raft yourself. A constant fierce pull is then applied to the cable by the buoyancy of the raft, amplified if there are waves on the lake. The sampler might be worked loose in this way. On the other hand, both the lost Colinvaux samplers went through this process, until the steel cables (3/32" wire rope — airplane cable) snapped. You can then go on hammering up until one of the connecting rod couplings also breaks underwater. This is the end of the road.

One other point about the use of shoes needs to be mentioned. Often when using a shoe, a drive of a full meter — ending with the piston satisfactorily locked into the tube holder at the end of the drive — will yield a sediment core less than a meter long. Despite having the piston held at the end of the drive, it is apparent that part of the core, usually 5–10 cm, has been lost out the bottom of the tube during recovery. The loss may be immediately apparent as escaping sediment, or you might discover the loss when opening or x-raying the tubes later when it appears as a gap or series of gaps in the core section. Our explanation of this phenomenon is that the shoe at the bottom of the borehole fits so tightly into the mud body that the seal between it and the mud does not break when one starts to pull the sampler out. Pressure inside the tube is so great that mud is expelled from the bottom. We are redesigning our pistons with grooves in the outer face in the hope that this will help to break the seal between shoe and borehole. If sediment should be lost from this cause, it is all of course lost from the bottom of the tube even though breaks can appear in the core section. In reconstructing the core stratigraphy, all the missing sediment should appear as a gap at the bottom of the meter section.

The best summary advice on the use of shoes and hammering is to proceed without shoes as long as possible. When you do encounter really stiff sediments, use a shoe and hammer with caution. Be prepared to hammer hard when you must, for there can be good sediments to be won in this way. The top 1.7 m of the Lake Pata cores (Figure 1.1) were raised without hammering, but we got a further 5 m below them by hammering until we reached the gravel lining of bedrock. We believe these extra 5 m (which are still being worked on at the time of writing but which we have already determined to be richly polliniferous) extend the Lake Pata record through most of the last glacial cycle.

To Extrude or Not to Extrude: The Compromise of Transparent Tubes

When the filled sample tube is safely on the raft a decision must be taken: should the core be sealed in the tube for later opening in the laboratory, or should it be extruded on the raft for inspection, before being wrapped in plastic for the journey to the laboratory? Different research teams have differing opinions, strongly held. We are somewhat uncompromisingly of the opinion that it is wrong to extrude in the field. But both schools of thought have valid reasons for their opinions.

A sediment core inside its sample tube is as safely packed as it ever will be. If the tube is closed securely, the core can be transported to the laboratory (however distant) essentially undisturbed. It can then be opened under controlled laboratory conditions when preparations for analyses have been made. The core can then be revealed by slitting open the tube instead of subjecting the sediments to the extra manipulation of extrusion (pushing the core out of the sampling tube). Moreover, it is far quicker (and thus a better use of field time in remote places) to close the tubes on the raft than to undertake the work of extrusion, description, and repacking in the field.

The reason given in support of extrusion is that one knows what one has got and can modify the rest of one's coring operation accordingly. We doubt that any advantage from having a quick look at your mud can justify the risks to your sediment inherent in exposing the mud to field conditions, in the risks of transport that result, and the time lost.

Nevertheless, there could be occasions when the sight of sediments (say spectacular banding, or a stratigraphic boundary half-lost at the bottom of a core section) should guide a new core. As a compromise we are investigating the use of transparent polycarbonate tubing to replace the aluminum core barrels we now use so that we can get sight of the sediments without extrusion.

Meanwhile we recommend for Amazon work the practice of sealing the core tubes in the field and shipping the sediments back to the laboratory safely encased in their aluminum or polycarbonate core tubes. If they are then stored in a 4°C cold room, they will last in apparently pristine condition for months — even for years if necessary — until you are ready to analyze them. Meanwhile, you can view the stratigraphy noninvasively and nondestructively by making x-radiographs of the sediments while still in their tubes. The standard x-ray facilities of an hospital are adequate for this purpose and we have found that it is usually possible to arrange for this with hospital management.

Our policy of keeping cores in the sample tubes is expensive in tubes (the aluminum tubes cost about \$20 each). This cost has to be evaluated, however, in light of the heavy investment represented by the whole coring operation. We believe the cost is a reasonable insurance premium to pay for so precious a cargo.

Sampling the Mud–Water Interface

A modified piston sampler can be used to sample accurately the very diffuse sediment of the top few centimeters and of the mud–water interface itself (Fisher *et al.*, 1992). This not only provides a voucher for the pollen rain falling on the lake from the modern vegetation, but it also allows close interval sampling of recent events so that the effects of contemporary environmental changes can be studied.

The principle of the surface sampler is to take a piston core that spans a length of water over the mud–water interface and the top few decimeters of soft sediment in a transparent tube. The tube is raised until near the surface, when the bottom of the tube is closed by reaching underwater to insert a laboratory rubber stopper that fits neatly into the tube like a second piston. The whole watery sample can be lifted out of the lake while keeping it vertical. The undisturbed mud–water interface can be seen in the transparent tube and then sampled by pushing the sediment from below and displacing the water until the mud surface is accessible at the top of the tube. All the sediment can be sampled as a series of slices as the sediment is repeatedly pushed up from below.

The important special property of the sampler that permits this upwards extrusion of the water/mud core is a bypass tube through the piston that can be opened to release pressure under the piston (Figure 3.10). With the bypass open, the piston can be pulled out by its cable without the water and mud underneath being disturbed. The piston-like rubber stopper that was used to block the bottom of the tube can now be used as a plunger to push the mud and water up through the tube. With our small rafts, it is best to take the filled tube ashore in the tender for this (being careful to keep the tube vertical all the way).[2]

This sampler can be made to fit on the extension rods used with the Livingstone sampler and should be used from the anchored raft, preferably before the casing is lowered with its danger of disturbing sediments under the raft. Taking the sample from off one side of the raft will keep it clear of the position of the main borehole. When visiting a lake to sample the sediment/water interface without raising a long core, however, this sampler — with its need for a string of rods and a stationary raft — is not very practical. However, it is simple to build a weighted version that can be lowered with a rope rather than pushed with rods. This is the best choice for a project to collect surface samples from lakes of a region, since all that is needed is the smallest, lightest kind of inflatable boat, the sampler, a coil of rope, and vials to store the sediment samples in.

The Casing

The casing is a simple pipe through the water. The cheapest, almost certainly the most widely used, is ordinary PVC pipe sold for plumbing and drainage work. It can be bought at any large hardware store, it is cheap, and thus expendable. With standard sleeve or flared couplings, lengths can be pushed together to make smooth joints that can be held together with duct tape. It does have the minor disadvantage that long lengths of it are flexible, which can be awkward. The flexibility problem is overcome by using aluminum pipe, but then a fresh problem arises in that metal pipe is usually joined with screw couplings, which seldom let the sections of pipe butt neatly together. The consequence is a series of rough ledges on the inside of what ought to be a perfectly smooth guide tube for the sampler. When lowering the sampler through this casing, it is likely to catch at the joins. The problem is even worse in the recovery when you are pulling up the core tube as gently as you can and it jams on one of the rough places. The solution is to use split-sleeve couplings. These couplings are made of short lengths of an outer aluminum tube (ours are 6 inches [15 cm] long) that slide easily over the casing. A narrow longitudinal slice is cut out of one side of this 'sleeve coupling' so that it can be compressed by hose clamps. The lengths of casing to be joined are slid together inside the sleeve-coupling to make a virtually seamless butt-joint and the sleeve is squeezed onto the joint by tightening the hose clamps with a screw driver. The join is tight, stiff, and virtually impossible to pull apart. The resulting casing is completely smooth on the inside. But our first field use revealed one serious flaw: prolonged vibration from hammering worked on the long casing pipe as it dangled from the raft so that some couplings slipped. Vibration and whiplash on the long string had overcome the couplings when a straight pull could not. The solution was to provide a small protrusion or lug on the casing that slides into a right-angled slot in the sleeve coupling before the hose clamps are tightened (Figure 3.11). Couplings of this design stay together, despite whatever one does to them, and the inside of the casing string remains completely smooth.

THE RAFT AND ITS USE

Any raft that you can stand on will do: oil drums, styrofoam, balsa logs, canoes lashed together, rubber boats, jettisoned fuel tanks from fighter aircraft. We recently saw a photograph of a most elegant raft built by Francis Mayle out of tropical hardwoods in a remote site in Amazonian Bolivia. But it is most convenient to have a standard portable raft that can be taken with you. This means the use of inflatable components.

We use two Avon Redstart inflatable yacht tenders, placed side by side like the two hulls of a catamaran. We place a wooden platform across the hulls and work between them (Figure 3.12). Each Redstart is 8 feet long (2.4 m) and is rated to carry three people when used with a motor. It weighs 35 pounds (15 kg) when dry, which makes it a reasonable packload for one person that can be carried long distances if necessary. This is the ideal size for our purpose — any larger and the boat becomes unmanageable on the trail. The Redstart is the smallest boat made by Avon. The next size up is nine feet long, which does not sound much more, but in fact the difference is tremendous. The nine-foot boat is heavier and has much larger air tubes — a beautiful yacht tender but impracticable for our purpose. There are, of course, other brands of high quality inflatable boats on the market, but whatever brand you choose, make sure that it is roughly 8 feet long (no more than 2.5 m), rated for three people, is a reasonable packload for one person and that it does not have a wooden floor, wooden transom or other heavy parts. Anything bigger than this will not do, at least not if you are going to man-haul your raft into remote places.

Inflatable boats of this quality have a number of tie points so it is an easy matter to lash them side by side. You want a gap between the hulls wide enough so that the casing passes through freely. The best way to arrange this is to fashion two spacer blocks out of styrofoam or wood that are set one each end of the hulls, which are then squeezed against them by the lashing.

The working platform is of wood and designed to straddle the narrow gap between the hulls. A large platform has obvious advantages as a comfortable working surface, and providing it with an understructure of poles 3 m long to span completely across both boats makes for great stability. But so large a platform is an extreme nuisance for transport, particularly for man carriage through the forest, as well as a nuisance as a piece of airline baggage. The Colinvaux platform has consistently shrunk over 30 years of experience. Now it is only a square yard (0.9 m), though he will concede that a meter square is acceptable. Even that small platform is made of two pieces which, together with two cross supporting pieces, can be dropped into a backpack or airline suitcase. Details of the design are shown in Figure 3.12. Use strong but light wood for construction (not plywood). To fasten anchor ropes or tools, and to tie the platform to the hulls, bore holes through the wood at suitable intervals. Never use cleats, ringbolts, etc. for this purpose as these would protrude dangerously in what will be a crowded work surface. Arrange a hole in the center big enough for the casing (plus a retrieving hand and arm) to reach through (Figure 3.12).

Raft Accessories

A bridge to hold the cable clamp over the casing is essential and a pair of wooden jaws to grip the casing, though not absolutely necessary is so useful that it quickly becomes to seem essential once you start using one.

The Colinvaux casing-jaws are 13 inches (33 cm) long and made of two pieces of pine 1.5 inches (4 cm) thick and 3.5 inches (9 cm) wide, hinged together at one end with a long strap hinge. Hemispherical slots in each side meet to make a round hole that fits round the casing but lets the wider couplings of the casing rest on top (Figure 3.13). Provision is made for tying the jaws securely closed, either with strong twine or with a catch.

Once on station with casing lowered to the full depth of the lake, the top of the casing is gripped with the jaws, which now straddle the hole through the raft. The top coupling, which in our standard model is six inches (15 cm) high, rests on the closed jaws. The opening into the casing must now be bridged to provide a slotted surface through which the piston cable can run and against which the cable clamp can rest securely during the drive. We have used a stout wooden box or sleeve, open at both ends and made to fit over the coupling as it protrudes through the jaws. A stout ledge, with a slot for the cable, covers about one third of the top opening, leaving space for the sampler and rods to pass through but also providing lodgement for the cable clamp (Figure 3.8). While this was being written, Jason Curtis had produced a better and simpler design in which the heavy wooden box is replaced by a cylindrical section of PVC pipe large enough to enclose the emergent coupling. The slot for cable and clamp is a half-moon-shaped piece of PVC reinforced by a steel plate epoxied to it.

A last and very essential raft accessory is a set of wooden fishtails to hold anchor ropes, other long ropes, and the piston cable (Figure 3.14). Attempting to coil ropes in a small rubber boat full of drill rods and human feet is not a rewarding undertaking. Accordingly, we wind all our ropes on fishtails. These are made of plywood. Marine plywood is best but commercial grade exterior plywood is quite OK. We tend to make ours out of whatever scrap plywood we happen to have.

Anchors and Anchoring

The raft has to lie to three anchors spread to divide a circle into three equal parts. All three anchor ropes must be pulled tight and taut, preventing the raft from moving in any direction. The anchors, therefore, must be very firmly sunk into the soft mud of the lake floor. Thus we need heavy mud-weights or the kind of small boat anchor that can dig deep into the mud, sometimes called sheet-metal anchors. Small, folding anchors will not do. By far the best plan for work at remote sites is to take along strong bags to be filled with rocks, gravel or sand from near the lake. Each anchor should be as heavy as can conveniently be managed by one person from a rubber boat, say 25 pounds (12 kg) or so.

In the Amazon lowlands, this simple plan of filling bags with rocks to make anchors is frequently frustrated by the fact that there are no rocks in the weathered bottom land. You can be faced with the regrettable necessity of packing rocks in from the nearest outcrop or the closest place you can reach by vehicle. Sheet metal anchors in the 10-pound (5 kg range) are then the answer, a nuisance to carry but better than a pack full of rocks.

An alternative to heavy anchors that is sometimes possible in small lakes is to tie to trees on the bank. To this end it is prudent to have anchor ropes that are 100 m long. Obviously this expedient will not work for lakes more than about 150 m across.

Anchor ropes must always be long so that the pull on the raft is as close to horizontal as possible: the working rule is that the anchor rope must be at least three times as long as the depth of the water. Lakes in the Amazon lowlands are mostly shallow, nearly all that we have encountered being less than 10 m deep, with most less than 5 m deep. If we have good reason to believe that we shall not encounter deeper water than this, so that we limit our depth range by taking only 10 m of casing, we take only 50 m of rope for each anchor, though we usually manage to have a spare coil of another 100 m. The best ropes to use are braided nylon, which are so strong that they can be quite thin.

An assault on deeper lakes requires foreknowledge of the depth likely to be encountered. Lake San Marcos in the Ecuadorian Andes was successfully cored in 37.5 m of water. This required not only anchor ropes each 150 m long but 40 m of casing and rods in proportion. Fortunately the field party was able to reach San Marcos by vehicle to carry this mass of equipment, but the point is that coring deeper lakes requires foreknowledge and appropriate planning. Send a reconnaissance field party to sound the lake before planning the coring operation.

Planting Anchors from a Tender

The actual act of anchoring a raft securely in open water is the part of the coring operation that most requires some developed skills, but these are the common skills of watermanship. Have as a team member someone who is used to 'messing about in boats' and all will go smoothly. The procedure is to plant the anchors from a separate boat. We use a third Avon Redstart for this purpose, so that our minimum field 'fleet' is three Avon Redstarts, two of them lashed together to make the raft and the third as our 'tender' to tow the raft, to plant anchors, and to ferry people and supplies between anchored raft and shore. For small lakes we rely on oars for propulsion, but in larger lakes we use a small outboard motor mounted on the tender.

Use the tender to tow the raft to the chosen spot in the lake, usually the deepest part. For this journey the anchors are carried on top of the wooden platform with their fishtails of rope tucked away in the hulls. Once on station, an anchor is passed to the person in the tender, its fishtail of rope being retained by someone who is standing on the platform. This person is in charge of the anchoring operation. The first anchor is taken obliquely up wind by the tender, which is slowly rowed in this direction while the person on the raft pays out rope from the fishtail. The idea is to get the line out quickly but without pulling the raft away from its station more than can be helped. (If the day is windy it can help to drop an extra anchor on a short rope directly from the raft to stop it from drifting off station while the three main anchors are emplaced.) When the tender has got far enough away the person on the raft orders the anchor to be dropped and ties the rope to the raft. The tender comes back for the next anchor.

Now comes the most tricky part of the operation. The second anchor must be placed obliquely upwind towards the side opposite to the first anchor, the object being to have the raft set between a 'V' of anchor ropes with the wind blowing straight down the 'V.' The person on the raft must keep this desired configuration in mind as he directs the tender to row away with the second rope and to drop the anchor. This is where a sense of watermanship comes in. Remember that the relative position of the raft can be moved after the first two anchors are planted by pulling or paying out one anchor rope or the other. If you do not get it right the first time, send the tender to pick up an anchor and move it (to do this, the tender of course has to get to the anchor by taking hold of the rope near the raft and pulling along hand over hand until above the anchor). Pulling up a 12-kg bag of rocks deeply buried in the mud is not an easy task from a rubber boat but any effort is worthwhile that gets the anchoring right.

When the first two anchors are placed satisfactorily, planting the third is simple: the tender collects the third anchor from the raft and is sent downwind (the direction opposite to the 'V' formed by the first two anchors). The three anchor ropes are now pulled taut, so taut that they twang when plucked like the strings of a guitar. They are tied to holes bored in the wood of the platform, never to the boats. When finished, the raft should be the center of a three rope star like the badge on a Mercedes automobile. If the ropes are taut, and the anchors heavy enough, the raft will remain on station for days if need be.

SUMMARY RUN-THROUGH OF A STANDARD CORING OPERATION

First, send your tender to sound the lake, to determine where to core (usually the deepest spot). Mark the chosen spot with an anchored float (a plastic bottle attached to a small rock with string is fine). Make a complete bathymetric map if possible. Take water samples (and plankton tows if practicable) to calibrate radiocarbon if for nothing else.

Build and launch the raft. Load the hulls with sufficient casing and core rods to cope with the depth of water. Add sample tubes, T-bar, piston and holder assembly with its cable on a fishtail, drop-hammer, anvil, shoes, jaws, duct tape, glass fibered tape, etc. Add a tool kit consisting of vice-grips, crescent wrench, cold chisel (for cutting steel cable if necessary), ball-peen hammer, and set of coarse files for use on aluminum: flat, triangular and rat-tail or round (these files are necessary to remove burrs or to reshape slots in core tubes).

Place the three anchors on the platform (the only things on it), with their attached fishtails of rope down in the hulls. Put two crew members aboard and have the tender tow the raft out to the core site.

Set the three anchors with the tender as described above. This is the most difficult part of the whole coring operation and requires teamwork. The placement is done by the person standing on the raft (though the perch is wobbly, the job cannot be done without standing; you need the height to be able to see clearly how the ropes lie in the water). With a total crew of three, the third person speeds up the operation by passing the heavy anchors to the tender as it comes alongside and by paying out from the fishtails.

Take the sediment/water interface sample from one side of the raft and send the tender to shore with the sampler for extrusion and subsampling.

Lower the casing; have a foot rope tied to the bottom section as an insurance against dropping it, but remember to give the foot rope plenty of slack at all times or it will distort the plumb (straight hanging) of the casing string. The wooden jaws are useful if the water is deep and the casing long because they can be secured round each length of casing as the next is fitted. One person manipulates the jaws while the second slides the tubes and coupling together and tightens the hose clamps on the couplings. Lock the jaws so that the last coupling sits firmly on it. Place the box-bridge over the top coupling.

Rig the sampler, with the piston locked in place in the mouth of the sample tube. The piston has to be pushed down the long tube with a rod, a proceeding in which the cable gets in the way. It is easy to modify the end of a broomstick to fit over the top of the piston and so be used as a pusher for this purpose, or you can use a short pusher that fits to a drill rod.

Lower the sampler into the casing, screwing on connecting rods as you go. Pause on the descent while the tube fills with water — one hears this as a musical gurgling, the most beautiful sound in the world. Continue adding extension rods until the sampler is poised at the bottom of the lake, paying out the piston cable as you go. Note that, if the piston cable is coiled on a fishtail as is our practice, you must roll the fishtail over to pay it out to avoid kinks in the cable. Do not pay off the cable as if it were rope.

Slide the piston cable into the slot in the box, take up any slack, and fit the cable clamp so that the cable is nicely taut against the box. Add a T-bar to the string of rods and apply a sharp jerk downwards to disengage the piston. Push down firmly and steadily until you feel the tube holder strike against the piston at the end of the drive. In the soft sediment at the start of a core like this you can usually feel a 'click' as the top of the piston locks into the holder.

Pull up the core. If the piston is properly locked into the tube holder there should be no need to maintain tension on the cable, but do not let it go slack or it will jam in the casing. Do not wind the cable back on its fishtail, but simply throw the slack over the side as it comes up, leaving the clamp in place (the clamp can go over the side too — it is fixed to the cable). Throwing the loose cable into the water lets it fall straight without kinks and is much the most satisfactory way of keeping it on a raft: you always have contact with both ends, one end on the fishtail kept in a hull and the other end attached to the sampler.

Remove the retaining screw securing the sample tube to the holder and pull out the piston-holder assembly (they should of course be locked together so that they come out as one unit; Figure 3.6). Sediment should be flush with the bottom of the tube, but some might have been pushed down by water pressure despite having locked the piston so that it projects from the tube. If so, sample it into a sealable plastic bag and make appropriate notes. Next, place a ring of glass-reinforced tape over the small hole near the bottom of the tube. This hole is there to take the projection of a retaining screw when a shoe is being used and needs to be sealed over to prevent both drying and escape of sediment.

Using glass tape rather than duct tape to close it, as it makes a rigidly strong cover.

The core tube must now be sealed at both ends. This can be done with corks but plastic caps are better. Put an 'outside' plastic cap over the bottom end of the core tube and reinforce its attachment with tape. Insert an 'inside' plastic cover into the top of the tube in the space left by the piston and work it down until it is contact with the sediment. You need a pinhole or cut in the face of the plastic cover to let air escape as the plastic disk is pushed down inside the tube to come to rest against the sediment. A hole stabbed with a marline spike against the raft surface does fine. Mark the core tube with a permanent, waterproof marker. Section one taken.

Unlock the piston from the holder, recock it and rig a fresh sample tube. Lower the sampler into the casing as before, but this time with the cable slack emerging from the lake as you go, until the section of the cable with the clamp on it appears. This clamp records the exact depth at which the core was started, but now you need to start at one meter deeper. Measure off one meter more cable from the old clamp position and move the clamp to this new position. Continue pushing the sampler down the hole until the clamp, in its new position, lodges against the top of the box. The head of the sampler is now at the bottom of the old hole. Take the second core section and proceed with it as for section 1.

It is probably wise to lower the casing by an extra meter at this stage, effectively jamming the casing bottom in the sediment. Add another meter section, therefore, remembering to pay out the foot rope and to replace the box over the top coupling as before. It does not matter that the casing will have disturbed the surface mud because you will push right past this with your locked piston blocking the mouth of the tube before you get to sampling depth.

Proceed, section by section, as long as you can. How long you can go on in this way in a particular lake nobody knows until they have tried. Often in the Amazon one or two meters pushed by hand is the limit, but in Ecuadorian Lake Kumpak[a] we pushed by hand for about 13 m and then hammered for 6 m more. This was unusual, but anything up to eight meters by hand is quite possible if a thick body of gyttja is present. But eventually, the coring will get harder. At first, you can progress by using your body as a hammer, applying your weight in jerks. If you weigh 170 pounds (77 kg) as I do, your body can make a powerful hammer if you use it right. But the time will come when you must try hammering with steel or bronze.

If the hammering necessity comes when a drive is part way down, it is too late to use a shoe for this section. Hammer anyway. Unscrew the top rod and the T-bar, screw an anvil in its place and screw a another rod to the anvil to act as a guide rod. It is good to be able to set the anvil at a convenient height, which is why we have a few half-meter rods in the storage hulls. Slide the drop hammer over the guide rod, have a crew member hold the rods steady, and hammer. Watch and feel constantly for progress (see our cautions on hammering in a preceding section). Be particularly alert for the feel of the piston locking into the sample tube holder which tells you that the drive is complete. Signals that you have got to the end of the drive include a thump on the wooden platform, or the spongy feel that come from the force of the blow being absorbed by the somewhat elastic cable because the piston is at the top of the tube. If you suspect you are feeling these phenomena, stop at once without chancing another blow.

Try to anticipate the need for hammering so that you can fit a shoe in good time. To fit a shoe, cock and lock the piston in the usual way, slip the mild steel shoe over the tube so that the tube rests snugly against the shoulder of the shoe, line up the projecting screw in the shoe with the slot in the tube and screw home. Be careful not to drive the screw in too far. We want the screw to project far enough to prevent the shoe falling off when you pull the sampler out of the hole, but the screw must not project so far that it interferes with the piston.

Always inspect the shoe after hammering a section. If the end is scratched or dented, this means that you have reached gravel or rock and it is time to quit. Very often prudence will tell you to stop hammering before you have got a complete core section. When this happens, you must remember to hold the piston cable tight throughout the pulling up, as the piston is not then locked into the holder.

Hammering is the one undertaking on our coring rafts that might be dangerous. Imagine a 5-kg chunk of metal being waved in the air over unprotected heads on a wobbly rubber boat. Be careful.

Once all the sections of the first core are satisfactorily taken, it is good practice to take a second core from the same station. This second core can be depth-offset so that all the section breaks in one core fall in the middle of a core tube in the second core. In this way, you can reconstruct the complete sediment stratigraphy from the cores without any uncertainty about what material might be missing where the sections change.

To proceed: raise the casing one or two meters to make sure the bottom is clear of the mud, then move the raft two or three meters by paying out an anchor rope on one side while pulling in on another. Make the anchor ropes taut again, lower the casing and take the second core. However when taking the first section of this core lower the sampler an extra half meter before beginning, then lock the cable clamp as before. From then on all core sections will be offset by half a meter from the first core.

NOTES

1. Piston samplers for lake deposits were invented by Daniel Livingstone in the early 1950s based on the principle used in the Kullenberg corers for deep sea sediments (Livingstone, 1955). The first sliding rod to hold the piston in place during descent through the water column was round and engaged the holder for the drive with protruding lugs (Vallentyne, 1955). Livingstone then improved this design by substituting a square rod for the round rod with lugs (Wright, 1967). All the variants, including ours with locking pistons, are properly called "Livingstone samplers."
2. Detailed instructions for building one of these samplers out of inexpensive materials are given in Fisher *et al.* (1992).

CHAPTER

4

Laboratory Protocols

This chapter is written for those coming newly to the techniques of pollen analysis. In writing it, we assume a researcher working in an institution where library resources are not large and where there is no prior tradition of palynological work. We concentrate on what is important to palynology in the Amazon. While we do expound some elementary fundamentals of palynology, intended for someone new to the subject, we do not pretend a comprehensive coverage. For this consult the standard works: Faegri and Iversen, 1989; Moore *et al.*, 1991; Salgado-Labouriau, 1973.[1]

DESCRIBING, STORING, AND DATING OF CORES

As soon as core sections are brought in from the field they should, if at all possible, be stored in a cold room at 4°C. A prime reason for this is to prevent the growth of mold fungi on the sediments, which can interfere with radiocarbon dating. To the extent that these fungi merely metabolize carbon compounds from the sediments their effect on dating is neutral, but unfortunately, fungi exchange CO_2 with the atmosphere in the course of their respiration, thus contaminating the sediments with modern carbon. So keep the sediments cool. Try also to exclude oxygen by ensuring that the core tubes are properly sealed before placing them in the cold room. If sediments have been extruded in the field (not recommended), see that they are well wrapped, preferably in *saran wrap* (which provides a gas-impermeable seal), rather than polyethylene (which allows slow gas exchange). Once cores have been opened, they must also be well-wrapped in *saran* for safe keeping.

X-Radiography Before the Cores Are Opened

X-radiographs of the sediments are easily made before the core tubes are opened. Not only will this provide a preview of the core stratigraphy, but the x-radiographs then serve as a permanent record of many of the features of the stratigraphy. X-radiography is worth doing because it is easy, as long as your institution has an x-ray facility or a friendly hospital is available. The core barrels are laid out directly on the x-ray film and are then irradiated from above. Because of the size of the sheets of film and the limited area irradiated, it is necessary to picture cores that are

1 m long in three sections, but 4–6 cores sections can be done at a time by laying them parallel on the film. The result is a print on a scale of nearly 1:1, with only minimal scale distortion due to the geometrical consequences of the fact that the irradiation comes from a point source. The typical x-ray transparency is a negative with the x-ray transparent materials (organic matter, water) showing as dark and x-ray opaque minerals as white. Gross stratigraphic features like volcanic ash layers show up brilliantly as white bands, but the x-radiographs can also show fine structure that is not apparent when looking at the exposed sediment itself. When this fine structure is present and shows up as bedding planes horizontal to the core tube, then this is good confirmation that the stratigraphy has not been disturbed in coring. Conversely, any disturbed sediment at the top of a core section can be identified and thus avoided in the subsampling program.

Opening Core Sections

If the sediments are soft, they can be extruded without damaging the core tubes. All that is needed is a plunger on a pole, such as can be made with a laboratory stopper fastened to a broom stick. The sediments are pushed out onto a sheet of *saran wrap*, perhaps with the *saran* supported by a half-round length of PVC, made by cutting PVC piping lengthwise into two halves. The core barrel can then be cleaned and used again. Such extrusion is widely practiced, yet we prefer to avoid the risk of distortion of the sediments inherent in this procedure. We recommend opening the core sections by slitting the aluminum alloy core tubes on both sides.

Description: Recording the Core Stratigraphy

As each core section is opened and described, a continuous record of the stratigraphy of the complete section is compiled and recorded. We recommend marking a depth-scale directly on the half-core tube in which the section rests. The depth-scale is continuous from section to section, thus describing the complete sediment stratigraphy from the mud–water interface down to the bottom of the core. If any sediment has been lost between core sections, this must appear as gaps so that depth-placement of core sections below the lost sediment remains accurate. When depth offset parallel cores have been taken as we recommend, the stratigraphy within any such gaps of lost sediment can generally be reconstructed from the parallel core. It is imperative that accurate depth placements be marked on the half-core barrels during the core description. This will allow subsequent resampling.

Clean the exposed face of the core with a spatula or knife to reveal fine stratigraphy.[2] At this first opening of the core tubes the sediment should still be anoxic. However carefully the core is afterwards wrapped for storage, colors will change as the sediment is oxidized; therefore, record the principal colors at first opening. The best written record of color is obtained with a Munsell soil color chart.[3] Color photography can also be useful, particularly with a digital camera system that will later allow scanning of such data as the incidence and intensity of fine banding. But for most cores used primarily for palynology, the stratigraphic analysis probably will be restricted to descriptions of sediment type, sediment color, and main stratigraphic boundaries. The really important data are depths in core of each stratum, and hence true placement of each sample used for analysis.

On Radiocarbon Dating

With the coming of AMS (Accelerator Mass Spectrometry) radiocarbon dating, our old reliance on dating of bulk sediments by beta-decay dating is ending. AMS dates cost more than the old beta-decay dates, and an insufficiency of laboratories equipped to do them can be a problem outside North America and Europe. But the advantages of AMS dates are so great that they should be obtained whenever possible. The advantage for lake sediment work comes from the small size of the samples needed. This means that the dated sample need no longer be an average of organic matter deposited over many years as with the old bulk samples. More important, however, is that the dated sample can be on discrete organic remains: leaf fragments from terrestrial plants, or body parts from zooplankton, for instance. Dating these fragments screened out from the sediment significantly reduces errors introduced by such contaminants of bulk sediments as graphite or coal particles.

As a matter of practical expediency, it might be prudent first to obtain a bulk beta-decay date from near the bottom of the core. This will suggest the age spanned by the sediments and can guide subsequent sampling for AMS dating at critical intervals.[4]

Stable carbon isotopic ratios ($^{12}C/^{13}C$) are commonly reported along with age estimates based on radiocarbon dating. The primary use of these ratios is to correct for isotopic fractionation so that radiocarbon dates from different laboratories can be compared. Although changes in stable isotope ratios do occur within a core, it is difficult to interpret what they mean. Carbon sources and sinks in lakes are extremely complex so that it is hazardous to interpret a change in the ratio as reflecting changes in the composition of surrounding terrestrial vegetation. Any paleovegetational postulates based on changes in the ratios of stable carbon isotopes in lake sediments or soils need testing by the pollen data.

EXTRACTING POLLEN FROM SEDIMENTS

The mass of pollen in lake sediments remains tiny compared to the mass of minerals or non-pollen organic matter, even though they are present at the rate of 100,000 grains per milliliter (cc).[5] For analysis, therefore, pollen grains have to be extracted from a relatively vast mass of dilutants, both organic and inorganic. We are able to do this because the outer coat of pollen grains is extraordinarily resistant to many chemical treatments except oxidizing agents. At its most simple, pollen extraction can consist of dissolving everything that is not pollen: humic substances in potassium hydroxide, cellulose in a sulfuric acid–acetic anhydride mixture (acetolysis), carbonates in hydrochloric acid, and silicate minerals in hydrofluoric acid. What is left after these digestions is a brown smear that contains the concentrated pollen that has survived this ferocious chemistry unharmed.

Unfortunately, however, these chemical preparations are never 'clean'; that is, they are never pure pollen, but always include both organic and inorganic particles that escape the treatment. The organic particles are the most intractable because they are chemically most similar to the pollen. Cell walls of some phytoplankton have compositions that are closely similar to pollen and have so far defied our efforts to destroy them without destroying the pollen as well (*Pediastrum*, *Botryococcus*).[6] Other phytoplankters have spores that resist the treatment, as do many fungal spores. A pollen preparation from lake deposits, therefore, is usually an extract of pollen plus assorted spores and cell parts of algae, fungi and fragments of particularly resistent lignified tissue. Typically these unwanted objects are significantly more numerous than the pollen, so that a pollen analysis is, in part, a hunt for pollen grains in a forest of non-pollen objects.

Some mineral matter also survives the typical digestion with hydrofluoric acid (HF). This can be partly because the sheer mass of mineral is hard to dissolve, but more particularly, it results from the presence of a minority of non-silicate minerals that will not dissolve in HF. Two alternative methods are available for separating pollen from these stubborn minerals — the pollen can be filtered out or floated out. Unfortunately, both methods risk losing some of the tiny pollen grains characteristic of the neotropics.

Filtering pollen from mineral matter became possible with the commercial availability of synthetic cloths with apertures between the fibers in the 5 micron range. The finely divided mineral particles in clay or silt are much smaller than this and will pass through these screens when finely dispersed with sodium hydroxide or detergents. Pollen grains are frequently in the 30 micron size range or bigger (Traverse, 1988). In the classical regions for pollen analysis like the north temperate belt, few plant species have pollen grains smaller than 18 microns, and even these are rare insect-dispersed types. In those circumstances, filtering offers a quick way to achieve a relatively clean separation of pollen from a mineral matrix (Cwynar *et al.*, 1979). But for us in the Amazon, filtering will not work because *Cecropia*, one of our most abundant pollen types (as well as many species of *Mimosa, Piper, Peperomia, Begonia*, among others), range from 6 to 10 microns on their long axes, and many others of the Moraceae other than *Cecropia* are not much bigger. Many of these grains would squeeze through the 5 micron mesh alongside the clay minerals and be lost. At the other size extreme for pollen, we can use a very coarse sieve (say 250 microns) to remove debris, but not the 100 micron sieve commonly used in temperate zones because our larger pollen grains could be removed.[7]

The specific gravity of pure silica (quartz) is 2.7 such that it will sink in a liquid whose specific gravity is 2.0. Other minerals are heavier so that all will sink in this liquid. But the carbon-based compounds of organic matter will all float on a specific gravity of 2.0. Thus it is possible to remove as much organic matter as possible by the chemical methods and then to float the rest off the mineral matrix in a 'heavy liquid' such as bromoform or zinc chloride solution. The one technical catch is that water (specific gravity 1.0) also floats on the heavy liquid so that water attached to any mineral can make it float up with the pollen. Therefore, to remove the silicate clay minerals and silts, it is necessary first to remove all water, including water of hydration and the water adsorbed by clay minerals. This can be done, though the task involves repeated washing with the alcohol or acetone used to remove the water. When done well, the addition of a heavy liquid separation to the protocol can produce an essentially mineral-free pollen extract.

The objection to using heavy liquids for Amazon samples is the same as the objection for using filters: the *Cecropia* and other small grains tend to be lost differentially during the numerous decantings required. This, coupled with the fact that the best heavy liquid, a bromoform–acetone mixture, is a dangerous chemical carcinogen has dampened enthusiasm for heavy liquid separations and they are not routinely used in our Amazon work.

Actual protocols for pollen extraction vary from laboratory to laboratory although the major steps and principles are the same. The proceedings can be arranged in any order and there is room for substitution in some steps, e.g., potassium hydroxide (KOH) can be replaced with sodium hydroxide (NaOH) or the HF digestion can be for a short time in a boiling water bath or overnight 'cold.' This is 'cookbook' science with everyone to his favorite recipe. The really vital and irreplaceable step is use of acetolysis mixture (1 part concentrated sulfuric acid to 9 parts acetic anhydride) to destroy cellulose. This 'acetolysis' not only destroys cellulose dilutants but it also cleans the pollen grains, in particular removing waxes, thus revealing the grain structure. Pollen grains in reference slides are also cleaned with acetolysis mixture so that all grains are of comparable appearance.

Pollen extracts are typically made on 0.25 or 0.5 cc samples. It is very important for subsequent pollen statistics that this volume be measured accurately. The best way is to pack sediment into containers of known volume.

Ours are made of stainless steel and are essentially a block of metal with a cylindrical cavity of the desired volume and a plunger that allows sediment to be expelled as a cylindrical pellet. We make them in three sizes, 0.25 ml, 0.5 ml, and 1 ml. The sediment is packed into the cylinder and then expelled with the plunger.

The equipment needed is fairly minimal, the most expensive item being a small, table top, clinical centrifuge that will hold six 15 ml tubes. Centrifuge tubes for the HF digestion must be polypropylene, not glass because glass is soluble in HF. Sometimes we have found it convenient to do most of the extraction in graduated glass tubes, only switching to polypropylene tubes for the HF extraction (the polypropylene tubes are somewhat opaque with surface tension characteristics that can interfere with decanting the waste liquid safely). On the other hand, glass tubes sometimes shatter, causing the loss of the whole sample. Because of this, we recommend planning to use only polypropylene tubes. Apart from ordinary laboratory glassware, the only other equipment needed is a water-bath or a heat block to heat six or more centrifuge tubes at a time, and a fume hood. The dry heat block is a great convenience but the job can be done perfectly well with an improvised boiling water bath.

Addition of Exotic Marker Pollen

Pollen analysts now routinely use an exotic marker technique to let them calculate the total number of pollen grains present in the original measured volume of the subsample. The exotic markers most frequently used are preparations of either *Eucalyptus* pollen or *Lycopodium* spores (Stockmarr, 1971). In north temperate or arctic latitudes *Eucalyptus* pollen types do not occur naturally, making this a truly exotic marker. But in the Amazon basin, although *Eucalyptus* itself is not native, other members of the family Myrtaceae are important forest trees whose pollen cannot easily be separated from *Eucalyptus*. Although some species of *Lycopodium* are also found in tropical forests they are such minor constituents of the flora that the effect of native *Lycopodium* spores on the sum of exotic *Lycopodium* can reasonably be assumed to be negligible. Accordingly we use *Lycopodium* spores as our exotic marker. This is practicable because preparations (similar to pharmaceutical tablets) of *Lycopodium* spores of known concentration are available commercially. Tablets of exotic spores are added to each sample being processed.[8] In the final counting, the ratio of *Lycopodium* to all other pollen counted allows the calculation of the original pollen concentration. If the sedimentation rate has been measured accurately, it is then possible to calculate pollen accumulation rates (influx), taxon by taxon. Accumulation rates, unlike pollen percentages, are independent statistics.

Lycopodium is least useful as an exotic marker in the cloud forests of our region because the genus is well-represented in the local vegetation. A suitable alternative developed by Salgado-Labouriau is *Kochia scoparia* (Chenopodiaceae) which has the periporate pollen type of the chenopods that is abundant in dry deserts but not present in cloud forest. *Kochia* pollen preparations cannot be purchased commercially and you must make up your own concentrated suspensions of the pollen (Salgado-Labouriau and Rull, 1986). An alternative to using exotic pollen is to use synthetic microspheres manufactured for the purpose, but we have not yet tried this new technique on Amazon samples.[9]

The following is the protocol that, with minor variations, we have used in our Amazon studies, including all the preparations on which the plates in this book are based. If a different recipe is substituted it is important to supply evidence that *Cecropia* and other small grains have not been differentially lost.

POLLEN EXTRACTION PROCEDURE

All treatments are in 15-ml conical, graduated centrifuge tubes. It is simplest to use tubes made of polypropylene, since these will resist all the reagents used, including both acetolysis mixture and HF. After each treatment the tubes are spun in a centrifuge, after which the waste liquid is carefully poured out; carefully so as to lose no pollen. A typical run in a centrifuge is defined as 2500 rpm for 5 minutes. In the following table this proceeding is listed as SC&D (stir, centrifuge and decant).

Work with batches of six tubes at a time, each with its own polypropylene or teflon stir-rod: a rack for the tubes and a simple rack for the stir-rods laid out so as to avoid all possibility of putting the wrong rod in a tube.

1. One Lycopodium tablet to each tube, add 10% HCl a drop at a time to total of 5 ml. This to dissolve the carbonate matrix of the tablet. Careful, it will foam: wait for foaming to cease after the first few drops before adding the rest. Add sediment sample (usually 0.25 or 0.5 cc). SC&D.

2. Wash with 12 ml water. SC&D. Repeat washing until supernatant is clear.

3. Cold HF digestion. 8 ml conc. HF, leave overnight, SC&D.

4. Hot HF digestion. 8 ml conc. HF in boiling water bath or heater block at 100°C for 30 minutes. Stir frequently. SC&D. (This step is necessary only in mineral-rich samples.)[10]

5. Wash with 12 ml water SC&D. Repeat 2 or 3 times or until supernatant is clear.

6. Wash out residual fluoride with HCl. 8 ml 10% HCl. SC&D.

7. Wash with 12 ml water SC&D. Repeat 2 or 3 times or until supernatant is clear.

   NOTE: In very humic samples we insert a caustic digestion here. 8 ml 10% KOH, stir, water bath 5 min. SC&D. Repeat step 7.

8. Acidify with acetic acid to prepare for acetolysis mixture. 8 ml glacial acetic acid. Stir. SC&D.

9. Acetolysis. 8 ml acetolysis mixture = 9 parts acetic anhydride (CH_3COCH_3) by volume to one part concentrated sulfuric acid (H_2SO_4). Mix fresh in a graduated cylinder in a bed of ice (it heats rapidly from an exothermic reaction). Place acetic anhydride in first then run sulfuric acid down the side of the cylinder a drop at a time — this is a dangerous, explosive mixing process = CARE — eye protection, rubber gloves, etc. Add the 8 ml of the mixture cautiously, a drop at a time, with stirring. Heat in boiling water bath until the liquid turns an oily black (5–20 min depending on sediment). SC&D.

10. Wash with glacial acetic acid. 8 ml. stir. SC&D.

11. Wash with water. SC&D. repeat twice more.

12. Wash with absolute alcohol (C_2H_5OH). Stir. SC&D.

13. Transfer sample to 5 ml polypropylene vial with screw cap (Evergreen vials) using a wash bottle with a fine jet. The 5 ml vial will fit into the same centrifuge as the 15 ml tubes. SC&D.

14. Suspend in glycerine (glycerol). Add 5–6 drops of glycerol (can be stained with safranin). Stir, label, put in 6O° oven overnight to drive off water. Cap tightly for storage. Keep in dark.[11] Slides should be stored horizontally.

Slide Mounting

To make a slide for counting, stir the preparation in the vial with a fine glass rod and transfer one drop to a slide, add cover slip and spread by pressing on the cover slip. The glycerine is now a circular pool occupying not much more than half the area of the square coverslip, the outer edges of which must now be filled with paraffin. Put paraffin round the edges of the cover slip and heat gently on a hot plate, just sufficiently to melt the paraffin so that it runs under the edges of the coverslip to seal the glycerine in its bubble. Once the slide is cool, clean off the congealed paraffin from the edges of the cover slip and paint the edges with nail varnish. Label the slide and store in the dark. Slides prepared in this way are long-lasting, the glycerine is contained and will not take up moisture from the air, and yet the pollen can still be gently turned over by pressing down on the cover slip with a needle.[12]

The above assumes that a single drop of the pollen concentrate will yield a dense enough concentration of pollen for comfortable counting. Ideally there should be some multiple of ten grains in each traverse of the slide. Preparations from Lake Pata sometimes had an hundred grains per traverse. If there are too many (not a common occurrence) the answer is to dilute with more glycerine. If there are not enough, then the only answer is to tweak the extraction technique, perhaps starting with a larger sediment sample. But if you cannot concentrate the extract any more there is nothing for it but a long and tedious time at the microscope.

Alternative Mounting Media

The correct mounting medium to use is controversial, with different proponents favoring silicone oil, glycerine, or hard media like Canada balsam. We use glycerine as described above, preferably stained with the dye safranin. Being water miscible, glycerine is simpler to use (i.e., it is not necessary to extract water with a series of alcohols before suspending the pollen in silicone oil,[13] a series of steps pregnant with the possibility of losing smaller pollen grains). At the same time, glycerine remains fluid under a cover slip, unlike that excellent but rigid preservative, Canada balsam. A fluid medium allows grains to be turned over, or nudged out from under a piece of obscuring tissue, by pressure on the cover slip with a needle. With our diverse pollen flora it is often essential to be able to move a pollen grain in order to get more than one view.

Silicone oil became favorite among those who found it possible to refine generic identifications by size–frequency statistics because the oil proved to be the medium that best prevented pollen grains from altering shape or size while stored. Confident that the pollen on a slide had retained its original size (typically 'no swelling') pollen of *Betula, Picea, Pinus* and a few others could be separated into subgeneric groups of species, or a single species could sometimes be demonstrated to be present. These data added significantly to the resolution of northern pollen diagrams with their depauperate species lists. Doubtless size–frequency statistics will one day come into their own in the Amazon also, though not in the present infancy of Amazonian palynology.[14] We are probably better served by seeking our details in that splendid Amazonian record of animal pollinated taxa whose pollen is washed off the watershed.

POLLEN COUNTING AND POLLEN DATA

Pollen counting must be done with at least a 40× objective. The pollen analysis at Lake Pata was done entirely under oil immersion (100×) to ensure that no small grains were overlooked, but conventionally only a 40× lens is used for routine counting, with the oil immersion lens being reserved to identify difficult grains.

The Pollen Sum

The pollen sum is an arbitrary number at which the counting stops. In our recent Amazon work we have set this number at 500 but have not been precise. 500 grains is far too few to provide a statistically meaningful sample of the rare elements but is more than enough to establish the statistical relativity of the few abundant pollen taxa delivered to the lake by wind.

In Mark Bush's TWINSPAN and DCA analyses of surface spectra described in Chapter 1 the chosen sum was a mere 200 pollen grains and fern spores, which proved to be adequate to allow the multivariate statistics to detect community types. This is because the pollen sum is largely made up of a few abundant pollen types whose signal becomes clear at comparatively low counts. Percentage pollen statistics have so many independent variables (all the taxa counted) that impracticably large numbers (tens of thousands) must be counted to secure statistical significance for rare taxa, particularly those present at less than 1%. The few abundant taxa that account for most of the variation, however, become statistically significant at very low counts. A simple way of testing this is to plot % pollen sum against pollen sum for the abundant taxa, when it will generally be found that little variation in % occurs with sums beyond 200. A formal mathematical statement of these realities will be found in Mosimann (1965). These tests show that it is acceptable to count to a pollen sum of only 250 grains. The quantitative data provided by this count could then be amplified by searching the slides for other pollen taxa which would only be interpreted as "present." This should suffice when the rare elements are used only as index fossils or when the research questions posed can be answered with a taxon list.

It used to be the practice among some European palynologists to choose as an arbitrary pollen sum the number of tree pollen only, and then to go on counting all pollen until this sum of tree pollen was reached. When the ancient vegetation had changed from grassland or tundra to forest this produced the absurd result of counting thousands of grains in the herb zones in order to get the same sum of tree pollen as in the forest zones. This was a statistically foolish undertaking (see Mosimann, 1965). The pollen sum should include all the vascular plants present, the only flexibility permitted being the removal from the pollen sum of taxa that can be safely assigned to special habitats, the best example being aquatic plants. The pollen history of aquatic plants can be interesting in its own right, but not as percentages of fluctuating number of plants in that different habitat, the land. It is also appropriate to remove from the sum for the purposes of statistical analyses those taxa that are heavily overrepresented, an example being the *Cecropia* that were removed from the pollen sums in Mark Bush's surface sample analyses (Chapter 1).[15]

A nagging worry to the pollen analyst of conscience can be the lumping of the spores of ferns with the pollen of higher plants into a common sum, for the two kinds of *palynos* (spore) are functionally quite different. When the influx of fern taxa is of the same order as the influx of pollen taxa the practice is satisfactory if entirely arbitrary; we are using the *palynos* of different groups of vascular plants as proxies for the presence of the parent populations. The actual biological functions of the spores do not matter as long as the relationship of spore influx to parent plant population is of the same order. Experience suggests that in most neotropical communities this relationship does, in fact, hold. Amazonian pollen diagrams are not distorted by fern spores which are usually present in low percentages despite the prevalence of ferns in Amazonian plant communities.

But it is necessary to be alert for the overrepresentation of fern spores, particularly when this might be due to postdepositional effects. Fern spores seem to be strongly concentrated in samples that have been exposed to weathering, presumably because of the differential destruction of pollen grains. Attempts at pollen analysis on the famous hominid strata of the Olduvai Gorge in East Africa have come to very little because the enrichment with fern spores during subsequent erosion serves as a warning signal that the original pollen percentages have been fatally skewed by differential weathering. Marine deposits are also often enriched with fern spores weathered out of older deposits and differentially preserved, suggesting that it is usually wise to remove fern spores from the pollen sum in marine deposits.

Fern spores can also be grossly overrepresented in fern-filled swamps, or where tree-ferns surround water bodies, when they should be removed from the pollen sum for statistical prudence like other taxa that are locally overrepresented.[16] In general, therefore, although fern spores can be proxies for their parent populations that are roughly comparable to pollen grains they are not completely interchangeable with pollen and should always be used with care.

Pollen Identification and the Reference Collection

This manual provides keys and photographs as a guide to the identification of Amazon pollen taxa. No photographs, however, can replace the real thing, so serious Amazon palynologists will want to build their own reference collections. Reference slides are made by taking a flower, sometimes more than one flower, and processing it through acetolysis and KOH exactly as the fossil samples are processed (though there should be no need for removing minerals with HF). Preparation of herbarium material is described in Erdtman (1952) and Salgado-Labouriau (1973). The only material difference in the preparation is that fixed mounts can be made, usually in glycerine jelly.[17] These rigid mounts are acceptable for reference slides because they hold numerous grains of the same kind so that there should always be a grain oriented in the direction needed.

It is essential that plants used for building the reference collection be accurately identified. The best practice is to sample plants in herbarium collections where the identification has been done by experts. Copy onto the slide label the herbarium number of the specimen and the name of the person who made the identification. In this way it will always be possible to refer back to the original herbarium specimen. This information is particularly valuable when trading slides. Pollen reference collections are built partly by trading with other pollen analysts and the herbarium data provide a provenance and assurance of the taxonomy.[18]

Pollen Diagrams

Standard pollen diagrams, like Figures 1.1 and 1.3 are 'pollen percentage diagrams,' in which each pollen taxon is presented as relative (%) to all other taxa. These diagrams can be supplemented by the 'pollen concentration diagram,' made possible by the use of exotic marker pollen. The concentration of each pollen taxon is expressed as a function of the volume of sediment, and is thus not dependent on the relative presence of other pollen types as is the pollen percentage. Pollen concentration can thus be useful as a check on an inference drawn from the percentage diagram: "Does the increase in percent of taxon 'Y' mean a real increase in the population of 'Y'? or is the high percent actually no more than a reduction in the population of taxon 'Z'?" Checking to see if the concentrations of 'Y' and 'Z' change can help to answer questions of this kind. But concentration diagrams present their own difficulties of interpretation also as their data are also relative: relative that is to the volume of water or clay or organic matter that make up the matrix of the sediment. Pollen concentration of taxon 'Y' can change markedly from level to level as a result of changed sedimentation rate even though the population and pollen production of 'Y' actually remained constant. Pollen percentage and pollen concentration diagrams read in conjunction, however, allow many of these uncertainties to be resolved.

The only truly independent pollen statistic is the pollen 'influx,' sometimes called the 'pollen accumulation rate.' This can be calculated from the pollen concentration if the sedimentation rate is known accurately over every interval, simply by dividing the pollen concentration per cc by the time taken to deposit 1 cm of sediment. In lake sediments AMS dates at say 1000 year intervals can yield believable sedimentation rates suitable for this enterprise, but obviously at a large pecuniary cost. In marsh deposits, valley fill and the like the attempt is usually futile because rates of sedimentation vary too much throughout the section.

The chore of doing the numerous simple calculations, and the plotting of the results, are now handled on personal computers with widely available software. It takes a day or two longer to do the calculations on a calculator and plot the results on graph paper in pencil, but many of us still in the profession did that once. It is still possible to do so if a computer is not available. Otherwise we suggest using the software packages TILIA and TILIA.GRAPH written by Eric Grimm (see appendix).

MEASURING THE MODERN POLLEN RAIN

The modern pollen rain can be measured in basically two ways: with pollen traps, or by sampling natural pollen collectors like lake mud and moss polsters. The two methods sample different facies of the pollen rain and are useful for different purposes. Natural collectors are the best choice for calibrating fossil pollen spectra to vegetation or climatic variables but pollen trapping yields data on pollen as index fossils or community components that can scarcely be obtained in any other way.

Pollen Trapping

Figure 4.1 shows the parts of an effective pollen trap that is very cheaply made. The trap itself is a wad of fibrous material filling a glass or plastic funnel, which is lined with a Whatman glass microfiber filter (GF/D). The funnel is held upright by being glued into the neck of a plastic bottle, which can itself be held upright by being half-buried in the ground or by being attached to a post. A covering of wire mesh prevents the funnel from being choked by fallen leaves and other debris. Holes in the sides of the plastic bottle allow water to drain out.

Pollen falling into the open funnel is washed down in rainwater and is collected either within the mass of fiber or on the filter paper. At the end of six months or a year the funnels are collected, and replaced by fresh funnels. Pollen is then extracted from both the fibrous mass and the filter paper.

The type of fiber chosen is important because it must allow recovery of the pollen. Our first preference was a form of cellulose acetate fiber that is soluble in acetone. Dissolving the fibrous mass in acetone should leave only a residue of pollen and other airborne particles. Unfortunately suitable cellulose acetate fiber is hard to find. Accordingly we have used rayon fiber. This cannot be dissolved by readily obtainable solvents but the texture is such that pollen can easily be washed out of it, using a 5% solution of KOH or NaOH as a detergent.

Place the rayon fiber, which has probably adhered to the filter paper, in a glass beaker, then add enough 5% KOH to cover the mass. Dissolve one or more tablets of exotic marker (*Lycopodium* spores in our Amazon work) in a few drops of 10% HCl in a test tube and wash the residue into the beaker of fiber. Stir to separate the pollen. Leave for a number of hours for the KOH to dissolve the more soluble organic detritus, stirring at intervals. Getting all the pollen out is merely a matter of being careful and thorough in the washing. This whole procedure can be done in a thick plastic bag instead of a beaker, when the fibrous mass can be worked by squeezing the bag. The bulk of liquid holding the pollen is centrifuged and the waste liquid poured off. It helps to have a centrifuge that can handle large volumes of liquid, but the separation can be done easily enough with the clinical centrifuge and 15 ml tubes used in normal pollen extraction, though it will take longer. Once the extract is in a single 15 ml tube it is processed to clean and concentrate the pollen exactly as for the fossil pollen samples by acetolysis, etc. Sometimes the filter paper separates from the rayon fiber.

When placed in forests of known species composition these traps collect pollen from animal pollinated trees within 20–30 m of the trap so that an array of 20 or more traps spaced through the forest yields an excellent measure of the pollen signal of the forest. Because the local species list is known, the task of identifying the great diversity of pollen is simplified. In addition the pollen taxa learned in this way are types that are demonstrated to be deposited in quantities onto the watershed and thus likely to be found in sediments. Using arrays of these simple traps, therefore, is the best way to calibrate the local pollen signal from animal pollinated trees.

The traps also collect a regional pollen signal as wind-dispersed pollen from the surrounding country falls into the traps. This part of the signal can be interesting for studying year to year variations in the regional pollen rain but the trap data are less useful for calibrating the regional signal in lake sediments. This is because almost any sediment sample is an amalgam of the pollen fallout from several to many years. The sediment subsample of 0.5 cc used for a typical pollen extract represents the accumulation of perhaps ten years in sediment deposited at average rates to perhaps an hundred years or more in slowly accumulating mud like that at Lake Pata. The percentage composition of pollen in the extract is thus the average of many years. It must be calibrated against a surface sample that was likewise collected over several years. And the best way to do this is to sample the surface layer of lake sediment itself.

Natural Pollen Traps

By far the best surface samples for calibrating the regional part of the total pollen rain are mud–water interface samples of lakes and ponds. They are easily taken, as described in Chapter 3. One of the most valuable and pioneering research projects will be the gridding of the Amazon basin with mud–water interface pollen samples, keyed to local vegetation and climate parameters. All that a researcher needs is a rubber boat, a bottom sampler, a GPS (global positioning system) meter for an accurate measure of site coordinates, and a note book. So equipped, a few people can go exploring in the Amazon and make a major contribution to knowledge. If necessary, a designated laboratory could undertake the actual pollen analyses of the samples.

The grid of surface pollen percentage data that would result from this effort would allow transfer functions to be calculated between pollen percentages and measures of precipitation, temperature and seasonality. The way would then be open to quantitative reconstruction of past Amazon climates from pollen percentage data in lake cores that should be comparable to the reconstructions already available for Europe and eastern North America (Huntley and Birks, 1983; Birks and Birks, 1980; Prentice *et al.*, 1991).

Where there are no lakes or ponds, other natural pollen traps must be found: the choices are few. Moss polsters, that is soft green mats of moss, are the best. Moss polsters act very like our pollen traps, sometimes being natural versions of funnels full of fiber that have been left out for many seasons or years. In the north temperate world of wind-pollinated forests, moss polsters, like our pollen traps, can yield an approximate sample of the regional pollen rain, though the data are of course subject to local overrepresentation of pollen from plants growing alongside the moss. In the Amazon, suitable moss polsters are not common and, when they can be found, are much inferior to lake sediments. But where lakes are not available moss polsters can be the next best thing.

If there are no lakes and no moss polsters, the only expedient left is to sample the surface soil. This will inevitably yield a biassed pollen sample. Soil, like moss polsters or traps, inevitably collects pollen differentially from the plants actually growing at the point of sampling but it also has the added disadvantage that preservation of pollen is selective also. Soil can be aerobic, which allows oxidative destruction of pollen grains. Furthermore, soils are typically well supplied with a microfauna of scavengers like soil arachnids, and with bacteria and fungi able to destroy pollen. The result is that soils tend rapidly to lose the smaller and more fragile pollen grains, becoming enriched with the more robust pollen and fern spores. In the Amazon ecosystems, where the regional pollen signal is given by the tiny pollen grains of families like Moraceae, Urticaceae, Melastomataceae and others, therefore, sampling soils for pollen is generally unsatisfactory. When it has to be done, because there is nothing else, the best practice is to collect small pinches at random from over a significant area, say a 100 m square, and homogenize the sample.

NOTES

1. Other general references include Brown (1960), Erdtman (1969) and Traverse (1988).
2. As the lake coring community moves to investigate as many proxies for climate change as possible in addition to pollen, it will increasingly be the custom to archive sagittal sections of cores in the manner long practiced by paleooceanographers. Core sections are then split into two halves, one of which is stored while the other is used for immediate subsampling and analysis. When this is done the split face of the core sections provides the clean face necessary for core description. In North America a lake core handling facility has been established by the National Science Foundation at the Limnological Research Center of the University of Minnesota, where protocols for investigating lake cores by multi-proxy methods are being developed. The address is: LRC Core Lab Facility, CME 500 Pillsbury Drive SE, Minneapolis, MN 55455, USA (email lrc@maroon.tc.umn.edu).
3. Munsell soil color charts provide colored chips against which the sediment color is matched. The color chips have been formally classified in a number and letters system that can be used to record the color: an example being one of the dark brown colors possible in sediments classified as Munsell 10YR 2/2, meaning that 10YR is the "hue" and 2/2 means a "value" (essentially the lightness) of 2 and a "chroma" (measure of color saturation) of 2. For the user, the technical meaning of the notation is unimportant: what matters is that they provide a permanent reference for the color in the way that a pollen reference slide gives a voucher for the identification of a pollen grain. The colors in the fresh sediment can be accurately described, recorded and reported. Sources for Munsell charts are given in the appendix.
4. The radiocarbon time scale differs significantly from the solar time scale over parts of its range. This is because a primary assumption of radiocarbon dating is that production of ^{14}C by solar radiation is constant, but this assumption does not hold precisely over some intervals. In parts of the Holocene radiocarbon dates diverge from true ages by as much as a thousand years. The radiocarbon time scale has now been calibrated against solar years by dating tree-rings from temperate latitudes, as well as other annual periodic phenomena, by Minze Stuiver and his colleagues at the University of Washington, as well as by other laboratories. A calibration curve based on this work is usually provided with the reports of commercial radiocarbon laboratories.

5. Actual concentration data for the Lake Pata core are:
   minimum concentration = 88,076 grains/cc
   maximum concentration = 778,400 grains/cc
   mean concentration entire core
   = 247,627 grains/cc
   mean concentration Holocene section
   = 160,678 grains/cc
   mean concentration LGM = 320,086 grains/cc
   mean concentration prior to LGM
   = 249,326 grains/cc
6. Separation of pollen from organic detritus has been achieved by centrifugation in density gradients, though the process seems to be too time-consuming to have come into general use. Density gradients can also be used to separate pollen into density fractions that have taxonomic consistency (Forster and Flenley, 1993).
7. Large pollen grains that might be removed by fine sieving include: *Cayaponia* (Cucurbitaceae) 120 μm, *Manihot* (Euphorbiaceae) 325 μm, *Hibiscus* (Malvaceae) 180 μm, *Pavonia* (Malvaceae) 260 μm, *Dalechampia* (Euphorbiaceae) 130 μm, and *Inga* (Mimosaceae) 200 μm (all sizes are maxima).
8. The number of Lycopodium or other exotic spore tablets to be used depends on the number of spores in the tablets and on the pollen concentration in the sediment sample. The final ratio of exotic spores to counted pollen should be in the range 1:2–1:6. The Stockmarr (1971) tablets of Lycopodium spores have between 12,000 and 15,000 spores per tablet. In the Lake Pata analysis one tablet to 0.25 cc sediment samples worked nicely but in large samples it might be necessary to use more than one tablet. In our pollen trapping it was necessary to use 5 tablets for the pollen collected by each trap.
9. Microspheres are made of polystyrene, can be ordered in the 15 micron size range, and are used in liquid suspension (for suppliers see appendix).
10. Hydrofluoric acid is a lethal substance of which you must be very wary. Always use it inside a fume hood and never breath its fumes. Use rubber gloves and protective clothing. Prolonged contact on the skin can be extremely dangerous, even fatal. An anti HF burn creme can be made with 1 part magnesium oxide to 1.5 parts glycerine jelly (Moore *et al.*, 1991).
11. Dark storage avoids the bleaching that can occur if the tubes have prolonged exposure to light. The bleached grains then become translucent and thus difficult to see under the microscope. Staining with safranin also helps to keep the grains visible.
12. Some pollen analysts use rigid media for mounting their pollen, usually glycerine jelly. We believe this to be poor practice because in such media the pollen grains are held rigidly in place by the medium and can be viewed from only the side that happens to be uppermost. So many grains in our diverse flora can be identified only from characters that can be seen properly from a single direction that we must feel skepticism about many identifications made of grains imprisoned in solid media. Among the taxa that particularly need rotation to be sure of their identity are *Copaifera*, *Caryocar*, Myrtaceae, Sapindaceae, Combretaceae, many Meliaceae, and even some Melastomaceae, especially when the grains are flattened. Identification of these taxa in fixed media should always be suspect.
13. The alcohol series needed is 96% alcohol, 99% alcohol, and tertiary butyl alcohol.
14. The one probable exception is Gramineae. Already we note the sizes of large grass pollen, seeking suggestions that they might be of the bamboo *Chusquea*, and we are always on the lookout for the very large grains that might be *Zea*. Finding cereal-sized grass grains (80–120 microns) in the sediments of Lake Ayauch[i] in Ecuadorian Amazonia suggested the possibility of *Zea* in deposits dated to 6000 years ago, which was confirmed when distinctive *Zea* phytoliths were found in the same deposit by D. Piperno (Bush *et al.*, 1989). On a more subtle level, it would be splendid if grasses from dry vegetation types, floating mats, and forest communities could be separated by size frequency statistics. The outlook, however, is not bright. Salgado-Labouriau and Rinaldi (1990) found that it was not possible to separate most of the Gramineae taxa of the Venezuelan Andes based on size alone. Extending this work to the Amazon lowlands would require mounts in silicone oil, and it would require a large effort studying the reference material, but perhaps one day.

15. When one taxon is heavily overrepresented, as, for instance, Gramineae in savanna parts of central Brazil, saturation curves for the next most abundant grains would level only after very large numbers of the overrepresented taxon were counted. To avoid wasting time counting huge numbers of overrepresented taxa, therefore, it is sensible to stop counting the overrepresented taxon at a predetermined number (if necessary set by a saturation curve for that taxon alone), express that taxon as a percentage of all other grains counted so far, and then continue to count remaining taxa until the stipulated pollen sum is reached.
16. Two examples are the Cana swamp in the Darien of Panama, on which herbaceous ferns are one of the dominant plant populations, and a cloud forest lake in the Galapagos Islands that is surrounded by tree-ferns. In both of these wet habitats, fern spores were heavily overrepresented, the fern pollen in the Galapagos lake being actually 300% of total pollen (Bush and Colinvaux, 1994; Colinvaux and Schofield, 1976).
17. Those working with silicone oil for determining size frequency statistics must mount their reference slides in silicone oil also.
18. More precise pollen identifications are sometimes possible using electron microscopes. These have been used on fossil material in a few studies where identification at the species level, rather than just to genus, has been crucial (Traverse, 1988). The technology has not yet been adapted for general pollen analysis, however.

CHAPTER

5

POLLEN MORPHOLOGY

Like the spores of ferns and other cryptogams, pollen grains are the product of meiosis, the cell division that produces haploid cells from diploid parents. Haploid fern spores are the first cells of gametophytes, the free-living haploid stage of the life cycle in which gametes are eventually produced by simple mitosis, fuse into a zygote, and finally develop into the diploid sporophyte which is the adult fern. Pollen grains are homologous to fern spores and equally haploid, but the gametophyte generation has been suppressed so that pollen grains are effectively no more than carriers of the male gamete. Pollen grains disperse to flowers, adhere to the stigma, and grow a pollen tube through which the gamete nucleus moves to enter the ovum. Thus the ecological functions of spores and pollen are subtly different. Both represent the reduction division necessary for a sexual life cycle, but spores also are dispersal agents. Structural homology between the two, however, remains clear and is the basis of their principal structural characteristics.

The meiosis which produces both spores and pollen makes them four at a time: four grains pressed together in a structure called a tetrad. In a few plant families, notably Ericaceae, pollen grains never separate and are dispersed as tetrads. Usually, however, both pollen and spores separate from their tetrads and disperse singly (monads) or, rarely, in pairs (dyads), but the separated grains often retain visible relics of their origin in tetrahedral shapes. This can be seen most clearly in the many fern spores that carry a triangular marking (trilete) recording folds of closest packing when the original four spheres were pressed together.

THE CELL WALL

Both pollen and spores develop a very resistant cell wall, the exine, which is made up of sporopollenin, a carotenoid-like polymer, which is highly stable molecule, resistant to various acids and bases, thus allowing the use of palynomorphs in geology, paleoecology, stratigraphy, archaeology and other disciplines.

Historically, the structure of the sporoderm or cell wall of pollen grains and spores has been the subject of various interpretations. Potonié (1934) and Wodehouse (1935) divided it into two units: the exine (Latin *exter*, on the outside) and the intine. Faegri and Iversen (1950) later divided the exine into ektexine (outer) and endexine (inner), which were later subdivided by Erdtman (1952) as shown on Figure 5.1. It is the sexine, the outermost layer of the exine that contains the sculptural and ornamental features used by pollen analysts to identify pollen.

SHAPES OF POLLEN

An idealized pollen grain can be thought of as a spheroidal object to which some geographical terms can be applied. An imaginary polar axis passes through the pollen grain connecting both extremities, termed poles. The equatorial line divides symmetrical grains into two "hemispheres." Shape classes which were developed based on the relationship between the polar axis (P) and the equatorial diameter (E) (Erdtman, 1952) are given on Table 5.1 and Figure 5.2. Gymnospermous bisaccate grains are shown on Figure 5.3. These grains possess two vesicles filled with an internal spongy tissue (Traverse, 1988), which are interpreted as possible flotation devices or used for enhanced wind dispersal and the protection of the corpus against dehydration (Wodehouse, 1959).

APERTURES

The germination of the pollen tube is made possible by the presence of apertures, thinner regions of the sporoderm, which permit the exit of the pollen tube. Apertures can be found as ecto-, endo-, or mesoapertures depending on their position in the different layers of the exine. The ectoapertures are more superficial and are located in the ectexine/sexine, endoapertures in the endexine/nexine and mesoapertures in an intermediary position (Punt *et al.*, 1994). Inaperturate grains are those lacking apertures, having instead patches of thin and easily penetrated exine.

Determining the numbers, types and positions of apertures is immensely important in the classification and identification procedure. The three basic aperture types are the colpus, the pore and the colporus. The colpus is an elongated germinal furrow (Erdtman, 1947) or an elongated thin area of the exine (Faegri and Iversen, 1950). The pore is a rounded opening or thin section of exine, and the compound aperture, colporus, is made up of both a colpus and one or more pores (Punt *et al.*, 1994). The morphological terminology of pollen and spore

TABLE 5.1. Erdtman's (1952) Shape Classes of Grains Seen in Equatorial View

Shape classes in equatorial view	P/E	100 · P/E
Peroblate	<4/8	<50
Oblate	4/8–6/8	50–75
Subspheroidal	6/8–8/6	75–133
Suboblate	6/8–7/8	75–88
Oblate-spheroidal	7/8–8/8	88–100
Prolate-spheroidal	8/8–8/7	100–114
Subprolate	8/7–8/6	114–133
Prolate	8/6–8/4	133–200
Perporlate	>8/4	>200

Based on the relationship between the polar axis (P) and the equatorial diameter (E). These shapes are represented in Figure 5.2.

grains, based on type, number and location of apertures is facilitated by the use of prefixes. For example, mono-, di- or bi-, tri, tetra, penta-, and hexa- for one to six, and poly- for many apertures respectively. The location of the aperture in the grain is defined by the prefixes zono- (or stephano-), for apertures or features distributed around the equatorial region and, peri- (or panto-) for those spread all over the surface of the pollen or spore. In some cases as, for example, in Melastomataceae, three colpori are intercalated by three false colpi or pseudocolpi. Such a grain is termed heterocolpate. The most commonly found types of apertures, numbers and position found in most neotropical pollen grains are shown on Figure 5.3.

THE SCULPTURAL AND ORNAMENTAL FEATURES

The sexine is the exine layer that contains sculptural and ornamental features which are sought by the palynologist during the identification of grains. One of the basic ornamental features separating angiosperm pollen from that of gymnosperms is the presence of the tectate-columellate condition in the angiospermous pollen (Taylor and Taylor, 1993). The tectum is a roof-like structure of the exine supported by column-like rods named columellae. Other basic differences between angiospermous and gymnospermous pollen are: the prevalence of positive ornamental or sculptural features of exine in angiosperms; the presence of a spongy or alveolar outer exine in gymnospermous pollen; the presence of a homogeneous endexine in angiosperms as opposed to a laminate inner exine in gymnospermous pollen (Traverse, 1988).

Many pollen grains have minimal ornamentation, a condition called "psilate," but most have ornamental features that can be either negative or positive. Negative features are those "carved" into the exine, as opposed to positive features, which are those resting on the grain surface. Among the negative ornamentation categories are: (a) perforate or micropitted tectum surface possessing small pits (<1 μm in diameter); (b) foveolate: surface with more or less rounded depressions (>1 μm in diameter); (c) fossulate: surface with elongated and irregular grooves.

Positive ornamentation categories are: (a) scabrate: pattern composed of granules of any shape but with diameters smaller than 1 μm; (b) gemmate: with somewhat rounded bud-like structures (gemmae); (c) clavate: with column-shaped sculptural elements (clavae), in which the tip is broader than the base; (d) verrucate: having wart-like sculptural elements (verrucae), usually larger than 1 μm in diameter; (e) echinate: having spines larger than 1 μm; (f) rugulate: ornamentation made up of elongated elements, >1 μm long; (g) striate: with grooves between elongated sculpturing elements; (h) reticulate: ornamentation made up of elongated sculptural elements, more than 1 μm long, forming a network-like pattern made up of cavities (lumina) bordered by other narrower elements (Punt *et al.*, 1994; Kremp, 1965). These ornamental categories are shown on Figure 5.4.

SPORE MORPHOLOGY

The basic spore types are the trilete, monolete and the alete (Figure 5.5). The trilete spore has Y-shaped laesura, i.e., a triradiate tetrad scar (Erdtman, 1943), whereas the monolete has one elongated laesura, and the alete condition lacks such scar. The trilete condition was reported as early as the lower Silurian (Hoffmeister, 1959), supporting the paleobotanical evidence that he alete has arisen from the monolete and the latter from the trilete condition (Richardson, 1969; Chaloner, 1970). Although not as highly ornamented, spores may contain similar ornamentation patterns as pollen grains, thus permitting the use of the same sculpturing or ornamental terminology (Traverse, 1988).

Important features of spore morphology present in some neotropical taxa are: (a) a Y-shaped laesura as in Lycopodium; (b) a monolete condition, that is., the single line of laesura, as in Blechnum; (c) a cingulum, or an equatorial collar-like extension, also referred to as zona (for example, in *Dicksonia*); (d) perine or perisporium, a membrane comprised of sporopollenin surrounding some spores (Jackson, 1928), which in most cases is destroyed during acetolysis or during the fossilization process.

POLARITY OF SPORES

In quaternary sediments, spores are usually found dissociated from the original tetrad. The proximal face (that facing the center of the original tetrad) in a spore is where the laesura is found. The distal face in some spores may contain ornamental features as it does in various species of Lycopodium, for example. *Cyathea*, an important tree fern found in Amazonian quaternary sediments, has ornamentation features varying from psilate to verrucate, whereas *Anthocerus* normally has a spinulose sculpture in both proximal and distal faces.

CAPÍTULO

1

Análise Palinológica Neotropical

O registro palinológico da rica flora neotropical pode fornecer a história, não somente das mudanças de vegetação através do tempo e, portanto, de clima, mas também da dinâmica de comunidades. Por um longo tempo, os palinólogos negligenciaram a região tropical, particularmente a amazônica. Em parte, isso pode ser explicado pela dimensão proibitiva da flora polínica, ou pelas dificuldades inerentes a um trabalho em florestas tropicais. Mais importante, contudo, foi o fato da análise palinológica ter sido inventada na região temperada do hemisfério norte. Os primeiros palinólogos, assim como a maioria dos ecólogos, trabalhavam para universidades de pesquisa que, por motivos históricos, se localizavam na estreita faixa entre as latitudes 40° e 55° em volta do hemisfério norte. A análise palinológica foi criada com a finalidade de estudar a história das florestas do Pleistoceno tardio em altas latitudes.

As florestas do hemisfério norte possuem algumas propriedades particulares que simplificaram o trabalho dos palinólogos. A diversidade de espécies é tão reduzida, que um só país pode ser o lar de apenas algumas dezenas de espécies de árvores. No país insular da Grã-Bretanha, por exemplo existe somente, cerca de uma dúzia de espécies arbóreas nativas. A maioria das árvores das florestas do hemisfério norte são anemófilas, isto é, sua polinização é mediada pelo vento, e o pólen é produzido em grande quantidade. Automóveis estacionados nas proximidades de um pinheiro (*Pinus*) podem ficar amarelos com a deposição de seu pólen em apenas algumas horas. Ervas abundantes, como as das famílias Cyperaceae e Gramineae (Poaceae), são também anemófilas e sua produção polínica é tão alta que muitas pessoas, que vivem no cinturão de clima temperado, sofrem com alergias ou com a *"febre do feno"* nos Estados Unidos, quando respiram ar com alto teor polínico.

O vento carrega os grãos de pólen sobre amplas áreas e, no decurso de um ano, sua precipitação local em forma de "chuva polínica" representa a vegetação de uma área geográfica considerável. A concentração dos grãos de pólen dessa chuva polínica, em sedimentos lacustres e em pântanos, é alta. Nestas circunstâncias, o sinal polínico, para ambas a mudança climática e a composição da comunidade vegetal, é obtido através de contagens percentuais relativamente simples de pólen de algumas espécies de árvores e ervas que constituem a grande parte das chuvas polínicas preservadas nos sedimentos.

Na Floresta Amazônica, contudo, a produção e dispersão polínicas são muito diferentes. A diversidade de espécies é imensamente maior, tão maior que cerca de 300 espécies de árvores foram encontradas em um simples hectare (Gentry, 1988). Além disso, a grande maioria das árvores da Amazônia não perdem pólen para a atmosfera e dependem diretamente de animais — insetos, pássaros ou morcegos — para coletar e distribuir seu pólen. Esses fatos, por um longo tempo, desencorajaram os palinólogos que estavam habituados com essas suposições atribuídas ao pólen anemófilo de florestas simples, e que chegaram a questionar a viabilidade desses estudos em florestas tropicais (Faegri & van der Pilj, 1979).

De acordo com essa visão pessimista, as árvores da Floresta Amazônica, polinizadas por insetos, deveriam produzir poucos grãos de pólen, ou o suficiente para carregar seus vetores de polinização. Sua produção polínica rarefeita não permitiria que grãos excedentes escapassem a esse ciclo fechado e sedimentassem em lagos ou pântanos. Portanto, para um palinólogo treinado nas latitudes temperadas, como todos o eram, as florestas da Amazônia deveriam ser ricas em "espécies silenciosas" que não deixariam registro palinológico nos sedimentos de lagos e pântanos. Os próprios sedimentos tropicais eram tidos como pobres em pólen, pois a eles deveriam ter sido negados vastos influxos de pólen anemófilo. Essa conclusão lógica era devido à suposta ausência do pólen a ser examinado e a noção de que o pólen presente na amostra sedimentar produziria uma visão distorcida da vegetação que o produziu, resultando na inutilidade da técnica palinológica. Esses receios foram comprovados como errôneos e alguns estudos palinológicos mostraram que lagos da Amazônia contêm um amplo registro palinológico, que pode produzir uma notável imagem da floresta local ou de outras comunidades vegetais.

PRODUÇÃO DE PÓLEN NA AMAZÔNIA

Sedimentos lacustres da Amazônia possuem grandes concentrações polínicas. No Lago da Pata, na floresta tropical úmida do Alto Rio Negro, Brasil, as concentrações polínicas variam com uma média de 247.600 grãos/cm^3. No Lago de Kumpaka, na Província de Oriente do Equador, que recebe sedimentos de uma grande bacia de drenagem, as concentrações polínicas médias giram em torno de 100.000 grãos/cm^3.[1] Essas concentrações são favoravelmente comparáveis com aquelas encontradas em sedimentos lacustres em New Hampshire, EUA (Davis *et al.*, 1984).

A prolífica produção polínica das florestas da Amazônia é, também, demonstrada pelo nosso programa de coleta de chuva polínica dentro de florestas fechadas. Colocamos coletores simples (descritos no Capítulo 3), em grupos de 10–100, em quadrados delimitados dentro da floresta com composição florística conhecida: perto de Manaus, na Amazônia central,[2] na Amazônia do Oriente do Equador e na Ilha de Barro Colorado, no Panamá. Os coletores de pólen são coletados em intervalos de 6 meses a um ano, permitindo o cálculo do influxo polínico em grãos/cm^2/ano. Os nossos coletores retêm de 7.000 a 30.000 grãos de pólen/cm^2/ano (Bush, 1992). Esses resultados são comparáveis aos valores de influxo obtidos nas florestas temperadas dos Estados Unidos (10–15.000 grãos/cm^2/ano) em Mirror Lake (David *et al.*, 1984).

As associações polínicas da Amazônia são qualitativamente diferentes daquelas da região temperada, pois são caracterizadas por uma maior biodiversidade. A riqueza taxonômica total, representada por um simples coletor de chuva polínica na Amazônia, é muitas vezes maior do que aquela encontrada nas regiões temperadas do hemisfério norte. Mas essa alta riqueza taxonômica é constituída por vários táxons que aparecem em baixa freqüência, geralmente um grão encontrado em uma contagem polínica de 300 a 500 grãos. Estes tipos polínicos raros, quando adicionados a outros, podem constituir uma porcentagem significativa da soma total de pólen da amostra. Isto sugere que a alta diversidade de espécies de plantas polinizadas por animais, na Floresta Amazônica, é representada, na chuva polínica, por uma riqueza comparável de tipos polínicos em baixa concentração. Os resultados dos nossos estudos de chuva polínica confirmam uma das suposições dos palinólogos da zona temperada: o pólen disperso por animais não é carregado pelo vento. Encontramos tais tipos polínicos distantes até 50 m da árvore produtora, sendo esse valor o limite máximo de dispersão encontrado. A grande maioria dos tipos polínicos encontrada foi sedimentada em coletores colocados em baixo da copa das próprias árvores produtoras. Apesar disso, muitos desses tipos polínicos são encontrados em sedimentos lacustres. A conclusão, quase que inevitável, sugere que o pólen é carregado das árvores até as bacias de sedimentação através do fluxo de água corrente da superfície. Tempestades derrubam pólen e flores no chão; estes são redistribuídos na área da bacia de drenagem e, mais tarde, levados à sedimentação pelo escorrimento hídrico superficial (*run-off*). O sinal polínico das plantas polinizadas por animais, portanto, registra somente a comunidade vegetal que realmente está crescendo dentro da bacia ou da drenagem local. Dessa forma, a hesitação clássica dos palinólogos está justificada: o pólen de árvores polinizadas por animais não produz nenhum sinal da vegetação regional, pois registra somente comunidades locais de florestas. Para estudos de comunidades florestais, contudo, esse transporte de pólen a uma bacia de drenagem adjacente tem a grande vantagem de permitir a análise detalhada de estudos históricos de associações florestais, dentro dessa bacia de sedimentação.

Assim, a grande maioria das árvores da Amazônia que são polinizadas por animais não são "silenciosas" no registro polínico, mas, ao contrário, têm o potencial de fornecer registros locais detalhados. Sua produção polínica total é surpreendentemente alta. Possivelmente, isto reflete a competição por polinizadores na floresta, na qual o próprio pólen é parte de um sistema de "isca" e "recompensa" para o inseto ou outro animal vetor do transporte. Essa super-produção de pólen, contudo, pode ser simplesmente uma função das estruturas de floração que expõem flores e anteras às vicissitudes ambientais causadas pelas tempestades e chuvas. Uma das evidências informais que apoia essa idéia refere-se ao fato de que pólen de orquídeas (Orchidaceae) não é encontrado em sedimentos ou em coletores de chuva polínica. Possivelmente, os refinados mecanismos de polinização das orquídeas evitam as perdas polínicas sofridas por outras plantas, negando dessa forma aos palinólogos o sinal fornecido pela grande massa de grãos de pólen redundantes ou perdidos.

Os ecossistemas amazônicos também produzem quantidades, algumas vezes grandes, de pólen que pode ser disperso pelo vento, mesmo nas florestas. As famílias Moraceae, Urticaceae, Ulmaceae e Myrtaceae são representadas nas florestas das terras baixas amazônicas por muitas espécies, cuja produção polínica é grande e amplamente dispersa. O pequeno pólen da árvore invasora (pioneira) de *Cecropia*, por exemplo, pode representar 40% da chuva polínica em alguns lagos da Amazônia (Bush & Colinvaux, 1988).

Outros gêneros de árvores da Amazônia que aparecem em diferentes florestas ou outros tipos de vegetação arbórea são também anemófilos, particularmente os gêneros arbóreos *Ilex*, *Podocarpus*, *Bursera*, e Euphorbiaceae. Adicionalmente, as gramíneas (Gramineae/Poaceae) e as ciperáceas (Cyperaceae), são produtoras de quantidades copiosas de pólen anemófilo, tanto na Bacia Amazônica como em outras áreas.

A produção polínica em ecossistemas amazônicos, portanto, é ampla para os propósitos da análise polinológica, mas os sinais polínicos na Amazônia devem ser lidos com cautela, devido às suas diferenças daqueles das florestas temperadas do hemisfério norte, onde os métodos palinológicos foram desenvolvidos. A Amazônia produz um registro altamente diversificado de comunidades florestais dentro de bacias hidrográficas, mas os registros regionais devem ser inferidos através da combinação de dados provenientes de florestas estritamente locais, com registros regionais fornecidos pelo pequeno sub-grupo de gêneros cujo pólen, por algum motivo, é transportado pelo vento.

INTERPRETANDO ESPECTROS POLÍNICOS NA AMAZÔNIA

Começamos com o diagrama polínico de porcentagem do Lago da Pata, na Floresta Amazônica do Alto Rio Negro, Amazônia brasileira, mostrado na Figura 1.1. O Lago da Pata ocupa uma bacia endorreica (sem córregos de saída ou entrada), provavelmente pseudokárstica formada pela lenta dissolução de silicatos de rochas plutônicas. O lago está a uma elevação de apenas 300 m, sobre um "inselberg" localizado acima do grande domínio das terras baixas da Bacia Amazônica, com aproximadamente 100 m de elevação e coberto com floresta tropical de terra firme. De um dos cumes desse "inselberg," pode-se ver, até o horizonte, um "tapete" florestal contínuo. O "inselberg" também é coberto por floresta tropical, embora algumas vezes reduzida em altura e modificada em locais pelos solos rasos sobre rochas que resistiram à erosão, formando florestas edáficas. A pequena bacia do Lago da Pata é coberta por "facies" locais de floresta tropical, embora uma aparente ausência das exposições rochosas vistas em certos setores do "inselberg" sugira que a comunidade florestal tem um certo controle edáfico. Com um amostrador de pistão, tipo Livingstone, coletamos 7 m de sedimentos, sendo os primeiros 1,7 m correspondentes à sedimentação nos últimos 40.000 anos radiocarbônicos. A análise palinológica do topo desta seção fornece a história das florestas baixas da Amazônia, através do último ciclo glacial (Colinvaux *et al.*, 1996).

Os espectros polínicos superiores (mais recentes) do digrama da Figura 1.1, ilustram o sinal polínico de Floresta Amazônica das terras baixas, semelhante àquela existente atualmente na região. Quatro conclusões são óbvias nesta parte do diagrama polínico:

1. Presença de táxons anemófilos: Moraceae/Urticaceae, Euphorbiaceae (*Alchornea*) e outros.
2. Presença de quantidades significativas de um ou mais táxons de árvores polinizadas por animais (*Mauritia* [Palmae]; *Copaifera* [Leguminosae]).
3. Quando todo o pólen arbóreo, incluindo as baixas contagens de grãos raros encontrados (listados como "outros arbóreos") é somado, este chega a valores de até 90%.
4. Contagem muito baixa de pólen de ervas, especialmente de Gramineae, que é quase ausente.

Assim, esse sinal polínico fundamental, no interior do complexo florestal da Amazônia, é dado pelo pólen de árvores e por quantidades insignificantes de gramíneas e outras ervas. Dessa forma, o diagrama básico de uma área dentro de uma grande região de floresta tropical no Alto Rio Negro, na Amazônia brasileira, é caracterizado por pólen de árvores, com raras gramíneas e outras ervas.

As propriedades essenciais do diagrama de porcentagem do Lago da Pata também caracterizam as partes mais antigas (inferiores) do diagrama e mostram que a floresta tropical das terras baixas amazônicas tem ocupado sua paisagem, pelo menos, nos últimos 40.000 anos, através do tempo glacial e pós-glacial. Essa importante conclusão é derivada de uma parte muito pequena da alta diversidade na floresta. O esperado registro proveniente do *"run-off,"* caracterizado por pólen de árvores polinizadas por animais também se encontra representado. Sua presença é registrada na categoria "outros arbóreos" (*other arboreal*) que inclui gêneros de floresta tropical como *Apeiba, Cedrela, Cordia, Didymopanax, Guarea, Iriartea, Luehea* etc.

As Árvores do Lago da Pata Polinizadas por Animais

O táxon *Mauritia* é proeminente na seção holocênica do diagrama polínico, e não ocorre no Pleistoceno. *Mauritia*, uma palmeira de áreas alagadiças e pantanosas, parece ter polinização efetuada por animais. Grupos de *Mauritia* são, geralmente, comuns nas bordas de lagos o que favorece a presença de seu pólen em diagramas polínicos da Amazônia. Existe um agrupamento significante de *Mauritia* nas margens do Lago da Pata e o seu diagrama polínico sugere que esse táxon tem estado presente na paisagem por todos os 10.000 anos do Holoceno. Antes de 10.000 anos AP (antes do presente), *Mauritia* não era encontrada no Lago da Pata. Esse é um registro puramente local de uma pequena área do pântano dentro da bacia que pode ou não registrar o clima alterado. Aumento na precipitação poderia criar um pântano para *Mauritia*, mas também o poderia um represamento natural do lago ou um aumento do nível do lençol freático. A coincidência do aumento de *Mauritia* no Lago da Pata, com uma mudança estratigráfica nos sedimentos e o início do tempo pós-glacial dão apoio à hipótese de que o pântano em questão estava associado ao final da última glaciação devido ao aumento da precipitação e conseqüentemente da área pantanosa ao redor do lago. Uma conseqüência alternativa do final do período glacial, contudo, refere-se simplesmente aos lençóis freáticos que se tornariam mais elevados devido à elevação do nível do mar a seus valores atuais e que a elevação do nível do lençol freático teria inundado áreas que se tornariam em pântanos de *Mauritia*.[3]

Persistência da floresta tropical antes e depois do estabelecimento de *Mauritia* sugere que qualquer alteração no regime pluviométrico, da região do Lago da Pata, não poderia ter sido de grande magnitude. Assim, o registro do pólen de *Mauritia* sugere que o nível médio do lago subiu no final do período glacial, inundando a porção rasa da bacia, agora ocupada por um pântano de *Mauritia*. O mesmo diagrama polínico, contudo, não indica a causa do aumento do nível da água.

Em outros depósitos sedimentares da Amazônia o aumento do pólen de *Mauritia* é devido mais aos processos ecológicos do que à mudança climática. Isso é particularmente verdade para as histórias palinológicas da Amazônia, que são, essencialmente, registros de preenchimento de antigas bacias ou antigos canais fluviais. Nesses depósitos, os dados palinológicos registram o estabelecimento de populações de *Mauritia* nas regiões pantanosas, depois de terem sido suficientemente drenadas pela ação de pequenos rios.

O táxon *Copaifera* aparece no médio-Holoceno com valores de 10%. *Copaifera* é um gênero de árvores de grande porte da família Leguminosae. Vinte e cinco espécies são conhecidas na região neotropical. Esse gênero está representado no diagrama polínico do Lago da Pata. Nossos dados palinológicos sugerem um evento com população local de *Copaifera*, na bacia de drenagem do lago, perdurando por muitas gerações arbóreas, que começam por volta de 6.000 anos AP. Mais do que isso não podemos dizer, baseados na evidência disponível, mas o registro de *Copaifera* ilustra o tipo de detalhe inerente aos diagramas de pólen da Amazônia.

Outros táxons arbóreos polinizados por animais aparecem regularmente, embora em baixas porcentagens, que são esperadas para a grande maioria de árvores polinizadas por animais:

Caesalpiniaceae, Lecythidaceae, *Tapirira* [Anacardiaceae]. Dessas, *Tapirira* é um gênero sómente de árvores de florestas tropicais. Infelizmente ainda não podemos identificar o pólen de muitas Caesalpiniaceae e Lecythidaceae a nível de gênero e espécie, mas ambas as famílias são bem representadas entre as árvores da floresta tropical das terras baixas da Amazônia, e Caesalpiniaceae inclui lianas da floresta. Junto à *Copaifera*, a ocorrência desses táxons demonstra a presença de árvores da floresta tropical das terras baixas crescendo dentro da pequena bacia de drenagem do Lago da Pata.

Cecropia e Melastomataceae: O Problema dos Cosmopolitas

Cecropia (embaúba) é um gênero de árvores pioneiras pertencente à família Moraceae, tradicionalmente tida como anemófila. Essas árvores poderiam ser chamadas de "primeiras colonizadoras" ou "espécies oportunistas," "espécies fugitivas" ou até mesmo "espécies de estratégia r" nas várias formas de uso da Língua Inglesa na teoria ecológica. As embaúbas estão presentes em quase todos os tipos de vegetação neotropical, desde as terras baixas até cerca de 2000 m de elevação. A produção polínica é muito alta e seu pequeno grão característico (6–14 μm, biporado) aparece em, virtualmente, quase todos os diagramas polínicos neotropicais, mesmo naqueles das altas elevações dos Andes do Peru (4000 m), situação geralmente explicada pelo longo transporte do seu pólen pelos ventos ascendentes (Hansen *et al.*, 1984). A presença de uma porcentagem modesta de pólen de *Cecropia* contra um sinal típico de floresta tropical de terras baixas, como no diagrama do Lago da Pata, é esperada devido à presença normal de colonizadores de clareiras das florestas. Porcentagens similares e relativamente baixas de *Cecropia* em outros tipos de vegetação, contudo, devem ser interpretadas com cautela, pois, possivelmente, são resultantes de transporte à longa distância (como nos Andes do Peru e em regiões secas). Devido ao fato de margens de lagoas serem naturalmente abertas e perturbadas, é muito provável que *Cecropia*, algumas vezes, seja super-representada em sedimentos lacustres. No Lago Ayauchi, situado em uma cratera vulcânica, na floresta tropical amazônica do Equador, o pólen de *Cecropia* estava presente em porcentagens maiores que 70% da soma total de pólen, possivelmente mostrando o efeito da borda do lago (Bush & Colinvaux, 1988). Altas porcentagens de *Cecropia* também podem ser esperadas nos primeiros estágios do preenchimento de sucessões hidrárquicas, com o abandono progressivo de canais fluviais e com o desenvolvimento de agrupamentos pioneiros desse táxon. Algumas ilhas arenosas ou argilosas, dentro do sistema fluvial do Rio Amazonas mantêm agrupamentos praticamente mono-específicos de *Cecropia*, que com certeza, contribuem para a super-representação desse táxon na chuva polínica local.

A ubiqüidade do pólen de *Cecropia*, em muitos tipos de vegetação neotropical é, na verdade, um pequeno exemplo de um problema altamente difundido. Os táxons de ampla ocorrência, freqüentemente identificados ao nível de família tendem a incluir espécies adaptadas a muitos tipos diferentes de hábitats, ambientes ou comunidades. Um exemplo conhecido é a família Melastomataceae, cujos membros são amplamente encontrados como arbustos, pequenas e grandes árvores, em praticamente todas as partes da região neotropical, desde as terras baixas até as florestas nebulares dos Andes, ou até mesmo nas Ilhas Galápagos. Nesse arquipélago ocorre *Miconia robinsoniana*, uma espécie endêmica que forma um cinturão mono-específico na porção sul da Ilha de Santa Cruz (Lawesson *et al.*, 1990).

As melastomatáceas são polinizadas por insetos ou beija-flores (Cronquist, 1988). As flores, muitas delas com forma de sinos e as vistosas flores de outras espécies, não sugerem polinização anemófila, contudo o seu pólen muito característico, é encontrado, freqüentemente, nos sedimentos. Praticamente, em todo o diagrama polínico do Lago da Pata, o táxon Melastomataceae flutua entre 10% e 30% e, mesmo em suas porcentagens mais baixas, essa família mantém valores de 5% a 10%. Em outros sedimentos lacustres da Floresta Amazônica e em lagos andinos abaixo da linha de árvores, encontramos porcentagens de Melastomataceae comparáveis a essas encontradas no Lago da Pata. Os valores desse táxon são muito altos para plantas com polinização por animais. Uma possível explicação para os altos valores de Melastomataceae nos sedimentos refere-se à possível produção de pólen redundante ou produção polínica em excesso. Muitos grãos chegam ao solo sendo, mais tarde, transportados para as bacias de sedimentação pela água superficial. A avaliação do pólen como uma função dos elementos associados encontrados em uma certa comunidade vegetal, através de métodos estatísticos multivariados, é uma forma de medirmos seu significado paleoecológico. Uma explicação viável para a redução de Melastomataceae no Holoceno do Lago da Pata, é que suas espécies locais foram inibidas pelo aquecimento pós-glacial.

REAGRUPAMENTO DE TÁXONS FLORESTAIS EM REPOSTA A MUDANÇAS DE TEMPERATURA

A Figura 1.2 mostra espectros polínicos de superfície, ao longo de um gradiente de elevação, no flanco leste das montanhas dos Andes do Equador. Todas as amostras, com exceção de pântano de Puyo (Puyo Bog) e Mera (amostras de solo vegetal e massas — *moss polsters* — de briófitas), são da interface sedimento/água de lagos. A vegetação amostrada varia desde floresta tropical das terras baixas na elevação de 280 m até vegetação de páramo, desprovida de árvores, na altitude de 3.400 m. As contagens polínicas foram feitas, razoavelmente cedo, nas nossas pesquisas neotropicais, pois nossas coleções palinológicas de referência e o nosso conhecimento ainda eram um pouco deficientes, pois as somas de pólen não identificados (listados como "outros") eram relativamente altas (20–40%) em todas as elevações. Contudo, 60% a 80% dos táxons são facilmente identificados e correlacionados com as principais características da vegetação moderna ao longo do transecto (Bush, 1991).

A distribuição de *Cecropia*, com relação à altitude, é coincidente com o limite da floresta tropical na Amazônia. Na Floresta Atlântica do Brasil (Hueck, 1966), e na região do Pico da Neblina, no noroeste da Amazônia brasileira, o limite superior da floresta está perto dos 2.000 m (Radambrasil, 1976). Nos flancos andinos, oeste da Amazônia, onde nosso transecto foi tomado, o limite superior e convencional da floresta tropical é de 1.200 m. Na Figura 1.2, altas porcentagens de pólen de *Cecropia* são sinais da presença de floresta.

O pólen anemófilo da família Moraceae, à qual *Cecropia* também pertence, é proeminente na floresta tropical abaixo de 1200 m, na seção equatoriana. Suas baixas porcentagens nas altas elevações são, possivelmente, atribuídas à dispersão pelo vento ascendente.

O pico na porcentagem de Moraceae a 2600 m no lago Yambo (Equador), é consistente com essa hipótese de distribuição pelo vento, pois é uma região desértica no Planalto Inter-Andino com pouca produção polínica e o seu pólen trazido por transporte, à longa distância, aparece com alta porcentagem na chuva polínica total (Colinvaux *et al.*, 1988). A porcentagem praticamente constante de pólen de Melastomataceae em todas as elevações, certamente reflete a presença de diferentes espécies desta família em virtualmente todos os tipos de vegetação neotropical.

Nesse transecto, as maiores elevações são facilmente caracterizadas por altas porcentagens de pólen de gramíneas, ciperáceas e compostas (Compositae/Asteraceae). Obviamente, nenhuma dessas famílias está restrita às altas elevações e suas espécies podem ser encontradas em abundância em várias partes das terras baixas da Amazônia. Suas proeminentes porcentagens polínicas, em elevações acima de 2000 m ao longo deste transecto altitudinal, simplesmente refletem a substituição progressiva de floresta por vegetação aberta, principalmente as "savanas" de altitude, conhecidas nos Andes equatoriais como "páramo." O aparecimento de *Podocarpus, Alnus* e *Hedyosmum* acima de 3000 m, sugere, contudo, a restrição dessas espécies a grandes elevações, presumivelmente pela temperatura. O gênero *Podocarpus* tem muitas espécies na região neotropical; uma fonte sugere um total de 29 espécies (Buchholtz & Gray, 1948). A maioria de suas populações modernas estão confinadas às grandes altitudes como sugerido na Figura 1.2, mas no Brasil Central duas espécies, *P. sellowii* e *P. lambertii* crescem em elevações mínimas de 750 m em florestas úmidas de galeria (De Oliveira, 1992). Alguns indivíduos já foram encontrados ao nível do mar em alguns locais especiais como, por exemplo, nos solos de areias brancas da Amazônia (Gentry, 1993). Como a família Melastomataceae, o gênero *Podocarpus* tem indivíduos de espécies, que podem ser encontradas em quase todas as elevações na região neotropical, onde há condições de crescimento. Assim, *Podocarpus* não pode ser usado como um fóssil-indíce para a temperatura. Mas, ao contrário de Melastomataceae, seu pólen não é rotineiramente encontrado em quantidades significativas em todas as elevações. Para encontrar mais do que alguns grãos esparsos em amostras superficiais ou em coletores, que podem ser atribuídos ao transporte de longa distância, é necessário que a amostragem seja feita em locais com elevação superiores a 1000 m. Isto sugere que populações recentes e significativas estão confinadas às grandes elevações.

Descobertas recentes de porcentagens significativas de *Podocarpus* em terras baixas estão sendo feitas em sedimentos com idade glacial. Um exemplo é a ocorrência acima de 10% de *Podocarpus* em espectros polínicos datados de 18.000 a 14.000 anos AP no Lago da Pata (Figura 1.1). Nesse lago e em algumas localidades no Equador, as altas porcentagens de *Podocarpus* são acompanhadas por outros táxons polínicos que sugerem ambientes alto-montanos, como por exemplo *Hedyosmum*, ou mesmo nos pés dos Andes, *Alnus* e o gênero quase que exclusivamente montano, *Drimys* (Winteraceae). Nenhum desses táxons (com a possível exceção de *Drimys*) são fósseis índices satisfatórios para a temperatura. Contudo, o esfriamento glacial parece ser a hipótese mais simples e lógica para explicar a co-ocorrência desses táxons em florestas tropicais das regiões baixas.[4]

Embora os dados polínicos da Amazônia ofereçam poucos, se alguns, fósseis-índices para a temperatura, eles, com certeza, em combinação com outros tipos polínicos, fornecem um sinal para a temperatura. Esta conclusão espelha a experiência dos palinólogos na zona temperada, que têm usado funções de transferência de composição percentual de gêneros arbóreos para reconstruir a temperatura média anual nos últimos milênios. Essa técnica usa porcentagens de pólen como valores de importância que podem, então, ser correlacionados com a temperatura em amostras de superfície.[5] Porcentagens de pólen de comunidades do passado (geralmente diferentes) são então transferidas como temperaturas. O aparecimento de 10% de pólen de *Podocarpus* e outros táxons nas florestas da Amazônia da Idade do Gelo sugere que o mesmo raciocínio é válido e deve ser seguido na região neotropical. Mas antes, devemos ter uma base de dados de centenas de amostras superficiais através de toda a Bacia Amazônica, todas correlacionadas a comunidades vegetais e parâmetros climáticos modernos. Até que tenhamos esses dados, inferir paleotemperaturas, através de pólen fósseis neotropicais, permanecerá uma prática controversa.

SINAIS POLÍNICOS DE CERRADOS E SAVANAS

Todas as investigações palinológicas do Pleistoceno tardio da Amazônia estarão alertas à possibilidade de redução de precipitação no passado. Isto, em parte, é resultado da "Teoria dos Refúgios" de Jurgen Haffer e outros (Prance, 1982; Whitmore & Prance, 1987) e da poderosa evidência de aridez glacial em outras partes do cinturão tropical, particularmente na África. Uma Idade do Gelo mais árida se torna admissível pela evidência, cada vez maior, de esfriamento glacial na região neotropical e os trópicos mais frios necessariamente significariam trópicos mais secos. Na sua forma mais extrema, a hipótese de aridez glacial tem toda ou muito da Floresta Amazônica substituída por vegetação de cerrados ou savana. Felizmente, o sinal polínico de vegetação mais árida (cerradão, cerrados e savana) pode ser distinguido, sem dúvida, do sinal de florestas tropicais.

A Figura 1.3 mostra um diagrama polínico de porcentagem de sedimentos glaciais e pós-glaciais, no domínio fitogeográfico do cerrado, na região relativamente seca do planalto do Brasil central, em Cromínia, Goiás (Ferraz-Vicentini & Salgado-Labouriau, 1996). Esse trabalho é do laboratório da palinóloga pioneira brasileira M.L. Salgado-Labouriau e trata de um depósito de sedimentos orgânicos de uma "vereda," vegetação pantanosa onde *Mauritia* é o elemento arbóreo principal. A região estudada tem elevação de 500 a 1000 m, e uma estação seca prolongada que não permite o estabelecimento de floresta fechada, exceto nos vales fluviais. A atual vegetação de cerrado é descrita como "vegetação parecida com a de savana na qual o extrato inferior é dominado por gramíneas e possui arbustos e pequenas árvores tortuosas, pertencentes principalmente à familia Leguminosae, espalhadas pela paisagem." Em vegetações mais secas de savana, a proporção de arbustos e árvores é mais reduzida.

Este diagrama de pólen do cerrado é caracterizado por pólen de ervas pertencentes à Gramineae, com 50% nas amostras holocênicas e em amostras superficiais retiradas do local estudado. Nas amostras de idade glacial, o pólen de gramíneas alcança 80% da soma total. Os contribuintes mais importantes à soma de pólen, com exceção das gramíneas, são as famílias de ervas Compositae e Cyperaceae. O aumento de 80% de pólen de Gramineae em tempos glaciais sugere uma tendência da vegetação à savana atual, embora seja igualmente admissível que os 30% extra de gramíneas seja um artefato estatístico da remoção de Compositae, *Mauritia* e Melastomataceae da chuva polínica, sem a compensação por aumentos de táxons, outros que Gramineae.

Quase totalmente ausentes neste diagrama de pólen, estão as assembléias polínicas de árvores pertencentes à floresta tropical, com elementos transportados pelo vento ou por animais. Este diagrama de pólen de uma região seca do Brasil tropical é inteiramente diferente da "assinatura" polínica de uma floresta tropical das terras baixas amazônicas, ilustrada pelo registro do Lago da Pata. Se tal vegetação de clima seco tivesse substituído a floresta tropical nas terras baixas da Amazônia, o sinal polínico em um diagrama teria um grande impacto. Obviamente, não houve tal substituição no centro-oeste da Amazônia, onde se situa o Lago da Pata (Figura 1.1).

O registro de Cyperaceae neste testemunho, provavelmente, enfatiza plantas crescendo no pântano, assim como o faz o pólen de *Mauritia*. Aqui o pólen de *Mauritia*, como sempre, fornece o sinal de um hábitat úmido. sugerindo que sua ausência nos depósitos glaciais poderia indicar o ressecamento da área. Uma explicação alternativa, contudo, é que a "vereda" estava presente, mas o esfriamento ambiental era inibidor para *Mauritia*. Esse argumento é possivel porque populações de *Mauritia*, atualmente, não se expandem a latitudes superiores a 20°S (Lorenzi *et al.*, 1996). A sensibilidade de temperatura de *Mauritia* explicaria também seu aparecimento no Holoceno, mas não em tempos glaciais na vereda de Águas Emendadas, no Brasil Central (Figura 1.4) como é sugerido pela autora (Barberi-Ribeiro, 1994). Neste exemplo, uma explicação, baseada na aridez, para a ausência de *Mauritia* em tempos glaciais se torna improvável devido à presença de plantas aquáticas (Pontederiaceae) que demonstram a presença contínua da umidade.

Interpretando o Pólen de Gramineae em Espectros Polínicos da Amazônia

As "facies" secas da vegetação neotropical: estepe, cerrado e cerradão, caatinga têm um uma cobertura de gramíneas e produzem espectros polínicos nos quais o pólen de Gramineae é predominante, possivelmente com 50% ou mais da soma total. Assim, qualquer espectro polínico sem altas porcentagens de Gramineae não pode representar clima ou hábitat seco. A proposição oposta, contudo, que alta porcentagem de Gramineae sempre significa clima seco é, infelizmente, falsa.

Um típico hábitat para as gramíneas nas terras baixas e úmidas da Amazônia é aquele proporcionado pelos "colchões" flutuantes de vegetação que se formam em lagos rasos ou em riachos, cujo fluxo de água é lento, como em muitas várzeas. As espécies de gramíneas que formam muito da matriz desses "colchões" de vegetação aquática, embora pertencentes a outras espécies que não estão presentes na vegetação de savanas, não podem ser diferenciadas pelo seu pólen através da microscopia ótica: ambos os grupos de gramíneas aparecem nos diagramas de pólen, simplesmente como Gramineae. Outro hábitat comum para as gramíneas, ao longo dos rios amazônicos, são as pequenas ilhas e bancos de areia e argila expostos pelos movimentos dos canais fluviais, os quais são freqüentemente colonizados por grupos densos de gramíneas, parecidas com aquelas que crescem nas savanas. Ambos os "colchões" flutuantes de vegetação e as ilhas de areia ou argila dos canais fluviais injetam grãos de pólen de gramíneas no sinal polínico das florestas tropicais. Na interpretação de picos de gramíneas em um diagrama polínico, o princípio básico a ser seguido deve ser o da busca de evidência de mudança em outros elementos do diagrama. Se táxons florestais continuam ao lado do perfil expandido dos espectros de Gramineae, o pólen de gramíneas não deve e não pode ser o fator principal na evidência do aumento da aridez.

Outra interpretação equivocada de máximos de Gramineae, pode acontecer em diagramas polínicos de regiões pantanosas ou de turfeiras. Ao contrário de sedimentos lacustres profundos, os depósitos de pântanos e turfeiras são sujeitos ao ressecamento superior causado por pequenas mudanças na precipitação ou por mudanças nos padrões de drenagem na paisagem local. Um dos resultados desses fenômenos é que gramíneas e outras ervas que crescem densamente na superfície seca desses depósitos invariavelmente produzem pólen que se tornam super-representados nos espectros polínicos locais. A estatística, então, requer que as porcentagens de todos os outros táxons na soma total de pólen se tornem reduzidas para acomodar as porcentagens aumentadas de pólen de gramíneas e ervas derivadas de plantas que crescem na superfície de pântanos e turfeiras. Alguns palinólogos, vítimas de diagramas polínicos concentrados ou resumidos de tais áreas, chegaram a sugerir que as savanas ou a vegetação de campos abertos estavam em expansão à custa da redução de florestas, quando realmente a única mudança real na paisagem era o ressecamento da turfeira ou do pântano amostrado, possivelmente causado por uma mudança climática.

O pólen de Gramineae (ou a soma de pólen de ervas), portanto, não pode ser usado sózinho como fóssil-índice para a vegetação de savana ou para qualquer outro tipo de vegetação seca. Da mesma forma, não existem tipos polínicos de ervas que possam ser usados como fósseis-índices exclusivamente para savanas. Um candidato a fóssil-índice para a savana é o gênero *Cuphea* (Lythraceae), o qual tem a vantagem de ser um tipo polínico morfologicamente distinto e facilmente identificado. Apesar de ocorrer proeminentemente em savanas e seu pólen corresponder a 40% em alguns espectros polínicos da caatinga, no nordeste do Brasil (De Oliveira *et al.*, 1999, In Press), *Cuphea* compartilha com *Podocarpus*, Melastomataceae e muitos outros táxons polínicos, a propriedade de ter espécies adaptadas a muitos tipos diferentes de hábitats. Algumas espécies de *Cuphea* são arbustos do extrato inferior de florestas tropicais densas. Mais importante que esse fato, é que outras espécies de *Cuphea* são ervas pioneiras de margens de riachos ou de pântanos e turfeiras. Assim, uma turfeira com uma superfície seca, por qualquer motivo, provavelmente produzirá espectros polínicos ricos em Gramineae, apoiados por alguns valores percentuais de pólen de *Cuphea*, muitas vezes associados ao pólen de outras ervas tropicais como *Borreria* (Rubiaceae). Estes táxons, encontrados em grande número ou individualmente, não são fósseis-índices exclusivos da vegetação de savana.[6]

Embora uma alta porcentagem de pólen de Gramineae, provavelmente com ervas associadas, esteja sempre presente em perfis polínicos de regiões de savanas ou cerrados, altas porcentagens dessas ervas podem estar presentes por outros motivos, não caracterizando aridez regional. Quando esses tipos polínicos são produzidos por vegetação flutuante, como os conhecidos "colchões de gramíneas" ou em bancos ou ilhas de areia e argila nos canais fluviais, eles não podem, em nenhuma hipótese, implicar em aumento da aridez ambiental.

Sinais de Savana por Plantas Lenhosas

O pólen de três gêneros arbóreos são úteis como indicadores de cerrado ou savana, quando considerados no contexto do restante do espectro polínico no qual eles ocorrem. São *Curatella* (Dilleniaceae), *Stryphnodendron* (Mimosaceae) e *Caryocar* (Caryocaraceae). O gênero *Curatella* tem somente duas espécies na região neotropical, ambas de hábitats secos, sendo *C. americana* característica de cerrados. O pólen de *Stryphnodendron* ocorre em políadas, característica da grande família das mimosáceas, mas são morfológicamente distintos e identificados ao nível de gênero. Ambos os gêneros são bons indicadores de cerrado, se o resto do espectro polínico for também consistente com este tipo de vegetação. Mas *Caryocar* tem que ser usado com mais cuidado. Árvores de *Caryocar* são importantes elementos do cerrado, de tal forma que seu pólen é esperado nos sedimentos do cerrado. Contudo, esse gênero também contém espécies que são grandes árvores na Floresta Amazônica, de forma que só sua presença, não pode ser usada como indicativa de cerrado. É necessário interpretar com cautela a presença de todos os táxons no contexto geral do espectro polínico.

ESTATÍSTICA MULTIVARIADA OU AGRUPAMENTOS ECOLÓGICOS?

É óbvio que uma dificuldade fundamental da análise palinológica na Amazônia é a ubiqüidade de certos táxons abundantes nos mais variados tipos de vegetação. Gramineae, Melastomataceae, Moraceae, Urticaceae, *Cecropia*, Compositae e outros, contribuem com as chuvas polínicas desde as terras baixas até grandes elevações, e desde as regiões secas às superúmidas. Esses táxons são grandes produtores polínicos, muitos dos quais têm polinização mediada pelo vento (anemofilia). Alguns desses tipos polínicos (e.g., Urticaceae, Ulmaceae) são dificilmente diferenciados e, muitas vezes, aparecem não distinguidos de Moraceae, a nível de família permitindo que muitas espécies sejam agrupadas em cada uma dessas categorias, por uma taxonomia não refinada. Felizmente, essas famílias incluem uma irrisória minoria de todo o "pool" de espécies existentes na Amazônia.

Como já demonstramos, é possível distinguir os mais importantes tipos de vegetação, com base nos espectros polínicos dominados por estes táxons, através de critérios puramente subjetivos: a floresta tropical da Figura 1.1 não pode ser confundida com os cerrados e florestas de galeria da Figura 1.3, embora ocorra uma superposição na distribuição de muitos táxons. As dificuldades começam quando precisamos fazer distinções mais sutis, para separar a floresta estacional da floresta tropical úmida, várzea de igapó, florestas inundadas, em geral, das florestas de terra firme, savana de campos inundados de gramíneas, etc. A experiência adquirida por palinólogos do hemisfério norte, onde a biodiversidade é altamente reduzida e onde o pólen arbóreo é anemófilo, sugere que este problema pode ser resolvido pela reunião dos táxons polínicos em grupos ecológicos, e pela representação gráfica da soma das árvores, arbustos, elementos de hábitats úmidos e secos etc. Em alguns estudos palinológicos neotropicais publicados, essa técnica tem sido usada para identificar tais agrupamentos ecológicos como "grupo de savana," "grupo de floresta," "aquáticos," etc. Ela, quando aplicada à vegetação amazônica, é capaz de conduzir o palinólogo a sérios erros, pois requer que este subjetivamente atribua táxons de ampla ocorrência como Gramineae, Melastomataceae ou Moraceae a um grupo ecológico determinado. É, portanto, melhor requerer da análise estatística, o agrupamento ecológico.

Mark B. Bush, foi o primeiro a investigar, seriamente, a possibilidade de discriminar ecossistemas em espectros palinológicos neotropicais pela análise multivariada (Bush, 1991). Bush considerou somente tipos polínicos que aparecem com valores acima de 2% do pólen total da amostra, removendo da análise o diversificado grupo de plantas com pólen zoófilo tal como foi usado no Lago da Pata para confirmar a história da floresta tropical de terras baixas. Bush trabalhou somente com os bons produtores de pólen e sua taxonomia, ainda um pouco rudimentar, com o objetivo de discriminar as amostras de superfície baseadas nos mais diferentes hábitats neotropicais (Figura 1.5).

A análise de Bush começou com a exclusão de *Cecropia* das somas totais de pólen, apoiando-se no fato de que sua super-abundância deformaria as análises (em muitos amostras *Cecropia* compunha 40% da soma total polínica). Bush, então, aplicou a análise TWINSPAN ao restante dos tipos polínicos. TWINSPAN é a sigla para "*two way indicator species analysis*," um programa computacional que classifica espécies (neste caso espectros polínicos) através da comparação de seus caracteres (porcentagens polínicas). A Figura 1.6 mostra a classificação hierárquica resultante. Alguns espectros polínicos de superfície estão inequivoca e satisfatóriamente agrupados: floresta de terra firme, floresta semi-decídua, igapó e várzea. Quatro amostras superficiais (cerrado e Lago Geral, Lagoa Comprida e La Yeguada) não estão separados. Essas quatro amostras, contudo, têm em comum a evidência de uma estação seca. Cerrado é inerentemente seco. Comprida e Geral, são dois lagos entre Monte Alegre e Santarém, na Amazônia central brasileira, uma região com uma forte e prolongada estação seca, onde a agricultura atual foi estabelecida há dez anos. La Yeguada é um lago na região de floresta da costa do Oceano Pacífico do Panamá, cuja paisagem tem sido influenciada por atividades humanas, durante a maior parte do Holoceno e onde a arqueóloga D.

Piperno descreve a vegetação resultante como "cerrado antropogênico." Assim, TWINSPAN identifica denominadores comuns nessas quatro amostras superficiais, de um modo fácil de ser interpretado.

Distinções mais sutis entre agrupamentos de pólen podem ser feitos por "Detrended Correspondence Analysis" (DCA). DCA ordena grupos de dados ao longo de gradientes ambientais hipotéticos (eixos), que explicam grande parte da variância no conjunto de dados. Os eixos são regidos por táxons relativamente abundantes em uma amostra ou num grupo de amostras, mas ausentes ou raros em outras. Variações pequenas nos principais táxons não são importantes para o programa DCA. A Figura 1.7 mostra os resultados de uma DCA de amostras dos espectros polínicos da Figura 1.5. Assim como a classificação pelo TWINSPAN, os sinais polínicos dos principais tipos de vegetação são separados claramente, particularmente no primeiro eixo onde florestas de várzea e terra firme estão em lados opostos da ordenação. É interessante observar que nesse eixo, várzea está mais perto de locais de estudo onde ocorre uma estação seca definida ou cerrados, do que está de floresta úmida de terra firme, um resultado possivelmente devido aos sinais conflitantes dados por Gramineae e outros tipos polínicos de ervas.

A separação bem sucedida desses espectros polínicos de superfície serve para refutar o antigo receio de que o pólen anemófilo da Amazônia, sendo originário de espécies cosmopolitas e pioneiras (invasoras), forneceria pouca informação sobre a composição da floresta. As análises de Bush estão restritas a táxons polínicos presentes com porcentagens maiores de 2%, representando predominantemente espécies anemófilas e invasoras, mas os principais tipos de comunidade vegetal são claramente diferenciados. Assim, pode-se esperar que sedimentos amazônicos forneçam sinais regionais de seu pólen anemófilo e também de histórias de comunidades dentro da bacia de drenagem de um lago. Esses sinais são fornecidos pelo rico e diverso espectro polínico produzido por plantas zoófilas. Chuvas polínicas, na Amazônia, potencialmente produzem dois tipos de sinais climáticos: o sinal regional transportado pelo vento e o sinal local trazido pelas águas superficiais da área de captação polínica.

A descoberta de sinal regional proporciona grande estímulo à idéia de que o cálculo de funções de transferências entre as porcentagens de pólen neotropical e as variáveis climáticas poderão ser usadas na região neotropical, como o foram nas latitudes temperadas do hemisfério norte. Para isso, será necessário um grande grupo de amostras de superfície, calibradas com as variáveis climáticas e de vegetação. Quando isso ocorrer, será possível usar o pólen de uma rede de testemunhos sedimentares de lagos amazônicos para o mapeamento do clima da Bacia Amazônica, em intervalos de tempo do Pleistoceno Tardio.

NOTAS

1. O Lago da Pata tem uma pequena bacia de captação, resultando num *"run-off"* limitado (Colinvaux *et al.*, 1996). No Lago do Dragão, um lago similar ao Lago da Pata, encontramos a concentração média de 76.800 grãos/cm^3. O valor de 100.000 grãos/cm^3 no Lago de Kumpaka, provavelmente, registra a concentração de grãos em uma grande e bem drenada bacia de captação (Liu & Colinvaux, 1988).
2. A coleta de chuva polínica em Manaus foi uma cortesia do INPA em associação com o Projeto de Dinâmica Biológica de Fragmentos Florestais. A coleta de chuva polínica no Equador foi realizada na Província do Oriente, na Estação de Pesquisa de Cuyabeno sob a direção de E. Asanza.
3. O Lago da Pata está a uma altitude de 300 m, acima do nível do mar. Com o abaixamento eustático do nível do mar, em tempos glaciais (isto é, antes do aumento de *Mauritia*), o lago estaria pelo menos 400 m acima do nível do mar.

Isso efetivamente elevou a bacia em relação ao lençol freático.

4. As evidências desses espectros polínicos para depressão de temperatura em tempos glaciais é discutida no capítulo 4. As principais referências para a descida de populações de árvores em tempos glaciais do Equador são: Liu & Colinvaux (1985); Bush *et al.* (1990); Colinvaux *et al.* (1996b, 1997).
5. Veja, por exemplo: Huntley & Webb (1988); Berglund (1986).
6. Acreditamos que algumas interpretações do diagrama polínico do pântano do platô da Serra de Carajás, no centro-leste da Amazônia brasileira estavam equivocadas pelas seguintes razões: a superfície do pântano secou e isso foi mostrado, elegantemente, pelas análises mineralógicas dos testemunhos; a interpretação que sugere que os dados polínicos mostram substituições simultâneas de floresta tropical por savana não pode ser comprovada (Absy *et al.*, 1991; Van der Hammen & Absy, 1994).

CAPÍTULO

2

REVISÃO DA PALEOECOLOGIA DA AMAZÔNIA: A AULA DE NWO/HUYGENS

No dia 12 de novembro de 1997, Colinvaux proferiu em Nieuwe Kerk, em Haia, Holanda a aula anual de Huygens, da Academia Nacional de Ciência da Holanda (NWO), sobre o tema "A Idade do Gelo da Amazônia e o problema da diversidade." A aula foi seguida de um comentário do Prof. Henry Hooghiemstra com o título *"Floresta Tropical verso savana: dois lados de uma moeda preciosa?"* Este capítulo é uma versão resumida dessa aula que foi impressa como uma brochura pela NWO, da qual foi removido, na sua maior parte, o uso da primeira pessoa.

Algumas partes da aula, particularmente a conclusão que a floresta tropical das terras baixas amazônicas nunca se fragmentou pela aridez, durante qualquer tempo do último ciclo glacial, são controversas. Em seu comentário, o Prof. Hooghiemstra discute, particularmente, o trabalho de Van der Hammen & Absy (1994) sobre o registro paleoecológico de Carajás, e sugere que partes do centro-leste da Bacia Amazônica foram cobertas por savanas em tempos glaciais, ao contrário das conclusões de Colinvaux. Mas a controvérsia é o sangue vital da ciência e a controvérsia de hoje é a hipótese a ser testada amanhã. É fácil entender porque os pesquisadores mantêm diferentes opiniões sobre algo tão importante como a estabilidade e antigüidade das florestas da Amazônia: é porque temos muito poucos dados a seu respeito. Uma série adequada de histórias ambientais derivadas da análise palinológica de todas as partes da Amazônia deve resolver essa controvérsia. O nosso objetivo ao escrever este manual é de servir como estímulo à aquisição desses perfis polínicos.

INTRODUÇÃO: O PROBLEMA DA DIVERSIDADE

A nossa estatística favorita para a Amazônia é originária do trabalho de Al Gentry, sobre a floresta amazônica do Peru. Gentry contou e identificou todas as árvores presentes em um único hectare de floresta tropical úmida nas terras baixas. Foram encontradas 600 árvores nesse hectare, que pertenciam a 300 espécies. Isso equivale ao número de espécies de floras inteiras da Europa e América do Norte juntas (Gentry, 1988).

Ainda não temos muitas dessas contagens na Amazônia, pois o trabalho de censo florístico requer, não somente um notável conhecimento taxonômico, como também uma extraordinária perseverança. Gentry foi obrigado a subir em cerca de 600 árvores para identificá-las. Mais tarde, ele faleceria quando o seu avião, que estava sendo usado pelo RAP (Programa de Levantamento Rápido) para avaliar o valor da conservação da floresta amazônica do Peru, caiu sobre as árvores que ele estava inspecionando.

Contagens de espécies animais e vegetais na Amazônia, embora não tão esplêndidas como as de Gentry, são suficientes para mostrar que a diversidade total é muito grande: 80.000 espécies de plantas, talvez milhões de espécies de insetos, e muitas espécies de outros animais. A Floresta Amazônica é o mais rico ecossistema da Terra.

Por que existem tantas espécies diferentes de plantas e animais na Amazônia? Por que muito mais do que nas férteis e altas latitudes da Europa e da América do Norte? Estas são as mais intrigantes questões da teoria biológica.

Existem dois modelos hipotéticos para a alta diversidade biológica amazônica: "A Amazônia como um museu" e "A Amazônia como um motor."

No modelo teórico de "museu," a Amazônia é considerada um local seguro, onde a extinção natural é tão rara, que as espécies tendem a se acumular durante o tempo geológico: ausência de invernos rigorosos, nenhuma incursão do Mar do Norte, e acima de tudo, a ausência de geleiras que destroem as florestas com a freqüência das Idades do Gelo. A "Amazônia como um museu," mantém espécies de tempos passados, assim como as produzidas recentemente. A adição de ambas leva aos números colossais atuais. "A Amazônia como um museu" é um local agradável, quente, um paraíso tropical, onde prevalece a estabilidade ambiental.

O modelo da "Amazônia como um motor," quase o oposto, sugere que a Amazônia é, na verdade, um local cheio de acontecimentos. Árvores caem durante violentas tempestades tropicais, chuvas falham em chegar na época certa, rios transbordam. Acima de tudo, o comportamento de predadores e animais que se alimentam de sementes, não é controlado por invernos rigorosos. As populações são continuamente selecionadas pelo estresse da vida tropical, mas raramente pela extinção. O efeito prático é que populações raramente se tornam grandes o suficiente para dominar suas comunidades em meios que excluem outras espécies. A coexistência de muitas espécies é possível. A vida conta com recursos sempre disponíveis, contribuindo para que uma nova espécie encontre um nicho ecológico. Estilos de vida inovadores são recompensados, pois uma nova espécie surge, como idéias inovadoras em um local de mercado competitivo. Onde os modelos de "museu" vêem a Amazônia como segura e estável, os modelos de "motor" prevêem violência que mantém em cheque os valentes moradores e os invasores da terra, deixando espaço para os recém-chegados.

Ambos os modelos de "museu" e de "motor" dependem de propriedades especiais das baixas latitudes: há pouca variação da temperatura mesmo em grandes áreas. Um grau ou dois de latitude perto do equador, na Floresta Amazônica, causam pouca diferença em temperaturas ambientes, enquanto que a mesma mudança de latitude, no norte da Europa, pode levar você a uma outra zona climática. Existe constância climática na Amazônia, cujo enorme território é maior que o da União Européia (Figura 2.1). Portanto, todo habitat amazônico é copiado, como se fosse uma abundante coleção de habitats similares para os quais as espécies estressadas podem escapar.

"Museu" ou "motor," a diversidade da Amazônia depende da história. Vivemos em uma época de Idades do Gelo, quando todos os ambientes da terra sofrem periódicos choques climáticos. Em cada Idade do Gelo, geleiras avançam para ocupar as terras do norte por cinqüenta ou mais milhares de anos. Depois, as geleiras derretem e um intervalo de temperaturas altas domina a Terra, igual ao que estamos vivendo agora, seguido de outro tempo de glaciação. O ciclo tem se repetido, várias vezes, nesses últimos dois milhões de anos.

A existência de ciclos glaciais é do conhecimento geral; menos familiar é a idéia de que a condição glacial é a normalidade. Os tempos glaciais duram mais que os tempos quentes interglaciais, como o que estamos vivendo no presente. Assim, a Idade do Gelo tem sido a condição normal nestes últimos dois milhões de anos. O aquecimento de um período interglacial, como o nosso, é excepcional (Imbrie *et al.*, 1984).

As espécies vivem mais tempo que as glaciações, geralmente vários milhões de anos, dependendo do tipo de organismo. O tempo de vida de uma espécie é muito mais longo que um ciclo glacial. Portanto, praticamente todas as espécies contemporâneas passaram a maior parte de sua existência em uma Idade do Gelo. A diversidade atual é a diversidade da Idade do Gelo, atualmente acomodada ao clima interglacial incomum de nossos dias, comparativamente breve em relação ao glacial.

Portanto, para entender a diversidade da Amazônia, devemos primeiramente entender a Amazônia da Idade do Gelo. A última Idade do Gelo terminou por volta de, apenas, 10.000 anos atrás, um tempo tão recente, que somente quarenta gerações de grandes árvores passaram desde o último recuo do gelo.[1] Por comparação, apenas quarenta gerações humanas de 50 anos cada nos separam do tempo em que Júlio César cruzou o Rio Rubicon com seu exército, assinalando a destruição final da República Romana. A última Idade do Gelo foi extremamente recente.

A RAREFAÇÃO DO DIÓXIDO DE CARBONO

O clima da Idade do Gelo pode ter se diferenciado do atual em temperatura e precipitação. As latitudes nortes se tornaram, certamente mais frias do que atualmente, e algumas partes dos trópicos, certamente mais secas. Mas a mudança universal, da qual temos absoluta certeza, refere-se à redução da concentração do dióxido de carbono (CO_2) na atmosfera.

A concentração global do CO_2 na Idade do Gelo foi medida em bolhas de ar daquela época, aprisionadas no gelo da Antártida e da Groenlândia e era de dois terços da concentração desse gás no nosso período interglacial, no início da revolução industrial: 0.02% por volume em oposição ao valor atual de 0.03%. Esta é uma diferença profunda. O CO_2 é um dos materiais indispensáveis para o crescimento de todas as plantas verdes. A baixa concentração desse gás sozinho sugere que, a relativa abundância de muitas espécies vegetais, foi diferente na Amazônia durante a Idade do Gelo.

As plantas conhecidas como "C4" (plantas com metabolismo C4) e outras que chamamos de "CAM" (plantas com metabolismo do ácido crassuláceo) são equipadas com enzimas especiais que coletam CO_2, removendo-o da atmosfera, mesmo quando esse gás está em baixas concentrações. Os mecanismos metabólicos fotossintéticos C4 e CAM são adaptações para a vida em ambientes estressados, particularmente pela aridez ou altas temperaturas (Colinvaux, 1993a). Tipicamente, esse sistema é usado principalmente por cactos e bromélias epifíticas, e representa uma adaptação a uma aridez relativa: para o cacto a adaptação é para a vida em ambiente desértico; para a planta epífita, a adaptação é para a ausência de solo. O mecanismo fotossintético C4 é, particularmente, usado pelas gramíneas de savanas, permitindo sua adaptação às condições quentes e secas, onde essas plantas são geralmente encontradas. Outras gramíneas têm o mecanismo C3, não necessitando dessa adaptação para condições ambientais secas. Mas a química de plantas C4 e CAM possui uma peculiaridade que permite a absorção do CO_2 do ar deficiente desse gás, mais rapidamente do que podem as plantas C3, que são mais comuns. Assim, cactos, bromélias e gramíneas de savanas deveriam ter uma vantagem ecológica na época glacial, quando o ar possuía concentrações deficientes desse gás.

Vemos a Floresta Amazônica da Idade do Gelo mais carregada com bromélias epífitas do que a floresta contemporânea, e talvez com muito mais bromélias terrestres cobrindo o solo dessas matas. Áreas ricas em gramíneas de vales fluviais, sazonalmente inundados, teriam tido mais gramíneas de savanas, com metabolismo C4. Talvez, tenha ocorrido uma substituição de espécies de gramíneas C3 por C4 nas partes altas das montanhas, onde os efeitos da altitude teriam intensificado a deficiência do CO_2.

Estas mudanças decorrentes do empobrecimento do suprimento de CO_2, devem ter sido suficientes para garantir que as associações vegetais da Amazônia da Idade do Gelo, fossem diferentes das que conhecemos atualmente, ainda que o CO_2 alterado, seguramente, tivesse menos efeito que teriam outras mudanças, quer na temperatura, quer na precipitação.

O ESFRIAMENTO DA REGIÃO EQUATORIAL

O bom senso pode sugerir que houve um esfriamento na Idade do Gelo, mas, surpreendentemente, até recentemente, a opinião predominante era que as temperaturas equatoriais se alteraram muito pouco até a época atual. Essa hipótese de "esfriamento mínimo" foi uma interpretação da paleooceanografia, particularmente de um "grupo de pesquisa" conhecido pela sigla "CLIMAP."

O objetivo do CLIMAP era produzir mapas das temperaturas da superfície dos oceanos, durante o Último Máximo Glacial, definido por eles como há 18.000 anos atrás, usando dados de testemunhos sedimentares marinhos. Os "termômetros" do CLIMAP para a temperatura da superfície dos oceanos foram os fósseis foraminíferos encontrados nos sedimentos, calibrados pela informação de distribuição atual das espécies relacionadas à temperatura das águas nas quais elas são atualmente encontradas. A conclusão do CLIMAP foi que, embora as temperaturas dos oceanos das latitudes médias e altas do hemisfério norte estivessem seis ou mais graus mais baixos na Idade do Gelo, os oceanos tropicais praticamente não esfriaram mais do que 2°C, em relação a temperaturas atuais (CLIMAP, 1976).

Se os mares tropicais não esfriaram na Idade do Gelo, devemos concluir que as terras adjacentes, também não poderiam ter esfriado. Tal era o consenso, mesmo um paradigma, da ciência, por duas décadas que seguiram a primeira publicação dos resultados do CLIMAP, em 1976.

Todavia, os geólogos, há tempos, vêm mostrando evidências de que, geleiras nas altas montanhas tropicais, avançaram, simultaneamente, com as glaciações continentais. Isto foi demonstrado convincentemente para a África, quando Daniel Livingstone datou a última Idade do Gelo, pelo método do radiocarbono, em sedimentos de um lago glacial nos altos flancos das montanhas do Ruwenzori Range. Idades de radiocarbono para as morenas terminais, abaixo de geleiras montanas dos Andes, também mostraram geleiras expandidas em montanhas tropicais durante a última Idade do Gelo. Tudo isso indicava esfriamento nas montanhas tropicais, quando o CLIMAP sustentava que não teria havido esfriamento nas baixas latitudes (Livingstone, 1962).

Logo ficou claro que as geleiras tinham se expandido nas montanhas tropicais, em todos os continentes. A explicação mais simples para essas descidas das geleiras montanas nos trópicos, era de um esfriamento da ordem de 6 a 9°C. Mas o paradigma do CLIMAP, que indicava que os trópicos tiveram um esfriamento mínimo, permaneceu, e nós nos defrontamos com o paradoxo aparente de regiões altas mais frias sobre as terras baixas, que teriam permanecido quentes.

O controle que a "hipótese do esfriamento mínimo" teve em nossa ciência pode ser julgado pelas tentativas de se explicar como as geleiras avançaram sem esfriamento. Modelos matemáticos apareceram e mostraram que geleiras tropicais poderiam crescer sem esfriamento adicional, se a precipitação fosse aumentada. Em outras palavras: "forneça bastante neve nos topos e você poderá vencer o derretimento embaixo."

Nesse meio tempo, apareceram na literatura as notáveis descobertas na Colômbia, por T. Van der Hammen, H. Hooghiemstra e seus colaboradores. A cidade de Bogotá, situada próxima à linha de árvores nos Andes, foi construída sobre os sedimentos de um lago de uma caldeira extinta, que secou há cerca de 20.000 anos atrás. Um total de 300 m de sedimentos lacustres contém os últimos 3.000.000 de anos de história ambiental. Análises palinológicas mostraram que a porcentagem total de árvores flutuou, em sincronização com as glaciações do hemisfério norte e com geleiras montanas dos picos mais altos dos próprios Andes, sendo de 1.500 metros a faixa de amplitude vertical da linha de árvores.

Os dados palinológicos forneceram evidência de momentos frios, não somente nos picos gelados das montanhas mais altas, como também na vegetação localizada de um a dois mil metros abaixo. Os argumentos usados na defesa da idéia do avanço das geleiras, sem esfriamento adicional, são pouco convincentes, especialmente quando a vegetação mais abaixo responde ao esfriamento.[2]

Apesar disso, as conclusões do CLIMAP permaneceram aparentemente intactas. Biogeógrafos desenharam mapas fitogeográficos, mostrando os diferentes e atuais cinturões de vegetação, comprimidos em faixas estreitas nos flancos inferiores das montanhas, com surpreendentes e íngremes gradientes de temperatura, desde o nível do mar até o nível rebaixado de árvores. A hipótese da ocorrência de ar seco nas baixas altitudes e ar úmido na altas, insuficientemente apoiava a complicada física desses gradientes de temperatura.

Esse paradoxo de terras altas mais frias, estabilizadas sobre as terras baixas, ainda quentes, era profundamente insatisfatório. Ao contrário do modismo da época, começamos a suspeitar que as regiões tropicais baixas esfriaram simultaneamente com as altas. O teste, para resolver esse paradoxo, seriam registros de temperatura das regiões baixas. Começamos a procurar registros climáticos nas regiões equatoriais baixas, primeiro nas Ilhas Galápagos e depois na própria Bacia Amazônica. Afortunadamente, conseguimos um registro palinológico no sopé dos Andes, em um lugar chamado Mera, no Equador.

Kam-biu Liu encontrou a seção sedimentar de Mera. Estávamos usando uma pista de pouso de onde voávamos a locais remotos da Amazônia, mas devido às condições de tempo, tivemos que ficar em terra. Andamos um pouco para passar o tempo e o Dr. Liu obteve amostras de madeiras e outros pequenos fragmentos vegetais e antigos sedimentos lacustres, expostos por um corte de estrada, a uma elevação de aproximadamente 1100 m, ainda dentro do limite altitudinal da Floresta Amazônica. A idade medida por radiocarbono, das amostras de madeira, chegaram perto de 30.000 anos. Isso era idade glacial! A análise palinológica realizada por Liu mostrou não só a presença de pólen das florestas das regiões baixas, como também pólen de árvores como *Podocarpus* e *Alnus*, atualmente confinadas em altas elevações dos Andes, 1.000 metros acima de Mera. Os resultados foram publicados como uma nota na revista *Nature*, em 1985, mostrando que um esfriamento de mais de 4°C prevaleceu em todas as elevações, desde o topo das montanhas até as regiões baixas amazônicas (Liu & Colinvaux, 1985).

Mera foi uma abertura, mas não definitiva. Se paleoecólogos não gostaram dos resultados, razões para descrédito podiam ser encontradas facilmente: idades de radiocarbono dentro da faixa de 30.000 anos podem ser sempre questionáveis, "*obviamente contaminação, meu caro colega, as amostras devem ser muito mais antigas,*" etc. A melhor resposta aos descrentes seria mais seções palinológicas. Escrevemos uma proposta de pesquisa à Fundação de Ciências (NSF-EUA), requerendo fundos para buscar nos sopés dos

Andes mais uma seção. Incrivelmente conseguimos a verba e encontramos a seção. .

A segunda seção foi encontrada na floresta em San Juan Bosco, a 160 km de Mera. Mark B. Bush e Paulo E. De Oliveira encontraram-na, depois de Colinvaux ter sido deixado em Quito, letárgico por uma enfermidade contraída na floresta. As datações da seção de San Juan Bosco sobrepõem-se às de Mera, e a análise palinológica realizada por Bush forneceu os mesmos resultados, isto é, florestas das terras baixas invadidas por *Podocarpus*, *Alnus* e outros elementos montanos. Juntos, os dados do Equador e da Colômbia, indicaram que as geleiras desceram cerca de 1.500 m e, abaixo delas, a linha de árvores também desceu 1.500 m e, ainda, abaixo dela, algumas árvores da floresta andina alto-montana desceram também 1.500 metros e entraram na floresta das regiões baixas. Calculamos que o máximo desse episódio de esfriamento exigiu uma redução de 7.5°C nos valores atuais de temperatura (Bush *et al.*, 1990).[3]

Estes dois locais no Equador foram os primeiros estudados dentro das florestas tropicais úmidas da Amazônia. A descida de árvores sensíveis à temperatura, demonstrada nesses estudos, revelou que os movimentos simultâneos de linhas de árvores e as geleiras acima delas estavam associados aos declínios de temperatura e que, não deveriam ser atribuídos à redução do CO_2.[4]

Encontramos uma história similar em sedimentos lacustres da região da costa pacífica do Panamá. A pequena cidade de El Valle, no Panamá, está construída sobre sedimentos, naturalmente drenados, de uma antiga caldeira vulcânica, uma versão de Bogotá, em menor escala, mas dentro da floresta costeira, a apenas 500 m de altitude. A história da Idade do Gelo de El Valle, revelada pela análise palinológica, é paralela àquela de San Juan Bosco e Mera, com florestas de tempos glaciais sendo invadidas por árvores que desceram cerca de 1000 metros. Assim, como nas duas localidades no Equador, esta também é uma história de plantas adaptadas a climas mais frios invadindo as florestas das regiões baixas. Não se tratando de uma história de vegetação movendo-se em cinturões, mas o contrário, o pólen mostrou que a composição da floresta tropical continuou sendo alterada, embora ainda reconhecida como floresta tropical (Bush & Colinvaux, 1990).[5] O registro de El Valle é contínuo, com bom controle de radiocarbono em toda a seqüência. Portanto, tínhamos demonstrado que as florestas das regiões baixas, tanto do Panamá como do Equador, esfriaram na época glacial, ao contrário das conclusões do CLIMAP.

O registro de esfriamento foi também encontrado no sudeste do Brasil, através da evidência palinológica que demonstrou que a floresta de Araucária, hoje concentrada na região sul do Brasil, se expandiu cerca de 700 km ao norte, cruzando o Trópico de Capricórnio. O sinal para o esfriamento de, no mínimo, 6°C foi encontrado no registro polínico de *Araucaria* de Serra Negra (De Oliveira, 1992) e Salitre (Ledru, 1993), em Minas Gerais.

Até então, nós, os palinológos, ainda estávamos sozinhos com nossos registros de esfriamento glacial; apenas com alguns testemunhos e a inferência biológica, contra os dados compilados da temperatura das águas superficiais oceânicas pelo CLIMAP. Mas a ajuda estava chegando. Martin Stute encontrou um registro de temperaturas em águas subterrâneas da época da Idade do Gelo, no nordeste do Brasil. O "termômetro" usado por Stute foi a concentração de gases nobres em solução. Nas águas subterrâneas do nordeste do Brasil, datadas pelo radiocarbono como pertencentes ao último máximo glacial, a temperatura estimada foi de 5°C a menos que a atual (Stute *et al.*, 1995). Enquanto isso, pesquisadores procurando registros de paleotemperaturas em razões químicas e isotópicas, na matriz de recifes de corais, encontraram evidências que as temperaturas das águas superfíciais oceânicas tinham sofrido um decréscimo, apesar dos cálculos de CLIMAP (Guilderson *et al.*, 1994). Finalmente, em fevereiro de 1997, uma equipe trabalhando com modelos climáticos publicou na revista *Nature*, um modelo climático global que permitia o esfriamento dos oceanos tropicais e continentes, de 6°C sugeridos por nós (Webb *et al.*, 1997). Um correio eletrônico, enviado a Colinvaux por um dos autores, dizia: "*Na minha opinião, sua vindicação veio dos trabalhos sobre corais e águas subterrâneas, enquanto que nosso artigo fornecia um mecanismo ... muitas pessoas agora estão redirecionando a mim, o desgosto que você recebeu durante anos.*"[6]

Agora o paradigma está sendo transferido em direção à aceitação de que os trópicos são mais frios nos tempos glaciais. Não negamos que isto é uma sensação doce, depois de tantos anos de revisores (referees) anônimos desconsiderarem nossos manuscritos e propostas de pesquisas, referentes ao esfriamento das regiões tropicais baixas. Mesmo os cientistas têm prazeres humanos!

A AMAZÔNIA FOI ALGUMA VEZ ÁRIDA?

E o terceiro parâmetro de mudança climática, a precipitação? O paradigma predominante é que os trópicos foram áridos durante a Idade do Gelo. Na África, por exemplo, os níveis dos lagos foram, sem dúvida, mais baixos e o grande Lago Victória possivelmente secou por completo. Perto da Amazônia, em partes da América Central e no Caribe, evidências incontestáveis mostram secas violentas durante a última glaciação. Os melhores dados vêm da Guatemala, onde o pólen encontrado em sedimentos lacustres de idade glacial indica uma paisagem árida de savana, onde hoje predominam florestas tropicais úmidas (Deevey *et al.*, 1983).

Nossa demonstração de esfriamento glacial, em parte, fortalece a hipótese da aridez, pois o esfriamento deveria reduzir a evapotranspiração. Uma Amazônia mais fria deveria implicar em uma Amazônia, inevitavelmente, mais seca.

Mas, quanto, mais seca? Uma recente revisão geológica mostra um mapa de todas as regiões baixas da Amazônia como uma savana (Clapperton, 1993). Biólogos conscientes da necessidade de manter a diversidade através da Idade do Gelo são mais modestos, permitindo ilhas de florestas, chamadas por eles de "refúgios," persistirem nesse grande mar de savana. Esses mapas devem mais ao raciocínio indutivo do que a dados concretos (Figura 2.2).

Os mapas, mostrando uma Amazônia árida, servem como afirmação de uma hipótese que está aberta a testes pelo método científico. Esses mapas sugerem que as regiões baixas da Amazônia tiveram savanas na época glacial, uma hipótese facilmente testada pela análise palinológica de testemunhos de sedimentos lacustres da Amazônia. O Lago da Pata é adequado a esse propósito, pois está localizado dentro da floresta tropical úmida e densa das regiões baixas do centro-oeste da Amazônia, onde ambos os mapas: geológico e biogeográfico baseiam-se na hipótese de uma Amazônia mais árida e mostram a vegetação local, durante a Idade do Gelo, como savana.

O Lago da Pata ocupa uma bacia fechada, sem córregos de entrada ou saída, embora a água seja, aparentemente, perdida através de um sumidouro, bem como pela evapotranspiração. Acumulação sedimentar é extremamente baixa, e a idade de 18.000 anos do CLIMAP foi demonstrada, pela datação do radiocarbono AMS (espectrometria de acelador de massa), estar na profundidade de 0,70 m. Sedimentos mais profundos que um metro têm mais de 30.000 anos de idade.

Análises palinológicas mostram, incontestavelmente, que a região estava coberta por florestas durante todo o tempo de acúmulo de sedimentos (veja Figura 1.1 e a discussão no Capítulo 1). Cerca de 80% de todos os grãos de pólen encontrados podem ser atribuídos a elementos arbóreos da floresta tropical. Ainda mais decisivo é o fato de que a porcentagem de pólen de gramíneas está com valores mínimos em todos os níveis, ambos em tempos glaciais e pós-glaciais. Sedimentos coletados em regiões de savana, no planalto central do Brasil, têm porcentagem de pólen de gramíneas com valores de 40 a 80% da soma polínica total (veja Figura 1.3, do Capítulo 1). O fato de que o pólen de gramíneas é insignificante nos sedimentos glaciais do Lago da Pata, é uma demonstração incontestável de que vegetação de savana não estava presente nessa região na época glacial. Essa parte das terras baixas amazônicas foram sempre cobertas por florestas (Colinvaux *et al.*, 1996a).[7]

No entanto, o testemunho do Lago da Pata sugere uma precipitação reduzida, consistente com o esfriamento. De 30.000 a 14.000 anos atrás, a cor e a estrutura dos sedimentos depositados sugerem exposição intermitente ao ar, portanto, com níveis de água mais baixos. As associações polínicas encontradas mostram, contudo, que uma precipitação suficiente permaneceu na região, permitindo a permanência de vegetação de floresta. A precipitação local anual agora está dentro da faixa de 3.000 a 4.000 mm. A interpretação mais prudente dos resultados do Lago da Pata é que o esfriamento realmente reduziu essa abundante precipitação, mas nunca o suficiente para interromper a existência de florestas.

EVIDÊNCIA DO LEQUE ALUVIAL DO RIO AMAZONAS

O Lago da Pata com seus lagos irmãos Verde e Dragão são importantes, pois produziram um registro palinológico contínuo desde os tempos glaciais, dentro da floresta da região baixa da Amazônia. O longo registro palinológico de Carajás também contém idades glaciais, mas é de uma região montana coberta atualmente por vegetação edáfica e mais seca, e não floresta tropical das baixas altitudes (veja nota 10). As seções de Mera e San Juan Bosco, no Equador, contêm somente o intervalo correspondente ao estágio 3 de isótopos de oxigênio, da última glaciação. Desta forma o Lago da Pata e seus lagos irmãos são, ainda, os únicos a conter um registro contínuo desde os tempos glaciais. Assim, temos um longo e completo registro de somente uma localidade, dentro da imensidão de terras baixas amazônicas. O risco de uma extrapolação desta localidade para toda a bacia amazônica é óbvio. Mas um teste geral da hipótese de savana é permitido pela análise palinológica dos sedimentos do leque aluvial do Rio Amazonas.

O leque aluvial do Rio Amazonas é um grande acúmulo de sedimentos fluviais na profunda região marinha na costa do seu delta. Os sedimentos alcançam o leque somente durante glaciações continentais, devido a níveis do mar, rebaixados em cerca de 100 metros em relação ao nível atual, de forma que as águas, contendo sedimentos dessa época, são despejadas através de uma "garganta" cortada na rasa plataforma continental e daí para o mar. Os sedimentos são, finalmente, depositados além da plataforma, em profundidades de cerca de 4.000 metros, formando, assim, o enorme leque amazônico.

Com o alto nível do mar em um período interglacial, o Rio Amazonas descarrega sua carga sedimentar na plataforma continental. Portanto, fora do delta do Rio Amazonas, existem amostras separadas de sedimentos glaciais e interglaciais.

O programa internacional de remoção de sedimentos oceânicos (Ocean Drilling Program-ODP) amostrou sedimentos glaciais desse leque, e depósitos interglaciais da plataforma continental. Nosso colega, Simon Haberle, estava a bordo desse navio e amostrou os testemunhos para formar um registro sedimentar contínuo de todo o último ciclo glacial (Haberle, 1997; Haberle & Maslin, 1998).[8] Haberle, então, fez a análise palinológica desse longo registro para determinar mudanças na vegetação na Bacia Amazônica inteira, durante as mudanças de condições glaciais a interglaciais e vice-versa.

Enquanto esperávamos pelo término e conclusões da análise palinológica de Haberle, estávamos convencidos de que os resultados mostrariam alguma evidência da expansão de savanas, apesar da nossa rejeição da idéia da substituição das florestas tropicais. O sistema amazônico drena a bacia inteira, de forma que um aumento na vegetação rica em gramíneas (savana, cerrado, caatinga), em qualquer parte da bacia, deveria enviar seu sinal polínico de gramíneas até os sedimentos marinhos. Atualmente, partes da bacia, como os limites ao norte e sul e alguns planaltos elevados têm savanas ou outras faces de vegetação seca. O esfriamento e o tipo de precipitação reduzida, sugerida pelo sedimento físico do Lago da Pata, aparentemente deveria aumentar a área de ocorrência de vegetações secas na periferia, mesmo se toda a floresta tivesse se mantido intacta. Portanto, esperávamos que o pólen da Idade do Gelo incluiria uma quantidade consideravelmente maior de gramíneas, ou seja, um sinal incrementado de vegetação de savana. Mas estávamos errados. Não havia excesso de pólen de gramíneas no leque amazônico (Figura 2.3).

O diagrama de Heberle, de pólen depositado pelo sistema inteiro da Amazônia, é igual àquele do Lago da Pata, pois mostra mudanças mínimas entre tempos glaciais e interglaciais. Em particular, pólen de gramíneas, em tempos glaciais, representados nos sedimentos carregados pelos rios, não eram mais abundantes do que o são durante o nosso atual período interglacial.

Ainda é possível que o pólen de áreas mais remotas da bacia estejam sub-representados nos depósitos do leque, ou tenham sido destruídos, filtrados ou se tornado rarefeitos por outros tipos polínicos, durante a sua longa jornada até o mar, de forma que savanas poderiam ter se expandido em partes secas e distantes da Bacia Amazônica, sem ter deixado um sinal nos diagramas polínicos de Haberle. O que é seguro dizer, contudo, é que as savanas não poderiam ter ocupado a ampla faixa de terras baixas através das quais os rios passam, na última parte de sua jornada e que agora está coberta de floresta tropical úmida.[9] Haberle testou a hipótese que sugere que a floresta amazônica das regiões baixas foi substituída por savanas, durante a Idade do Gelo e descobriu que essa hipótese é falsa.

CONTROVÉRSIA SOBRE A ARIDEZ DA AMAZÔNIA

É nossa conclusão que as regiões baixas da Amazônia permaneceram cobertas por florestas através das glaciações. As florestas nunca foram substituídas por savanas, e as regiões baixas da Amazônia nunca foram áridas, como a palavra árida é normalmente entendida. Mas a probabilidade de redução significativa da precipitação em tempos glaciais permanece ainda alta.

A melhor evidência direta para a redução de precipitação está no testemunho de um pântano na Serra de Carajás (Pará), tomado no topo da serra, na parte centro-leste das terras baixas da Amazônia brasileira (Absy *et al.*, 1991). O alto da Serra de Carajás está entre altitudes de 700 e 900 m e, em uma parte relativamente mais seca (mais sazonal) da Amazônia. Essa região possui várias das facies de vegetação mais seca, incluindo savanas. Os excelentes dados sobre a mineralogia dos sedimentos mostram que o nível de água, no sistema do lago pantanoso estudado, foi significativamente mais baixo em tempos glaciais. Enquanto nossas observações no Lago da Pata sugerem níveis mais baixos de água até 18.000 anos atrás, baseando-se na aparência grosseira dos sedimentos, o trabalho de Carajás demonstrou, convincentemente, que aquele planalto tornou-se mais seco. Esses são os únicos dois registros, bem datados, de seca dentro da Bacia Amazônica nesse intervalo de tempo. Embora esses dois registros estejam separados por 2.000 km, é razoável correlacioná-los e mostrar uma redução generalizada das chuvas (Absy *et al.*, 1991).[10]

Observação de solos amazônicos e suas características como as tão mencionadas linhas de pedras (*stone lines*) têm sido, por muito tempo, usadas como argumento para uma Amazônia, não somente árida, como também desprovida de vegetação em alguns lugares. Isso não pode ser verdade. Tais fenômenos, como as linhas de pedras, necessitam ser reavaliados sob a luz da poderosa evidência que a Bacia Amazônica permaneceu coberta de vegetação. Partículas de pó encontradas nos testemunhos de gelo do Andes peruanos, da mesma forma, não podem ser atribuídas às regiões baixas da Amazônia (Thompson *et al.*, 1995).

Dunas fossilizadas no Orinoco, no médio Rio São Francisco, e em várias partes do nordeste brasileiro, estão todas fora da Bacia Amazônica. Provavelmente, a aridez limitada sugerida pelos registros do Lago da Pata e Carajás, estava associada à formação e mobilização de dunas nessas regiões mais secas, fora da bacia. Mas, estudos atuais no sistema de dunas fósseis do médio Rio São Francisco, agora coberto por vegetação de caatinga, mostram que a ativação dessas dunas foi mais alta nos últimos 4.000 anos, que durante a transição Pleistoceno/Holoceno, entre 11.000 e 9.000 anos atrás (De Oliveira *et al.*, 1999, In Press).

Portanto, a preponderância de evidências mostra que a maior parte das regiões baixas da Amazônia permaneceram cobertas por florestas na Idade do Gelo, embora a precipitação tenha sido reduzida. Temos a expectativa de mudanças sutis na composição da vegetação em tempos mais secos.

Este, no momento, é o nosso modelo, proposto para o ambiente amazônico durante a Idade do Gelo. As regiões baixas nunca foram áridas, mas mantiveram-se cobertas de vários tipos de floresta tropical, em um clima mais frio e com seca moderada.

MUSEU OU MOTOR EVOLUCIONÁRIO?

O modelo de "museu" para a diversidade amazônica refere-se à manutenção segura de espécies da floresta tropical. Essa "manutenção segura" é fornecida pela Amazônia da Idade do Gelo, como nós e nossos colegas a reconstruímos. Negamos que a aridez eliminou muito da floresta em tempos glaciais. Ao contrário, a vasta floresta tropical das regiões baixas ofereceu espaço para a vida de muitas espécies da floresta amazônica, em todos os estágios do último ciclo glacial, o que torna razoável a conclusão que isto também foi verdade para os ciclos glaciais precedentes. Um armazém seguro existiu para o museu, desde, pelo menos, o início da época pleistocênica, há dois milhões de anos atrás.

Mas o modelo do "motor" também recebe um grande apoio, devido ao modo pelo qual longos períodos de normalidade glacial são interrompidos por breves interglaciais, caracterizados por temperaturas mais quentes e enriquecimento do CO_2 atmosférico. Deve ser enfatizado que os interglaciais quentes é que são incomuns e não as idades do gelo. Biólogos tropicais deveriam visualizar um interglacial como o atual, como os "maus tempos," ou "época de problemas," quando longos períodos de plácida normalidade da Idade do Gelo são caracterizados por um aflitivo calor.

Os dados palinológicos mostram que espécies vegetais tropicais podem sobreviver através desses ciclos, tanto na Amazônia como em outros locais das regiões tropicais. Mas não pode haver dúvidas que diferentes espécies vegetais toleram diferentemente essas flutuações. Cada vez que a normalidade glacial foi interrompida pelo aquecimento e enriquecimento de CO_2 de uma fase interglacial, deveria ter ocorrido um novo arranjo de espécies nas comunidades vegetais da Amazônia.

Nossos dados palinológicos nos fornecem uma idéia remota desse novo arranjo em progresso da biota. O registro palinológico de florestas da Idade do Gelo, do Lago da Pata, mostra que gêneros arbóreos como, por exemplo, *Podocarpus* e *Hedyosmum* estavam incluídos no que era, então, uma floresta tropical das regiões baixas. Estas são espécies arbóreas, cujas populações modernas são encontradas cerca de 1.000 metros mais altas, sendo o ponto mais perto do Lago da Pata, as montanhas do complexo do

Pico da Neblina, centenas de quilômetros ao norte. Nós interpretamos suas presenças, na floresta das terras baixas, como uma indicação do esfriamento, um dado que traz o nosso registro de esfriamento para dentro das regiões baixas da Amazônia. Mais importante que isso, são os espectros polínicos de idades glaciais, no qual a presença de plantas não tolerantes ao calor está sobreposto ao sinal moderno de floresta tropical. É a evidência direta do novo arranjo das comunidades vegetais.

O maior triunfo da análise palinológica tem sido a demonstração, por Margaret Davis, Tom Webb e outros trabalhando no leste da América do Norte, que comunidades e associações de espécies florestais arbóreas nunca são permanentes. (Davis, 1986; Huntley & Webb, 1988). Costumava-se imaginar que, espécies vegetais viviam em "associações" que persistiam através do tempo, e que na América do Norte e Europa comunidades vegetais inteiras migraram ao sul, com suas composições florísticas intactas, quando as geleiras avançaram, retornando depois ao norte com a retração das geleiras. Essa mesma idéia foi aplicada às montanhas onde se acreditava que a vegetação existia em cinturões que caminhavam para cima e para baixo, como comunidades vegetais intactas. O triunfo da Palinologia foi mostrar que isso não ocorreu. Ao contrário, comunidades florestais dissolveramse antes do avanço do gelo, pois cada espécie procurou sua própria rota para a sobrevivência. Uma malha de cerca de 300 diagramas polínicos mostrou que as florestas, comparativamente ricas do leste da América do Norte, foram formadas lentamente por migrações seqüenciais, praticamente durante os últimos dez mil anos.

Nossos dados palinológicos do Lago da Pata mostram que árvores das comunidades florestais amazônicas obedecem às mesmas regras. Uma floresta tropical úmida pode incluir mais árvores de *Podocarpus* se esfriada de alguns graus centígrados, sem deixar de ser uma floresta tropical. As comunidades florestais alteramse à medida que o ambiente se altera. As espécies mantêm-se em algum lugar da grande área amazônica, mas as comunidades de espécies são efêmeras e estão em constante alteração.

O modelo de "motor" requer perturbação ambiental que não seja catastrófica; perturbação que impeça competidores agressivos ou predadores de se tornarem demasiadamente numerosos, o que potencialmente eliminaria outras espécies. O registro histórico da Amazônia agora mostra que esse nível de perturbação opera em escalas de tempo glacial de milhares de anos, assim como dentro dos ciclos de vida de plantas individuais. Jogos de sobrevivência na Amazônia seguem regras que mudam sutilmente com o ritmo das idades do gelo; isso alimenta o motor. Mas a mesa do jogo é grande e permanente, possibilitando a fuga, que abaixa a penalidade de perder o jogo, mas permite que o museu seja construído.

O PROBLEMA DO ENDEMISMO AMAZÔNICO

A resposta para outro problema da diversidade — o dos endemismos locais — também se encontra na época glacial da Amazônia. Gêneros de aves e de borboletas existem como espécies distintas em diferentes partes da Bacia Amazônica, de tal forma que suas distribuições geográficas mapeadas parecem ilhas pequenas e separadas num mar verde de floresta. Além disso, algumas das comunidades isoladas de borboletas e pássaros têm, aproximadamente, distribuições concêntricas, ocupando os mesmos agrupamentos no mar contínuo de floresta. Mapas desses agrupamentos endêmicos assemelham-se a mapas de um arquipélago no oceano, mas nesse caso não há um oceano (Figura 2.2).

Os pássaros e borboletas desses agrupamentos isolados são parentes próximos e, portanto, espécies que certamente evoluíram de ancestrais comuns dentro da floresta. Mas tal evolução deveria ser possível somente quando populações fossem, efetivamente, impedidas de cruzamentos genéticos por um longo período de tempo. Sugerimos que deve existir "vicariância"; ou seja, aquelas populações separadas por uma barreira física. Na floresta amazônica contínua, fronteiras geográficas que separam populações para a necessária vicariância, não são imediatamente identificadas.

Em um trabalho clássico, publicado em 1969, Jurgen Haffer observou que populações endêmicas isoladas se agrupam na periferia da Bacia Amazônica, em locais mais elevados da floresta, os quais o levaram, pois ele é um geólogo, a sugerir que as florestas dessas áreas elevadas poderiam ter sido fragmentadas entre si no passado geológico recente, embora elas agora façam parte integral da floresta. As idades do gelo foram os únicos acontecimentos geológicos que pareceram drásticos e suficientemente recentes. Logicamente, nenhum dado atual da época glacial da Amazônia era disponível até 1969, mas a hipótese da aridez tropical já estava em moda. Haffer sugeriu que as remanescentes manchas florestais subsistiram em áreas elevadas, e que as terras baixas intermediárias foram ocupadas por savanas; assim sendo as manchas florestais teriam sido efetivamente isoladas. Aqui, então, estava a vicariância: novas espécies evoluíram de manchas florestais separadas, por milhares de anos na época glacial, por savanas áridas. Linda hipótese![11]

Se estivermos corretos em concluir que as regiões baixas da Amazônia foram sempre cobertas de florestas, a hipótese de Haffer, agora testada, é falsa: não houve nenhum isolamento geográfico pela aridez que promovesse a vicariância necessária. Mas o princípio da hipótese de Haffer, "vicariância na Idade do Gelo" pode estar correto. Necessitamos substituir esfriamento glacial e reduzida concentração de CO_2 pela aridez de Haffer.

Esfriamento e concentração de CO_2 deveriam ter sido sempre mais importantes em regiões elevadas, onde seus efeitos são intensificados pela altitude. Assim, esses efeitos não podem ser vistos nas áreas separadas do "mar de savana" de Haffer, mas nos próprios isolamentos de espécies. A Idade do Gelo não criou o mar de savanas como Haffer acreditava, mas criou as "ilhas." A vicariância é preservada pelas descobertas de esfriamento glacial e concentrações reduzidas de CO_2, os quais não eram conhecidos quando Haffer apresentou sua hipótese.[12]

NOTAS

1. O cálculo de quarenta gerações de árvores, desde o fim da última Idade do Gelo usa a premissa arbitrária de que, se não morrerem jovens, as espécies arbóreas da floresta vivem cerca de 250 anos. 250 anos, embora, seja uma idade comum para uma árvore, agora é considerado um valor conservador. Não podemos determinar a idade das árvores da floresta pela dendrocronologia, isto é, através da contagem dos anéis de crescimento, pois árvores da Amazônia não enfrentam invernos e, por isso, não possuem esses anéis. Algumas espécies da várzea amazônica produzem anéis de crescimento, mas esses são correlacionados somente com períodos secos e flutuações do nível de água dos rios. Contudo, recentemente, grandes árvores amazônicas foram datadas pelo método do ^{14}C, que resultou para algumas, idades de até 1.500 anos (Chambers, 1998). Portanto, somente existiram sete gerações de árvores de 1.500 anos de idade, desde o final da Idade do Gelo.

2. Em algumas circunstâncias, é possível que a linha de árvores também seja sensível a concentrações de CO_2, de forma que os dados polínicos possam refletir somente a redução desse gás na atmosfera glacial. Isso foi recentemente sugerido para as linhas de árvores do leste da África, e defendido pelo uso de um modelo de simulação computacional do bioma com pronunciada aridez, como umas das condições limite (Jolly & Haxeltine, 1997). O modelo é para um bioma de floresta baixa na África, e não para as linhas de árvores mais úmidas dos Andes. Nossos dados

(descritos abaixo) para o Equador mostram que o fenômeno registrado nos lados das montanhas, não se trata do movimento de biomas, mas de populações individuais de espécies, muitas em elevações mais inferiores do que a linha de árvores. Essas são plantas C3, vivendo em florestas para as quais o modelo do bioma não se aplica. Outro modelo (Street-Perrott *et al.*, 1997) sugere que, devido ao fato de serem plantas C3, a depressão da linha de árvores em tempos glaciais reflete sua inabilidade de se manter nas altas elevações das montanhas, em condições de concentrações reduzidas de CO_2. Isso implica na questão sobre limites fisiológicos de árvores (Tranquilini, 1979) e pode ser discutida em detalhe. Contudo, esse assunto é duvidoso para os flancos andinos onde as populações arbóreas, bem abaixo da linha de árvores, têm descido simultaneamente com o avanço das geleiras, como foi mostrado pelos dados das seções de Mera e San Juan Bosco.

3. O registro de Mera, assim como o de San Juan Bosco, foram diretamente contestados na literatura por K. Heine, que essencialmente usou do argumento que as datações de radiocarbono de 26.000 a 30.000 anos poderiam, possivelmente, ser o resultado da contaminação de madeiras pleistocênicas muito antigas, por um pouco de carbono recente. Heine, erroneamente, descreveu os sedimentos lacustres de Mera como depósitos de lignita e declarou que, o intemperismo do depósito, parecia indicar uma idade mais antiga. Ele tentou datar os detritos vulcânicos de Mera, pelo método do Urânio-Tório, que lhe forneceu idades de 8.000 a 86.000 anos. Estas idades variaram, presumivelmente, dependo das histórias das amostras, antes destas terem sido encaixadas por detritos vulcânicos. Dessa forma, Heine não foi capaz de contestar as idades de radiocarbono exceto por declaração (Heine, 1994). A demonstração subseqüente de descidas similares de vegetação em San Juan Bosco, na costa pacífica do Panamá, no Lago de Pata, juntamente como outras evidências de esfriamento glacial, apoiam a interpretação original do registro de Mera como uma história de esfriamento glacial, de datação entre 26.000 e 30.000 anos atrás, dentro do tempo do estágio 3 de isótopos de oxigênio.
4. Obtivemos, recentemente, um registro sedimentar localizado perto da linha de árvores, 1.000 metros diretamente acima da seção de San Juan Bosco, que fornece evidência palinológica para a descida dessa cinturão arbóreo no Equador (Colinvaux *et al.*, 1997).
5. A cratera de El Valle foi drenada no início do Holoceno e, afortunadamente, encontramos uma completa seqüência contendo a porção do glacial tardio e o Holoceno, no registro de La Yeguada, próximo a El Valle. Este diagrama de porcentagem é mostrado na Figura 2.4. Observe que o sinal de Urticaceae/Moraceae de floresta tropical está presente em toda a seqüência, apesar dos sinais de esfriamento dados por *Quercus* e outros tipos polínicos na época glacial tardia. As altas porcentagens de Gramineae do Holoceno indicam a savana antropogênica (veja amostras superficiais no capítulo 1); Gramineae tem valores baixos na floresta tropical mais fria do glacial tardio (Bush *et al.*, 1992).
6. Modelos climáticos adicionais estão aparecendo na literatura e fornecem evidência para o esfriamento tropical, como por exemplo, Ganopolski *et al.* (1998).
7. Neste trabalho publicamos os dados palinológicos somente dos primeiros 1,7 m da coluna sedimentar, que contém os últimos 42.000 anos. A extrapolação sugere que a coluna sedimentar completa de 7 m contém o último ciclo glacial na sua totalidade. Uma análise preliminar desses sedimentos mostra que florestas nunca foram substituídas por savanas nesse longo período.
8. Haberle amostrou os sedimentos que foram datados segundo os critérios do projeto ODP, até a seção equivalente ao estágio 6 dos isótopos de oxigênio (a penúltima glaciação). Haberle também amostrou a carga sedimentar moderna do Rio Amazonas, em um transecto de oeste a leste, com o objetivo de calibrar as associações polínicas fósseis contra as atuais produzidas pela vegetação moderna.
9. Objeções a esta conclusão, embora não publicadas, têm sido oferecidas. Essencialmente, elas sugerem que o pólen nos sedimentos do leque representam "somente" as margens dos rios, e que florestas de galeria existiam ao longo dos rios e não nos interflúvios. É completamente

irracional a idéia de que nuvens de pólen de gramíneas de enormes áreas de savana poderiam ter sido camufladas por faixas estreitas de vegetação ciliar ao longos dos rios. Essa sugestão vai além da imaginação. Compare os diagramas polínicos de floresta do Lago da Pata com os do cerrado (Figuras 1.1 e 1.3).

10. Diferenciamos desses autores e de Van der Hammen & Absy (1994) quanto às nossas interpretações dos registros palinológicos. O diagrama polínico de Carajás mostra, em toda a sua extensão, altos valores de gramíneas e de componentes herbáceos e está, virtualmente, sem os tipos polínicos que associamos com floresta tropical, que no Lago da Pata constituem 80% ou mais da soma polínica total, assim como nas contagens de nosso extenso programa de chuva polínica, na floresta tropical moderna. O pólen arbóreo, do diagrama de Carajás, nos parece inteiramente de árvores, típicamente associadas às várias formas de vegetação aberta e savana de gramíneas, do tipo encontrado atualmente no topo da Serra de Carajás. A razão pólen herbáceo/pólen arbóreo, certamente aumentou no clima mais seco da última Idade do Gelo, mas em nossa opinião isto é um resultado do limite imposto por condições edáficas na própria localidade, que não fornece nenhuma informação sobre a floresta tropical de terras baixas, 600 m abaixo do platô (veja nota 6 no Capítulo 1). O Prof. Hooghiemstra discorda de nós neste ponto, e aceita que a porcentagem reduzida de pólen arbóreo, em tempos glaciais em Carajás, requer fragmentação da floresta abaixo do platô. Sua reconstrução da Idade do Gelo da Amazônia mostra, conseqüentemente, uma faixa larga, possivelmente de savana, estendendo-se de norte a sul através da parte leste da Bacia Amazônica. Esta reconstrução, contudo, como a nossa, mantém um setor imenso de floresta no oeste da Bacia Amazônica.

11. A hipótese do refúgio de Haffer agora tem uma vasta literatura. A referência principal é Haffer (1969). Elaborações dessa hipótese encontram-se em trabalhos publicados em Prance (1982) e em Whitmore & Prance (1987).

12. A alternativa óbvia para uma hipótese de vicariância pleistocênica tem sido aquela que sugere que endemismo, pelo menos o de pássaros, é mais antigo que o Pleistoceno. Essa possibilidade recentemente se tornou provável, com os trabalhos modernos moleculares sobre as relações entre as aves da América do Norte. Pesquisadores, há muito tempo, sugeriram que o padrão de distribuição encontrado entre diferentes espécies de pássaros encontradas entre o leste e o oeste dos Estados Unidos se desenvolveu no Pleistoceno, quando populações ancestrais foram fragmentadas pelas geleiras. Resultados moleculares recentes, contudo, sugerem, convincentemente, que essas separações ocorreram há muito mais tempo. Se as espécies endêmicas da Amazônia também são da era terciária (mais de três milhões de anos), então necessitamos somente explicar como elas se mantiveram separadas durante todo o Pleistoceno. Essas espécies endêmicas estão na periferia de uma vasta floresta que, segundo nossas reconstruções, persistiu durante o Pleistoceno. Isto sugere que o ambiente local (áreas elevadas de Haffer) são suficientemente diferentes, permitindo a essas populações permanecerem isoladas.

CAPÍTULO

3

GUIA PARA AMOSTRAGEM, COM PISTÃO, DE SEDIMENTOS LACUSTRES

Um palinólogo, em busca de registros longos e antigos da Bacia Amazônica, ficará contente em aceitar qualquer depósito polinífero: pântanos, perfis em cortes de estradas, paleocanais, perfis de paleosolos e, até mesmo, escavações arqueológicas. Mas lagos, particularmente, aqueles com extensão entre 100 m e 1000 m de comprimento, têm vantagens que os tornam os melhores alvos do paleoecólogo. Em lagos com dimensões dentro dessa faixa, a chuva polínica regional (parte do sinal polínico transportada pelo vento) está mais bem representada, com chances mínimas de super-representação local como ocorre nas turfeiras, sedimentos de pântanos e áreas semelhantes. Geralmente, a preservação do pólen é excelente em sedimentos lacustres. Taxas de sedimentação são, geralmente, constantes no decorrer de longos períodos e podem ser precisamente calculadas com um número adequado de datações por carbono 14 (Capítulo 4), preferivelmente do tipo AMS que, por sua vez, permite o cálculo do influxo polínico. Sedimentos lacustres são, ao contrário de outros tipos de sedimentos quaternários, menos suscetíveis a distúrbios pós-deposicionais. Outros sinais paleoecológicos, tais como algas diatomáceas, química dos sedimentos, pigmentos vegetais, intensidades magnéticas, razões de isótopos, etc., tendem a ter maior valor em depósitos lacustres do que em turfeiras, pântanos ou paleosolos. Na Amazônia, sedimentos lacustres são, praticamente, o destino final de pólen transportado nas bacias de drenagens pelo escorrimento de água superficial (*run-off*).

Coletar as melhores amostras para a análise palinológica, portanto, envolve escolher um lago e coletar testemunhos completos de seus sedimentos. Para a realização dessa atividade recomendamos o uso de barcos infláveis, que acoplados formam a base flutuante de uma plataforma de madeira, adequadamente estabilizada. A plataforma é posicionada no meio do lago ou na sua região com sedimentos, previamente evidenciados por batimetria ou outro método, e é estabilizada por três âncoras. Uma tubulação-guia (*casing*), feita de seções de 1 m cada e acopláveis, é formada e descida na coluna de água. Este tubo-guia conecta a plataforma a um ponto fixo no fundo do lago e guia os tubos amostradores (*core tubes*). Os testemunhos (*cores*) são, então retirados em seções, com um tubo amostrador contendo no seu interior um pistão. Cada amostra sucessiva reentra no velho canal formado pela coleta, pois ela é direcionada pelo tubo-guia (Figura 3.1).

AMOSTRADORES DE PISTÃO

Os sedimentos lacustres orgânicos (gyttja) são, quando sob a água, macios, maleáveis e fácilmente penetrados por um tubo. Mas a obtenção de um testemunho encontra a séria dificuldade causada pelo bloqueio que o sedimento tende a causar no final do tubo. Se esse bloqueio ocorrer, a ação contínua, meramente empurra os sedimentos para os lados, em vez de deixá-los entrar no tubo amostrador. Isto acontece porque a fricção entre o sedimento e o interior do tubo se torna maior que a força necessária para empurrar o sedimento de lado. Em testemunhagem de lago, superamos esta fricção interna pela ação do pistão na sua extremidade, que permite a entrada dos sedimentos no tubo amostrador por toda a sua extensão.

Amostradores de pistão funcionam como bombas de bicicletas ao reverso. Quando o ar é bombeado no pneu da bicicleta, o pistão é empurrado ao longo de um tubo, para expulsar o ar do seu fundo. Em um amostrador de sedimentos, o pistão começa no fundo do tubo, o qual é empurrado contra o pistão. O pistão estando bem selado em relação às paredes internas do tubo amostrador que desce contra os sedimentos, faz com que a pressão no seu interior possa ser igualada com a exterior, permitindo a entrada de sedimentos tão depressa quanto o tubo é empurrado contra eles. O peso da água no topo dos sedimentos (pressão hidrostática), pressionando a superfície dos sedimentos, iguala o peso da água no topo do pistão, enquanto os sedimentos enchem o tubo. Deste modo a pressão hidrostática é usada para superar a fricção dos sedimentos dentro do tubo. O processo pode ser visualizado mais claramente quando se diz que o sedimento é sugado para dentro do tubo: uma bomba de bicicleta ao reverso.

Assim, as características essenciais do amostrador de pistão são o tubo amostrador, sistema de hastes (*rods*) e um conector (*holder*) para abaixá-las através da coluna d'água, e um sistema que permita segurar e manter o pistão em seu lugar quando o tubo é empurrado contra ele. O pistão é mantido em posição por um cabo de aço flexível, cujo final é preso na plataforma, por um chave de pressão modificada (*vice-grips*). O contínuo empurrar das hastes propulsiona o tubo amostrador ao longo do pistão estacionário (Figura 3.1).

Para o correto funcionamento do amostrador com pistão, este não deve se mover antes do tempo. O amostrador deve ser abaixado e reentrar no velho canal feito anteriormente, e o pistão deve permanecer na sua posição basal inicial. Sómente, então, deve ser liberado o cabo de aço preso na plataforma, e o tubo amostrador empurrado contra os sedimentos. Podem existir fortes pressões que tendem a mover o pistão de sua posição inicial e correta, especialmente durante a sua descida dentro do canal formado nos sedimentos.

É absolutamente essencial que o pistão não escorregue e fique fora de sua posição basal até que a profundidade dentro da coluna sedimentar, na qual a próxima coleta deve começar, seja alcançada. Dois métodos alternativos são usados para segurar o pistão em sua posição inicial até sua liberação. Um método consiste em ter hastes quadradas (*square rods*) dentro do tubo amostrador, que fisicamente impedem que o pistão seja empurrado para cima pela pressão hidrostática. Quando na profundidade de amostragem, o pistão é liberado pela haste quadrada, que é puxada para cima. A segunda alternativa é um pistão que se trava na parte terminal do tubo amostrador, até que seja liberado pela tensão no seu cabo de aço.

O amostrador de haste quadrada representa a primeira alternativa. Neste modelo, somente a última haste é quadrada, em seção cruzada. Esta haste passa por uma abertura quadrada dentro do tubo amostrador até que o seu final conecte com o pistão. O tubo amostrador e o pistão são empurrados simultaneamente até que a profundidade de amostragem seja alcançada; nesse momento a haste é puxada para cima. O cabo do pistão é, então, preso na plataforma, não permitindo que o pistão se aprofunde nos sedimentos. A coluna de haste é empurrada contra o pistão que agora está estável.[1]

Este sistema de haste quadrada tem dado bons resultados em muitos lagos, mas possui certas características que achamos insatisfatórias. Para trabalho e transporte em locais remotos da floresta tropical o sistema de haste quadrada é pesado, pois essa haste é de metal sólido e tem cerca de 1.2 m de comprimento. Além de ser sensível a danos físicos, uma pequena flexão dessa haste pode torná-la presa dentro do orifício quadrado. Mais séria ainda, acreditamos, é a dificuldade de martelar a tubulação, o que permite que o tubo coletor amostre sedimentos mais grosseiros ou lâminas de cinzas vulcânicas, pois não há conexão rígida entre o final da linha de hastes e a "cabeça" ou segurador (*holder*) do tubo amostrador. Assim, eliminamos a necessidade de uma haste quadrada pela construção de um pistão, que se acopla na base de tubos amostradores e que é liberado somente pela tensão no seu cabo de aço.

O pistão travável (*locking piston*) foi desenhado pelo engenheiro mecânico Vince Vohnout e é mostrado nas Figuras 3.2 e 3.3. O princípio básico desse modelo é composto de dois anéis de borracha, que são conectados por uma haste sólida e separados por um came (*cam*), peça circular com movimento de rotação quando pressionada. Os anéis de borracha são pressionados contra a parede do tubo amostrador (flechas na Figura 3.4a). Um aperto moderado do seu parafuso e o pistão se torna fortemente fixo em posição, mesmo quando à seção é imposta a ação de um martelo gravitacional (*drop hammer*). Contudo a pressão na borracha é liberada pela rotação do came, que se estabiliza em uma superfície horizontal (Figura 3.4b). O selo hermético proporcionado pela borracha readquire seu formato e o pistão está livre para se mover. A rotação do came exige somente um forte repuxo no cabo do pistão.

Para usar o amostrador, o pistão é inserido dentro do tubo em posição de "disparo" onde o came (girado) permite que sua face convexa seja pressionada contra a "porca" de metal (*washer*), que segura os anéis de borracha. O pistão é então empurrado contra o tubo amostrador até que atinja o seu final, quando então apertamos o parafuso do pistão, tornando-o imóvel na extremidade basal do tubo amostrador (Figura 3.4a). O amostrador está agora pronto para penetrar o lago ou o antigo canal, pela conexão de hastes até que a profundidade desejada seja alcançada. Você também pode afundar o pistão através de sedimentos que não devem ser coletados, pois ele estará imobilizado na extremidade basal do tubo. O cabo do pistão é então conectado na plataforma. Um forte solavanco nas hastes tensiona o cabo do pistão, o came gira e o pistão não mais se encontra selado contra a parede do tubo, você pode, então, empurrá-lo firmemente contra o pistão estacionado, mas livre, usando pressão nas hastes (Figura 3.4b).

Uma vantagem óbvia deste sistema sobre o do de haste quadrada refere-se à rigidez do tubo amostrador em relação à haste, que permite que o impacto da ação de um martelo seja transferido da haste para o tubo amostrador com uma perda mínima de impacto. Nós usamos um martelo de 10 libras (5 kg), que possui um orifício por onde uma haste de metal pode passar e que causa um grande impacto em uma peça metálica hexagonal acoplada às hastes. O martelo também pode impactar essa peça de baixo para cima, caso seja necessária a remoção do tubo amostrador de um sedimento pegajoso ou muito resistente (Figura 3.5).

Este modelo de pistão tem uma vantagem adicional: permite que o mesmo seja travado dentro do tubo amostrador durante sua descida nos sedimentos. No final da operação, o tubo amostrador está cheio de sedimentos e o pistão se localiza, nesse estágio, na parte superior do tubo amostrador contra o qual está pressionado. Mas agora você tem que retirá-los, o que causa uma forte sucção que tende a puxar ambos, o pistão e os sedimentos para fora do tubo. Para impedir que isso ocorra, uma pessoa começa a retirar as hastes puxando-as para cima, enquanto outra pessoa no mesmo compasso, puxa o cabo de aço. Isto é o que precisa ser feito com o sistema de hastes quadradas. No nosso sistema, temos uma peça acopladora (*locking device*) para o pistão, que o mantém preso quando atinge a parte superior do tubo amostrador. Isto permite que ambos, o pistão e o tubo amostrador sejam retirados simultaneamente pela ação das hastes. Usamos conectores de encanamento, com uma porção "macho," no topo do pistão e a parte "fêmea" desse conjunto, no segurador do tubo (Figura 3.6). O pistão agora, no final de cada operação, está preso ao segurador e não pode ser empurrado para baixo pela pressão hidrostática. O perigo de se perder o testemunho torna-se muito reduzido. O alívio proporcionado aos operadores do sistema, que agora não necessitam segurar o cabo de aço, é também considerável.

Presilhas para o Cabo e Anotações de Campo

Com o amostrador abaixado até a profundidade correta e o pistão em sua posição, bloqueando a boca do tubo, o cabo de aço do pistão deve ser preso na plataforma. Duas são as formas de se fazer isso: esticar e enrolar o excesso de cabo em volta de um "rabo de peixe" de madeira, ou usar um tipo de chave (*clamp*). Nós recomendamos o uso de um sistema de chave de pressão e um orifício (Figuras 3.7 e 3.8). A chave não pode passar pela abertura do orifício que forma uma "ponte" sobre o topo da tubulação-guia (Figura 3.8). Isto faz com que o cabo fique apertado diretamente sobre o orifício e verticalmente sobre o amostrador. Mais importante, contudo, é a posição da chave no cabo, que fornece um registro preciso da profundidade na qual a amostragem começou, com uma vantagem imensa quando do preparo para a retirada da próxima seção.

Quando o trabalho é realizado em profundidades de 10, 20 ou mesmo 30 m de água, não é possível sentir, através das hastes, quando o amostrador está entrando nos sedimentos ou quando ele alcança o fundo do canal. Portanto, deve-se manter um bom registro da profundidade alcançada através da medição desde a plataforma ou da superfície da água. Sem uma chave (do tipo inglesa ou de pressão) no cabo do pistão, esta medição se torna cansativa, pois é necessário manter registro preciso do número de hastes que já entraram no orifício. Tendo uma chave firmemente presa ao cabo do pistão, no ponto onde a operação precedente começou, esse problema é eliminado.

A chave marca a profundidade na qual o pistão ficou posicionado no começo de cada operação anterior. Durante a operação, o pistão foi liberado pela tensão no cabo de aço; depois disso o tubo coletor passa pelo pistão, pela distância de um metro. O pistão não se move durante a operação e acaba ficando na parte superior do testemunho, que agora contém 1 m de sedimento. Para começar a nova seção, o novo tubo carregando o pistão na sua extremidade deve penetrar exatamente um metro adicional. Isto é fácilmente realizado pela liberação da chave seguida do seu deslocamento no cabo de aço por um metro. Isto significa que haverá mais um metro de cabo de aço disponível para penetração nos sedimentos. Toda unidade, com um novo tubo amostrador preparado com o pistão em sua extremidade e conectado ao cabo de aço, tem agora um metro adicional em sua extensão. O pistão travado está agora no fundo do antigo canal formado pela remoção do metro anterior de sedimento. E a nova operação pode começar. Este procedimento é mais simples e mais preciso do que manter uma contagem do número e de comprimentos parciais de hastes, que entraram tanto na água como nos sedimentos.

Persiste ainda o problema de se encontrar a profundidade da interface sedimento–água, onde a primeira seção de um metro do testemunho deverá começar. A amostragem dos primeiros centímetros de sedimentos não consolidados requer o uso de um amostrador especial (descrito em outra seção deste manual). Mas a superfície sedimentar aproximada para o início da remoção do testemunho deve ser encontrada. Esta profundidade não pode ser sentida através das hastes, mas sua posição aproximada pode ser encontrada através do uso de um disco de 12 pol. (25 cm) de acrílico (ou material similar) preso a uma corda. Quando da confecção de mapas batimétricos, usamos um batímetro para ver a descida do tubo amostrador em direção aos sedimentos, e interrompíamos sua descida quando o seu contato com os sedimentos estivesse registrado nesse aparelho. No caso do primeiro metro amostrado, o final da tubulação-guia é mantido deliberadamente acima dos sedimentos e o tubo amostrador pode ser visto com uma projeção abaixo do final do tubo-guia.

Uma vez que o cabo de aço do pistão está firmemente preso com uma chave para a primeira operação, uma marcação precisa da profundidade relativa é facilmente mantida com movimentos sucessivos da chave ao longo do cabo, pela distância equivalente em profundidade de sedimento amostrado na operação anterior. Geralmente, essa distância é de um metro completo, mas algumas vezes a operação pode ter sido realizada parcialmente; então, o comprimento do sedimento removido no tubo amostrador é medido e a chave é movida ao longo do cabo pela distância equivalente, em vez de um metro completo.

O Uso de um Cortador em Sedimentos Mais Coesos

Sedimentos orgânicos e maleáveis (sapropel ou gyttja) são facilmente penetrados por um simples empurrar da linha de hastes com seu tubo amostrador: o peso de 110 libras (50 kg) de uma pessoa é suficiente para o trabalho. Siltes mais pegajosos e argilas, contudo, são mais resistentes à perfuração e com certeza encontrados na maioria das bacias lacustres. Muitas vezes esses sedimentos são encontrados na parte basal da seção estratigráfica adjacente ao embasamento. Quando a penetração não pode ser mais realizada devido à resistência sedimentar, recomenda-se impactar a unidade com o martelo gravitacional. Esta operação pode se tornar mais fácil com a redução da fricção por uma estrutura metálica, denominada "cortador" ou "sapato" (*shoe*).

A séria resistência oferecida por sedimentos mais coesos não é causada pela necessidade de se deformar o corpo sedimentar para acomodar o amostrador, e sim pela resistência causada pela fricção contra os lados do tubo amostrador. Argilas pegajosas aderem ao tubo amostrador como a lama nos seus sapatos, quando se caminha em um pântano. Para penetrar com sucesso sedimentos coesos, é necessário reduzir a fricção. O método mais prático é aumentar o diâmetro da parte basal do tubo amostrador. O cortador faz exatamente isso (Figura 3.9). Os cortadores que usamos são, essencialmente, tubos curtos (5 cm) de alumínio, afinados na sua extremidade, o que permite a fácil penetração nos sedimentos. Eles não devem ser demasiadamente afiados, evitando assim que essas partes possam agir como pontas de facas em barcos infláveis. O cortador é facilmente acoplado ao final do tubo amostrador, que se apóia a uma linha saliente no próprio cortador. Um orifício no cortador projeta-se num orifício similar no tubo amostrador, e um parafuso o mantém em sua posição durante a recuperação dos sedimentos (Figura 3.9).

É importante notar que o propósito do cortador de aço não é proporcionar um corte melhor do que o faria o tubo amostrador sozinho, e sim reduzir a fricção criada com a criação de um canal sedimentar de maior diâmetro do que aquele feito com o tubo amostrador. Teoricamente, a única fricção encontrada deveria então ocorrer ao longo do comprimento do cortador, liberando assim o tubo amostrador da fricção causada pelos sedimentos. Mas, muitas vezes, uma certa quantidade de sedimentos entra em contato com o tubo, mesmo nesse canal de maior diâmetro, mas a penetração pode ser facilitada pela ação simultânea do cortador e o martelo gravitacional.

Os limites práticos de coleta em sedimentos coesos são determinados, não pela dificuldade de penetração, mas pela dificuldade de remover o tubo amostrador, dos sedimentos. A unidade pode ser aprofundada nos sedimentos coesos durante a coleta pelo impacto do martelo gravitacional, mas o operador pode ter dificuldades durante a remoção do tubo amostrador e suas hastes conectantes. Em seus 37 anos de testemunhagem de lagos, Colinvaux perdeu dois amostradores de pistão dessa forma: um, em um grande lago na região ártica do Alaska, e outro, na Amazônia central do Brasil. Eles permanecem nesses locais como relíquias submersas, troféus potenciais para um arqueólogo industrial do futuro.

A força que usamos para extrair amostradores presos nos sedimentos é limitada pela flutuação da plataforma. É comparativamente simples montar um sistema de roldana na plataforma de madeira, mas a aplicação desse tipo de força nas hastes, quando o amostrador está preso na lama, meramente empurra toda a plataforma para debaixo d'água. Nossa experiência mostra que a força de duas ou três pessoas fortes, é suficiente para a liberação da unidade da força de coesão dos sedimentos. Não existe outra possibilidade de se exercer mais força de extração quando em plataformas flutuantes, exceto talvez com a ação inversa do martelo gravitacional.

Martelar de baixo para cima libertará o amostrador preso nos sedimentos dada a rigidez do nosso sistema de conexão das hastes de alumínio ao tubo amostrador, mas o método tem restrições. Os conectores das hastes podem se romper e quebrar antes que o amostrador seja liberado. A ação do martelo gravitacional, de cima para baixo, funciona bem, pois as hastes são empurradas de "costa a costa." Mas para retirar a linha de hastes, estas estão ligadas por uma conexão de parafusos. Cada conector está seguro pelo seu "parafuso pai," mas este pode ser rompido e a conexão perdida. O melhor conselho a ser oferecido é martelar com cuidado. Se a cada batida, o amostrador descer em profundidade, é seguro continuar até que um metro completo seja amostrado. Mas o operador deve estar preparado para remover menos que um metro de sedimentos, caso o amostrador não penetre com batidas adicionais do martelo. Quando o objetivo da pesquisa sugere que sedimentos mais coesos devem ser coletados, é sempre uma boa idéia usar tubos amostradores de meio metro de comprimento, ao invés de tubos de um metro. O corte do tubo original de um metro pode ser feito com uma simples serra manual. Sempre use o cortador; a redução resultante de fricção pode ser mais importante na remoção do que na penetração. Se a unidade ficar presa nos sedimentos e se o uso de força não solucionar este problema, use o esforço humano em combinação com o uso do martelo gravitacional, impactando a unidade de baixo para cima. Essa ação combinada de esforços deve ser suficiente para a liberação da unidade amostradora dos sedimentos. Num eventual fracasso dessas opções, aconselha-se encher os barcos com água, causando um razoável afundamento dos mesmos. Depois, com a mesma força da equipe, puxar a linha de haste e segurar o cabo de aço e amarrá-lo a um local firme da plataforma com a chave. Depois retirar a água dos barcos, retirar a equipe dos barcos, permitindo assim que os mesmos subam em relação ao nível da água. Essa ação causará uma tensão permanente na haste e no cabo de aço. Ambos coletores de Colinvaux foram perdidos dessa forma, pois as hastes de aço perderam suas estruturas conectantes e o cabo de aço escapou da chave, mantendo a unidade permanentemente presa nos sedimentos.

Um outro ponto sobre o uso do cortador deve ser mencionado. Freqüentemente, quando o cortador é usado, a operação para coleta de um metro de sedimentos, termina com o pistão satisfatoriamente travado no dispositivo acoplador dentro do tubo amostrador, resultando em testemunho sedimentar com menos de um metro. Embora o pistão seja mantido em sua posição basal durante a descida, é aparente que uma parte basal do testemunho, em geral 10 cm, é perdida para o exterior do tubo durante a recuperação do tubo amostrador. A perda pode se tornar imediatamente aparente pela visualização dos sedimentos durante a abertura dos tubos e mesmo através da radiografia por raio x. A perda é geralmente representada por um ou por uma série de hiatos na seção do testemunho. Nossa explicação para este fenômeno é que o cortador na extremidade basal do tubo amostrador está tão fixo no fundo do orifício do canal amostrado, que os sedimentos não se rompem quando a unidade é puxada para cima. A pressão, dentro do tubo e abaixo do pistão, é tão grande que o sedimento é expelido do seu fundo. Estamos atualmente redesenhando nossos pistões com ondulações na sua superfície exterior, na esperança, de que isso quebre o selo hermético entre o pistão e o fundo do canal sedimentar amostrado. Se sedimentos são perdidos por esta causa, então sua perda está ocorrendo na parte basal do tubo amostrador, embora descontinuidades possam aparecer na seção testemunhada. Na reconstrução da estratigrafia do testemunho, todas as seções ausentes devem aparecer como hiatos.

Um conselho final sobre o uso de cortadores e do martelo gravitacional — a penetração deve prosseguir tanto quanto possível. Quando sedimentos muito coesos são encontrados, use o cortador e o martelo com cuidado. Muitas vezes, o operador terá que acionar com muita força o martelo, pois sedimentos importantes para o estudo podem ser mais coesos que os sedimentos mais recentes e mais maleáveis. Os primeiros 1,7 m dos testemunhos do Lago da Pata (Figura 1.1) foram retirados sem o uso do martelo gravitacional, mas cinco metros adicionais foram amostrados através do uso do martelo, até que uma unidade de seixos, cobrindo o embasamento, foi encontrada. Esses 5 metros adicionais, que estão sendo analisados palinológicamente no momento da editoração deste trabalho, são ricos em pólen e extendem o registro paleoecológico do Lago da Pata através de grande parte do último ciclo glacial.

Extrair ou não Extrair os Sedimentos: Vantagens e Desvantagens do Uso de Tubos Transparentes

Quando o tubo amostrador, cheio de sedimentos, está seguro na plataforma, uma decisão deve ser tomada: o testemunho deverá ser selado para posterior abertura no laboratório, ou deverão os sedimentos ser extraídos na plataforma para inspeção, antes de serem embalados em plástico para depois serem levados ao laboratório? Diferentes equipes de pesquisas possuem diferentes opiniões, mantidas fortemente por suas convicções. Temos a opinião unânime que é errado extrair, ainda no campo, os sedimentos. Mas ambas as escolas de pensamento têm razões válidas para as suas opiniões.

Os sedimentos dentro de seu tubo estão acondicionados da melhor maneira possível. Se o tubo está seguramente selado, o testemunho pode ser transportado para o laboratório sem sofrer perturbações. Assim ele poderá ser aberto sob condições controladas. O tubo amostrador é aberto, por um corte na sua parede de alumínio, em vez de sujeitar os sedimentos à uma manipulação de extrusão, que empurra os sedimentos para fora do tubo. Selar os tubos na plataforma é mais rápido e o tempo no campo é usado mais eficientemente, especialmente em locais remotos, ao contrário das atividades de extrusão, descrição e acondicionamento dos sedimentos para o transporte até o laboratório.

A vantagem da extrusão no campo é que a quantidade de sedimentos obtidos é determinada rapidamente e a equipe pode modificar a operação de acordo com os resultados.

Duvidamos que alguma vantagem em poder observar os sedimentos após a remoção, possa justificar os riscos envolvidos ao expô-los às condições do campo, os riscos de transporte e o tempo gasto.

Mesmo assim, podem existir ocasiões em que o rápido exame dos sedimentos no campo é importante, especialmente quando da presença de laminações espetaculares ou de um limite estratigráfico. Para isso estamos atualmente investigando o uso de tubulação transparente em substituição aos tubos de alumínio. Esses tubos permitirão o exame visual dos sedimentos no campo sem necessidade de sua extrusão.

Recomendamos, para trabalhos na Amazônia, a prática da selagem dos tubos de testemunhos no campo e seu envio ao laboratório, no seus próprios tubos amostradores. Se os testemunhos são armazenados em salas frias (4°C) e escuras eles manterão suas condições inalteradas por muitos anos. Nesse meio tempo, a estratigrafia de seus sedimentos, pode ser visualizada pela técnica não-destrutiva da radiografia por raios x, antes de serem abertos. A infra-estrutura de radiografia de um hospital é adequada para este propósito e em geral é possível contar com o apoio da direção de hospitais para este tipo de análise.

Nossa política de manter os sedimentos em seus tubos amostradores é relativamente cara, pois o custo de cada tubo de alumínio é de cerca de US$ 20,00. Esse custo tem que ser avaliado, contudo, sob a luz do alto investimento representado por toda a atividade de remoção dos mesmos no campo. Acreditamos que esse custo (o preço de um seguro) é razoável para uma carga tão preciosa.

Amostragem da Interface Sedimento–Água

Um amostrador de pistão pode ser usado para amostrar corretamente os sedimentos inconsolidados dos primeiros centímetros da coluna sedimentar, e da própria interface sedimento–água (Fisher *et al.*, 1992). Isto não só proporciona amostras da chuva polínica recente no lago e um sinal da vegetação moderna, como também a amostragem, em intervalos curtos, de mudanças ou eventos recentes. Essas amostram permitem que os efeitos de mudanças ambientais contemporâneas possam ser estudadas.

O princípio do amostrador de sedimentos superficiais baseia-se em um tubo transparente com um pistão, e que tenha o comprimento equivalente a profundidade entre os sedimentos mais superficiais e os mais compactados. O tubo é trazido quase até a superfície da água, e sua extremidade basal é vedada por um tipo de rolha de borracha, cujo diâmetro permite sua entrada e movimento dentro do mesmo, como se fosse um pistão. Com o seu fundo vedado, a unidade pode, então, ser retirada do lago, mantida sempre em posição vertical. A interface sedimento–água, não perturbada pela ação, pode ser vista através do tubo coletor transparente. A amostragem dos sedimentos é realizada empurrando cuidadosamente a rolha de borracha (pistão) para cima, até que a superfície dos sedimentos não consolidados cheguem à extremidade superior do tubo amostrador. Toda a coluna sedimentar pode ser amostrada por uma série de fatias, seqüencialmente através do deslocamento consecutivo e, para cima, do pistão.

A propriedade importante desse amostrador que permite a extrusão vertical do testemunho de sedimentos não consolidados, é um tubo que passa pelo pistão e que pode ser aberto em sua extremidade superior para liberar a pressão embaixo (Figura 3.10). Com essa abertura, o pistão pode ser puxado para fora do tubo pelo seu cabo de aço sem que a água e os sedimentos superficiais sejam perturbados. A "rolha" de borracha, semelhante a um pistão que foi usado para bloquear a extremidade basal do tubo amostrador pode ser usada agora como um "desentupidor," que empurra os sedimentos e a água através do tubo. A melhor forma de realizar isso é empurrar o tubo para baixo, contra uma haste vertical apoiada na madeira da plataforma ou contra o solo. Com uma pequena plataforma e barco extra, como os nossos, é mais aconselhável levar toda a unidade, em posição vertical, para a margem do lago.[2]

Esse amostrador pode ser adaptado ao amostrador de Livingstone e, como este, deve ser usado para trabalho em uma plataforma flutuante, devidamente estabilizada por âncoras. A amostragem do sedimento superficial deve ser realizada antes da remoção dos sedimentos mais consolidados pelo amostrador de Livingstone. Isso evita a perturbação da estratificação dos sedimentos não consolidados. Na visita a um lago somente para a amostragem do interface sedimento/água, sem a remoção de um testemunho completo, este amostrador não é muito prático, pois requer o uso de hastes e cabos. Contudo, é muito fácil contruir uma versão mais pesada que pode ser abaixada por uma corda, do que por hastes. Esta é a melhor opção para um projeto de coleta de sedimentos superficiais de uma região, pois só requer um pequeno barco inflável, o amostrador, cordas e frascos para armazenagem das amostras.

A Tubulação-Guia

A tubulação guia (*casing*) nada mais é que um simples tubo, que atravessa a coluna d'água do lago a ser amostrado. Certamente, a forma mais econômica e a mais utilizada de tubos-guia é feita de pvc comum. Esse material, assim como seus conectores, podem ser comprados em praticamente qualquer loja de materiais de construção. É um material razoavelmente pesado, mas seu peso não chega a ser proibitivo e tem custo moderado, o que permite seu uso sem muitas considerações financeiras. As conexões podem ser unidas por "silver tape" ou outro adesivo resistente. A conseqüência disso é uma série de projeções horizontais eliminando a possibilidade de um interior completamente livre e liso, por onde correrão os tubos amostradores, suas hastes e o cabo de aço que segura o pistão. O abaixamento dessas peças em um interior não liso fatalmente será problemático, pois estas poderão ficar aprisionadas nas junções da tubulação. Esse problema é ainda maior na recuperação do tubo amostrador com suas hastes. A solução é usar presilhas ou braçadeiras (*coupling*) de meia-seção, feitas de seções curtas de tubulação de alumínio de 6 pol. (15cm), que correm facilmente pelo exterior da tubulação-guia como se fossem braceletes. Um corte estreito e longitudinal existe em um lado dessa presilha, permitindo assim sua compressão e ajustamento à tubulação-guia, por cintas de metal. As unidades da tubulação-guia a serem acopladas deslizam dentro dessas braçadeiras do tipo "presilhas com cintas" (sleeve-couplings), fazendo uma conexão muito firme. As braçadeiras são apertadas por meio de um sistema de parafusos laterais, com uma chave de fenda. A junção dos tubos é firme e, virtualmente, impossível de se separar e a tubulação-guia resultante é totalmente lisa no seu interior. Contudo, nosso primeiro trabalho de campo com esse equipamento revelou uma falha séria: vibração prolongada resultante da ação do martelo gravitacional sobre a tubulação-guia, fato que causou o escorregamento de algumas braçadeiras, embora as partes não tivessem se separado. A solução para este problema foi providenciar pequenas intrusões, nos tubos-guia individuais, que se encaixassem em um corte em forma de letra L (90°) antes que as cintas fossem apertadas (Figura 3.11). Presilhas com este modelo permanecem juntas, mesmo quando submetidas à ação das batidas e vibrações causadas pelo trabalho do martelo gravitacional.

A PLATAFORMA E SEU USO

Pode ser usada qualquer plataforma flutuante que permita o trabalho da equipe em pé: tambores de óleo, troncos de árvores formando uma balsa, canoas juntas, barcos infláveis, etc. Recentemente vimos a foto de uma elegante plataforma construída por Francis Mayle feita de madeira tropical em um remota região da Amazônia da Bolívia. É importante que sua plataforma seja portátil e possa ser levada, sem dificuldades, ao local de estudo. Isto significa que componentes infláveis devem ser usados.

Nós usamos dois barcos infláveis Avon Redstart, de produção inglesa, colocados lado a lado, como se fossem as partes flutuantes de um catamarã. Sobre os barcos colocamos uma plataforma de madeira com um orifício, por onde passa toda a tubulação (Figura 3.12). Cada barco tem a dimensão de 8 pés (2.4 m) de comprimento e capacidade para até 3 pessoas, quando usado com um motor. Seu peso é de 35 libras (15 kg) quando seco, o que equivale a um pacote de peso razoável para um carregador. Esse é o tamanho ideal para os nossos propósitos e qualquer barco com dimensões maiores e mais pesado, torna-se inadequado para o transporte manual. O Redstart é o menor tipo de barco inflável produzido pela Avon. O próximo tamanho tem nove pés de comprimento e, embora não pareça ser muito maior, tem uma grande diferença: é bem mais pesado e possui tubos de ar também maiores, chegando a ser impraticável o seu uso para os nossos propósitos. Existem outras marcas e estilos de barcos e infláveis de alta qualidade no mercado, mas qualquer que seja o tipo selecionado, o paleoecólogo tem que estar seguro que ele deve ter, aproximadamente 8 pés de comprimento (não mais que 2,5 m), capacidade para 3 pessoas, um peso razoável para ser transportado por uma pessoa, e que não possua um chão de madeira e outras partes pesadas.

Barcos infláveis desta qualidade têm vários pontos de amarração e, em geral, são fácilmente conectados lado a lado. Deve existir um espaço entre os barcos por onde passará a tubulação. Uma das melhores maneiras de se criar esse espaço entre os barcos, vital para a tubulação, é usar dois blocos de isopor ou madeira resistente que são colocados entre cada uma das extremidades dos barcos. Estes podem ser, então, amarrados, com os dois blocos entre si.

A plataforma de madeira cobre o espaço criado pelos dois blocos de isopor. Em seu centro está um orifício por onde passará a tubulação. Uma grande plataforma tem, obviamente, suas vantagens, sendo a estabilidade uma delas. Mas uma plataforma grande causa incômodos durante o transporte, especialmente quando tem que ser transportada pela floresta densa. Também é alto o custo do seu transporte aéreo. A plataforma de Colinvaux teve suas dimensões diminuídas nos últimos 30 anos de experiência nesse tipo de trabalho e, atualmente, ela é um quadrado de 90 cm, embora ele admita que um metro quadrado é aceitável. A plataforma é constituída de duas partes iguais, unidas por duas peças de apoio, e podem ser empacotadas sem problemas para eventuais transportes aéreos. Detalhes do modelo da plataforma encontram-se na Figura 3.12. A madeira para a confecção da plataforma deve ser resistente e leve e não deve ser compensada. Para amarrar âncoras, ferramentas ou mesmo para amarrar a plataforma aos barcos, alguns orifícios são feitos na sua madeira, em intervalos adequados. Não use peças de metais para este propósito, pois elas formarão pontas protuberantes naquilo que será uma superfície de trabalho um pouco apertada. O orifício central deve ser suficientemente grande para a passagem da tubulação-guia e também para o espaço de uma mão e um braço, no caso de uma eventual recuperação da tubulação-guia (Figura 3.12).

Acessórios da Plataforma

Uma ponte para segurar a chave sobre a tubulação-guia e uma guilhotina modificada, denominada "jaws" para segurá-la, embora não seja absolutamente necessária é tão útil que rapidamente se tornou essencial, pois é um excelente dispositivo de segurança.

A guilhotina da tubulação-guia de Colinvaux tem 13 pols. (33 cm) de comprimento e é feita de dois pedaços de madeira de pinho de 1,5 pols. (4 cm) de espessura e 3,5 pols. (9 cm) de largura, unidos por uma dobradiça. Orifícios hemisféricos em cada lado se juntam para formar um orifício circular, que circunda a tubulação-guia e permite que as braçadeiras (mais largas) descansem em seu topo (Figura 3.13). A guilhotina deve ser equipada com uma trava de segurança que permite seu fechamento ininterrupto, feita com dois pregos presos por uma linha ou barbante bem resistente, ou com uma presilha.

Uma vez na sua posição desejada no lago, e com a tubulação-guia abaixada até a superfície dos sedimentos, o topo da tubulação-guia é preso pela guilhotina que agora cruza o orifício central da plataforma. A braçadeira superior, que em nosso modelo tem seis polegadas (15 cm) de altura, se apoia sobre a guilhotina fechada. A abertura deve existir para o livre deslocamento do cabo de aço do pistão. Nós temos usado uma caixa de madeira sem fundos, cujas dimensões permitem cobrir a braçadeira, que está sendo presa pela guilhotina. Um corte horizontal na placa metálica da caixa fornece o espaço para o cabo de aço e fornece também um apoio para a chave que o segurará (Figura 3.8). Enquanto este trabalho estava sendo escrito, Jason Curtis criou um modelo mais simples e melhor, no qual a caixa de madeira (pesada) sem fundos é substituída por uma seção cilíndrica de pvc, suficientemente grande para conter em seu interior a última braçadeira da seção de tubos-guia. O corte para o cabo de aço e para a chave é um pedaço de pvc com a forma de uma meia-lua e reforçado por uma placa de aço aderida por Epoxy.

Finalmente, um acesório muito importante é um grupo de "rabos de peixe," que são feitos de madeira que guardam as cordas das âncoras e do cabo de aço do pistão (Figura 3.14). Tentativas de se enrolar cordas em um barco inflável cheio de hastes e outros equipamentos não é uma tarefa agradável. Por isso enrolamos as cordas das âncoras em "rabos de peixe," que são feitos de madeira compensada e podem ser molhados.

Âncoras e Seu Posicionamento

A plataforma deve ser estabilizada por três âncoras posicionadas em ângulos de 120°. Todas as três cordas das âncoras devem ser bem esticadas depois do lançamento, impedindo assim o seu deslocamento pela ação do vento. Elas devem permanecer firmemente presas nos sedimentos do fundo do lago, e devem ser, portanto, razoavelmente pesadas. O melhor plano de trabalho em locais remotos tem sido o uso de sacos ou bolsas resistentes de náilon, que são preenchidas com rochas, cascalhos ou até mesmo areia provenientes das redondezas do lago. Cada uma deve ser tão pesada, por exemplo 25 libras (12 kg), quanto possa ser convenientemente manuseada por uma pessoa da equipe, trabalhando em um barco inflável.

Nas terras baixas da Amazônia este simples plano de encher sacos com rochas e usá-los como âncoras é freqüentemente frustado pelo fato de que, muitas vezes, rochas têm que ser transportadas de uma exposição rochosa num local acessível por veículo. Neste caso, é recomendável o uso de peças de metal encontradas em sucatas. Um pedaço de ferro velho com peso de 5 kg afunda facilmente nos sedimentos lacustres e, na maioria das vezes, esse tipo de âncora não é tão volumoso.

Uma alternativa para âncoras pesadas, algumas vezes possível em pequenos lagos, é usar árvores das margens dos mesmos. Neste caso, recomenda-se ter cordas com até 100 m de comprimento. Obviamente, essa técnica não pode se aplicada em lagos cuja maior dimensão é superior a 150 m.

As cordas das âncoras devem ser sempre longas de forma que o seu repuxe forme uma linha horizontal, tanto quanto possível. As âncoras devem ser lançadas a uma distância três vezes maior do que a profundidade da coluna d'água, no local a ser amostrado. Lagos nas terras baixas da Amazônia são, em geral, rasos e quase todos aqueles encontrados por nossa equipe não tiveram profundidade superior a 10 m, e muitos não passavam de 5 m de profundidade. Se temos bons motivos para acreditar que lagos mais profundos não serão encontrados, definimos assim a profundidade limite ao levarmos somente 10 m de tubulação-guia, 50 m de cordas para as âncoras e um "rabo-de-peixe" extra contendo outros 100 m de corda. As melhores cordas a serem usadas, nesse tipo de trabalho, são aquelas de náilon trançado que, devido à sua alta resistência, podem ser finas.

Trabalho em lagos mais profundos requer o conhecimento prévio da profundidade a ser encontrada. O Lago San Marcos, nos Andes Equatorianos, foi testemunhado com sucesso, sob 37,5 m de água. Este trabalho exigiu não somente cordas de âncoras com até 150 m de extensão, como também 40 m de tubulação-guia e hastes em proporção. Felizmente, a equipe de trabalho foi capaz de alcançar o Lago San Marcos com um veículo que transportou o equipamento. É importante ressaltar que a testemunhagem de lagos profundos requer um conhecimento prévio e um planejamento apropriado. O ideal é realizar o reconhecimento da área da lagoa e sua profundidade, tanto na época seca como na úmida, antes da amostragem.

Fixação das Âncoras com Barco

O ato de ancorar uma plataforma seguramente sobre a superfície de um lago é a parte da operação que mais requer habilidades comuns náuticas. O procedimento tem por objetivo fixar as âncoras com um terceiro barco separado, que no nosso caso também é um Avon Redstart, amarrado à plataforma durante seu deslocamento até o local de amostragem. Para essa jornada, as âncoras são transportadas sobre a plataforma de madeira e os seus "rabos-de-peixe" são seguramente colocados no chão de um dos barcos. Uma vez no local desejado, as âncoras são passadas uma de cada vez, para a pessoa que está no terceiro barco. Ela será responsável pelo procedimento de ancorar a plataforma. A primeira âncora é levada pelo terceiro barco em direção contrária ao vento predominante no local e a pessoa na plataforma libera gradualmente a corda do "rabo-de-peixe" para o terceiro barco, que aos poucos se afasta da plataforma. O objetivo desta operação é liberar a corda da primeira âncora o mais rápido possível, sem puxar a plataforma nem removê-la do local determinado para a remoção do testemunho sedimentar. Se o dia estiver ventoso é uma boa idéia lançar uma âncora extra (amarrada em uma corda curta) diretamente da plataforma, a fim de impedir que esta seja arrastada pelo vento a outra posição no lago, enquanto as três âncoras estão sendo fixadas. Quando o terceiro barco se encontra suficientemente distante da plataforma, a pessoa que libera a corda do "rabo-de-peixe" na plataforma ordena o lançamento ao lago da âncora, pela pessoa no terceiro barco e amarra a corda da âncora lançada em um dos orifícios da plataforma de madeira. O terceiro barco retorna, então, à plataforma para pegar a próxima âncora e repetir a operação.

Chegamos aqui a uma parte importante dessa operação. A segunda âncora deve ser fixada obliquamente em relação ao vento predominante, e do lado oposto onde foi fixada a primeira âncora. O objetivo dessa parte da operação é de fixar a plataforma, como se fosse o ponto central da letra "Y." A pessoa na plataforma deve manter esta configuração desejada, enquanto libera a corda da segunda âncora e direciona o barco rebocador em sua jornada, carregando-a. É aqui que as habilidades náuticas são importantes. É importante lembrar que a posição relativa das primeiras duas âncoras fixadas no lago podem ser reajustadas através de uma nova fixação, da liberação de mais corda ou de sua amarração mais apertada na plataforma. Se a pessoa direcionando essa operação da plataforma não estiver satisfeita com o resultado, a pessoa no barco rebocador deverá remover a âncora fixada, remar novamente para o local desejado e arremessá-la novamente, segundo direções fornecidas pela pessoa na plataforma. Muitas vezes, a coleta de uma âncora de 12 kg, fortemente fixada nos sedimentos de um lago, pode ser uma tarefa árdua, mas o posicionamento correto delas reverterá em uma excelente estabilização da plataforma no local desejado para a amostragem.

Quando as duas primeiras âncoras são fixadas corretamente, a fixação da terceira é, geralmente, um procedimento simples: o barco rebocador pega a terceira da plataforma de madeira e é enviado a favor do vento, na direção oposta aos dois braços da letra "Y" formado pelas duas primeiras âncoras. As três âncoras são então puxadas da plataforma, até que se tornem tão esticadas que possam ser dedilhadas como uma corda de violão. As âncoras devem ser sempre amarradas na madeira da plataforma e nunca nos barcos. Ao final da operação de fixação das âncoras, a plataforma formada pelos dois barcos de apoio deve estar no centro de uma letra "Y," ou seja, no ponto de conexão dos três eixos. Se as cordas das âncoras estiverem tensas e forem suficientemente pesadas, a plataforma permanecerá na posição desejada por alguns dias, caso seja necessário.

RESUMO DE UMA OPERAÇÃO PADRÃO DE AMOSTRAGEM DE SEDIMENTOS

Primeiramente, envie o barco rebocador para determinar a profundidade do local com sedimentos a ser amostrado. Marque a posição dessa estação (geralmente a parte mais profunda do lago) com uma bóia colorida (pode ser uma garrafa plástica) amarrada a uma pedra por um barbante ou corda. Faça um mapa batimétrico completo, se possível. Colete amostras da água para análise com redes de plâncton. O plâncton poderá poderá ser usado para calibração dos resultados radiométricos de datação.

Monte e lance a plataforma ao lago. Carregue-a com suficiente quantidade de tubos-guias e hastes, para realização da coleta de sedimentos sob altura já determinada da coluna d'água. Inclua tubos amostradores, barras-T, pistão e seu final acoplador já conectados ao cabo de aço enrolado em seu "rabo-de-peixe," peça hexagonal de apoio para o martelo gravitacional, cortador, guilhotina, fita plástica adesiva (silver tape), etc. Tenha em mãos uma caixa de ferramentas contendo chave de boca, martelo e um jogo de limas grossas para alumínio. Limas planas e redondas são necessárias para ajustar, em alguns casos, orifícios nos tubos amostradores onde se acoplam o cortador e o segurador do pistão.

Coloque as três âncoras sobre a plataforma de madeira (as únicas coisas sobre ela), amarradas pelas suas cordas em "rabos-de-peixe," colocadas no chão de um dos barcos da plataforma. Dois membros da equipe, irão sentados nos lados dos barcos da plataforma. Uma pessoa, no barco rebocador, levará a plataforma até o local desejado no lago.

Fixe as três âncoras. Esta é a parte mais difícil de toda a operação com este equipamento e requer trabalho em equipe. A coordenação do posicionamento das âncoras é feita por uma pessoa, em pé, na plataforma de madeira. Embora a plataforma não esteja ainda estável, esse trabalho requer que a pessoa responsável pela coordenação das âncoras, fique em pé para poder ver claramente se a fixação delas está seguindo o padrão de letra "Y." Com uma equipe de três pessoas, a terceira pessoa acelera a operação com a entrega das âncoras pesadas para a pessoa no barco rebocador. Com o distanciamento do barco rebocador, essa mesma pessoa, na plataforma, libera a corda do "rabo-de-peixe" para esse barco.

Obtenha uma amostra da interface sedimento/água de um lado da plataforma e envie o barco rebocador à margem com o amostrador para extrusão e sub-amostragem.

Abaixe com cuidado a tubulação-guia na coluna de água e por segurança amarre uma corda fina a essa tubulação, que a protegerá contra uma eventual perda, caso uma de suas partes se dissocie. Essa corda de segurança deverá estar sempre frouxa para evitar que a tubulação saia da orientação vertical desejada. A guilhotina de madeira é útil, especialmente em lagos profundos. A tubulação-guia longa e pesada necessita de ser presa com firmeza para adição de suas unidades. Uma pessoa deve manipular a guilhotina enquanto a outra adiciona as unidades do tubo-guia e afrouxa os parafusos da braçadeira. Após essa fase, a pessoa operando a guilhotina abre esse dispositivo, permitindo assim a descida da tubulação-guia com suas braçadeiras. Quando a tubulação chegar à superfície dos sedimentos, trave a guilhotina. A última braçadeira deverá se apoiar na guilhotina agora travada, impedindo que a tubulação-guia escape e se aprofunde nos sedimentos. Prepare o primeiro tubo amostrador com o pistão, que ficará preso na sua extremidade basal. O pistão, em posição travada, é inserido juntamente com seu cabo de aço no tubo amostrador através de uma haste. É fácil ajustar uma das hastes somente para este fim, através de um pedaço de plástico que possa ser rosqueado na sua extremidade e que contenha um corte que permita o encaixe da parte superior do pistão. Conecte o dispositivo segurador do pistão à parte superior do tubo amostrador através de seu próprio parafuso que passará pelo orifício superior do tubo amostrador.

Coloque a caixa de madeira sobre a guilhotina e a última presilha que repousa sobre sua superfície. Desça o tubo amostrador dentro da tubulação-guia, rosqueando hastes conectoras adicionais. Uma pausa na descida é necessária para que o tubo amostrador se encha de água. Você ouvirá através de um som musical de gargarejo, o som mais lindo do mundo! Continue adicionando extensões de hastes e liberando o cabo de aço do pistão até que o tubo amostrador alcance a superfície dos sedimentos. Observe que o cabo de aço do pistão está enrolado em um "rabo-de-peixe" e este deverá ser girado para a liberação do cabo. Evite liberar o cabo de aço manualmente, pois o mesmo acabará se enrolando e perdendo sua linearidade. Não o manuseie como se fosse corda das âncoras.

O cabo de aço deverá correr dentro do corte linear da caixa de madeira e não poderá estar frouxo. Ajuste a chave de forma que o cabo esteja esticado contra a caixa de madeira. Adicione uma barra "T" a essa linha de hastes e aplique um forte empurrão para baixo (em direção aos sedimentos) para liberar o pistão. Continue empurrando toda a unidade (hastes + tubo amostrador) agora, gentilmente, contra os sedimentos, até que a operação chegue ao seu final com o acoplamento da parte superior do pistão ao seu segurador. Nos sedimentos friáveis superficiais, no início da operação, um "click" resultante desse acoplamento pode ser sentido.

Traga, verticalmente, toda a unidade para cima, dissociando simultâneamente suas hastes. Se o pistão estiver conectado ao segurador não será necessário manter a tensão no seu cabo de aço, embora deva ser lembrado que um cabo muito solto pode ocasionar o aprisionamento do tubo amostrador e hastes dentro da tubulação-guia. Não é necessário enrolar o cabo de aço no rabo-de-peixe. Com a subida de toda a unidade (tubo amostrador + hastes), simplesmente deixe o cabo de aço cair na água, ao lado da plataforma; isso impedirá que sejam formados nós em sua extensão. A chave fixada ao cabo de aço também pode ir para a água, permitindo que ele permaneça livre de nós, pois esse é o modo desejado de se ter cabo de aço em uma plataforma.

Remova o parafuso que mantém firme o segurador do pistão e os separe; ambos deverão estar conectados o que permite a sua saída como uma única peça — Figura 3.6. Os sedimentos deverão estar no mesmo plano horizontal que a parte basal do tubo amostrador, mas alguns sedimentos podem ser puxados para baixo pela pressão da água, causando sua projeção para fora do tubo. Neste caso, recomenda-se guardá-los em uma embalagem plástica adequada e anotar todas as informações pertinentes a esta amostra, tais como sua altura em cm, etc. Em seguida, com uma fita adesiva plástica sele os orifícios (do cortador e do segurador do tubo) do tubo amostrador, agora contendo os sedimentos. Isto impedirá que os sedimentos se tornem ressecados dentro do tubo amostrador ou dele escapem. Ponha uma "tampa" plástica sobre a extremidade basal do tubo amostrador e reforce sua fixação com uma fita adesiva plástica ou com "silver tape." Uma "tampa" plástica é inserida internamente na extremidade superior do tubo, no espaço deixado pela unidade pistão-segurador do pistão. Empurre essa tampa interna com uma haste até que fique em contato com os superfície superior dos sedimentos. Um pequeno orifício ou corte a ser feito nessa tampa interna, deixará o ar escapar, à medida que ela está sendo empurrada em direção aos sedimentos no tubo. Anote as informações referentes ao nome da lagoa, profundidade amostrada etc.

Destrave o pistão do seu segurador, ajuste-o na sua configuração de descida e o insira em um novo tubo amostrador. Desça com a unidade (tubo amostrador + hastes) dentro da tubulação-guia, como mencionado acima. O cabo de aço será retirado aos poucos da água, até que a chave a ele afixada apareça. Esta chave marca a exata profundidade onde os sedimentos começaram a ser removidos, mas agora, a operação deve começar um metro mais abaixo. Meça um metro adicional de cabo de aço e mova a chave para essa nova posição. Continue empurrando o amostrador para baixo até que a chave chegue até o topo da caixa. A parte basal do tubo amostrador está agora no fundo do antigo canal de perfuração nos sedimentos. Tome a segunda seção e proceda com ela da mesma forma que na seção 1.

Uma boa precaução é abaixar a tubulação-guia por um metro adicional nessa etapa, aprofundando efetivamente sua extremidade basal nos sedimentos. Adicione, portanto, uma outra seção de um metro, lembrando de soltar mais um metro da corda de segurança da tubulação-guia e de recolocar a caixa de madeira sobre a última braçadeira, como anteriormente mencionado. Nesse estágio, não mais importa que a tubulação-guia perturbe os sedimentos superficiais, o tubo amostrador e as hastes podem passar por esses sedimentos superficiais perturbados, pois o pistão estará travado e bloqueando a abertura basal do tubo amostrador. Além disso, os sedimentos superficiais foram coletados sem perturbação e estão na primeira seção obtida.

Proceda seção por seção, tanto quanto for possível. O tempo de coleta não é conhecido até que a operação seja realizada. Com freqüência, na Amazônia, um ou dois metros obtidos manualmente é o limite, mas no lago equatoriano de Kumpaka, o uso do martelo gravitacional foi necessário antes do décimo terceiro metro amostrado. Uma profundidade de até 8 metros de sedimentos pode ser retirada manualmente, se uma grossa camada de sedimentos orgânicos (gyttja) estiver presente. No início da operação o peso do corpo da pessoa amostrando os sedimentos pode ser usado como um martelo para liberar o pistão. Se o operador pesar 170 libras (77 kg), seu peso pode ser usado como um martelo poderoso.

Se o uso do martelo gravitacional se torna uma necessidade, quando o tubo amostrador penetrou somente metade do que deveria coletar, é demasiado tarde para se usar um cortador. Desrosqueie a haste superior e a barra "T," rosqueie a peça hexagonal em seu lugar e uma outra haste que servirá como guia para o martelo. É uma boa idéia ter a capacidade de colocar a peça hexagonal em uma altura conveniente, motivo pelo qual existem algumas hastes de apenas meio metro de comprimento. Deslize o martelo gravitacional cuidadosamente pela haste guia, enquanto um outro membro da equipe deverá segurar as hastes abaixo da peça hexagonal. Logo após essa etapa, a ação do martelo gravitacional poderá ser iniciada. Observe e tente sentir se a unidade está se aprofundando nos sedimentos (veja nossas recomendações sobre o uso do martelo gravitacional na seção anterior). Em particular, o operador deve estar alerta à sensação e até ao som causado pelo acoplamento do pistão ao seu segurador, o que indica o final da remoção daquele metro de sedimentos. Outros sinais indicadores desse acoplamento referem-se a uma vibração brusca na plataforma de madeira e a uma vibração diferente na haste, causada pela força de uma batida sendo absorvida pelo cabo, assim que o pistão chega ao topo do tubo amostrador. Se o operador perceber a ocorrência de qualquer desses fenômenos, a ação de penetração deve ser interrompida.

Tente antecipar a necessidade do uso do martelo gravitacional, de maneira que o cortador possa ser conectado ao tubo amostrador antes do inicio da operação de remoção. Para colocar o cortador, ponha o pistão na posição de disparo e prenda-o na extremidade do tubo da forma usual, coloque o cortador de aço no tubo permitindo que o tubo amostrador se apoie na sua linha de apoio interior. Alinhe o orifício do tubo amostrador com o correspondente orifício do parafuso do cortador, rosqueando-o. Todo cuidado deve ser tomado para que esse parafuso não penetre além do necessário, pois sua função é a de penetrar o suficiente, para manter o cortador conectado ao tubo e prevenir sua perda durante a retirada da unidade amostradora do canal sedimentar. Esse parafuso não deve interferir com a ação do pistão dentro do tubo amostrador.

O cortador deverá ser inspecionado depois do uso do martelo gravitacional. Se a extremidade do cortador estiver riscada, torta ou levemente danificada, isto significa que o tubo amostrador atingiu cascalho ou até mesmo rocha. Nesse momento, a operação de penetração deve ser interrompida. Muitas vezes, a prudência sugere a interrupção da penetração, mesmo quando uma seção de um metro não foi alcançada. Quando isto ocorrer, o operador deve lembrar que o cabo do pistão deve estar seguro em sua posição, em conjunto com as hastes, durante a operação de retirada da unidade (tubo amostrador + hastes) da tubulação-guia. Caso contrário, a pressão dentro do tubo expulsará os sedimentos do tubo amostrador.

O uso do martelo pode ser considerado perigoso, pois um pedaço de metal de 5 kg levantado a uma altura para caber dentro da haste, pode ser acidentalmente derrubado e atingir a cabeça desprotegida de algum membro da equipe trabalhando em uma plataforma, que está sofrendo ondulações causadas pelo vento e ondas no lago.

Uma vez que todas as seções do primeiro testemunho foram satisfatoriamente removidas, é uma boa prática tomar um segundo testemunho na mesma estação. Esse segundo testemunho pode ser iniciado com uma profundidade diferente da do primeiro, de modo que os sedimentos iniciais de uma seção correspondam aos sedimentos de um outro testemunho coletado na segunda estação. Desta forma, o paleoecólogo poderá reconstruir a completa estratigrafia sedimentar de testemunhos, sem nenhuma incerteza sobre qual material pode estar faltando e onde ocorre mudança nas seções.

Levante a tubulação-guia de um a dois metros, garantindo assim que os sedimentos superficiais da lagoa não estejam sendo afetados, mova a plataforma dois a três metros com a liberação suficiente de corda do "rabo-de-peixe" amarrada a uma das âncoras, enquanto que as outras são puxadas à mesma distância. As cordas das âncoras deverão ser bem amarradas garantindo assim tensão nas mesmas. Abaixe a tubulação-guia e comece a remoção dos sedimentos. É importante lembrar que antes do começo da operação, meio metro deverá ser adicionado a tubulação-guia. Depois conecte a chave. A partir desse instante, todas as seções obtidas estarão compensadas por meio metro do primeiro testemunho.

NOTAS

1. Mostradores de pistão para depósitos lacustres foram inventados por Daniel Livingstone no início dos anos 50 e foram baseados no princípio usado nos amostradores Kullenberg, usados, então, para sedimentos marinhos profundos (Livingstone, 1955). O primeiro tubo cilíndrico de haste, usado para segurar o pistão em posição basal durante sua descida na coluna de água, era redondo e segurava o segurador por protuberâncias metálicas (Vallentyne, 1955). Livingstone, então, aperfeiçoou aquele modelo pela substituição da haste quadrada pela cilíndrica com protuberâncias metálicas. Todas as variações, incluindo as nossas, são propriamente denominadas "Amostradores de Livingstone."
2. Instruções detalhadas sobre a construção de amostradores de Livingstone são encontradas em Fisher *et al.* (1992).

CAPÍTULO

4

METODOLOGIA PALINOLÓGICA: PROTOCOLO E MATERIAIS

Este capítulo foi escrito especialmente para aqueles que estão iniciando suas atividades na análise palinológica. Ao escrevê-lo, imaginamos um pesquisador trabalhando em uma instituição onde os recursos bibliotecários não são amplos e onde não há tradição em trabalhos palinológicos. Nos concentramos no que consideramos importante para a Palinologia da Amazônia. Expusemos alguns fundamentos de Palinologia, direcionados para pesquisadores novos na disciplina e o nosso objetivo não foi produzir um documento abrangente sobre o assunto. Para este intento, o leitor deverá consultar alguns trabalhos padrões: Faegri & Iversen, 1989; Moore *et al.*, 1991; Salgado-Labouriau, 1973.[1]

DESCRIÇÃO, ARMAZENAMENTO E DATAÇÃO DE TESTEMUNHOS SEDIMENTARES

Assim que as seções de testemunhos são transportadas do campo ao laboratório, devem ser mantidas, se possível, em salas frias com temperatura de 4°C, para evitar o crescimento de fungos e bactérias nos sedimentos. Esses organismos podem interferir na datação do carbono-14. Fungos podem metabolizar compostos de carbono dos sedimentos e podem trocar CO_2 com a atmosfera, durante o processo de respiração, contaminando assim os sedimentos com carbono recente. Os sedimentos, portanto, devem ser refrigerados. Logo após a remoção dos sedimentos, as extremidades dos tubos devem ser seladas com fita plástica, antes de serem armazenadas em sala fria. Se os sedimentos forem expelidos do tubo ainda no campo (o que não é recomendado), devem ser cuidadosamente embrulhados em plástico fino impermeável e não em plástico de polietileno, que permite trocas gasosas.

Radiografia Antes da Abertura dos Testemunhos

A radiografia dos sedimentos é facilmente realizada antes da abertura dos testemunhos. Esta técnica, além de fornecer uma visão geral dos sedimentos, produz um registro permanente de algumas das características estratigráficas. A técnica de raios-x é fácil de ser realizada em um hospital ou em uma clínica particular de raios-x. Os tubos cilíndricos de alumínio contendo os sedimentos são postos diretamente sobre o filme de raio-x, sendo irradiados de cima para baixo. Devido ao tamanho desse filme (chapa), é necessário usar três filmes para a radiação de 4 a 6 seções diferentes de 1 metro cada, que são postas paralelamente sobre o filme. O resultado é uma foto em uma escala praticamente 1:1, com um mínimo de distorção, devido ao fato de que a irradiação é originária de uma fonte pontual. A típica transparência de raio-x é um negativo que mostra a matéria orgânica como área escura e minerais opacos como áreas brancas. Características estratigráficas grosseiras como lâminas de cinzas vulcânicas aparecem na chapa como bandas brancas. Além disso, as radiografias podem mostrar também estruturas estratigráficas delicadas que não são aparentes durante a inspecção visual dos sedimentos. Quando essas estruturas finas estão presentes e apresentam-se como planos horizontais dentro do tubo amostrador, é uma boa confirmação que a estratigrafia não foi perturbada durante a amostragem. Da mesma forma, qualquer distúrbio sedimentar no topo da seqüência pode ser identificado e evitado durante a subamostragem dos sedimentos.

Abertura dos Testemunhos

Se os sedimentos são moles, eles podem se expelidos do tubo sem que ocorra danos à estratigrafia. Para isso, é necessário um êmbolo feito com um cabo de madeira e um tampão de borracha em sua extremidade. Os sedimentos são empurrados para fora do tubo, sobre um filme plástico apoiado por uma meia-seção de pvc, para manter a forma cilíndrica dos sedimentos. Desta forma, o tubo amostrador pode ser lavado e reusado. Embora esta técnica seja usada em vários laboratórios, preferimos evitar o risco de distorção dos sedimentos inerentes a esta pratica. Recomendamos a abertura dos tubos por um torno mecânico, que os corta com uma incisão em cada lado.

Descrição: Anotando a Estratigrafia

Com a abertura de cada seção, os sedimentos são descritos e um registro contínuo é produzido com as anotações no laboratório. Recomendamos o uso de uma fita métrica metálica, para determinar a profundidade obtida de sedimentos em cada seção e anotar essa informação nos próprios tubos de alumínio. A escala de profundidade deve ser contínua de seção para seção, descrevendo a estratigrafia sedimentar completa, desde a interface sedimento–água até a base do testemunho. Se algum sedimento for perdido entre as diferentes seções, esta informação deve aparecer como hiato permitindo assim a determinação da profundidade correta das seções encontradas abaixo do hiato. Quando testemunhos paralelos são coletados começando o trabalho com um desvio de 0.5–1 m, normalmente os hiatos são compensados com o testemunho paralelo. É importante salientar que anotações sobre as profundidades sejam feitas diretamente nos tubos de alumínio, facilitando assim futuras subamostragens.

Limpe a face exposta dos sedimentos com uma espátula ou faca, para revelar a estratigrafia mais fina.[2] Durante a abertura dos sedimentos, estes devem ser ainda anóxicos. Com a exposição dos sedimentos durante o trabalho, suas cores originais são alteradas. O melhor registro de cor dos sedimentos é obtido com o uso do tabela de cores de Munsell.[3] Fotografias a cores também podem ser muito úteis, especialmente quando feitas com uma câmera fotográfica digital, que pode mais tarde permitir análises mais detalhadas, por computadores, da intensidade e incidência de bandas sedimentares. Na maioria dos testemunhos usados principalmente na palinologia, a análise estratigráfica provavelmente ficará restrita às descrições de tipo, cor e limites estratigráficos dos sedimentos. O dado importante, nesta etapa, é a profundidade correta de onde foram retirados os sedimentos para a análise.

Datação Pelo Método do Radiocarbono

Com o advento da datação radiocarbônica por AMS (Accelerator Mass Spectrometry) ou espectrometria de acelerador de massa, nossa dependência da datação de sedimentos volumosos pelo declínio de emissão de partículas beta está chegando ao fim. Datações por AMS custam mais que as convencionais e a insuficiência de laboratórios equipados para esse fim pode ser um problema, fora da América do Norte e Europa. As vantagens desse novo tipo de datação são tão grandes, que deveria ser usado sempre que possível. A vantagem mais óbvia para trabalhos com sedimentos lacustres refere-se à pequena quantidade de sedimentos requeridos para a datação. Isto significa que ela não precisa mais ser uma média da idade da matéria orgânica depositada durante vários anos, como acontecia com as amostras volumosas. Além disso, o mais importante é que a datação possa ser obtida de pequenos remanescentes orgânicos: fragmentos de folhas de plantas terrestres ou partes do corpo de animais do zooplâncton, por exemplo. Datação desses fragmentos orgânicos, separados dos sedimentos, reduz de forma significativa os erros introduzidos por contaminantes como o grafite e partículas de carvão natural, que podem ocorrer em algumas localidades.

Por conveniência, recomendamos primeiramente a obtenção de uma datação basal pelo método de ^{14}C convencional. Esta informação dará ao palinólogo uma idéia do limite de idade da sedimentação amostrada e poderá guiar as subamostragens subseqüentes para datação AMS em intervalos críticos.[4]

Razões entre isótopos estáveis de carbono ($^{12}C/^{13}C$) são geralmente fornecidas com as estimativas de idade pelo laboratório de datação. O objetivo principal dessas razões é de corrigir o efeito do fracionamento isotópico permitindo assim que datações obtidas em diferentes laboratórios possam ser comparáveis. Embora mudanças nas razões entre isótopos estáveis possam ocorrer dentro dos tubos, é difícil interpretar o seu significado. A dinâmica entre as fontes e sumidouros de carbono em lagos é muito complexa, e muito cuidado é requerido para a interpretação dessas razões e a sua relação com a vegetação terrestre que circunda o lago. Qualquer interpretação paleovegetacional baseada em razões entre isótopos estáveis obtidas em sedimentos lacustres ou mesmo em solos, deve ser contrastada com a interpretação derivada da análise palinológica.

EXTRAÇÃO DE PÓLEN DE SEDIMENTOS

A massa polínica em sedimentos lacustres permanece pequena comparada à massa de minerais ou de matéria orgânica de origem não polínica, embora grãos de pólen possam ser encontrados em concentrações de até 100.000 grãos por centímetro cúbico(cm^3).[5] Portanto, para a análise palinológica, os grãos de pólen precisam ser extraídos de uma massa relativamente vasta de diluidores orgânicos e inorgânicos. A técnica palinológica é possível devido a extraordinária resistência da parede celular dos palinomorfos a muitos tratamentos químicos, com exceção de agentes oxidantes. De uma forma simplista, podemos afirmar que a extração de pólen consiste em dissolver da amostra a ser estudada, tudo que não é pólen: substâncias úmicas por meio do hidróxido de potássio, celulose pela mistura de acetólise (anidrido acético e ácido sulfúrico) e minerais silicáticos pela ação do ácido fluorídrico. Tudo que resistir a essas três digestões é um resíduo que contém o pólen concentrado, que sobreviveu sem alterações a essa química feroz.

Infelizmente, contudo, essas preparações químicas não produzem um resíduo "limpo," pois em geral o resíduo final não é constituído somente de pólen, mas inclui partículas orgânicas e inorgânicas quimicamente muito similares ao pólen. Paredes celulares de alguns elementos do fitoplâncton têm composições químicas que são intimamente similares às das dos grãos de pólen, e têm desafiado nossas tentativas de eliminá-las das preparações palinológicas, sem que ocorra a destruição do conteúdo polínico. Um exemplo dessas algas são *Pediastrum* e *Botryococcus*.[6] Outros elementos fitoplanctônicos produzem esporos que resistem ao tratamento químico mencionado acima, como também o fazem os numerosos esporos de fungos. Uma preparação palinológica de sedimentos lacustres, portanto, é um extrato de pólen, contendo os mais variados esporos e partes de células de algas, fungos e fragmentos resistentes de tecidos lignificados. Tipicamente, estes objetos indesejados são significativamente mais numerosos que o pólen, de forma que a análise palinólogica é, em parte, uma busca de grãos de pólen em uma "floresta" de elementos não-polínicos. A busca é conduzida em um microscópio ótico com aumento de 400X a 1000X.

Alguma matéria mineral também sobrevive à típica digestão com ácido fluorídrico (HF). Em parte, isso é conseqüência da presença de uma notável massa mineral que não é facilmente dissolvida, particularmente minerais não-silicosos que não sofrem dissolução pelo HF. Dois métodos alternativos são disponíveis para a separação do pólen desses minerais persistentes: os grãos de pólen podem ser separados por filtração ou por flotação. Infelizmente, ambos os métodos oferecem o risco de se perderem grãos pequenos e característicos da região neotropical.

Separar pólen da matéria mineral pela filtração se tornou possível com a disponibilidade comercial de tecidos sintéticos, com aberturas das malhas fibrosas dentro de um intervalo de 5 micra. As partículas minerais da argila ou do silte são muito menores que este valor e passam por esses filtros quando dispersos com hidróxido de sódio ou detergentes. Os grãos de pólen, freqüentemente, têm tamanho, dentro da faixa, de 30 micra ou mais (Traverse, 1988). Nas regiões clássicas de análise palinológica, especialmente o cinturão temperado do hemisfério norte, poucas espécies vegetais possuem grãos de pólen menores que 18 micra, e destes, poucos representam tipos polínicos dispersos por insetos. Nestas circunstâncias, a filtração oferece um meio rápido de se obter uma separação de pólen, de uma matriz mineral relativamente limpa (Cwynar *et al.*, 1979). Filtração, em sedimentos amazônicos, não é recomendada devido ao tamanho reduzido de *Cecropia*, um dos tipos polínicos mais abundantes nessa região. Assim também muitas espécies de *Mimosa*, *Piper*, *Piperomia*, *Begonia*, entre outras, têm o seu maior comprimento variando de 6 a 10 μm, e muitos membros da família Moraceae não são muito maiores que *Cecropia*. Muitos grãos passariam juntos com minerais de argila, pela malha de 5 μm, perdendo-se do resíduo a ser analisado. Por outro lado, podemos usar peneiras mais grossas como, por exemplo, as com malhas de 250 μm, para remover fragmentos orgânicos, mas não as de 100 μm, geralmente usadas nos países temperados, pois nossos maiores grãos podem ser removidos.[7]

A gracidade específica da sílica pura (quartzo) é 2.7, portanto ela sedimentará em um líquido cuja gravidade específica é 2.0. Outros minerais são mais pesados de forma que todos sedimentarão nesse líquido. Mas, os compostos de carbono encontrados na matéria orgânica flutuarão quando dispersos em líquidos com gravidade específica 2.0. Portanto, é necessário remover tanta matéria orgânica quanto for possível pelos métodos químicos e depois flotar o restante, separando-os da matriz mineral em um "líquido pesado." Um problema técnico surge quando levamos em consideração que a água, cujo peso específico é 1.0, também flutua sobre o líquidopesado, trazendo consigo minerais silicosos de argila. É necessário remover toda a água, incluindo a água de hidratação e a água adsorvida pelos minerais de argila. Isto pode ser feito, através de repetidas lavagens do material analisado com álcoois ou acetona. Se esta etapa for bem feita, a separação através da adição de líquidos pesados ao protocolo pode produzir um extrato polínico essencialmente livre da fração mineral.

A objeção ao uso de líquidos pesados em amostras da Amazônia é a mesma objeção feita ao uso de filtros; os pequenos grãos de *Cecropia* e de outros táxons, tendem a ser perdidos diferencialmente durante as numerosas decantações requeridas pelo protocolo. Este problema, acoplado ao fato de que o melhor líquido pesado empregado (uma mistura de bromofórmio e acetona) é um composto químico, reconhecidamente cancerígeno, diminuiu o entusiasmo dos palinólogos com as separações por líquidos pesados, não sendo usados, rotineiramente, em nosso trabalho na Amazônia.

Os protocolos para a extração de pólen variam de laboratório a laboratório, embora as estapas principais sejam as mesmas. Alguns procedimentos podem ser planejados em qualquer ordem, e há espaço para substituição de algumas etapas. Por exemplo, o hidróxido de potássio (KOH) pode ser substituído pelo hidróxido de sódio (NaOH), e a digestão com o ácido fluorídrico (HF) pode se feita através de sua fervura em banho-maria por 30 minutos ou usando-o a frio por 8 horas. Há uma receita para cada gosto nesta *"ciência de livro de receita."* A única etapa vital e insubstituível é a do uso da mistura de acetólise (1 parte de ácido sulfúrico concentrado, para 9 partes de anidrido acético) que destrói a celulose. A acetólise não só destrói os componentes celulósicos, como também limpa os grãos de pólen, removendo, em particular, as ceras, revelando, assim, a estrutura morfológica dos grãos de pólen e de outros palinomorfos. Grãos de pólen em lâminas de referência também são limpos com a mistura de acetólise, de modo que todos os grãos tenham aparência comparável.

Extrações de pólen são feitas em tubos cônicos de centrífugas. O volume de cada amostra sedimento deve ser medido com precisão. Nós usamos um medidor de volume, feito de aço inoxidável, produzindo pastilhas de sedimentos em três tamanhos: 0,25, 0,5 e 1 cm^3. O sedimento é colocado dentro da cavidade cilíndrica desse amostrador, e depois expelido por uma haste ou êmbolo. O volume subamostrado é agora uma pastilha cilíndrica (Figura 4.1).

O equipamento requerido é mínimo, sendo a centrífuga clínica para 15 tubos de 15 ml cada, o item mais custoso. Os tubos de centrífuga para a digestão com HF devem ser de polipropileno, e nunca de vidro, pois o vidro é solúvel nesse ácido. Muitas vezes achamos conveniente fazer a extração em tubos de vidro graduado, transferindo o material para tubos de polipropileno somente quando da extração por HF. Os tubos de polipropileno são um pouco opacos em condições de tensão superficial e podem interferir com a decantação segura do líquido residual. Contudo, um risco a ser corrido é a quebra desses tubos na centrífuga, causando a perda da amostra. Com exceção da vidraria usual de laboratório, o único outro equipamento necessário é um banho-maria ou um aquecedor de bloco, com espaço para 6 ou mais tubos de centrífuga por vez. O aquecedor de bloco "a seco" é de grande conveniência, mas o trabalho pode ser feito com um banho-maria tradicional (100°C) ou improvisado.

Adição do Pólen Marcador Exótico

Os palinológos usam, rotineiramente, a técnica do marcador exótico para o cálculo do número total de grãos de pólen, presentes originalmente no volume medido da subamostra. Os marcadores exóticos mais freqüentemente usados são preparações ou de pólen de eucalípto (*Eucalyptus* sp.) ou de esporos de *Lycopodium* (Stockmarr, 1971). Nas regiões temperadas do hemisfério norte e nas latitudes árticas, o pólen de *Eucalyptus* não ocorre naturalmente, tornando-se um verdadeiro marcador exótico. Embora *Eucalyptus* não seja nativo na Bacia Amazônica, outros membros da família Myrtaceae são importantes táxons arbóreos de suas florestas, cujo pólen não é facilmente separado do de *Eucalyptus*. Embora algumas espécies de *Lycopodium* sejam também encontradas em florestas tropicais, em geral elas constituem uma pequena fração de suas floras e o efeito dos esporos nativos de *Lycopodium*, na soma dos esporos exóticos pode ser considerada como insignificante. Por isso usamos esporos de *Lycopodium* como marcadores exóticos em nossos trabalhos. Isso é possível, pois preparações (similares a pastilhas farmacêuticas) de esporos de *Lycopodium* de concentrações determinadas são encontradas comercialmente. Pastilhas do marcador exótico são adicionadas a cada amostra a ser processada.[8] Na contagem final, a razão de esporos marcadores de *Lycopodium* e grãos de pólen contados permite o cálculo da concentração polínica original. Se a taxa de sedimentação for medida corretamente, então tornase possível calcular as taxas de acúmulo de pólen (influxo) de cada táxon. Taxas de acumulação, ao contrário das porcentagens polínicas, são estatísticas independentes.

O *Lycopodium* não é recomendado como marcador nas florestas nebulares da nossa região, pois este gênero é bem representado nesse tipo de vegetação. Uma alternativa, desenvolvida por Salgado-Labouriau é a de *Kochia scoparia* (Chenopodiaceae), que possui o tipo polínico periporado, típico das chenopodiáceas que são abundantes em regiões desérticas e ausentes na vegetação de florestas. Pastilhas de *Kochia scoparia* não são vendidas comercialmente, e o palinólogo deve preparar suas próprias suspensões desse pólen (Salgado-Labouriau & Rull, 1986). Uma alternativa ao uso de pólen como marcador exótico é usar microesferas sintéticas produzidas para este fim. Ainda não experimentamos essa técnica em amostras da Amazônia.[9]

A seguir, apresentamos um protocolo que, com algumas pequenas variações, temos usado em nossos trabalhos na Amazônia, incluindo todas as preparações usadas para a confecção das pranchas de pólen deste livro. Se um protocolo diferente é usado, é importante fornecer evidência de que *Cecropia* e outros grãos pequenos não foram perdidos diferencialmente.

PROCEDIMENTO PARA A EXTRAÇÃO DE PÓLEN

Todos os tratamentos químicos são feitos em tubos de centrífuga de 15 ml, de preferência com graduações. O mais fácil é usar somente tubos de propileno, pois são resistentes a todos os reagentes a serem usados, incluindo a mistura de acetólise e o HF. Depois de cada tratamento os tubos são centrifugados e o líquido residual é cuidadosamente decantado, para evitar a perda desnecessária de pólen. Em cada etapa do processo, a centrifugação é definida como 2500 rotações por minuto (rpm) durante 5 minutos. Na tabela a seguir, esse procedimento é indicado pelas letras M,C & D (misture, centrifugue e decante). Trabalhe com grupos de 6 tubos por vez, cada um com seu próprio bastão-misturador feito de teflon; um porta-tubos e um porta-bastão deverão ser usados para evitar toda possibilidade de contaminação que pode ocorrer com o uso do bastão-misturador errado.

Protocolo Palinológico

1. Uma ou mais pastilhas de *Lycopodium* são usadas; adicione uma gota por vez de uma solução de 10% HCl até atingir um total de 5 ml. O ácido irá dissolver a matriz carbonática da pastilha. Tenha cuidado, pois ocorrerá uma reação efervescente; espere até o cessar da efervescência para adicionar o resto de HCl. Adicione o sedimento (geralmente 0,25, 0,5 e 1,0 cm^3). M,C&D. Adicione água M,C&D.

2. Lave com 12 ml de água destilada. M,C&D. Repita a lavagem até que o supernatante esteja claro.

3. Digestão a frio com HF 8 ml de HF concentrado. Deixe por uma noite. M,C&D.

4. Digestão a quente com HF 8 ml de HF concentrado em um banho-maria ou em um bloco aquecedor, a 100°C por 30 minutos. Mexer freqüentemente, C&D. (Esta estapa é necessária somente se a amostra for rica em minerais).[10]

5. Lave com 12 ml de água destilada, M,C&D. Repita a operação 2 ou 3 vezes, ou até que o supernatante esteja claro.

6. Elimine o resíduo do ácido com HCl 8 ml de 10% HCl. M,C&D.

7. Lave com 12 ml de água destilada, M,C&D. Repita 2 ou 3 vezes ou até que o supernatante esteja claro.

   Nota: em amostras muito ricas em ácidos úmicos, inserimos uma digestão cáustica aqui. 8 ml de 10% KOH, banho-maria 5 minutos. M,C&D. Repita a etapa 7.

8. Acidificar com ácido acético, em preparação para a acetólise. 8 ml de ácido acético glacial. M,C&D.

9. Acetólise. 8 ml de mistura de acetólise = 9 partes de anidrido acético (CH_3COCH_3) por volume para uma parte de ácido sulfúrico concentrado (H_2SO_4). Misture esses componentes em uma proveta graduada, mais tarde envolvida em gelo (a mistura se aquece rapidamente devido a uma reação exotérmica). Coloque primeiro o anidrido acético, depois o ácido sulfúrico (lentamente), pois esta é uma reação potencialmente explosiva. Proteja os olhos com óculos de proteção, use luvas de borracha, etc. Adicione os 8 ml dessa mistura cuidadosamente e, lentamente, misture com o bastão. Aqueça em banho-maria ou no bloco aquecedor, até que o líquido se torne escuro (5–20 minutos, dependendo do sedimento). M,C&D.

10. Lave com 8 ml de ácido acético glacial. M,C&D.

11. Lave com água. M,C&D. Repita esta etapa duas vezes.

12. Lave com álcool absoluto (C_2H_5OH). M,C&D.

13. Transfira a amostra para tubos de 5 ml de polipropileno com tampas (recomenda-se tubos da Evergreen).

14. Suspensão em glicerina (glicerol). Adicione 5–6 gotas de glicerina corada com safranina. Misture bem, coloque a etiqueta correspondente a cada tubo de 5 ml, coloque em uma estufa a 60°C, por uma noite, para eliminar a água residual. Tampe e guarde. Mantenha no escuro.[11] Lâminas devem ser mantidas na posição horizontal.

Montagem de Lâminas para a Microscopia

Para montar lâminas para análise microscópica dos grãos de pólen e esporos, assim como de outros palinomorfos, misture bem o resíduo final com um bastão de vidro e transfira uma gota desse resíduo a uma lâmina de vidro. Coloque uma lâminula sob a gota de resíduo e pressione gentilmente. A glicerina forma um círculo que cobre mais da metade da área da lamínula; e suas bordas devem ser preenchidas com parafina. Derreta a parafina em um pequeno béquer. Coloque as lâminas a serem preenchidas com parafina em um placa aquecedora e comece a adicionar a parafina às bordas da lamínula. O calor do prato-aquecedor derreterá a parafina que, automaticamente, preeencherá os espaços vazios, circundando a glicerina com o resíduo a ser analisado. Uma vez removida do aquecedor, a glicerina ficará selada com a solidificação da parafina, e a lâmina deverá ter suas margens seladas por esmalte de unha transparente. Aplique a etiqueta a cada lâmina identificando o nome do local estudado, o número de identificação do testemunho e a profundidade em cm no testemunho do material processado. Lâminas preparadas dessa forma duram muitos anos, e a glicerina não mais absorverá a umidade do ar. O pólen, então, poderá ser girado gentilmente através de um simples toque de uma agulha sobre a lamínula.[12] O texto acima sugere que, uma simples gota do resíduo polínico final, irá fornecer uma concentração densa o suficiente para uma boa contagem. O número ideal de grãos de pólen deve ser algum múltiplo de 10 em cada transecto examinado da lamínula. As preparações dos sedimentos do Lago da Pata, muitas tiveram mais que 100 grãos de pólen por transecto. Se por algum motivo, o número de grãos não for suficiente para a contagem, recomenda-se uma nova extração de pólen dos sedimentos, dessa vez com uma amostra maior dos mesmos. Caso o extrato final não possa ser mais concentrado, o palinólogo terá que enfrentar uma contagem mais vagarosa e mais tediosa ao microscópio.

Meio de Montagem Alternativo

O correto meio de montagem a ser usado é controverso, pois diferentes pesquisadores preferem o óleo de silicone, glicerina ou um meio sólido como o bálsamo do Canadá. Nós usamos glicerina, como descrito acima, preferivelmente corada com safranina. Como a água se mistura por completo, a glicerina é mais fácil de ser usada, pois não é necessário extrair água com uma série de álcoois antes da suspensão do resíduo no óleo de silicone,[13] etapas que podem, potencialmente, implicar na perda de grãos pequenos. Ao mesmo tempo, a glicerina mantém-se fluida sob a lamínula, ao contrário do excelente e rígido meio do bálsamo do Canadá. Um meio fluido permite que o analista gire os grãos durante o exame microscópico, ou os retire debaixo da cobertura de fragmentos orgânicos ou inorgânicos da preparação, através da aplicação de uma pequena pressão na lamínula com uma agulha. Com a diversificada flora neotropical é sempre muito importante ter a capacidade de girar grãos para a visualização de mais de uma de suas vistas.

O óleo de silicone tornou-se favorito entre aqueles que acham possível refinar identificações genéricas, através da estatística de freqüência de tamanho, pois este meio provou ser o melhor em prevenir a alteração do tamanho e formatos de pólen armazenados em uma preparação. A segurança de que o pólen retido em uma lâmina conserva seu tamanho e forma originais, permitiu a separação em grupos subgenéricos de espécies de *Betula, Picea, Pinus* entre outros, que puderam ser separados em grupos sub-genéricos de espécies. Assim a ocorrência de espécies únicas pôde ser comprovada. Esses dados colaboraram, significativamente, com a resolução dos diagramas polínicos do hemisfério norte, com suas listas empobrecidas de espécies. Sem dúvida, a estatística de freqüência de tamanhos um dia será também usada na Amazônia, embora não na atual infância da sua palinologia.[14] Provavelmente seja mais produtivo, atualmente, buscar detalhes no esplêndido registro amazônico de táxons polinizados por animais, cujo pólen é transportado às bacias sedimentares pela água superficial.

CONTAGEM DE PÓLEN E DADOS PALINOLÓGICOS

A contagem de pólen deve ser feita com objetivas, com um aumento mínimo de 40 X. A análise palinológica do Lago da Pata foi feita inteiramente sob lente de imersão a óleo (100X), para assegurar que nenhum grão pequeno fosse ignorado. Convencionalmente, o aumento de 40X é usado rotineiramente nas análises, enquanto que as lentes de imersão são reservadas para a identificação de grãos problemáticos.

A Soma de Pólen

A soma de pólen é um número arbitrário, que estabelece o final da contagem. Em nossos trabalhos recentes na Amazônia estabelecemos esse limite em 500 grãos, mas não fomos precisos. Quinhentos grãos não são suficientes para fornecer uma amostra estatisticamente significativa dos elementos raros, mas é mais do que suficiente para estabelecer a relatividade estatística dos poucos grãos abundantes transportados ao lago pelo vento.

No estudo estatístico de Mark Bush, usando TWINSPAN e análises DCA dos espectros polínicos de superfície (Capítulo 1), a soma escolhida foi 200 grãos de pólen e esporos, o que mostrou ser adequado para a análise multivariada para a detecção de tipos de comunidades. Isto se deve ao fato de que a soma total de pólen é, quase na sua totalidade, constituída de poucos, mas abundantes tipos polínicos, cujo sinal se torna claro em contagens comparativamente baixas. A estatística polínica de porcentagem tem muitas variáveis independentes (todos os táxons contados) e números impraticáveis (dezenas de milhares) devem ser contados para assegurar um significado estatístico para os elementos raros, particularmente aquele presentes em valores menores que 10%. Contudo, os poucos táxons abundantes, responsáveis pela maior parte da variação, se tornam estatisticamente significativos em contagens muito baixas. Uma maneira simples de testar isso é através da representação, em um gráfico, da relação entre % da soma polínica e a soma dos táxons abundantes. Esse gráfico, geralmente, mostrará que existe pouca variação em porcentagem com somas acima de 200 grãos. A expressão matemática dessas realidades são encontradas em Mosimann (1965). Esses testes indicam que somas totais de pólen de, no mínimo, 250 grãos são aceitáveis. Os dados quantitativos fornecidos por essa contagem podem ser amplificados pela procura, nas lâminas, de outros tipos polínicos que seriam interpretados somente como "presentes." Isto deve ser suficiente, especialmente quando os elementos raros são usados somente como fósseis-índices, ou quando as perguntas estabelecidas no projeto de pesquisa possam ser respondidas com uma lista de espécies.

Costumava ser prática comum entre os palinólogos europeus, escolher uma soma arbitrária de pólen arbóreo, e continuar com a contagem de todo o pólen até que essa soma de pólen arbóreo fosse alcançada. Quando a antiga vegetação mudava de floresta para campo de gramíneas ou tundra, isto forçava o resultado absurdo de se contar milhares de grãos em uma zona polínica de ervas, a fim de se alcançar a mesma soma de pólen arbóreo encontrada nas zonas polínicas de floresta do diagrama. Isto foi um empreendimento estatisticamente insensato (veja Mosimann, 1965). A soma de pólen deve incluir todas as plantas vasculares presentes. A única flexibilidade permitida é a remoção da soma total de pólen de táxons que podem, com segurança, ser designados a hábitats especiais. O melhor exemplo disso são as plantas aquáticas. A história palinológica das plantas aquáticas pode ser interessante por si só, mas não como porcentagens flutuantes do número de plantas naquele hábitat tão diferente, a terra. Deve-se também remover da soma, para os propósitos da análise estatística, aqueles táxons que são super-representados. Um exemplo é Gramineae, ou *Cecropia* que foi removida das somas de pólen nas análises das amostras superficiais, realizadas por Mark Bush (Capítulo 1).[15]

Uma preocupação irritante para a consciência do palinólogo pode ser a inclusão de esporos de pteridófitas ou de algas à soma de pólen de plantas superiores, pois ambos os tipos de palinomorfos são, funcionalmente, muito diferentes. Quando o influxo de táxons de pteridófitas ou de algas é da mesma ordem do influxo de tipos polínicos, essa pratica é satisfatória embora inteiramente arbitrária. Nós usamos os palinomorfos de diferentes grupos de plantas vasculares como "equivalentes" (*proxies*) para a presença de populações dessas plantas. As reais funções biológicas dos esporos não importam desde que a relação entre o influxo de esporos e a sua população produtora seja da mesma ordem. A experiência sugere que, na maioria das comunidades neotropicais, essa relação se mantém de fato. Os diagramas de pólen da Amazônia não são distorcidos ou afetados pelos esporos de pteridófitas, os quais estão sempre em baixas porcentagens, apesar da prevalência de pteridófitas em comunidades vegetais amazônicas.

É necessário que o palinólogo mantenha-se alerta ao problema da super-representação de esporos de pteridófitas, particularmente quando este fenômeno é conseqüência de efeitos pós-deposicionais. Os esporos de pteridófitas parecem estar fortemente concentrados em amostras que foram expostas ao intemperismo, presumivelmente como conseqüência da destruição diferencial dos grãos de pólen. Tentativas de se usar a análise palinológica nos famosos estratos sedimentares com fósseis hominídeos de Olduvai Gorge na Africa Oriental não foram produtivas, devido ao enriquecimento das amostras com esporos de pteridófitas durante eventos erosivos subseqüentes. Isto deve funcionar como um aviso pois as porcentagens polínicas originais têm se mostrado tendenciosas pelo intemperismo diferencial. Depósitos marinhos também se tornam enriquecidos com esporos de pteridófitas erodidas e removidas de depósitos mais antigos e diferencialmente preservadas. Isto sugere que é mais aconselhável remover os esporos da soma total polínica em depósitos marinhos.

Esporos de pteridófitas também podem ser grosseiramente super-representados em sedimentos de pântanos onde essas plantas são abundantes, ou em sedimentos lacustres, quando fetos arborescentes vivem nas margens das bacias de sedimentação. Neses casos, eles devem ser removidos da soma estatística como todos os outros táxons que são localmente super-representados.[16] Portanto, pode-se dizer que, embora os esporos de pteridófitas possam ser "proxies" ou equivalentes dessas populações vegetais, eles são aproximadamente comparáveis aos grãos de pólen, mas não são completamente equivalentes e, portanto, devem ser usados com cuidado.

Identificação Dos Tipos e o Uso de Coleçes de Referência

Este manual fornece chaves de identificação e várias microfotografias, como um guia na identificação de táxons polínicos da Amazônia. Nenhuma fotografia, contudo, pode substituir o pólen, de forma que um palinólogo trabalhando com pólen amazônico, terá interesse em montar sua própria coleção de referência. Lâminas de referência são feitas a partir de uma ou mais flores, que são processadas com acetólise e KOH, exatamente como amostras fósseis, embora não haja neste caso, necessidade da eliminação de minerais com HF. A metodologia para processamento de material de herbário está descrito em detalhe em Erdtman (1952) e em Salgado-Labouriau (1973). A única diferença nesta preparação é que a montagem das lâminas é feita, em geral, com gelatina glicerinada.[17] A montagem rígida, por este meio, é aceitável para lâminas de referência, pois estas contém numerosos grãos de pólen da mesma espécie, permitindo a visualização do grão em várias posições.

É absolutamente essencial que, as plantas usadas para a formação de uma palinoteca, sejam precisamente identificadas. A melhor forma de se fazer isso é obter material polínico em herbários, onde a identificação das plantas foi feita por especialistas em taxonomia. Copie a informação das exsicatas em uma etiqueta, que identificará cada lâmina, como por exemplo, o número de referência do herbário da exsicata amostrada e o nome do identificador. Desta forma, será sempre possível conferir com o material original do herbário. Esta informação é particularmente valiosa na troca de lâminas entre palinotecas. As coleções de referências são montadas, principalmente, pela troca de lâminas entre os palinólogos, e os dados obtidos das exsicatas garantem a identificação.[18]

Diagramas Polínicos

Diagramas polínicos padres, como os das Figuras 1.1 e 1.3 são "diagramas polínicos de porcentagem," nos quais cada táxon polínico é representado como porcentagem relativa a todos os outros táxons. Estes diagramas podem ser suplementados por um "diagrama polínico de concentração," obtido a partir da técnica do marcador exótico de pólen. A concentração de cada táxon polínico é expressa como uma função do volume de sedimento e, portanto independente da presença relativa de outros tipos polínicos como o é o diagrama de porcentagem. O aumento da porcentagem do táxon "Y" significa um aumento real da população de "Y," ou a alta porcentagem é nada mais do que uma redução na população do táxon "Z"? A observação de mudanças nas concentrações de "Y" e "Z" podem ajudar a responder questes desse tipo. Mas os diagramas de concentração também apresentam suas próprias dificuldades de interpretação, devido ao fato de seus dados serem também relativos ao volume de água, argila ou matéria orgânica que constituem a matriz sedimentar. A concentração polínica de um táxon "Y" pode se alterar notavelmente de um nível sedimentar a outro, como resultado de mudanças na taxa de sedimentação, mesmo quando a população e a produção polínica de "Y," presente no momento, se mantenha constante. Contudo, diagramas polínicos de porcentagem e concentração, quando lidos em conjunto, permitem que muitas dessas dúvidas sejam esclarecidas.

A única variável verdadeiramente independente da estatística palinológica é o "influxo" polínico, muitas vezes denominado de "taxa de acumulação polínica" (pollen accumulation rate). Esta taxa pode ser calculada a partir da concentração polínica, se a taxa de sedimentação for determinada anteriormente com precisão para cada intervalo de amostragem, pela simples divisão da concentração polínica por cm^3, pelo tempo necessário para a deposição de 1 cm de sedimento. Em sedimentos lacustres, datações mas em intervalos, por exemplo de 1.000 anos, podem produzir taxas de sedimentação para este tipo de pesquisa, mas, obviamente, com um custo relativamente alto. Em depósitos sedimentares de regiões pantanosas, vales fluviais e paisagens semelhantes, as tentativas de se obter dados de influxo polínico não são bem sucedidas, devido às taxas de sedimentação, nessas regiões, serem inconstantes em praticamente toda a coluna sedimentar.

Fazer numerosos cálculos simples, assim como representar os resultados por meio de um gráfico, são tarefas agora realizadas por computadores que usam vários programas disponíveis comercialmente. Normalmente, o tempo requerido para se fazer manualmente todos os cálculos com uma calculadora e representações gráficas dos resultados é da ordem de vários dias. Atualmente, os diagramas são feitos por computadores, usando-se os programas Tilia e TiliaGraph, escritos por Eric Grimm (veja apêndice).

MEDINDO A CHUVA POLÍNICA MODERNA

A chuva polínica moderna pode ser medida basicamente de duas maneiras: com coletores artificiais de pólen, ou com coletores naturais como sedimentos superficiais lacustres e pequenos agrupamentos ou tapetes de briófitas, que chegam a reter muita umidade como se fossem verdadeiras esponjas. Ambos os métodos amostram "facies" diferentes da chuva polínica. Coletores naturais são, geralmente, a melhor escolha para se calibrar os espectros polínicos fósseis com a vegetação ou as variáveis climáticas, mas coletores artificiais de pólen fornecem dados sobre tipos polínicos considerados fósseis-índices ou componentes de comunidades que, raramente, podem ser obtidos de outra forma.

Coletando Pólen

A figura 4.1 mostra partes de um eficaz coletor de pólen que pode ser construído economicamente. A armadilha usada é constituída de uma massa fibrosa que preenche um funil de vidro ou plástico revestido, no seu interior, por com um filtro de microfibras de vidro (Whatman glass microfiber filters GF/D). O funil é mantido em posição vertical depois de ser colado ou amarrado ao "pescoço" de uma pequena garrafa plástica (300 ml ou menos). Essa garrafa também é mantida verticalmente enterrada no chão, até sua metade, ou pode ser amarrada a um poste ou tronco, sempre em posição vertical. Uma tela de náilon recobre a abertura do funil, protegendo-a contra o acúmulo de folhas ou outros fragmentos vegetais. Perfurações laterais na garrafa de plástico permitem a drenagem da água coletada com a precipitação.

O conteúdo polínico acumulado no coletor é trazido pelas águas das chuvas e retido na massa fibrosa embora alguns grãos escapem dessas fibras, sendo retidos posteriormente pelo filtro de fibras de vidro. Ao final de seis meses ou um ano, dependendo do tipo de informação necessária no projeto a ser desenvolvido, os funis são coletados e substuídos por novos. O pólen é, então, extraído tanto da massa fibrosa como do filtro.

O tipo da fibra a ser usada é importante, pois deve permitir a recuperação do pólen. Nossa primeira preferência é uma forma de fibra de acetato de celulose, solúvel em acetona. A dissolução da massa fibrosa na acetona permite recuperar totalmente o resíduo contendo pólen e outras partículas transportadas pelo vento. Infelizmente, é difícil encontrar comercialmente fibras de acetato de celulose. Conseqüentemente, temos usado em nossos coletores fibras de raiom, que são indissolúveis pela maioria dos solventes usados rotineiramente. Contudo a sua textura permite a lavagem e a retirada do pólen de sua massa, usando uma solução 5% de KOH ou NaOH como detergente.

Depois da coleta, coloque a massa de fibra de raiom, que provavelmente está aderida ao filtro de microfibras de vidro, em um béquer, adicione em seguida suficiente quantidade de 5% KOH para cobri-la. Dissolva uma ou mais pastilhas do marcador exótico (esporos de *Lycopodium clavatum*, no caso de nosso trabalho na Amazônia) em algumas gotas de HCl em tubo de ensaio e transfira o seu resíduo, para o béquer contendo o raiom. Misture bem com um bastão de vidro a fibra, para que o pólen e outros palinomorfos se separem das fibras. Deixe por algumas horas, o que permitirá a dissolução do detrito orgânico mais solúvel; misture com o bastão em certos intervalos. A remoção de todo o pólen requer cuidado e uma uma boa lavagem das fibras. Todo esse procedimento pode ser feito também com um saco plástico de boa resistência, ao invés de se usar um béquer. Isso permite que a massa fibrosa possa ser espremida. O líquido, agora contendo pólen e outros palinomorfos, é centrifugado e os palinomorfos sedimentados. O líquido residual é descartado. Nesse tipo de trabalho recomenda-se uma coroa de centrífuga que segure tubos de 50 ml, o que reduziria o tempo necessário para todas as lavagens. Caso contrário, tubos de 15 ml podem ser usados sem problema, aumentando, naturalmente, o tempo de centrifugação. Uma vez terminada a lavagem, o resíduo final é transferido para tubos de 15 ml onde serão realizados os processos químicos para extração do pólen e outros palinomorfos. Algumas vezes, durante a coleta dos funis, percebemos que o filtro se separa da massa fibrosa de raiom. Se isto acontecer, reuna-os no béquer e processe-os juntos.

Quando colocados em florestas com composições vegetais conhecidas, esses coletores recolhem pólen de árvores e plantas polinizadas por animais, situadas entre 20 e 30 m dos coletores, de forma que um grupo de 20 ou mais coletores, distribuídos pela floresta, produzem uma excelente medida do sinal polínico por ela produzido. Pelo fato das espécies locais serem conhecidas, o trabalho de identificacão ao nível palinológico dessa grande biodiversidade é simplificado. Os tipos polínicos encontrados nos coletores indicam que os mesmos são produzidos em quantidades significativas, o que explica seu aparecimento em sedimentos lacustres. O uso de grupos desses coletores, portanto, é a melhor maneira de se calibrar o sinal palinológico de plantas zoófilas.

Os coletores também registram o sinal polínico regional, pois o pólen anemófilo, proveniente de vegetação circundante da floresta, também é precipitado nesses recipientes. Esta parte do sinal polínico pode ser de grande interesse para estudos de variações anuais, mas os dados provenientes desses coletores não são tão úteis para a calibragem do sinal regional em sedimentos lacustres. Isto se deve à natureza de qualquer amostra sedimentar conter uma amálgama de elementos precipitados ao longo de alguns ou vários anos. O volume de subamostras de 0.5 cm^3 usado em uma típica extração palinológica representa a acumulação de, talvez, dezenas de anos de sedimentos depositados, até taxas médias de centenas de anos ou mais, especialmente em sedimentos como os da Lagoa da Pata. A composição porcentual de pólen em extrações palinológicas é, portanto, uma média de muitos anos. Isto deve ser calibrado pela amostra dos sedimentos superficiais que também foi formada em um período de vários anos. A melhor maneira de se fazer isso, é através da amostragem dos próprios sedimentos superficiais lacustres.

Coletores Naturais de Pólen

As melhores amostras superficiais para a calibragem de uma fração regional de chuva polínica são as superficiais da interface sedimento/água de lagos e lagoas. Essas amostras são facilmente obtidas, como foi descrito no capítulo 2. Um dos projetos mais valiosos e pioneiros será uma amostragem palinológica de sedimentos superficiais da interface sedimento/água, direcionada por vários transectos, dentro da bacia amazônica, e a correlação da vegetação local com os parâmetros climáticos. Tudo que um pesquisador necessita para tal trabalho é um barco inflável, um amostrador de sedimentos superficiais, um aparelho de GPS (Global Positioning System), isto é um sistema de posicionamento global para medição precisa das coordenadas dos locais amostrados e um caderno de anotações. Com esse simples equipamento, os palinólogos podem explorar a Amazônia e fazer uma das maiores contribuições ao seu conhecimento. Se necessário, um laboratório pode se encarregar das análises palinológicas das amostras.

Os dados palinológicos percentuais oriundos de uma amostragem planificada com uma rede de amostras, como acima mencionado, permitiriam o cálculo de funções de transferência entre porcentagens polínicas e medidas de precipitação, temperatura e salinidade. Quando isto acontecer o caminho estará aberto para a reconstrução quantitativa dos climas pretéritos da Amazônia, a partir de dados de porcentagem polínica em sedimentos lacustres, que poderão ser comparados com as reconstruções paleoambientais já disponíveis para a Europa e o leste da América do Norte (Huntley & Birks, 1983; Birks & Birks, 1980; Prentice *et al.*, 1991).

Onde não existem lagos e lagoas, outros coletores naturais de pólen devem ser encontrados; as escolhas são poucas. Tapetes ou pequenos "colchões" de briófitas (moss polsters) são os melhores. Estas plantas funcionam como os nossos coletores de pólen, sendo algumas vezes verses naturais de funis cheios de fibras que foram acumuladas por pólen de muitas estações de crescimento ou anos. Na região temperada das florestas de árvores anemófilas da América do Norte, aglomerados de briófitas, assim como nossos coletores, podem produzir uma amostra da precipitação polínica regional, embora os dados estejam naturalmente sujeitos à super-representação de pólen, oriundo de árvores crescendo em proximidades dessas briófitas. Na Amazônia, aglomerados satisfatórios de briófitas não são comuns e, quando podem ser encontrados, são muito inferiores a amostras de sedimentos lacustres superficiais. Mas onde não ocorrem lagos e lagoas, aglomerados de briófitas são a melhor opção.

No caso de ausência de lagos, lagoas e aglomerados de briófitas, o único recurso é amostrar o solo superficial. Este procedimento, inevitavelmente produzirá uma amostra palinológica tendenciosa. O solo, assim como aglomerados de briófitas ou coletores, inevitavelmente acumulam diferencialmente pólen das plantas que estão crescendo nas proximidades da área amostrada e, além disso, amostras de solo sofrem pela desvantagem de terem uma preservação polínica também seletiva. O solo pode ser aeróbico, o que permite a destruição oxidante dos grãos de pólen. Outra desvantagem refere-se ao fato de os solos serem bem providos de uma microfauna saprófoga como alguns aracnídeos, fungos e bactérias que destroem o pólen, esporos e outros palinomorfos. O resultado é que os solos tendem a perder rapidamente os grãos menores e os mais frágeis, tornando-se gradualmente enriquecidos com os grãos de pólen mais resistentes e esporos, principalmente de fungos e de algumas pteridófitas. Em alguns ecossistemas amazônicos, onde o sinal palinológico regional é fornecido por pequenos grãos das famílias Urticaceae, Moraceae, Melastomataceae, por exemplo, a amostragem de solos para análise palinológica é, em geral, insatisfatória. Contudo, essa amostragem tem que ser realizada por causa da ausência de outros tipos de amostras. Recomenda-se coletar pequenas quantidades de solo, ao acaso, de uma área estatisticamente significativa, como por exempo 100 m^2, e homogeneizar a amostra antes da extração dos palinomorfos.

NOTAS

1. Outras referências são: Brown (1960), Erdtman (1969) e Traverse (1988).
2. Com uma maior tendência à interdisciplinaridade, a comunidade científica busca cada vez mais o uso de outros "proxies," fósseis ou indicadores equivalentes para a determinação de mudanças climáticas. Portanto, está se tornando cada vez mais comum, arquivar seções de testemunhos, da mesma forma que os paleooceanógrafos. As seções sedimentares são cortadas ao meio, uma metade é arquivada e a outra usada nas análises. Na América do Norte, um centro de arquivo de sedimentos foi estabelecido pela Fundação Nacional de Ciência dos Estados Unidos (NSF), no Centro de Pesquisas Limnológicas (Limnological Research Center) da Universidade de Minnesota, onde protocolos para estudos de sedimentos lacustres com métodos "multi-proxies" (uso de vários indicadores e fósseis) têm sido desenvolvidos. O endereço é: LRC Core Lab Facility, CME 500 Pillsbury Drive SE, Minneapolis, MN 55455, USA (correio-eletrônico: lrc@maroon.tc.umn.edu).
3. Tabelas de cores de solo de Munsell fornecem retângulos coloridos com os quais a cor do sedimento é comparada. Esses retângulos coloridos são classificados por um sistema de números e letras: um exemplo refere-se às cores marrons encontradas em sedimentos e classificadas como Munsell 10YR 2/2, onde 10YR é o matiz e 2/2 significa um valor 2 para o brilho e outro valor 2 para o "croma," uma medida de saturação da cor. Para o usuário, o significado técnico desses números não é importante; o importante é que esse sistema permite uma referência permanente para a cor, da mesma forma que uma lâmina de referência para o pólen fóssil identificado. As cores do sedimento fresco, são assim anotadas e relatadas. Fontes comerciais para as tabelas de Munsell são dadas no apêndice.
4. A escala de tempo do método do radiocarbono difere significativamente da escala de tempo solar, sobre muitas partes do seu domínio. Isto se deve porque a principal premissa da datação pelo radiocarbono é que a produção de ^{14}C pela radiação solar é constante, mas esta premissa não

se mantém constante em alguns intervalos de tempo. Em partes do Holoceno, datações de radiocarbono divergem das verdadeiras idades por alguns milhares de anos. A escala de tempo do radiocarbono tem sido calibrada por anos solares através da dendrocronologia em regiões temperadas, assim como por outros fenômenos anuais periódicos, por Minze Stuiver e seus colaboradores, na Universidade de Washington, e por outros laboratórios. Uma curva de calibragem baseada neste trabalho é, geralmente, fornecida pelos laboratórios comerciais de datação por radiocarbono.

5. Dados de concentração polínica para os sedimentos da Lagoa da Pata são:
   a. concentração mínima: 88.076 grãos/cm^3
   b. concentração máxima: 778.400 grãos/cm^3
   c. concentração média do testemunho analisado: 247.627 grãos/cm^3
   d. concentração média da seção holocênica: 160.678 grãos/cm^3
   e. concentração média durante o último máximo glacial: 320.086 grãos/cm^3
   f. concentração média para o período glacial antes do último máximo: 249.236 grãos/cm^3
6. A separação de pólen, dos detritos orgânicos tem sido realizada por centrifugação em gradientes de densidade, embora o processo requeira um razoável consumo de tempo para se tornar de uso geral. Gradientes de densidade também podem ser usados para separar pólen em frações de densidade com consistência taxonômica (Forster & Flenley, 1993).
7. Grãos de pólen, relativamente grandes, que podem ser acidentalmente removidos por peneiras finas incluem: *Cayaponia* (Cucurbitaceae) 120 µm, *Manihot* (Euphorbiaceae) 325 µm, *Hibiscus* (Malvaceae) 180 µm, *Pavonia* (Malvaceae) 260 µm, *Dalechampia* (Euphorbiaceae) 130 µm e *Inga* (Mimosaceae) 200 µm (todos valores são de tamanho máximo).
8. O número de pastilhas de *Lycopodium* ou de outros marcadores exóticos a ser usado depende do número de esporos em cada pastilha e da concentração polínica nas amostras de sedimentos. A razão final de esporos exóticos e grãos de pólen contados deve estar dentro de uma faixa de variação de 1:2 a 1:6. No método de Stockmarr (1971), a concentração varia de 12.000 a 15.000 esporos por pastilha. Na análise da Lagoa da Pata usamos uma pastilha em 0.25 cm^3 de sedimentos, mas em algumas amostras pode ser necessário usar mais de uma pastilha. Nas nossas amostragens de chuva polínica na Amazônia, usamos 5 pastilhas por coletor.
9. Microesferas são feitas de polysterene, seu tamanho deve estar dentro de uma faixa de 15 µm, e são usadas em suspensões líquidas (veja suprimentos e materiais, no apêndice).
10. O HF (ácido fluorídrico) é uma substância letal, que deve ser usada com muita precaução. Sempre use o HF dentro de uma capela exaustora e nunca respire os vapores produzidos. Use roupas de proteção, luvas de borracha e proteção para os olhos. Contato prolongado com a pele pode ser extremamente perigoso e mesmo fatal. Um creme contra queimaduras provocadas pelo HF é feito de 1 parte de óxido de magnésio para 1,5 parte de gelatina glicerinada (Moore *et al.*, 1991).
11. A armazenagem de sedimentos (ou lâminas) no escuro evita a descoloração de grãos que pode ocorrer se expostos à luz por muito tempo. Essa descoloração pode tornar os grãos translúcidos e dificultar o exame palinológico. A solução é usar safranina ou outro corante para tornar os grãos mais visíveis.
12. Alguns palinólogos usam meios sólidos para a montagem de suas lâminas, geralmente a gelatina glicerinada. Acreditamos que essa é uma prática não recomendável, pois nessas condições os grãos se tornam imóveis e só podem ser visualizados em uma face. Muitos grãos da diversificada flora amazônica só podem ser identificados quando vistos em várias faces. Por isso, duvidamos de identificações feitas em lâminas onde os grãos estão aprisionados pelo meio sólido. Alguns dos vários táxons que particularmente necessitam de rotação para a sua identificação precisa são: *Copaifera, Caryocar,* Myrtaceae, Sapindaceae, Combretaceae, muitas espécies de Meliaceae, e mesmo algumas Melastomataceae, especialmente quando os grãos estão achatados lateralmente.
13. A série de alcóois usados no método do óleo de silicone são: álcool (96%), álcool (99%), e TBA (álcool terciário butílico).

14. Provavelmente a única exceção é Gramineae. Durante o desenvolvimento do trabalho palinológico, o analista dever anotar o tamanho exato dos maiores grãos de gramíneas encontradas, buscando evidências que permitam a sua identificação como pólen de bambu ou a exclusão dessa possibilidade. Ao mesmo tempo, o palinólogo deve estar atento à possibilidade de que grãos grandes de Gramineae pertençam a *Zea*. A descoberta de grãos de pólen de cereais (80–120 micra) nos sedimentos da Lagoa Ayauchi, na Amazônia do Equador, sugeriu a possibilidade de ocorrência de *Zea* em depósitos datados de 6000 anos antes do presente, o que foi confirmado pelos específicos fitólitos de *Zea*, encontrados nos mesmos depósitos, por D. Piperno (Bush *et al.*, 1989). Seria ideal se as gramíneas de vegetações secas, de massas flutuantes dessas ervas, e de comunidades florestais pudessem ser separadas por estatística de freqüência do tamanho de grãos de pólen. Contudo, a perspectiva dessa possibilidade não é das melhores. Salgado-Labouriau & Rinaldi (1990) verificaram que não é possivel separar os táxons de Gramineae, dos Andes da Venezuela baseando-se somente no tamanho dos grãos.
15. Quando um táxon se encontra super-representado como, por exemplo, Gramineae em regiões de cerrado do Brasil Central, curvas de saturação para os próximos tipos abundantes somente se estabilizam depois que um grande número de grãos do elemento super-representado é contado. Para evitar esse grande investimento de tempo, recomendamos, terminar a contagem dos elementos super-representados quando um valor pré-determinado (se necessário, estabelecido por uma curva de saturação daquele elemento sozinho) for atingido, expressar a porcentagem do elemento super-representado em relação aos elementos já contados, e depois continuar a contagem dos tipos presentes, ignorando o super-representado, até que a contagem estipulada de pólen seja alcançada.
16. Dois são os exemplos: a. o pântano de Cana, na região de floresta tropical do Darien, no Panamá, no qual as pteridófitas herbáceas formam a população vegetal dominante; b. um lago na floresta nebular nas Ilhas Galápagos, que é margeado por fetos arborescentes. Em ambos habitats úmidos, esporos de pteridófitas estavam super-representados nos sedimentos, chegando a 30% da soma total de pólen (Bush & Colinvaux, 1994; Colinvaux & Schofield, 1976).
17. Palinólogos trabalhando com estatísticas do tamanho de grãos, devem fazer a montagem de lâminas em óleo de silicone.
18. Identificações mais precisas de tipos polínicos são possíveis em alguns casos pelo uso de microscópios eletrônicos, principalmente o microscópio eletrônico de varredura (MEV). Esta técnica tem sido usada na identificação de materiais fósseis, essencialmente em casos onde a identificação de espécies é crucial (Traverse, 1988). No entanto, esta tecnologia ainda não foi adaptada para a análise palinológica geral.

CAPÍTULO

5

MORFOLOGIA DE PÓLEN E ESPOROS

Assim como os esporos das pteridófitas e de outras plantas inferiores, os grãos de pólen são o produto da meiose, divisão celular que produz células haplóides de plantas mães diplóides. Os esporos haplóides das pteridófitas são as primeiras células do gametófito, o estágio independente do ciclo de vida dessas plantas, no qual gametas são eventualmente produzidos pela mitose e, unindo-se formam um zigoto que finalmente desenvolve-se em um esporófito diplóide, que é a planta adulta das pteridófitas. Grãos de pólen são homólogos aos esporos das samambaias e outras pteridófitas, e igualmente haplóides, mas a geração do gametófito foi suprimida, permitindo que os grãos de pólen não passem de simples transportadores dos gametas masculinos. Os grãos de pólen dispersam-se e chegam às flores, aderindo-se ao estigma, e germinando um tubo polínico, através do qual o núcleo da gameta se move em direção ao óvulo. Assim, as funções ecológicas dos esporos e pólen são sutilmente diferentes. Ambos representam a divisão redutiva (meiose) necessária para o ciclo de vida sexuado. Os esporos também representam agentes de dispersão. A homologia estrutural entre pólen e esporos, contudo, é evidente e é a base das suas principais diferenças estruturais.

A meiose produz ambos pólen e esporos, em quatro unidades. Essas quatro unidades, bem compactadas entre si, resultam em uma estrutura denominada tétrade. Em algumas famílias botânicas, notadamente a família Ericaceae, os grãos de pólen nunca se separam e são dispersos em tétrades. Contudo, o pólen e os esporos geralmente se separam de suas tétrades e se dispersam individualmente (mônades), ou em alguns casos, em pares (díades), mas os grãos separados, muitas vezes retêm sinais da sua origem em formas tetraédricas. Isto é visto particularmente em muitos esporos de pteridófitas que possuem a marca trilete, sinal do antigo contato dos membros da tétrade, quando estavam juntos.

A PAREDE CELULAR DE PÓLEN E ESPOROS

Históricamente, a estrutura da esporoderma ou parede celular, dos grãos de pólen e esporos, tem várias interpretações. Potonié (1934) e Wodehouse (1935) a dividiram em duas unidades principais: a exina (do Latim *exter*, do lado de fora) e a intina, a camada mais interior da parede celular.

Faegri & Iversen (1950) mais tarde subdividiram a exina em ectexina (camada exterior) e endexina (camada interior), que também foram posteriormente subdivididas por Erdtman (1952), como mostra a Figura 5.1. É a sexina, a camada mais exterior da exina, que contém os elementos esculturais ou ornamentais, usados pelos palinólogos para a identificação dos tipos polínicos.

FORMAS DO GRÃO DE PÓLEN

O grão de pólen pode ser imaginado como um objeto esferoidal, ao qual alguns termos geográficos podem ser aplicados. Um eixo imaginário atravessa o grão de pólen longitudinalmente, conectando ambas as extremidades, denominadas "pólos." A linha equatorial divide o grão em dois "hemisférios."

Classes de formas de grãos de pólen e esporos de Erdtman (1952), baseadas na relação entre o eixo polar (P) e o diâmetro equatorial (E) são dadas na Tabela 5.1 e mostradas na Figura 5.2.

Grãos bisacados das gimnospermas são mostrados na Figura 5.3. Esses grãos possuem duas vesículas preenchidas com um tecido interno esponjoso, que são interpretadas como possíveis estruturas para flutuação na água (Traverse, 1988), ou aumento da sua capacidade aerodinâmica, ou ainda para a manutenção da água durante a desidratação dos grãos (Wodehouse, 1959).

ABERTURAS

A germinação do grão de pólen é facilitada pela presença de aberturas, regiões delgadas da esporoderma, que permitem a saída do tubo polínico. Aberturas podem ser ecto, endo ou mesoaberturas, dependendo de sua posição nas diferentes camadas que formam a exina.

As ectoaberturas são mais superficiais e são localizadas na camada ectexina/sexina, as endoaberturas estão posicionadas na camada endexina/nexina e as mesoaberturas em uma posição intermediária entre essas camadas (Punt *et al.*, 1994). Grãos inaperturados são aqueles desprovidos de aberturas e que possuem camadas de exine, fáceis de serem rompidas.

A determinação dos números, tipos e posições das aberturas é imensamente importante na classificação e identificação palinológica. Os três tipos básicos de aberturas são: poro, colpo e cólporo.

TABELA 5.1. Classes de pólen quanto à forma, baseadas na relação entre o eixo polar (P) e o diâmetro equatorial (E) (Erdtman, 1952)

Forma em vista equatorial	P/E	100 · P/E
Peroblato	<4/8	<50
Oblato	4/8–6/8	50–75
Subesferoidal	6/8–8/6	75–133
Suboblato	6/8–7/8	75–88
Oblato-esferoidal	7/8–8/8	88–100
Prolato-esferoidal	8/8–8/7	100–114
Subprolato	8/7–8/6	114–133
Prolato	8/6–8/4	133–200
Perprolato	>8/4	>200

O colpo é uma invaginação germinal (Erdtman, 1947) ou uma área delgada da exina (Faegri & Iversen, 1950). O poro é uma abertura, de forma arredondada a elíptica, ou simplesmente seções delgadas da exina, enquanto que a abertura composta denominada cólporo é constituída de um ou mais poros (Punt *et al.*, 1994).

A terminologia morfológica empregada para pólen e esporos, baseada no tipo, número e localização das aberturas, é facilitada pelo uso de prefixos. Por exemplo: *mono*, *di* ou *bi*, *tri*, *tetra*, *penta*, e *hexa*, são usados para grãos contendo de uma a seis aberturas respectivamente, e o prefixo *poli*, para grãos com muitas aberturas.

A localização da abertura em um grão é definida pelo uso dos prefixos *zono* (ou *estéfano*) para aberturas ou ornamentações distribuídas em volta da região equatorial e, *peri* (ou *panto*) para aquelas distribuídas por toda a superfície do grão de pólen ou esporo. Em alguns casos como, por exemplo, em Melastomataceae, três cólporos encontram-se intercalados por três pseudocólporos. Este tipo polínico das Melastomataceae recebe o nome de heterocolporado. Os tipos de aberturas, números e posições mais comumente encontrados em grãos neotropicais são mostrados na Figura 5.3.

AS CARACTERÍSTICAS ESCULTURAIS OU ORNAMENTAIS

A sexina é a camada da exina que contém os elementos de escultura ou ornamentação, os quais são procurados pelos palinólogos durante o processo de identificação dos grãos. Uma das características ornamentais que diferenciam grãos de angiospermas dos grãos das gimnospermas é a presença da condição tectado-columelado (Taylor & Taylor, 1993).

O tecto é uma estrutura similar a um teto, apoiado por elementos colunares e cilíndricos, chamados columelas. Outras diferenças básicas entre o pólen das angiospermas e das gimnospermas são: a). prevalência de elementos esculturais positivos na exina das angiospermas; b). presença nas gimnospermas de um tecido alveolar esponjoso na camada exterior da exina; c). a camada mais interior da exina nas gimnospermas é laminar, enquanto que nas angiospermas, a endexina é homogênea (Traverse, 1988).

Muitos grãos de pólen possuem um mínimo de ornamentação, condição que é chamada psilada. A grande maioria das ornamentações podem ser negativas ou positivas. Ornamentação negativa é aquela onde as estruturas se encontram na parte interior da exina, em oposição às estruturas positivas, que estão localizadas sobre a superfície da parede celular. Entre as categorias negativas de ornamentação encontram-se: a). **perforada**, no qual o teto, microperforado, possui pequenas cavidades menores que 1 μm de diâmetro; b). **foveolada**, ornamentação caracterizada por cavidades ou depressões maiores que 1 μm; c). **fossulada**, caracterizada pela presença de sulcos alongados e irregulares.

Categorias positivas de ornamentação são: a). **escabrada**, quando o padrão de ornamentação é composto de grânulos de qualquer formato, mas com diâmetros menores que 1 μm; b). **gemada**, ornamentação com estruturas arredondadas (gemas); c). **clavada**, ornamentação com elementos esculturais alongados com diâmetro da parte superior maior que o da base, denominados clavas; d). **verrucada**, quando a ornamentação é composta de elementos não ponteagudos, semelhantes a verrugas, geralmente com diâmetros maiores que 1 μm; e). **espinhosa**, ornamentação composta de espinhos maiores que 1 μm; f). **rugulada**, ornamentação constituída de elementos alongados, produzindo um padrão escultural intermediário entre a condição estriada e a reticulada (Iversen & Troels-Smith, 1950); g). **estriada**, ornamentação onde existem sulcos separando elementos esculturais salientes; h). **reticulada**, ornamentação em forma de malha, constituída de cavidades (lúmens) cercadas de elementos mais estreitos que os lúmens (Punt *et al.*, 1994; Kremp, 1965). As diferentes categorias de ornamentação são mostradas na Figura 5.4.

MORFOLOGIA DE ESPOROS

Três tipos morfológicos básicos são encontrados nos esporos das plantas pteridófitas: trilete, monolete e alete (Figura 5.5). O esporo trilete tem uma abertura em forma de letra "Y," ou seja uma lesão (laesura) trífida ou triradiada (Erdtman, 1943), enquanto que o monolete possui uma lesão alongada e estreita; o esporo alete é desprovido de tal estrutura.

A condição trilete apareceu no Siluriano inferior (Hoffmeister, 1959), apoiando a evidência paleobotânica que reconhece o esporo alete como derivado do monolete e este último da condição trilete (Richardson, 1969; Chaloner, 1970). Embora não tão ornamentados quanto o pólen, os esporos podem conter padrões similares de ornamentação, permitindo assim o uso da mesma terminologia (Traverse, 1988).

Importantes características da morfologia dos esporos encontradas em táxons neotropicais são: a). lesão trífida ou trirradiada como no esporo trilete de *Lycopodium*; b). lesão única e alongada, como no esporo monolete de *Blechnum*; c). cíngulo, ou uma extensão equatorial também denominada de *zona* como, por exemplo, em *Dicksonia*; d). perina ou perispório, uma membrana constituída de esporopolenina que envolve alguns esporos (Jackson, 1928) e que, na maioria dos casos, é destruída pela fossilização ou pela acetólise.

POLARIDADE NOS ESPOROS

Em sedimentos quaternários, esporos são geralmente encontrados dissociados da tétrade original. Face proximal em um esporo é sempre aquela onde a lesão é encontrada. A face distal, em alguns esporos pode conter elementos de ornamentação como em várias espécies de *Lycopodium*. *Cyathea*, um importante gênero de fetos arborescentes encontrado com certa freqüência em sedimentos quaternários da Amazônia, tem ornamentação que varia de psilada a verrugada em ambas as faces, enquanto que *Anthocerus*, um importante gênero de briófitas, tem uma ornamentação espiculosa em ambas as faces.

APPENDIX: SOURCES OF SUPPLY

Most of the supplies and equipment needed for a program of core drilling and pollen analysis are routine and obtainable from any laboratory supply house or camping equipment company. The coring equipment itself, and some specialist needs of pollen analysts, however, are more difficult to obtain. We list only the suppliers that we ourselves routinely use, noting that other first-rate suppliers of many items (e.g., rubber boats, radiocarbon dates) exist. Our list is offered only for the purpose of helping research teams beginning their studies in Amazonian paleoecology.

SPECIALIST SUPPLIES FOR POLLEN ANALYSIS

Tablets of Exotic Spores (*Lycopodium*) or Pollen

Thomas Persson
Department of Quaternary Geology
University of Lund
Tornavagen 13
S-22363
Lund, Sweden
Internet: **http://www.geol.lu.se/kvg/ehome.xtm**

Polystyrene Microspheres

DuPont New England Nuclear Research Products
Customer Services
549 Albany Street
Boston, MA 02118, USA
Phone: 1-800-551-2121

Synthetic Cloths (Spectra/Mesh)

Spectrum
1100 Rankin Road
Houston, TX 77073-4716, USA
Phone: 1-800-634-3300 (toll-free U.S. & Canada)
Phone: 1-713-443-2900 (all other areas)
Fax: 1-713-443-3100

Pollen Diagram and Statistics Software (Tilia)

Eric C. Grimm
Illinois State Museum
Research and Collections Center
1011 East Ash Street
Springfield, IL 62703, USA
email: **grimm@museum.state.il.us**
Internet: **http://www.ngdc.noaa.gov/paleo/tilia.html**

Statistics Packages (DCA)

Microcomputer Power
111 Clover Lane
Department C8
Ithaca, NY 14850, USA
Attention: Richard E. Furnas
email: **76556.3444@compuserve.com**

Global Pollen Database

This database is rapidly becoming a source for pollen statistics from all parts of the world. Its value is enhanced if all raw pollen data are submitted to the database when pollen diagrams and conclusions are published. The database also accepts unpublished pollen data.

World Data Center-A for Paleoclimatology
Department 955, E/GC
Boulder, CO 80303, USA
Attention: Mildred England
Internet: **http://www.ngdc.noaa.gov/paleo/gpd.html**

Radiocarbon Dating

Beta Analytic Inc.
University Branch
4985 SW 74 Court
Miami, FL 33155, USA
email: **beta@analytic.win.net**
Internet: **http://www.win.net/~analytic**

Munsell Soil Color Charts

Macbeth Division
Kollmorgen Instrument Corporation
Munsell Color
P.O. Box 230
Newburgh, NY 12551, USA

SPECIALIST FIELD EQUIPMENT

Coring Equipment

Our coring equipment has been developed from original designs of Professor Daniel A. Livingstone, who introduced piston coring into studies of lake sediments. The version we use was designed by V. Vohnout and P.A. Colinvaux, and can be made to order by:

GeoCore
3448 Trabue Road
Columbus, OH 43204, USA
Attention: Vincent Vohnout
email: **vohnout.1@osu.edu**

Raft Units

Any good quality rubber boats of suitable size (see text) can be used. We use the REDSTART model made by Avon in Britain.

Avon Inflatables Limited
Dafen, Llanelli
Dyfed SA14 8NA United Kingdom
email: **avon@celtic.co.uk**
email (North America): **info@avonmarine.com**
Internet: **http://www.avonmarine.com/contact.htm**

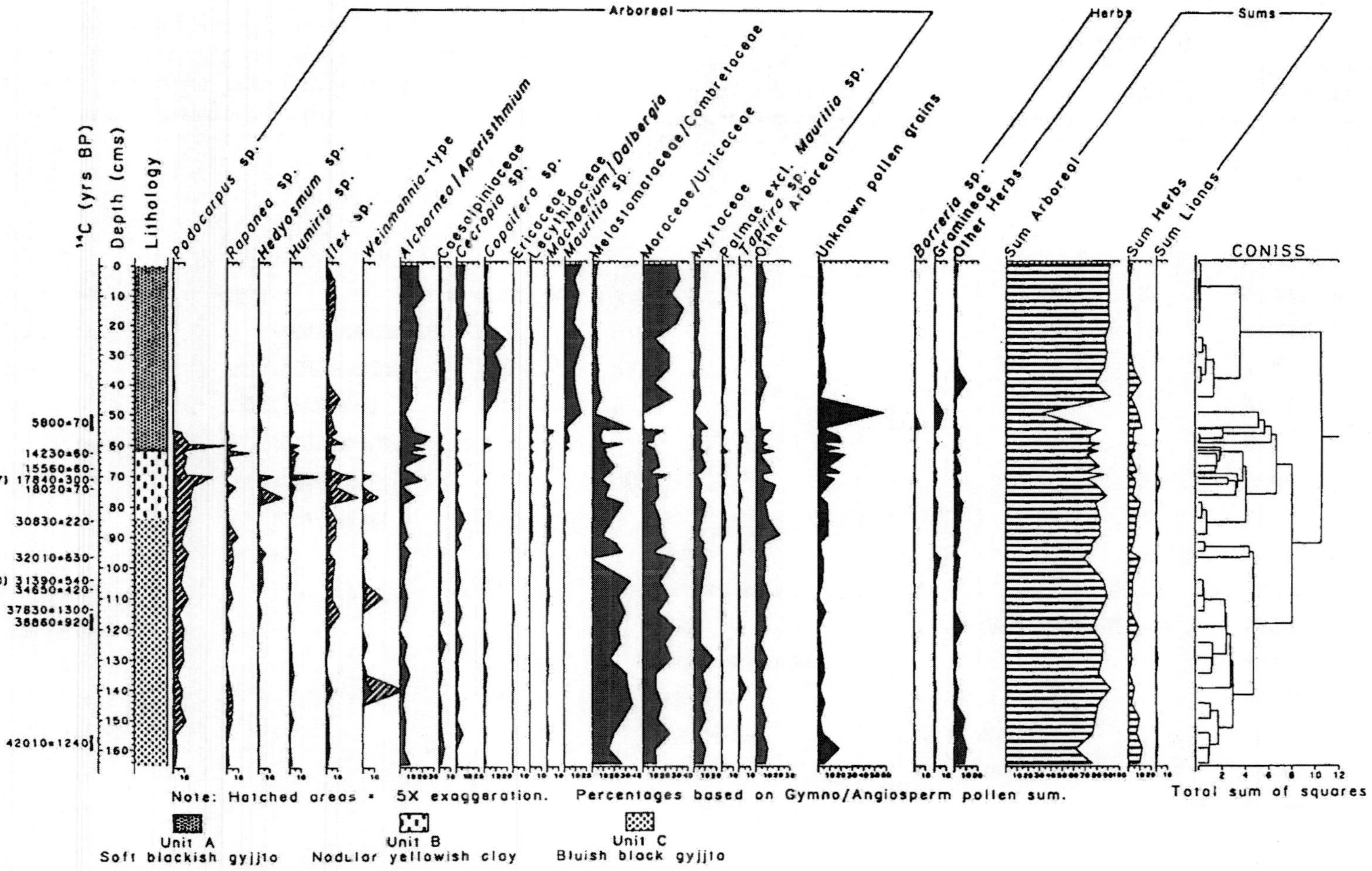

Figure 1.1
Lake Pata pollen percentage diagram. The site is at 300 m elevation in the lowland Amazon forest. The pollen record is one of lowland tropical rain forest throughout the 40,000 years spanned by the diagram. (From Colinvaux *et al.*, 1996a.)

Figura 1.1
Diagrama polínico do Lago da Pata. O lago está a uma elevação de apenas 300 m nas terras baixas da Bacia Amazônica. O registro palinológico é indicativo de floresta tropical das terras baixas amazônicas durante os últimos 40.000 anos (veja Colinvaux *et al.*, 1996a).

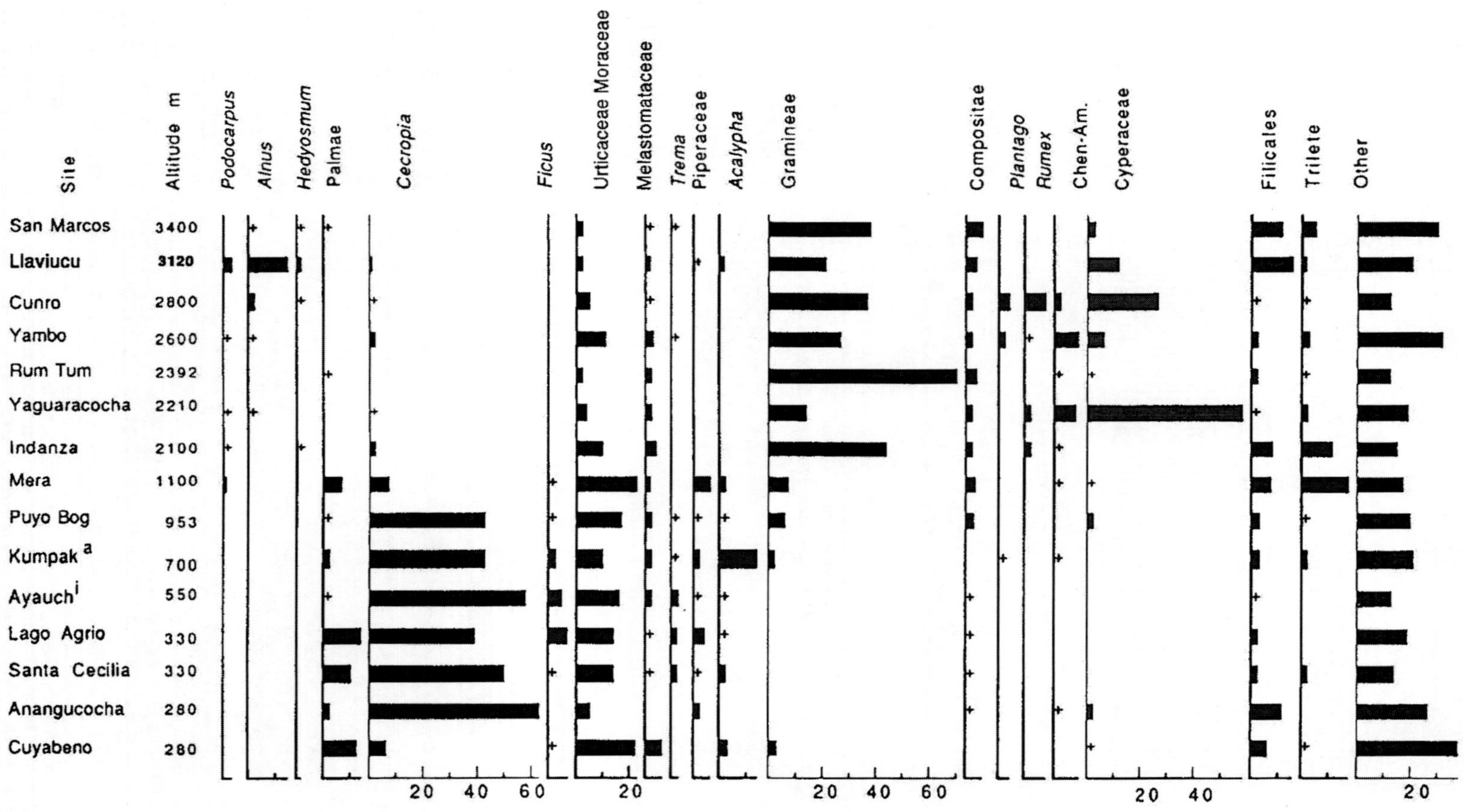

Figure 1.2
An altitudinal gradient of surface pollen spectra from the eastern flank of the Andes in Ecuador. The elevational limit of tropical rain forest in Ecuador is about 1200 m, which can be seen as a major boundary in the pollen diagram. (From Bush, 1991.)

Figura 1.2
Espectros polínicos de superfícies, ao longo de um gradiente de elevação, no flanco leste das montanhas dos Andes do Equador. O limite superior da floresta tropical do Equador é de 1200 m, que pode ser identificado como um notável limite no diagrama polínico (veja Bush, 1991).

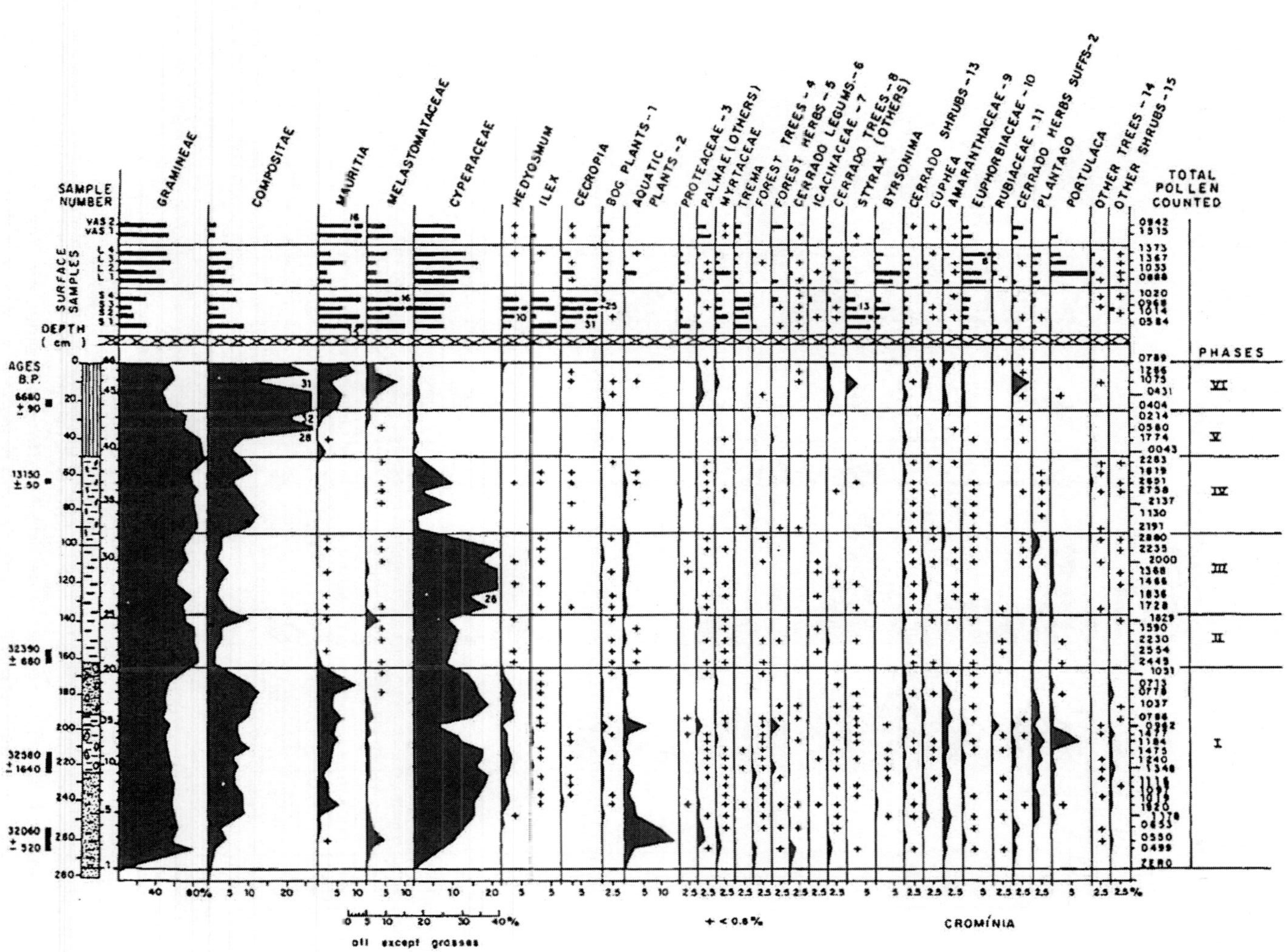

Figure 1.3

Pollen percentage diagram from a palm swamp in savanna vegetation on the Central Brazilian plateau. The diagram shows that the pollen rain from savanna is dominated (60–80% in most samples) by Gramineae pollen, with other herbs and the *Mauritia* palms growing actually on the swamp making up nearly all of the rest of the pollen. These pollen spectra are utterly different from the forest spectra of Figure 1.1. (From Ferraz-Vicentini and Salgado-Labouriau, 1996.)

Figura 1.3

Diagrama polínico de porcentagem de uma vereda, dentro da área nuclear do cerrado, no Planalto Central do Brasil. O diagrama mostra que a chuva polínica das savanas é dominada (60–80% na maioria das amostras) pelo pólen de Gramineae. Pólen de palmeiras de *Mauritia* e de ervas que crescem em ambiente de vereda perfazem o resto dos espectros polínicos. Esses são nítidamente distintos daqueles de floresta tropical da Figura 1.1 (veja Ferraz-Vicentini & Salgado-Labouriau, 1996).

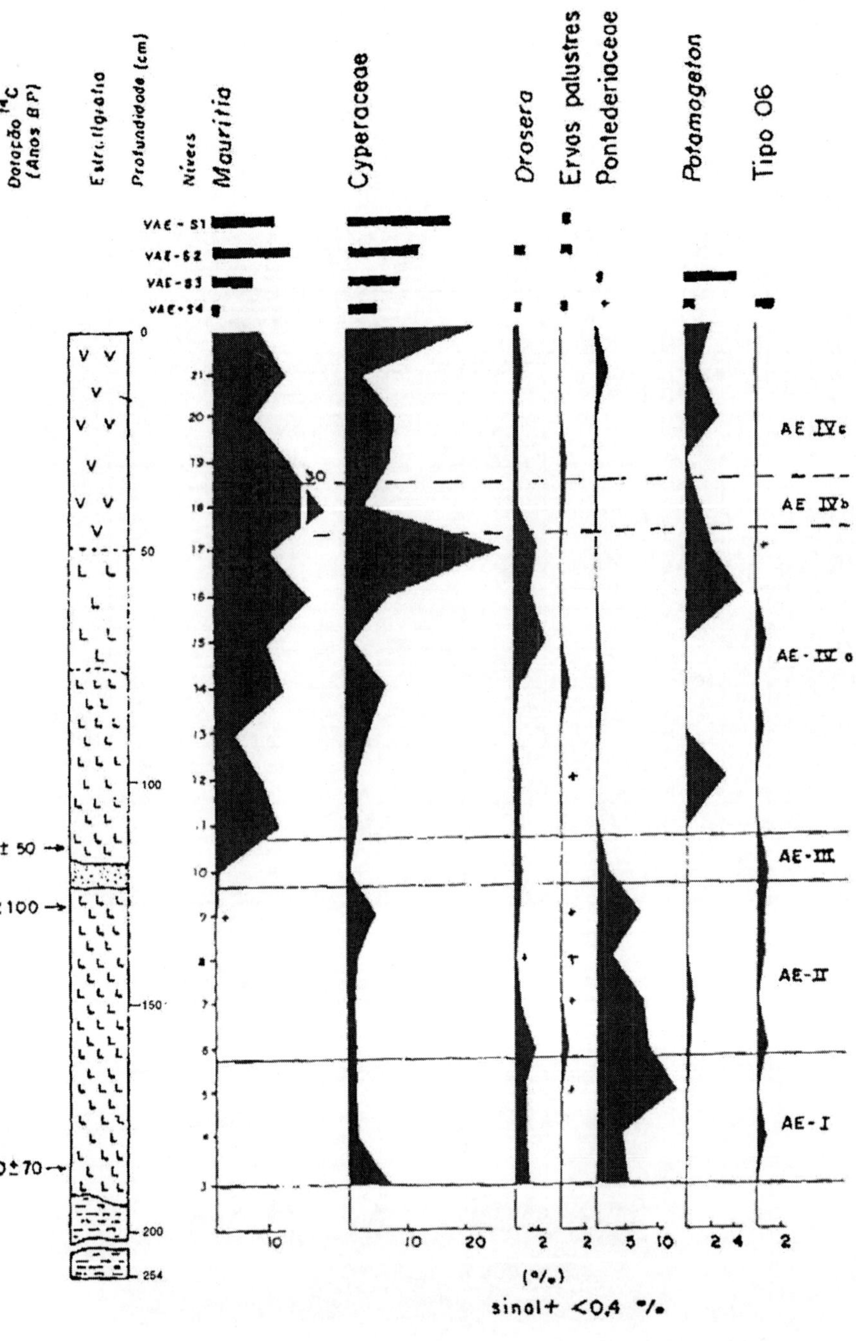

Figure 1.4
Partial pollen percentage diagram from a central Brazilian marsh showing the history of aquatic plants. The marsh palm *Mauritia* occurs only in the Holocene, although the marsh was present in glacial times also, to 25,000 years ago. The spread of *Mauritia* in the Holocene might be a response to the warming of postglacial times. (From Barberi-Ribeiro, 1994.)

Figura 1.4
Diagrama polínico parcial de porcentagem de uma vereda do Brasil Central, mostrando a história de plantas aquáticas. A palmeira *Mauritia* ocorre somente no Holoceno, embora a vereda estivesse presente em tempos glaciais há 25.000 anos. A expansão de *Mauritia* no Holoceno pode ser atribuída ao aquecimento ambiental pós-glacial (veja Barberi-Ribeiro, 1994).

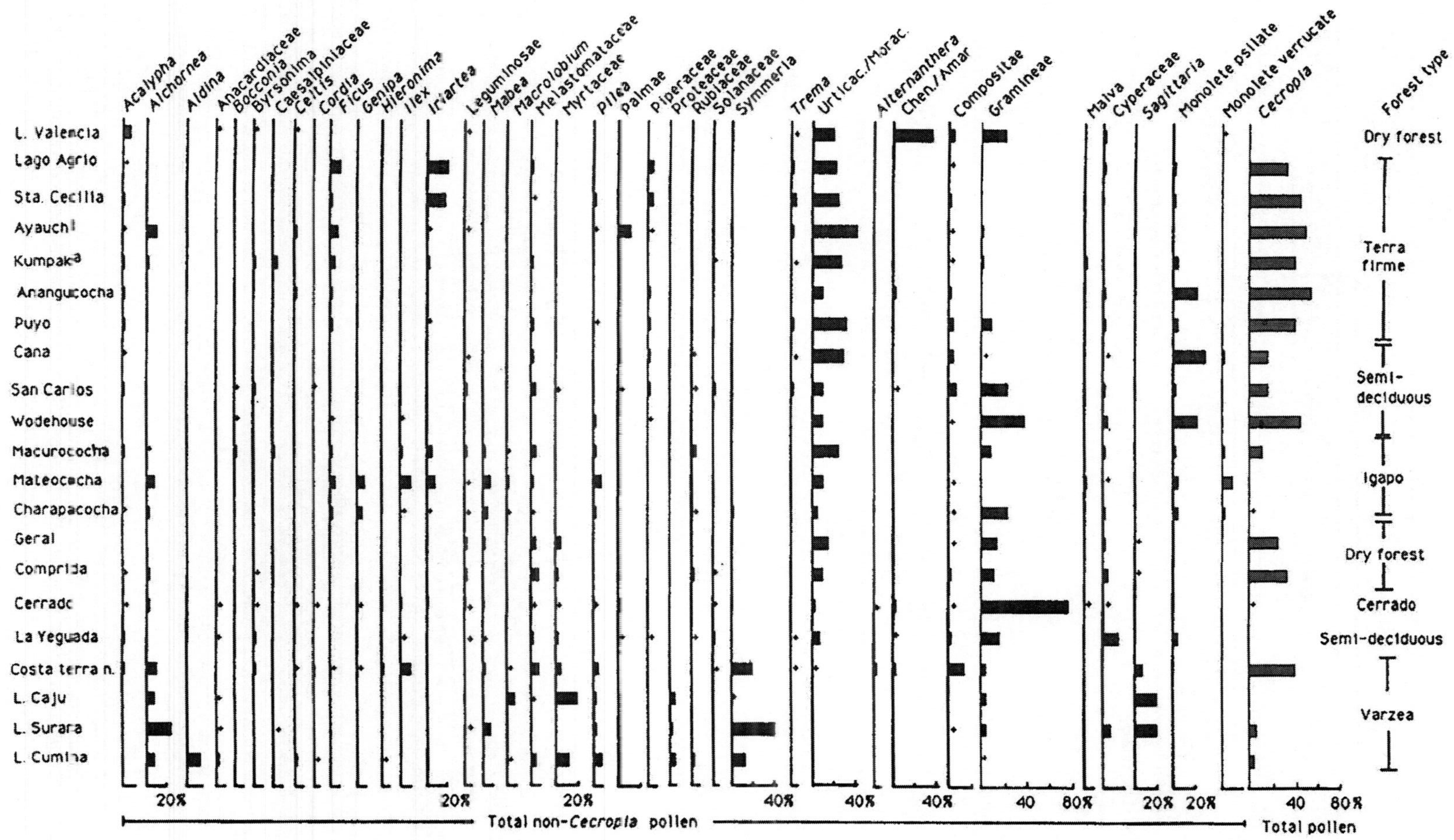

Figure 1.5
Pollen percentage spectra from surface samples of various vegetation types in the neotropical lowlands. The larger components of these pollen spectra are of a few wind-pollinated taxa, but multivariat analyses allow clear separations of the vegetation types from the pollen data, as shown in Figures 1.6 and 1.7. (From Bush, 1991.)

Figura 1.5
Espectros polínicos de porcentagem de amostras superficiais de vários tipos de vegetação de terras baixas neotropicais. Os componentes mais expressivos desses espectros pertencem a poucos táxons anemófilos, mas análises multivariadas permitem uma nítida separação dos tipos de vegetação a partir de dados palinológicos, como é mostrado nas Figuras 1.6 e 1.7 (veja Bush 1991).

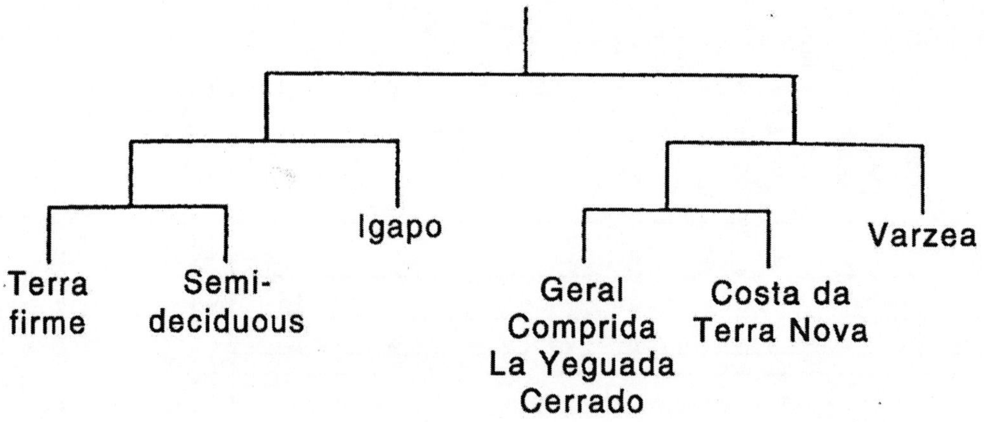

Figure 1.6
TWINSPAN classification of pollen spectra in Figure 1.5. (From Bush, 1991.)

Figura 1.6
Classificação pelo TWINSPAN dos espectros polínicos da Figura 1.5 (veja Bush, 1991).

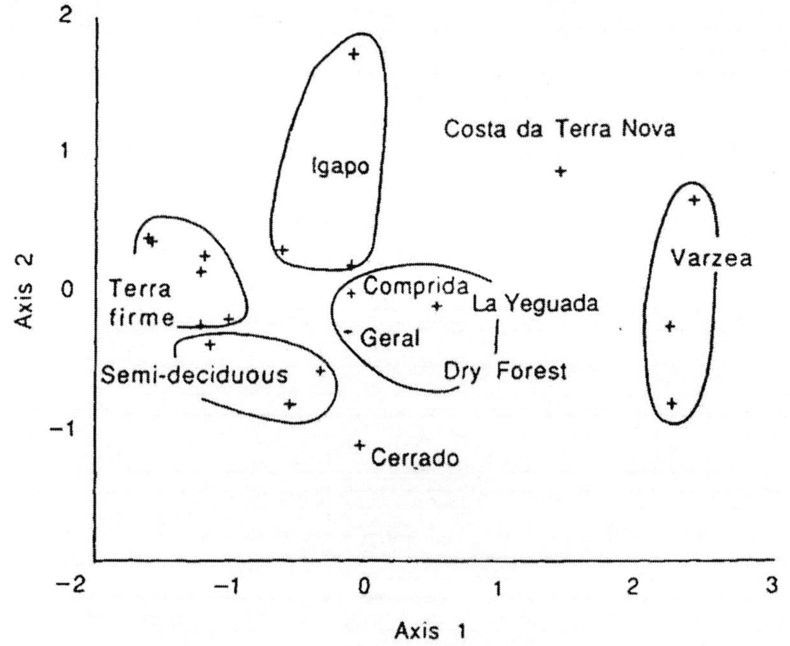

Figure 1.7
DCA ordination of pollen spectra in Figure 1.5. Samples are grouped according to the TWINSPAN classification of Figure 1.6. (From Bush, 1991.)

Figura 1.7
Ordenação DCA de espectros polínicos da Figura 1.5. As amostras são agrupadas de acordo com a classificação TWINSPAN da Figura 1.6 (veja Bush, 1991).

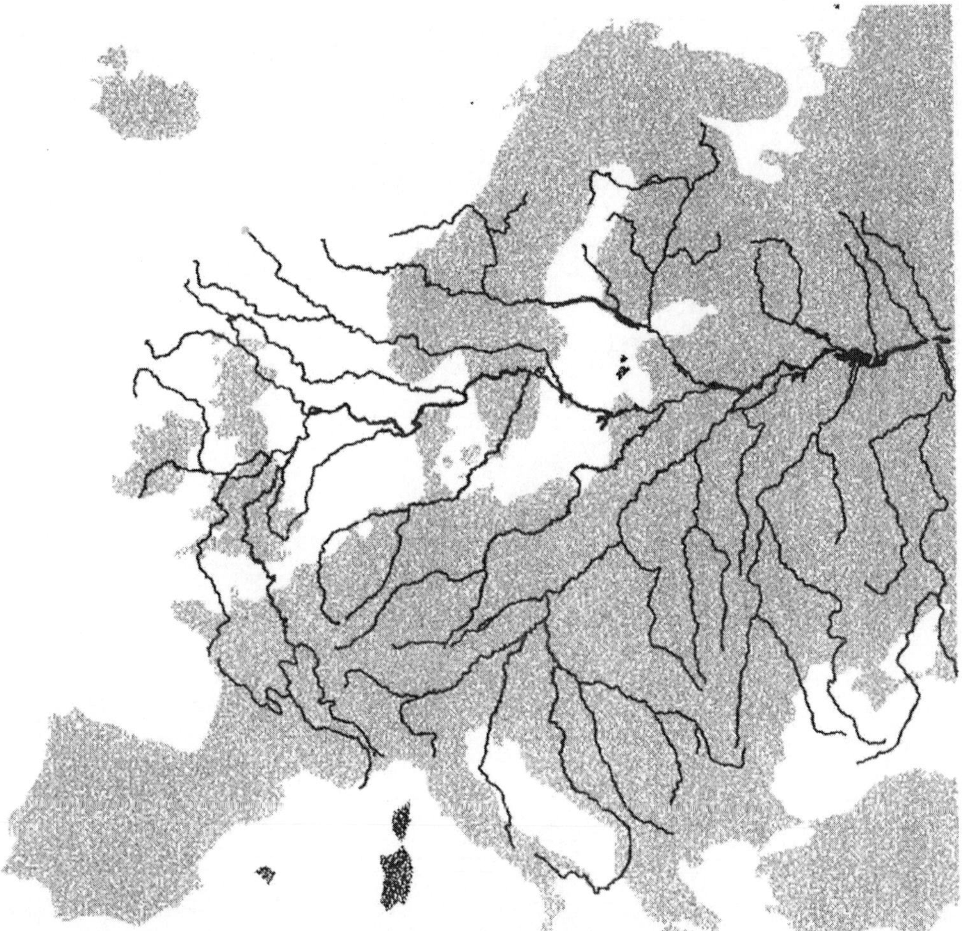

Figure 2.1
The Amazon drainage system superimposed on a map of Europe drawn to the same scale.

Figura 2.1
O sistema de drenagem do rio Amazonas superposto num mapa da Europa.

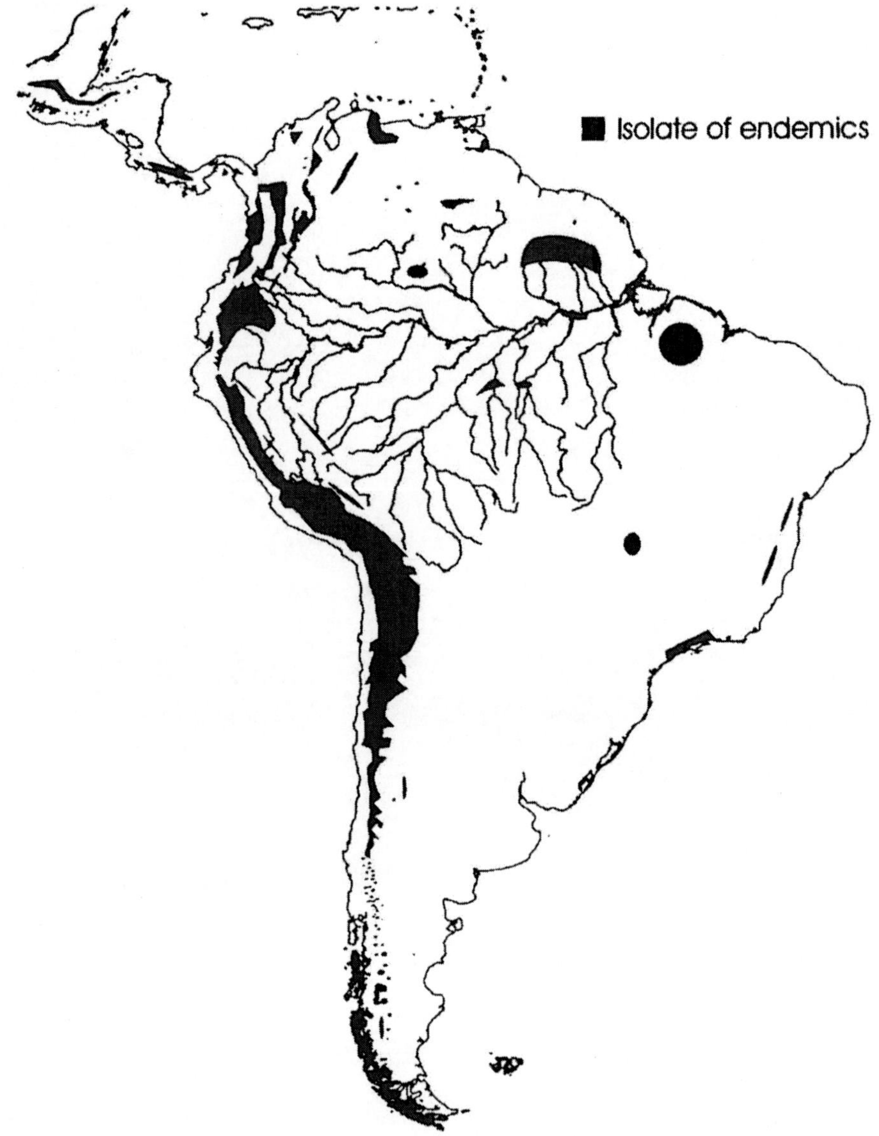

Figure 2.2
Endemic bird distribution in the Amazon basin. Each region of endemicity was postulated by Haffer (1969) to have been a forest refuge in ice-age time, each isolated from the others by a "sea" of savanna. Our interpretation of the pollen data is that the postulated savanna sea did not exist, the forest being continuous throughout glacial cycles. We postulate instead either that the endemic isolates had their origins much earlier than the Pleistocene, or that the endemic isolates record areas of maximum environmental change in glacial times because most were affected by lowered temperature and carbon dioxide. Haffer's model is thus reversed, with the isolates being regions of disturbance in an unchanging "sea" of forest.

Figura 2.2
Distribuição endêmica de pássaros na Bacia Amazônica. Cada região de endemismo foi postulada por Haffer (1969) como um refúgio florestal em épocas glaciais. Nossa interpretação dos dados palinológicos é de que o "mar de savanas" postulado não existiu e que a floresta manteve-se através dos ciclos glaciais. Postulamos, em vez disso, que as áreas endêmicas isoladas tiveram suas origens antes do Pleistoceno, ou que elas registram áreas de mudança ambiental máxima em tempos glaciais, pois a sua maioria foi afetada pela redução da temperatura ambiental e do gás carbônico. Portanto, o modelo de Haffer é invertido sendo os refúgios, áreas de perturbação em um "mar" intacto de floresta.

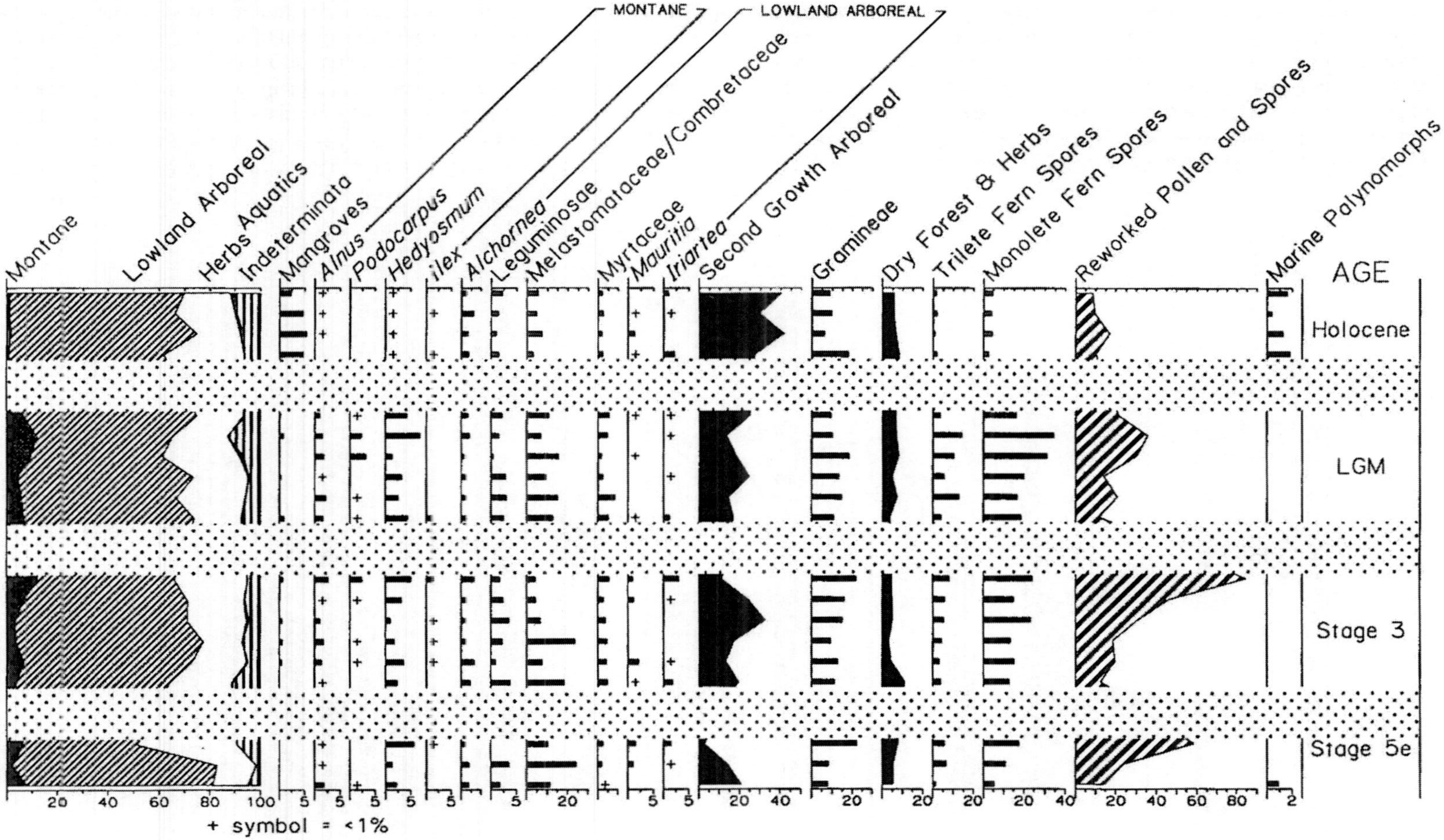

Figure 2.3
Pollen diagram from the Amazon fan and shelf. At no time in a glacial cycle was there significantly more grass (Gramineae) pollen coming down the Amazon system than in postglacial times. This is taken as evidence that the lowlands were never covered by savanna (Haberle, 1997).

Figura 2.3
Diagrama polínico do leque aluvial e da plataforma continental do Rio Amazonas. Em nenhuma época do último ciclo glacial houve mais pólen de gramíneas (Gramineae) descendo o sistema fluvial amazônico do que em épocas pós-glaciais. Isto é uma evidência de que as terras baixas da Amazônia nunca foram cobertas por savanas (veja Heberle, 1997).

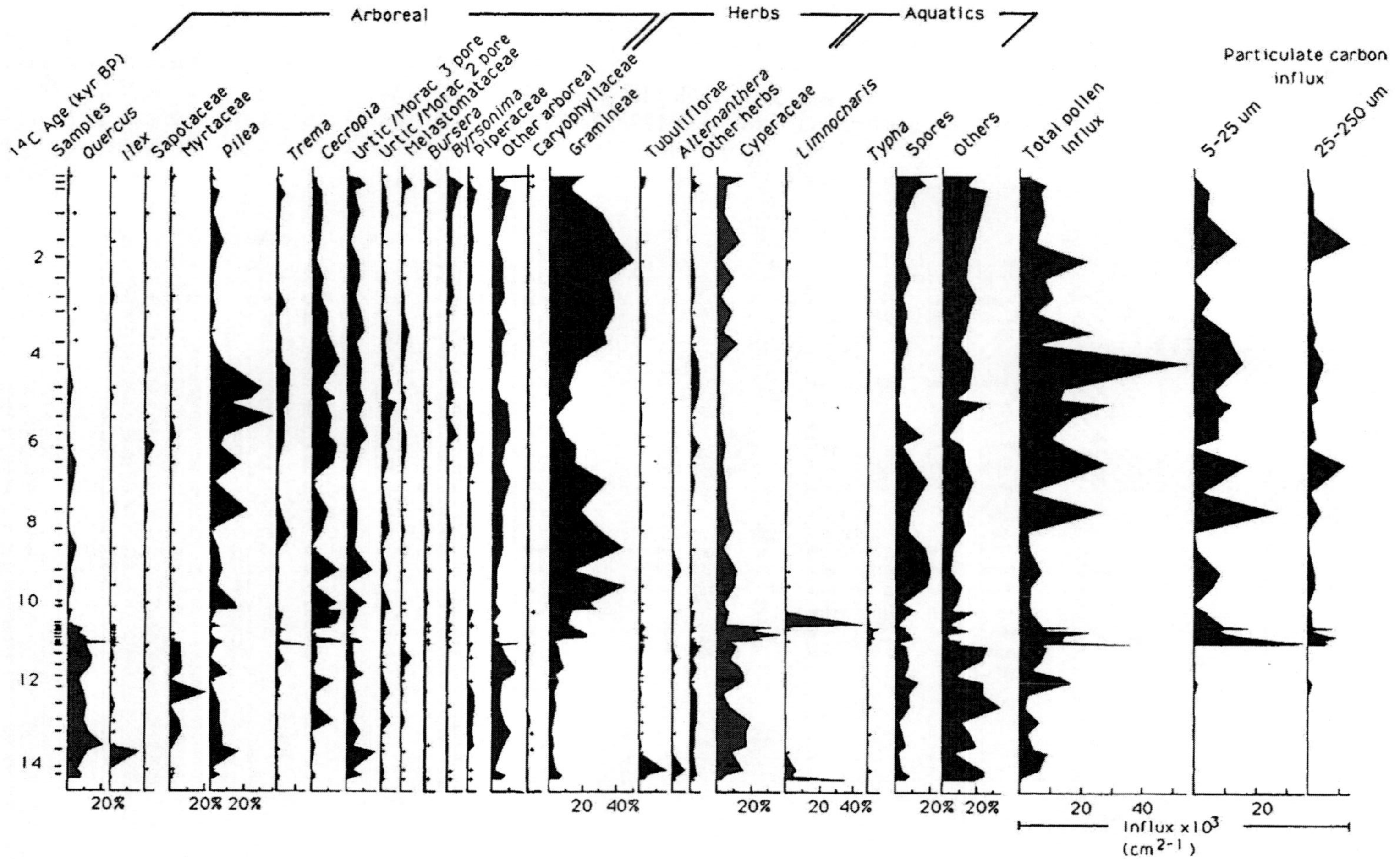

igure 2.4
ollen diagram from the Pacific coast of Panama. The record from Lake La eguada spans 14,000 years, thus beginning in late-glacial times. At all ages in the diagram can be seen the signal of tropical forest in the pollen f Moraceae, Urticaceae, etc. But in the glacial portion of the record an dmixture of *Quercus* and other pollen demonstrates invasion of the low-nd forest by plants from higher altitudes, which we take as a signal for ooling. The record is not one of replacement of one vegetation type by nother but rather of a reassortment of species in different environmental onditions. There is a rise in Gramineae pollen after 11,000 years (Bush *et al.*, 992).

Figura 2.4
Diagrama polínico da costa do Pacífico do Panamá. O registro do Lago La Yeguada contém 14.000 anos de história, começando, portanto, em tempos glaciais tardios. Em todos os setores do diagrama pode ser visto o sinal de floresta tropical fornecido pelo pólen Urticaceae/Moraceae, etc. Mas na porção glacial do registro, uma mistura de pólen de *Quercus* e outros táxons demonstra a invasão da floresta das terras baixas por plantas das altas elevações, que associamos com o sinal de esfriamento ambiental. Este registro, não mostra uma substituição de um tipo de vegetação por outro, mas sim de redistribuição de espécies em condições ambientais diferentes. Há um aumento no pólen de Gramineae depois de 11.000 anos AP (Bush *et al.*, 1992).

A

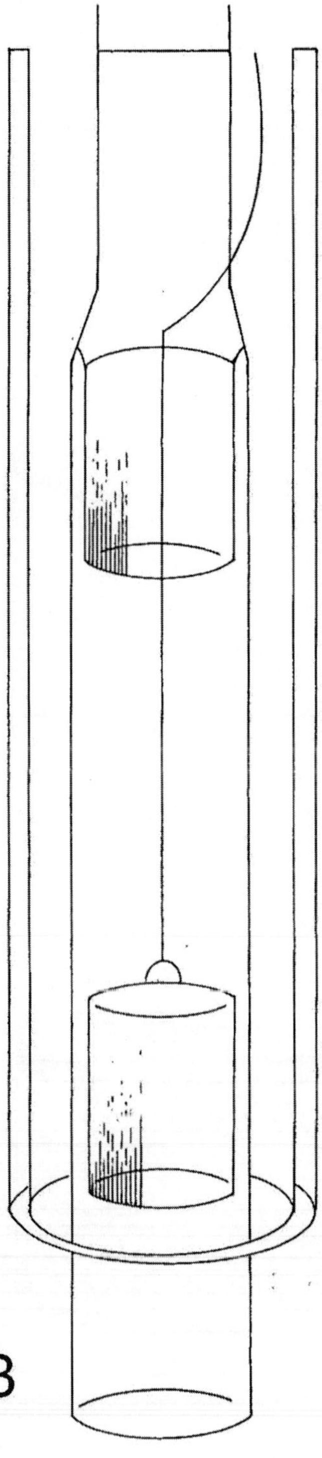

B

Figure 3.1
Piston sampler (B) in use from raft of rubber boat (A).

Figura 3.1
Amostrador de pistão (B) em uso em uma plataforma sobre barco inflável (A).

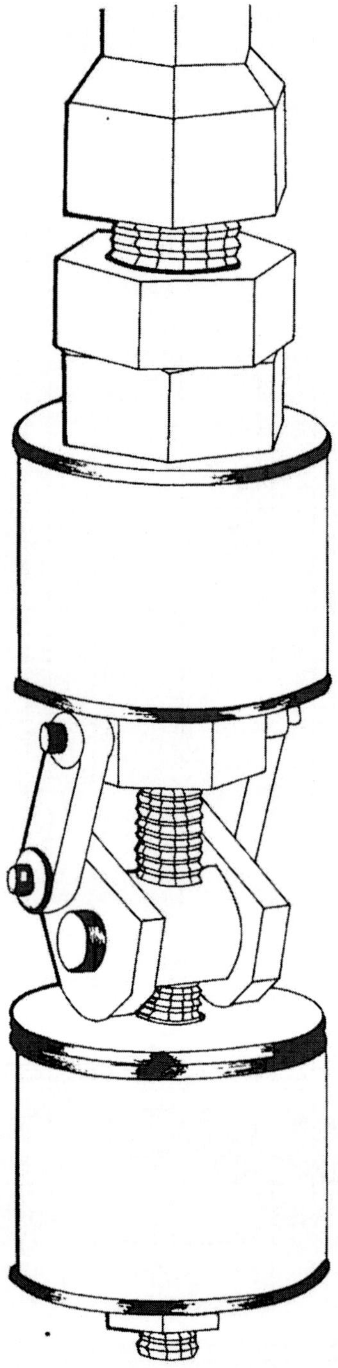

Figure 3.2
Vohnout locking piston. Original design, with the cam in the cocked position.

Figura 3.2
Pistão travável Vohnout. Desenho original do pistão onde o came está na posição de disparo.

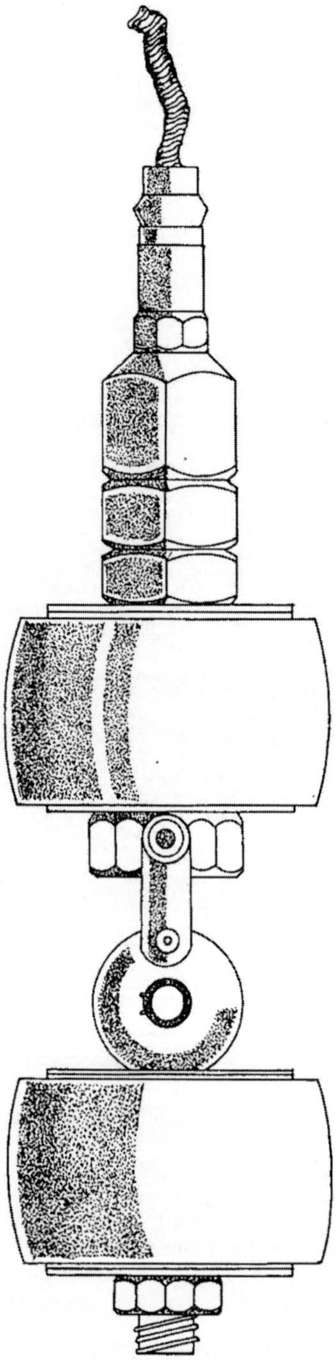

Figure 3.3
Vohnut locking piston. Modified cam design, with cam in the uncocked position (piston free to move).

Figura 3.3
Pistão travável Vohnout. Desenho do came modificado, que está em posição para se mover livremente dentro do tubo amostrador.

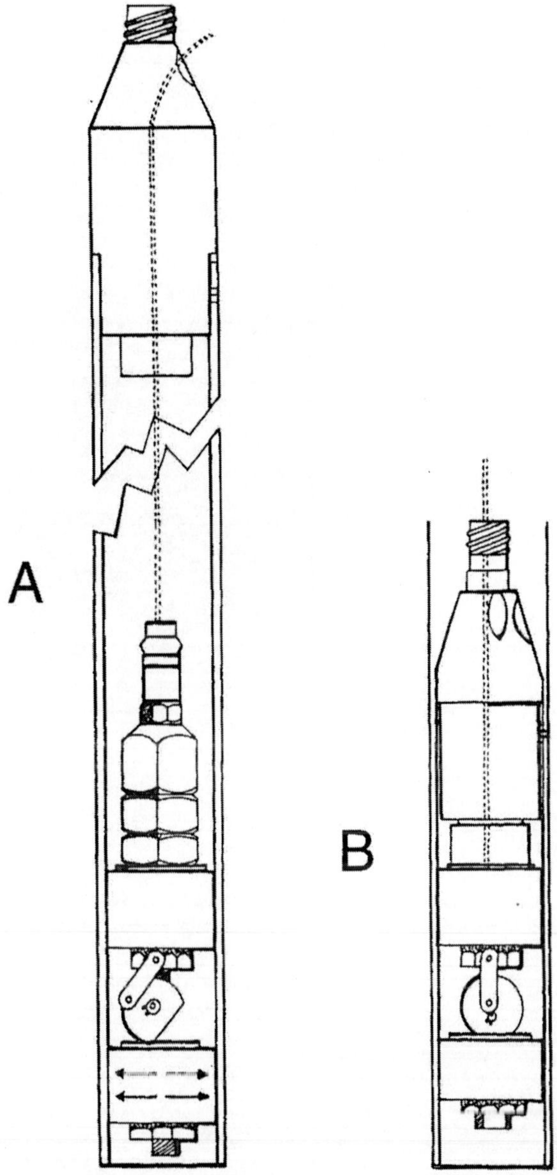

Figure 3.4
Piston at start and end of the drive. In the locked position the leading rubber seal is squeezed between the round face of the cam and the bottom nut and exerts pressure against the inside of the sample tube walls, as shown by arrows (a). In the unlocked position (b), the cam was rotated by the pull on the piston cable, to present its cutaway flat surface to the rubber seal. The seal is no longer squeezed against the sample tube walls, letting the sample tube be pushed down past it. The piston in (b) is shown at the end of the drive when locked into the holder.

Figura 3.4
Pistão no começo e no final do tubo amostrador. Na posição de "disparo," a face convexa do came é pressionada contra a porca de metal que segura os anéis de borracha. Estes são pressionados contra a parede do tubo amostrador (A). Na posição livre (B), um forte repuxo no cabo do pistão gira o came e a borracha não mais se encontra pressionada contra a parede do tubo, deixando o tubo ser empurrado contra o sedimento. O pistão em B está agora na parte superior do tubo e acoplado ao seu segurador, no final de uma coleta.

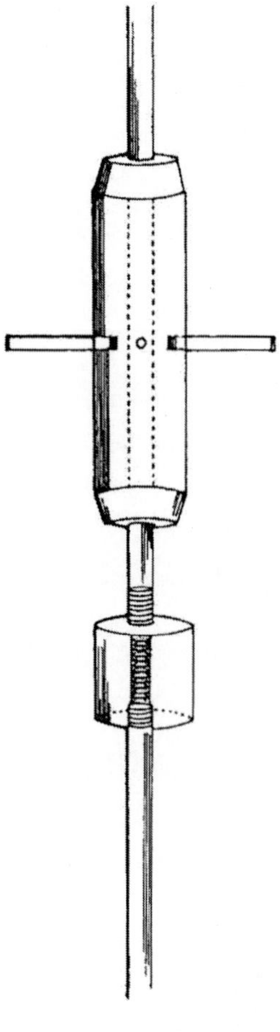

Figure 3.5
Sketch of hammer and anvil assemly.

Figura 3.5
Esboço do martelo e da peça metálica receptora do impacto.

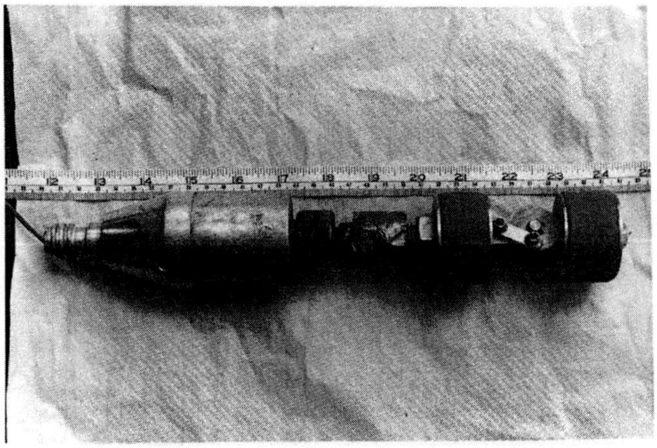

Figure 3.6
Photograph of piston and holder locked together. This is how the piston and holder are joined together inside the sample tube at the end of a drive. Notice that the piston is in the unlocked position (Figure 3.5b). The piston is coupled to the holder by the two parts of a hose coupling so that the piston cannot slip back down the tube as the sampler is withdrawn from the mud.

Figura 3.6
Fotografia mostrando o acoplamento do pistão com o seu segurador. É desta forma que eles são unidos dentro do tubo amostrador no final da amostragem. Observe que o pistão está na posição de pós-disparo. O pistão é unido ao segurador por duas partes de uma junção metálica que não permitem mais sua separação, após a conexão estar formada, evitando assim sua separação, quando da amostragem.

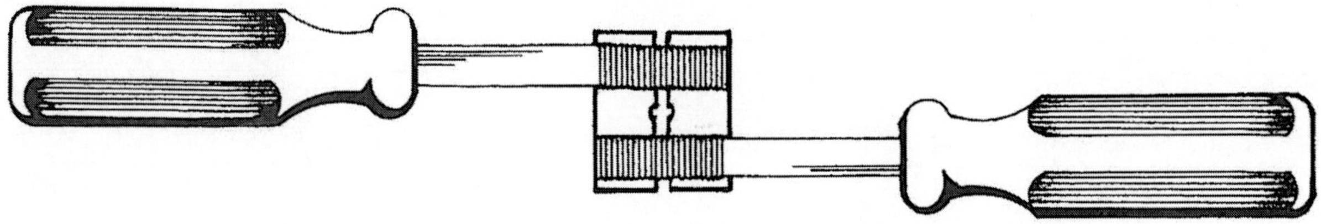

Figure 3.7
Sketch of cable clamp. The cable is gripped in the 'v' groove between two steel blocks that are pressed together by hand screws. The cable could be held by standard locking pliers instead of this cable clamp, but pliers crush and 'kink' the cable. This cable clamp holds the cable securely without damaging it.

Figura 3.7
Desenho das presilhas para o cabo de aço. O cabo é pressionado pelos dois blocos de aço por meio de duas chaves. O cabo também pode ser segurado por meio de uma chave de pressão, mas esta pode danificá-lo.

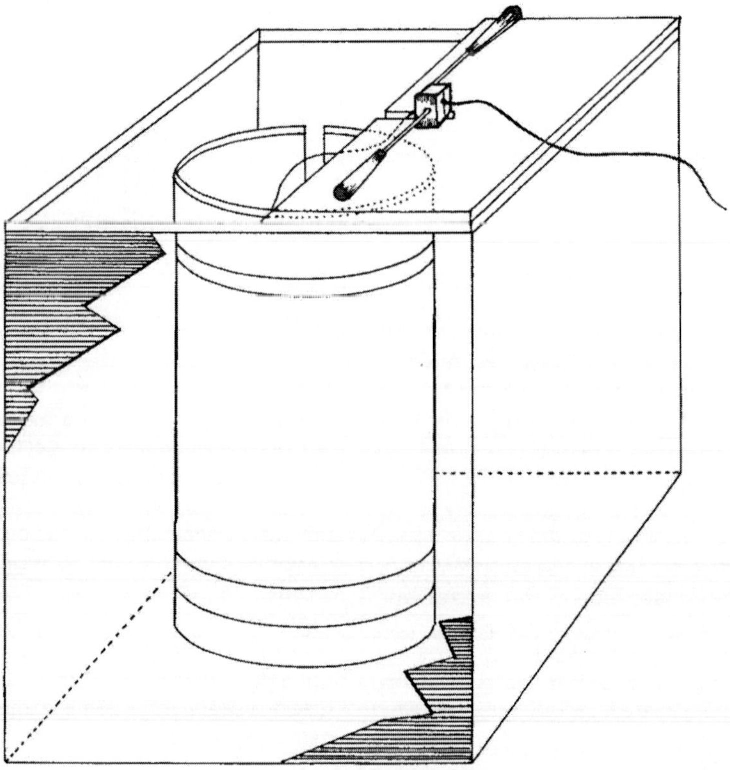

Figure 3.8
Cable clamp in use on bridge over the casing.

Figura 3.8
As presilhas em uso sobre a ponte acima da tubulação-guia.

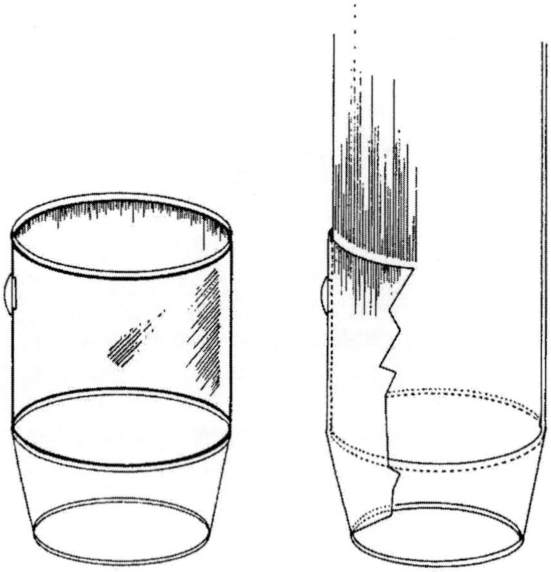

Figure 3.9
Sketch of mild steel shoe for use in stiff sediments. The shoe slides over the bottom end of the sample tube so that the tube presses against the inside shoulders, and the shoe is held in place by the tip of a steel screw that is threaded through a hole in the side of the steel shoe and projects inside to engage a hole drilled in the aluminum alloy sample tube. The shoe is tapered for ease of entry into the mud. The function of the shoe is not to 'cut' through the mud but to squeeze out a hole of larger diameter than the sample tube, thus reducing friction between the sample tube and the mud.

Figura 3.9
Cortador para sedimentos mais coesos. O cortador é acoplado ao final do tubo amostrador, que se apoia em uma linha saliente no próprio cortador. Um orifício no cortador projeta-se num orifício similar no tubo amostrador. O propósito do cortador não é proporcionar um corte no sedimento, mas sim reduzir a fricção criada com a formação de um canal sedimentar de maior diâmetro do que aquele feito com o tubo amostrador.

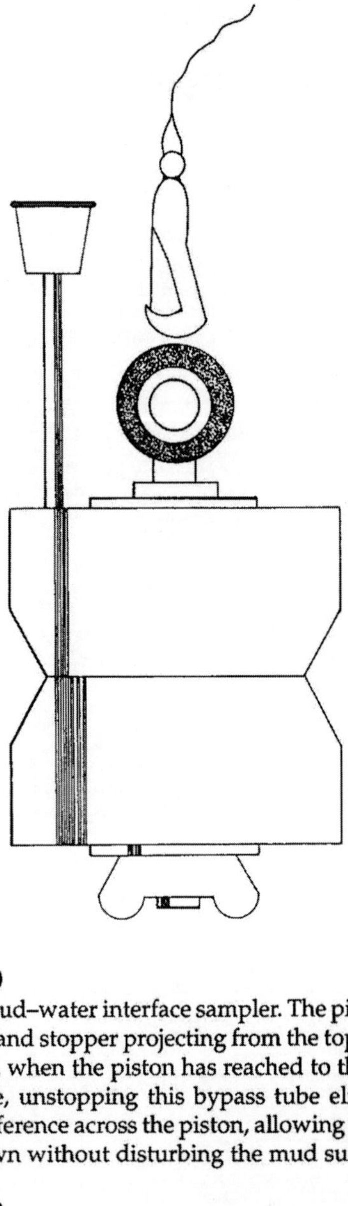

Figure 3.10
Piston for mud–water interface sampler. The piston is fitted with a tube and stopper projecting from the top. At the end of the drive, when the piston has reached to the top of the sample tube, unstopping this bypass tube eliminates the pressure difference across the piston, allowing the piston to be withdrawn without disturbing the mud surface.

Figura 3.10
Amostrador de pistão da interface sedimento/água. A rolha de boracha é empurrada para cima, até que a superfície dos sedimentos não consolidados cheguem à extremidade superior do tubo amostrador.

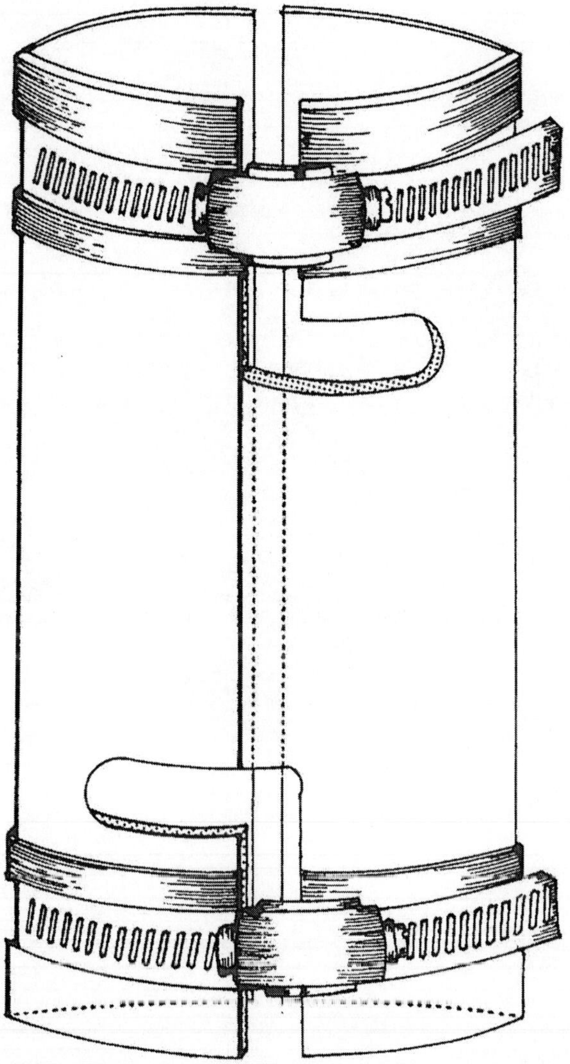

Figure 3.11
Coupling for aluminum casing. In (b) a couping is being locked onto the next length of casing by tightening the hose clamps.

Figura 3.11
Braçadeira para a tubulação-guia. A braçadeira é acoplada à tubulação-guia através de um sistema de parafusos que são apertados com uma chave de fenda.

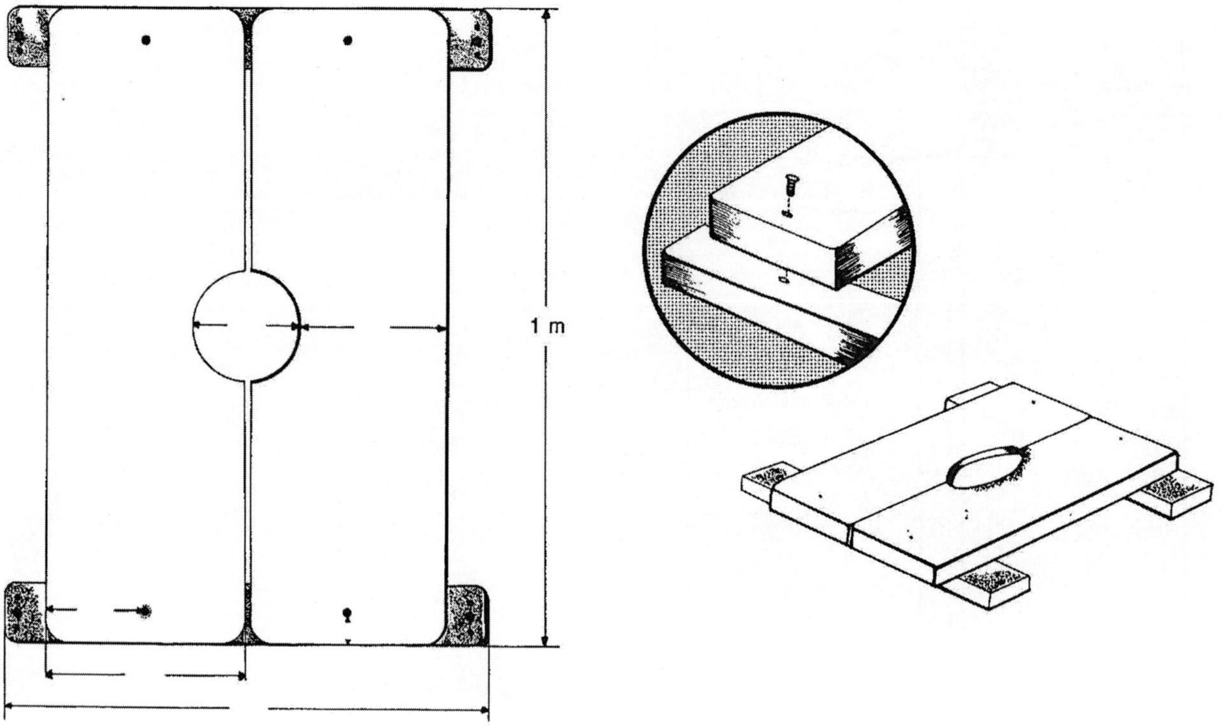

Figure 3.12
Plan of small portable wooden platform.

Figura 3.12
Plano para uma pequena plataforma portátil.

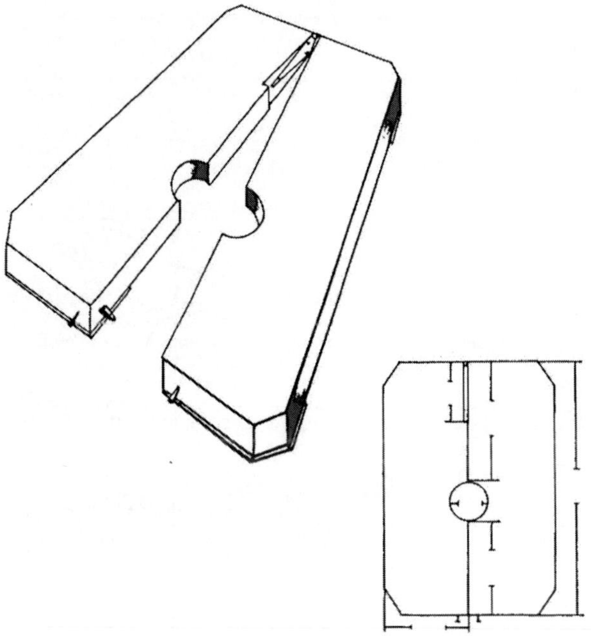

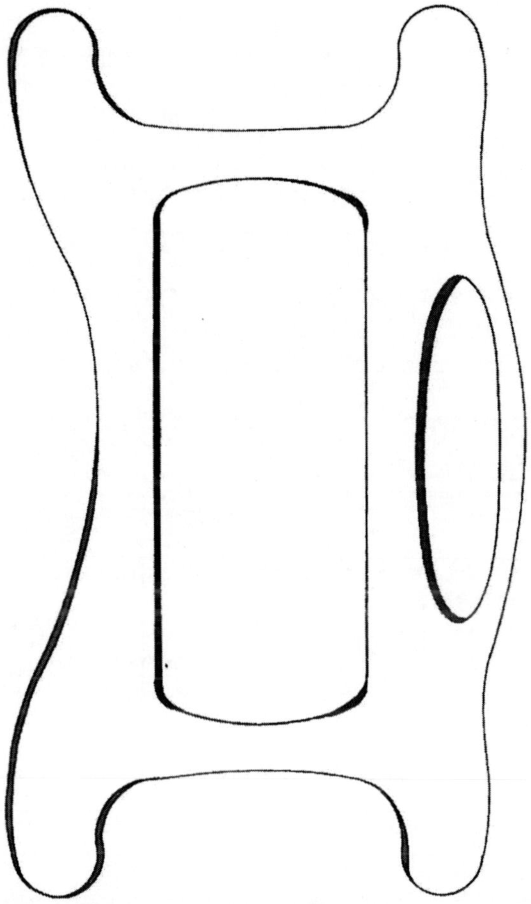

Figure 3.13
Wooden jaws for holding casing. See also Figure 3.8b for photograph of jaws in use.

Figura 3.13
Guilhotina (jaws) de madeira para segurar a tubulação-guia.

Figure 3.14
Anchor line coiled on a "fishtail."

Figura 3.14
Cordas das âncoras enroladas em um "rabo de peixe."

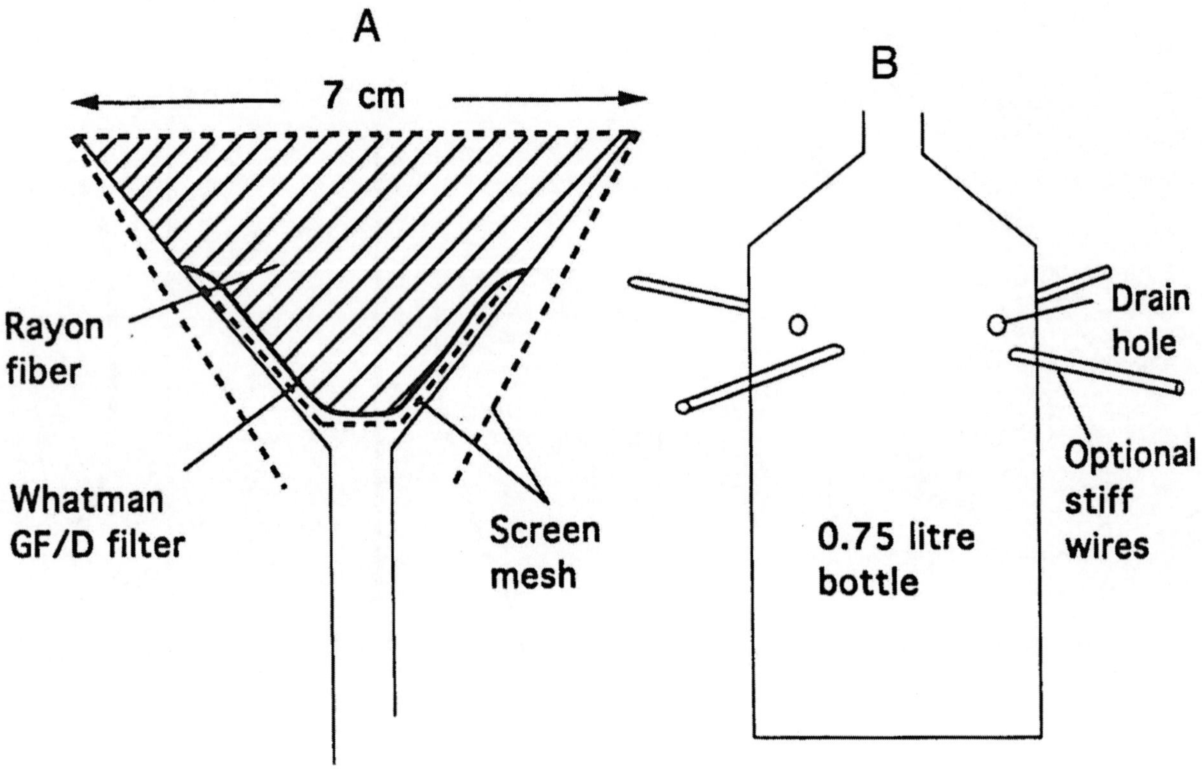

Figure 4.1
The parts of an Oldfield trap for airborne pollen. A funnel holding rayon fiber supported on filter paper, or a glass microfiber filter, is placed in the neck of a plastic bottle. A screen of wire mesh keeps large particles out of the funnel. Holes in the plastic bottle allow water to escape (Bush, 1992).

Figura 4.1
As partes de um coletor Oldfield para pólen anemófilo. Fibras de raiom são colocadas sobre um filtro de microfibras de vidro inserido num funil, que tem, colada na sua base, uma pequena tela de arame ou de náilon. Esta unidade é apoiada no gargalo de uma garrafa plástica com orifícios para o escoamento da água. Uma tela sobre a maior abertura do funil impede a penetração de folhas ou outros fragmentos vegetais (Bush, 1992).

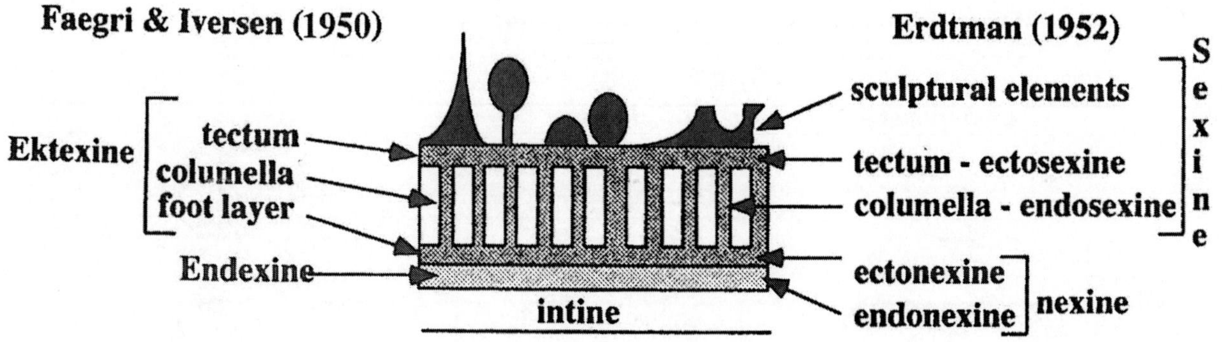

Figure 5.1
Sporoderm stratification according to Erdtman (1952) and Faegri and Iversen (1950).

Figura 5.1
Estratificação da esporoderma de acordo com Erdtman (1952) e Faegri & Iversen (1950).

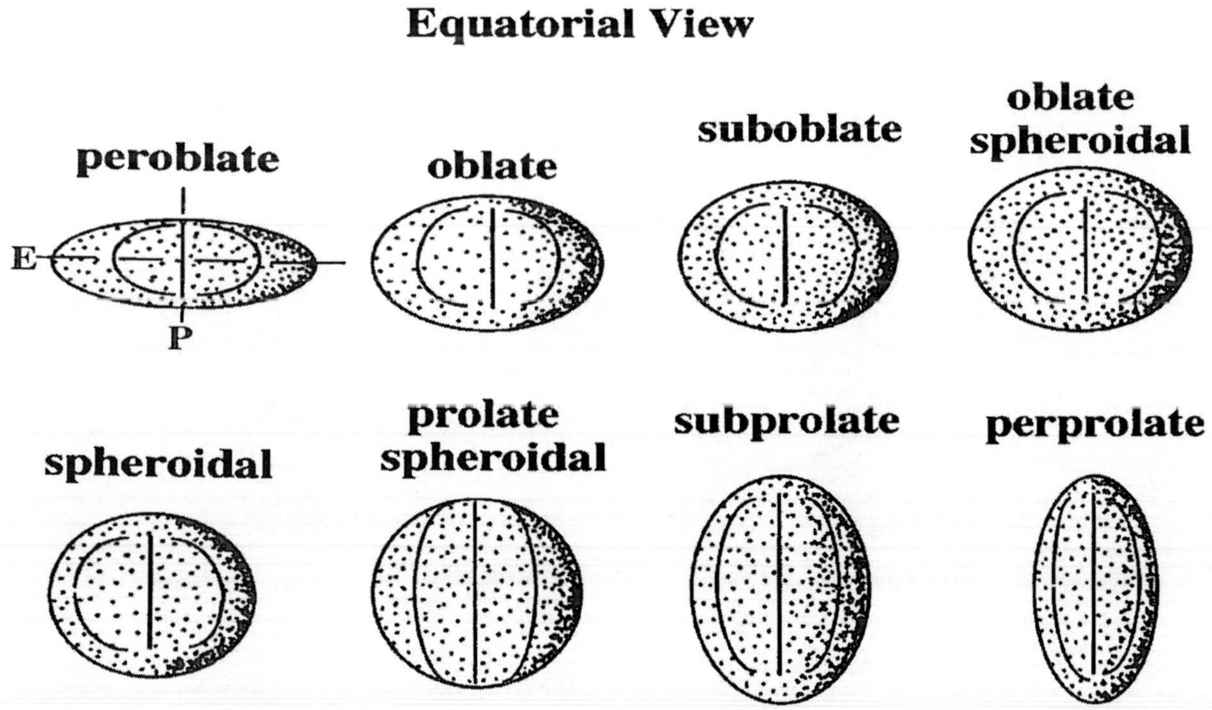

Figure 5.2
Erdtman's (1952) shape classes based on the relationship between the polar axis (P) and the equatorial diameter (E).

Figura 5.2
Classes de pólen quanto à forma, baseadas na relação entre o eixo polar (P) e o diâmetro equatorial (E) (Erdtman, 1952).

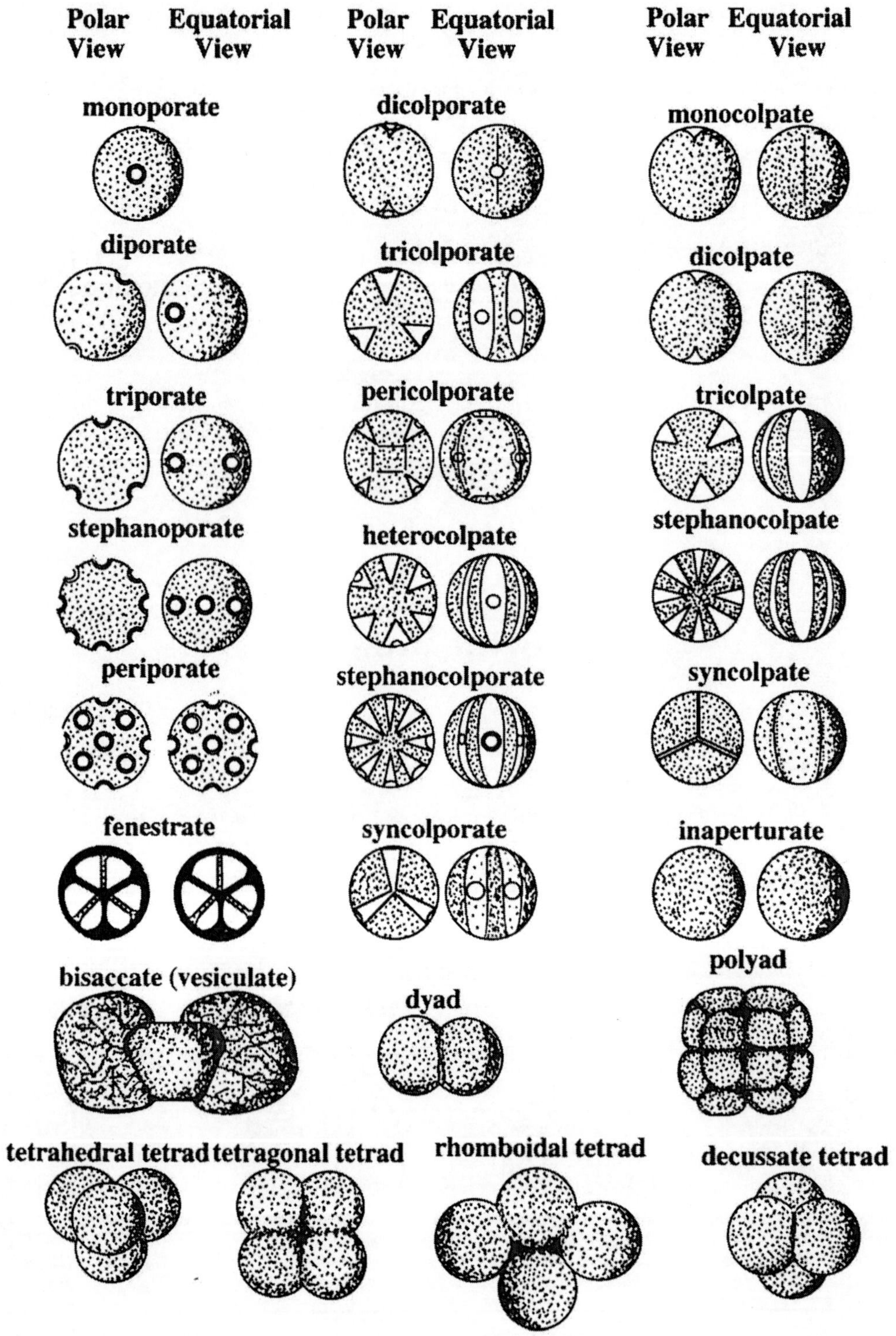

Figure 5.3
Apertural types, number, and position, most commonly found in netropical pollen grains.

Figura 5.3
Tipos de aberturas, número e posição mais comumente encontrados em pólen neotropical.

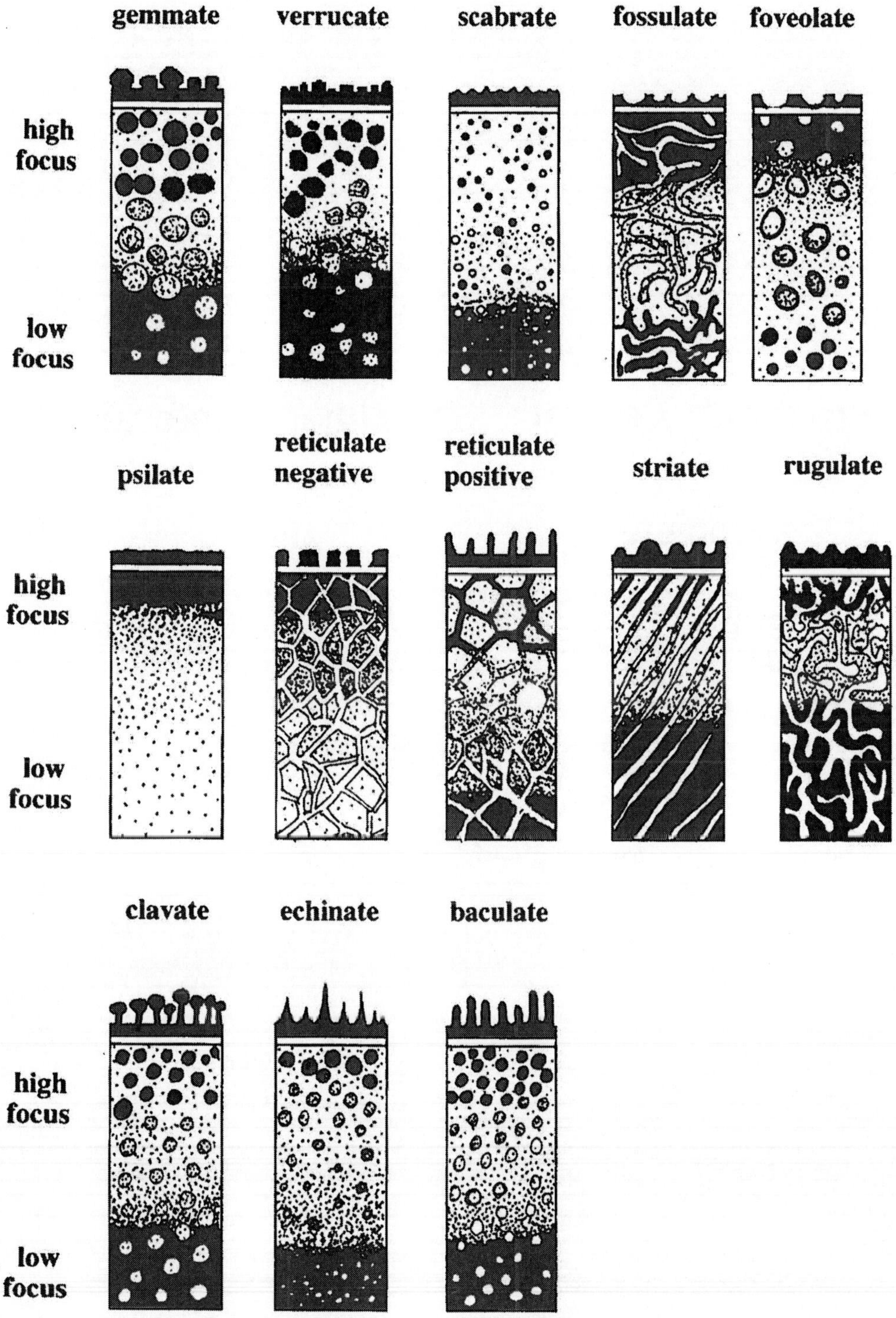

Figure 5.4
Different types of positive and negative ornamentation.

Figura 5.4
Categorias de ornamentação positiva e negativa.

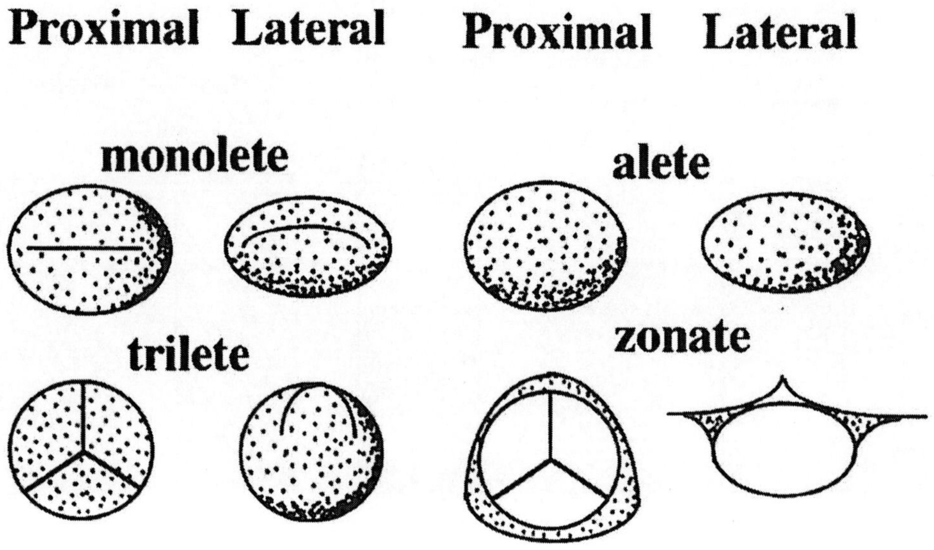

Figure 5.5
Spore morphology.

Figura 5.5
Morfologia de esporos.

Part II

Amazon Pollen Atlas

Introduction to Part II

We describe and illustrate 421 Amazon pollen taxa, spread among 91 families and 246 genera. We provide two dichotomous keys to help identify these taxa. The first key assigns pollen and spores to their morphological group, whereas the second key starts with the morphological group and finds the botanical taxon within that group. The keys are followed by a section providing formal descriptions of all 421 taxa, arranged alphabetically by family in the order: ferns, gymnosperms, dicots, and monocots. All 421 taxa are then illustrated with photomicrographs. A botanical index identifies the plate number and page on which the taxon is illustrated and described.

The photomicrographs are reproduced on a scale of 1 mm to 1 micron, so that the dimensions of the grains in microns can be read from the plates with an ordinary metric system ruler. A few very large grains are printed at a reduced scale, but these are all identified with an asterisk and the scale multiplier is given at the bottom of the plate.

All the photomicrographs are of subfossil pollen grains and spores in sediment cores from Lake Pata and its sister Lake Dragão (see Chapter 1). The grains are thus shown as they appear in pollen extracts from sediments. The smaller grains were photographed using a 100× oil immersion lens of low numerical aperture, giving the highest resolution obtainable with a light microscope, the larger grains with lower power lenses. Low ASA film was used throughout to retain the finest resolution (Ilford Pan F 50, Agfa Pan 25, Kodak TMX 100). The pollen and spores in the photographs were identified by comparison with reference slides in our collection, most of which were made from authenticated herbarium specimens at The Field Museum of Natural History in Chicago (F), The New York Botanical Garden (NY), and the United States National Herbarium (US).

Although we have drawn for our examples on pollen in sediments from lakes of the lowland tropical rain forest, our taxon list includes rain forest representatives of most of the pollen taxa prominent in other vegetation types of tropical America, so that the atlas can be used to identify most of the pollen types likely to be encountered in the whole Amazon basin. This atlas can, however, be supplemented by a number of excellent regional atlases from the neotropics: from the savanna region of Central Brazil, Salgado-Laboriau (1973), Absy (1975), the serial in *Anais*

Acad. Brasil ("Pollen grains of plants of the cerrado") and the serial in *Mem. Inst. Oswaldo Cruz* ("Catálogo sistemático dos pólens das plantas arbóreas do Brasil Meriodional") by the Brazilian palynologists Monika Barth, Terezinha Melhem and their associates; from Chile, Heusser (1971); from Argentina, Markgraf and D'Antoni (1978); from Mexico, Chavez *et al.* (1991) and the serial "Catalogo Polinologico para la Flora de Veracruz"; from Panama, Roubik and Moreno (1991); from Colombia, Hooghiemstra (1984), and for useful and cultivated plants of the Colombian Amazon, Herrera and Urrego (1996). For fern spores, an important source is Murillo and Bless (1974, 1978).

Our pollen keys are followed by diagnostic descriptions of all the taxa we recognize. These identify groups (monads, tetrads, etc.), polarity, symmetry, apertures, exine structure and ornament, etc. and show the plate number on which the photomicrograph appears. Measurements refer in order to longitudinal and equatorial axes respectively. A glossary of all technical terms used in the book is provided.

Our taxonomy is conservative. We use the suffix *aff.* (affinity) whenever we suggest specific names within a genus. Botanical and authors' names were obtained from the databases of the Royal Botanical Gardens (Kew and Edinburgh, UK) and the Missouri Botanical Garden, supplemented by the Gray Herbarium Index, Gentry (1993), Mabberley (1993), and Maas and Westra (1993). Because Amazonian plant taxonomy is under continuous revision the names used herein may not always reflect the most recent nomenclatural changes.

The pollen descriptions include a statement of probable habit (tree, herb, liana, etc.) for convenience when using the ecological groupings function of the software program TILIA and TILIA.GRAF. However, we repeat the caution about assumptions of ecological affinity that we gave in Chapter 1. Very many pollen taxa (genera or families) in our region include species of different habit so that one genus can include species that are rain forest trees and are lianas or bushes. Use ecological groupings only with extreme care.

INTRODUÇÃO

Descrevemos e ilustramos neste manual, 421 táxons polínicos amazônicos pertencentes a 91 famílias e 246 gêneros. Duas chaves dicótomas facilitam a identificação desses táxons: a primeira arranja o pólen e os esporos de acordo com o seu grupo morfológico, enquanto que a segunda começa com o grupo morfológico chegando até o táxon botânico. As chaves são seguidas por uma seção contendo descrições formais de todos os 421 táxons, ordenados alfabeticamente por família, seguindo a ordem: pteridófitas, gimnospermas, dicotiledôneas e monocotiledôneas. Um índice botânico identifica o número da prancha e a página na qual o táxon é ilustrado e descrito.

As fotomicrografias estão reproduzidas com escala de 1 mm para 1 mícron, que permite a determinação direta das dimensões dos grãos com uma régua métrica simples. Alguns grãos grandes estão reproduzidos em uma escala reduzida, mas estão identificados com um asterisco e a escala correspondente encontra-se na parte inferior da prancha.

Todas as fotomicrografias são de grãos de pólen e esporos fósseis encontrados nos sedimentos do Lago da Pata e do seu lago irmão, o Lago do Dragão (veja capítulo 1). Os grãos são, portanto, mostrados como aparecem nas extrações polínicas dos sedimentos. Os grãos pequenos foram fotografados com lente de imersão de 100X a óleo, de baixa abertura numérica, a maior resolução disponível com um microscópio ótico, enquanto que os grãos grandes foram fotografados em menor aumento. Usamos filmes fotográficos de baixa velocidade para a obtenção de maior resolução (Ilford 50 Pan F 50, Agfa Pan 25, Kodak TMX 100). O pólen e os esporos em todas as fotografias foram identificados por comparação com lâminas de referência de nossa coleção, cuja maioria foram obtidas de excicatas dos herbários do Field Museum of Natural History de Chicago, EUA (F), do New York Botanical Garden, EUA (NY), e do United States National Herbarium, EUA (US).

Embora as pranchas polínicas mostrem, exclusivamente, o pólen encontrado em sedimentos lacustres da região de floresta tropical úmida equatorial amazônica, nossa lista de táxons inclui representantes botânicos proeminentes em outros tipos de vegetação da América tropical, permitindo o uso deste atlas na identificação da maioria dos tipos polínicos encontrados em toda a Bacia Amazônica. Recomendamos o uso deste atlas em conjunto com outros excelentes atlas regionais neotropicais: para o Brasil (Salgado-Labouriau, 1973), Absy (1975), De

Oliveira (1992), a série de Palinologia de Plantas do Cerrado, e Catálogo Sistemático dos Pólens das Plantas Arbóreas do Brasil Meridional, publicado nos Anais da Academia Brasileira de Ciências e pelo Instituto Oswaldo Cruz pelas palinólogas brasileiras, Monika Barth e Therezinha Melhem e associados; Chile (Heusser, 1971), Argentina (Markgraf & D'Antoni, 1978), Mexico (Chaves *et al.*, 1991; Catálago Palinologico para la Flora de Veracruz), Panamá (Roubik & Moreno, 1991), Colômbia (Hooghiemstra, 1984; Herrera & Urrego, 1996). Uma importante fonte para esporos de pteridófitas é Murillo & Bless (1974, 1978).

Nossas chaves de identificação são seguidas por descrições diagnósticas de todos os táxons reconhecidos, que identificam em seqüência os grupos polínicos (mônades, tétrades, etc.), polaridade, simetria, aberturas, estrutura e ornamentação da exina e o número da prancha na qual as microfotografias aparecem. As medidas referem-se aos eixos longitudinais e equatoriais respectivamente. Apresentamos também um glossário de todos os termos técnicos usados nas descrições palinológicas.

Nossa taxonomia é conservadora. Usamos o sufixo *aff.* (afinidade) quando sugerimos nomes de espécies dentro de um gênero. Os nomes botânicos e das autoridades correspondentes foram obtidos dos bancos de dados do Royal Botanical Gardens (Kew e Edinburgh, Inglaterra), do Missouri Botanical Garden, EUA, suplementados pelo Índice do Gray Herbarium, EUA, Gentry (1993), Mabberley (1997) e Maas & Westra (1993). A taxonomia de plantas da Amazônia está sob constante revisão e portanto, os nomes usados aqui nem sempre refletem as mudanças mais recentes de nomenclatura.

As descrições polínicas incluem informações sobre os prováveis hábitos (árvore, erva, liana, etc) dos táxons polínicos, facilitando o uso da função de agrupamentos ecológicos do programa TILIA e TILIA.GRAPH. Contudo, reafirmamos que é preciso cautela nas suposições sobre a afinidade ecológica discutida no capítulo 1. Muitos tipos polínicos (gêneros ou famílias) da Amazônia incluem espécies encontradas com diferentes hábitos e um só gênero pode incluir espécies de árvores da floresta tropical, lianas e arbustos. Use os agrupamentos ecológicos com muito cuidado.

KEY TO MAJOR POLLEN AND SPORE GROUPS

1a. Grains appearing joined forming groups, normally as multiple of 4, showing approximately a symmetrical arrangement 2
1b. Grains appearing isolated, free (monads) 3

2a. Groups with only 4 grains . . . . . **Tetrad**
2b. Groups with 6 or more grains . . . . . **Polyad**

3a. Grains with two globose sacci (wing-like) . . . . . **Vesiculate**
3b. Sacci absent 4

4a. Grains apparently without visible aperture . . . . . **Inaperturate**
4b. Grains displaying scars (pseudoaperture) or true aperture 5

5a. Single aperture 6
5b. Multiple aperture 9

6a. Grains displaying pore type (circular aperture) . . . . . **Monoporate**
6b. Grains displaying furrow type (elongated aperture) 7

7a. Furrow triradially shaped (Y split mark) on proximal face . . . . . **Trilete**
7b. Furrow linear 8

8a. Furrows (laesurae) on proximal face; exine with indistinct stratification; aperture with frequently conspicuous and thick margo . . . . . **Monolete**
8b. Furrows (colpus) on distal face; exine stratified; margo absent or if present, very thin . . . . . **Monocolpate**

9a. Aperture circular (pore) or meridionally elongated (colpus tranversalis) 10
9b. Longitudinally elongated furrows (colpi, colpori or both conditions) 14

10a. Aperture as lacunae (surrounded by ridges) . . . . . **Fenestrate**
10b. Aperture as pore or colpus transversalis (not as above) 11

11a. Two opposite pores present . . . . **Diporate**
11b. More than two pores 12

12a. Three pores equidistant, arranged equatorially . . . . **Triporate**
12b. More than three pores 13

13a. 4 to 5 pores restricted to equator . . . . **Stephanoporate**
13b. More than 5 pores irregularly arranged on the surface . . . . **Periporate**

14a. Exclusively colpi condition 15
14b. Colpi with pores (colpori) or colpi alternating with colpori 18

15a. Two opposite colpi . . . . **Dicolpate**
15b. More than two colpi 16

16a. Three colpi arranged equatorially 17
16b. More than three colpi (4 to 16) restricted to equator . . . . **Stephanocolpate**

17a. Colpi ends free at poles . . . . **Tricolpate**
17b. Colpi ends joined at poles, occasionally forming small triangle (parasyncolpate condition) . . . . **Syncolpate**

18a. Colpi seen as pseudoaperture alternating with colpori . . . . **Heterocolpate**
18b. Colpi always with incorporated pores (colporate condition) 19

19a. Two opposite colpori . . . . **Dicolporate**
19b. More than two colpori 20

20a. Three colpori arranged equatorially 21
20b. More than three colpori 22

21a. Colpi ends joining at poles . . . . **Syncolporate**
21b. Colpi ends free at poles . . . . **Tricolporate**

22a. 4 to 8 colpori restricted to equator . . . . **Stephanocolporate**
22b. 4 to 6 colpori scattered on surface . . . . **Pericolporate**

KEY TO FAMILIES AND GENERA BY POLLEN AND SPORE GROUPS

VESICULATE

1a. Grains with corpus trapezoid-like . . . . Podocarpaceae (*Podocarpus* form 3)
1b. Grains with corpus elliptic to rounded . . . . Podocarpaceae (*Podocarpus* forms 1&2)

MONOLETE

1a. Spores displaying perine with spinulose undulating ridges 2
1b. Perine absent; spores showing psilate, gemmate, foveolate or verrucate sculptural patterns 7

2a. Spores <50 μm; sclerine <3 μm thick; margo inconspicuous 3
2b. Spores >50 μm; sclerine ≥3 μm thick; margo conspicuous 4

3a. Spores 38 μm . . . . Aspleniaceae (Dryopteridaceae) (*Asplenium* form 4)
3b. Spores 48 to 50 μm . . . . Aspleniaceae (Dryopteridaceae) (*Asplenium* form 2)

4a. Ambitus concave-convex to plano-convex (lateral view); spores >70 μm . . 5
4b. Ambitus elliptic-biconvex (lateral view); spores <60 μm . . . . Thelypteridaceae (*Thelypteris* form 2)

5a. Spores 70 to 75 μm; always showing monad condition; spinule fine . . . 6
5b. Spores 144 μm; occasionally displaying dyad condition; spinule up to 20 μm . . . . Aspleniaceae (Dryopteridaceae) (*Asplenium* form 3)

6a. Ambitus concave-convex . . . . Thelypteridaceae (*Thelypteris* form 1)
6b. Ambitus plano-convex . . . . Aspleniaceae (Dryopteridaceae) (*Asplenium* form 1)

7a. Sculpture psilate or foveolate 8
7b. Sculpture gemmate or verrucate 10

8a. Spores >55 µm; sclerine foveolate 9
8b. Spores <50 µm; sclerine psilate . . . . . Gleicheniaceae (*Dicranopteris*)

9a. Spores 67 µm; sclerine 3.5 µm thick; ambitus elliptic . . . . . Filicopsida form 3
9b. Spores 56 µm; sclerine 1 µm thick; ambitus concave-convex . . . . . Schizaeaceae (*Schizaea aff. pennula*)

10a. Spores less than 45 µm; sclerine gemmate . . . . . Blechnaceae (*Blechnum*)
10b. Spores more than 45 µm; sclerine verrucate . . . . . Polypodiaceae (*Polypodium*)

TRILETE

1a. Spores psilate 2
1b. Spores ornamented, with positive or negative sculpture 5

2a. Spores >40 µm; margo wide, thick 3
2b. Spores <40 µm; margo narrow, thin 4

3a. Ambitus subtriangular-plano (proximal face) . . . . . Dicksoniaceae (*Dicksonia* form 3)
3b. Ambitus subtriangular-concave (proximal face) . . . . . Cyatheaceae (*Cyathea* forms 3&4)

4a. Spores 26 µm; "Y" split mark conspicuous . . . . . Cyatheaceae (*Alsophila*)
4b. Spores 18 µm; "Y" split mark masked, inconspicuous . . . . . Selaginellaceae (*Selaginella* sp. form 1)

5a. Spores displaying positive ornamentation (scabrae, spinule, bacula, verrucae or gemmae) 6
5b. Spores displaying negative ornamentation (foveolate, rugulate or reticulate) . 20

6a. Spores finely scabrate 7
6b. Spores not as above 8

7a. Spores <40 µm; margo inconspicuous or absent . . . . . Filicopsida 2
7b. Spores >40 µm; margo conspicuous . . . . . Dicksoniaceae (*Dicksonia* forms 1&2)

8a. Spores baculate 9
8b. Spores spinulose, verrucate or gemmate 10

9a. Spores 80 µm; margo 4 µm wide . . . . . Hymenophyllaceae (*Hymenophyllum aff. ciliatum*)
9b. Spores 48 to 50 µm; margo 2 µm wide . . . . . Hymenophyllaceae (*Trichomanes aff. elegans*)

10a. Spores displaying spinulose sculpture 11
10b. Spores verrucate or gemmate, or with both conditions 14

11a. Spores >35 µm 12
11b. Spores <35 µm 13

12a. Spores 45 to 67 µm; spinule long, occasionally bifurcate . . . . . Selaginellaceae (*Selaginella aff. arthritica*)
12b. Spores 35 to 45 µm; spinule short, acute . . . . . Selaginellaceae (*Selaginella aff. asperula* forms 1&3)

13a. Sclerine 1.5 µm thick . . . . Lycopodiaceae (*Lycopodium aff. linifolium*)
13b. Sclerine 3 µm thick . . . . Selaginellaceae (*Selaginella* sp. form 2)

14a. Spores >50 µm 15
14b. Spores <50 µm 17

15a. Sclerine <5 µm thick 16
15b. Sclerine >5 µm thick . . . . Pteridaceae (*Pteris*)

16a. Verrucae 2 µm height; ambitus subtriangular-convex . . . . Grammitidaceae (*Grammitis*)
16b. Verrucae 3 to 4 µm height; ambitus subtriangular-semiconcave . . . . . . Cyatheaceae (*Cyathea* forms 2&5)

17a. Spores <35 µm; verrucae well developed, conspicuous 18
17b. Spores >35 µm; verrucae flat 19

18a. Laesurae thin, narrow; margo conspicuous . . . . Selaginellaceae (*Selaginella aff. asperula* form 2)
18b. Laesurae wide; margo inconspicuous . . . . Cyatheaceae (*Cyathea* form 1)

19a. Sclerine displaying gemmate pattern . . . . Selaginellaceae (*Selaginella aff. exaltata*)
19b. Sclerine displaying psilate pattern . . . . Selaginellaceae (*Selaginella aff. asperula* forms 4&5)

20a. Spores reticulate or rugulate; always displaying monad condition 21
20b. Spores foveolate; displaying dyad condition . . . . Lycopodiaceae (*Lycopodium aff. subulatum*)

21a. Spores <40 µm; sclerine <4 µm thick 22
21b. Spores >80 µm; sclerine 5 to 6 µm thick . . . . Filicopsida 1

22a. Spores rugulate, 30 µm . . . . Lycopodiaceae (*Lycopodium aff. trianae*)
22b. Spores reticulate, 35 to 40 µm . . . . Lycopodiaceae (*Lycopodium clavatum*) [Exotic marker]

POLYAD

1a. Polyad more than 45 µm; grains reticulate, heterobrochate 2
1b. Polyad less than 45 µm; grains psilate to scabrate 3

2a. Polyad 150 to 160 µm; grains symmetrically arranged; exine >3 µm thick . . . . Leguminosae – Mimosoideae (*Parkia aff. multijuga*)
2b. Polyad 46 to 50 µm; grains grouped in 4 tetrahedral tetrads; exine <2 µm thick . . . . Celastraceae (Hippocrateaceae) (*Hippocratea aff. volubilis*)

3a. Polyad with 16 grains 4
3b. Polyad with 6 to 12 grains 5

4a. Polyad >35 µm; grains scabrate . . . . Leguminosae – Mimosoideae (*Inga*)
4b. Polyad <35 µm; grains psilate . . . . Leguminosae – Mimosoideae (*Acacia*)

5a. Polyad with 6 or 8 grains 6
5b. Polyad with 12 grains . . . . Leguminosae – Mimosoideae (*Mimosa aff. rufescens*)

6a. Polyad with 6 grains 7
6b. Polyad with 8 grains . . . . Leguminosae – Mimosoideae (*Mimosa* forms 1&2)

7a. Polyad more than 30 µm . . . . Leguminosae – Mimosoideae (*Mimosa* sp.)

7b. Polyad less than 15 µm . . . . Leguminosae – Mimosoideae (*Mimosa aff. punctulata*)

8a. Polyad 17 × 13 µm; grains 7 × 10 µm . . . . Leguminosae – Mimosoideae (*Mimosa aff. extensa*)
8b. Polyad 13 × 11 µm; 5 × 7 µm . . . . Leguminosae – Mimosoideae (*Mimosa aff. bimucronata*)

TETRAD

1a. Multiplanar tetrahedral or decussate tetrad; grains tricolporate or periporate . . . 2
1b. Uniplanar tetragonal tetrad; grains triporate . . . . Leguminosae – Mimosoideae (*Mimosa aff. pteridifolia*)

2a. Tetrahedral tetrad; colpori with pores lalongate or, in periporate condition, pores inconspicuous <1 µm diameter 3
2b. Decussate tetrad; colpori with pores circular 5µm diameter . . . . Leguminosae – Caesalpinioideae (*Dinizia*)

3a. Tetrad more than 40 µm; grains tricolporate, scabrate to baculate 4
3b. Tetrad less than 20 µm; grains periporate, psilate . . . . Leguminosae – Mimosoideae (*Mimosa aff. micropteris*)

4a. Exine more than 2 µm thick; sculpture granulate to baculate 5
4b. Exine less than 2 µm thick; sculpture scabrate . . . . Ericaceae (*Agarista / Gaylussacia*)

5a. Exine >4 µm thick, intectate; bacula 3 µm length . . . . Gentianaceae (*Schultesia*)
5b. Exine <3 µm thick, tectate; bacula <1 to 1.5 µm length . . . . Ericaceae (*Gaultheria*)

INAPERTURATE

1a. Grains displaying reticulate pattern 2
1b. Grains with gemmate, baculate or clavate pattern 4

2a. Reticulum homobrochate; brochi <1 to 1 µm wide 3
2b. Reticulum heterobrochate; brochi variable 3 to 5 µm wide forming isolated patches (areolae) . . . . Bignoniaceae (*Memora*)

3a.	Grains >20 μm; exine semitectate	Cyperaceae (*Fimbrystylis*)
3b.	Grains <20 μm; exine tectate	Araceae (*Anthurium*)
4a.	Grains >30 μm; baculate or clavate	5
4b.	Grains <25 μm; gemmate	Boraginaceae (*Tournefortia*)
5a.	Grains displaying clavate pattern; ambitus circular	6
5b.	Grains displaying baculate pattern; ambitus elongated, irregular	7
6a.	Grains >50 μm; clavae 4 μm length, displaying "croton" pattern	Euphorbiaceae (*Croton*)
6b.	Grains <50 μm; clavae 1.5 to 3 μm length, scattered on surface	Chloranthaceae (*Hedyosmum*)
7a.	Grains <70 μm	8
7b.	Grains >80 μm	Annonaceae (*Diclinanona aff. tessmanii*)
8a.	Grains <40 μm	Annonaceae (*Unonopsis aff. pittieri*)
8b.	Grains >40 μm	Annonaceae (*Annona*)

MONOPORATE

1a.	Sexine echinate; pore with irregular borders; margo absent	2
1b.	Sexine psilate to scabrate; pore circular, surrounded by conspicuous margo	Poaceae (Gramineae)
2a.	Grains >35 μm; exine 1.5 to 4 μm thick; pore >5 μm diameter	3
2b.	Grains <35 μm; exine 1 μm thick; pore 3 μm diameter	Arecaceae (Palmae) 1
3a.	Grains 60 to 70 μm; exine 3 to 4 μm thick; pore 8 μm diameter	Arecaceae (Palmae) (*Mauritiella aff. aculeata*)
3b.	Grains <60 μm; exine 1.5 to 2 μm thick; pore 12 to 15 μm diameter	Arecaceae (Palmae) (*Mauritia aff. flexuosa*)

DIPORATE

1a.	Pores annulate	2
1b.	Annulus absent	3
2a.	Grains >20 μm, scabrate; pore 8 to 10 μm diameter; annulus 5 to 6 μm thick	Rubiaceae (*Coussarea*)
2b.	Grains <20 μm, psilate; pores 1 μm diameter; annulus 2.5 μm thick	Onagraceae (Oenotheraceae)
3a.	Grains <15 μm	4
3b.	Grains >20 μm	Moraceae (*Brosimum*)
4a.	Sexine psilate	5
4b.	Sexine scabrate	Cecropiaceae (*Cecropia* forms 1&3)
5a.	Grains oblate 12 × 15 μm; pore 2.5 μm diameter	Moraceae 1
5b.	Grains subprolate 9 × 6 μm; pores <1 μm diameter	Cecropiaceae (*Cecropia* form 2)

TRIPORATE

1a.	Grains psilate	2
1b.	Grains ornamented with positive or negative sculpture	3
2a.	Grains oblate; pores circular 6 µm diameter; ambitus triangular	Proteaceae (*Roupala* form 2)
2b.	Grains peroblate; pores lolongate 4 µm length; ambitus circular	Apocynaceae (*Lacmellea*)
3a.	Grains displaying reticulate condition	4
3b.	Grains displaying scabrate or echinate conditions	12
4a.	Reticulum homobrochate	5
4b.	Reticulum heterobrochate	10
5a.	Pores circular 3 to 8 µm diameter	6
5b.	Pores slightly lolongate or appearing as pseudoapertures, inconspicuous	8
6a.	Grains >30 µm; pores common type	7
6b.	Grains <25 µm; pores vestibulate type	Tiliaceae (*Mortoniodendron*)
7a.	Exine 2 µm thick; pores 8 µm diameter; ambitus triangular-plano	Proteaceae (*Roupala* form 1)
7b.	Exine 3 µm thick; pores 3 µm diameter; ambitus triangular-concave	Proteaceae (*Euplassa*)
8a.	Grains <35 µm; annulus absent; ambitus circular or irregular	9
8b.	Grains >35 µm; pores annulate; ambitus inter-subangular	Bombacaceae 1
9a.	Exine 2.5 to 3 µm thick; columellae (bacula) densely distributed on the surface	Magnoliopsidae 2
9b.	Exine 1 µm thick; columellae (bacula) scattered	Cyperaceae (*Rhynchospora*)
10a.	Grains tectate >40 µm; ambitus circular	11
10b.	Grains semitectate <40 µm; ambitus triangular	Rubiaceae (*Alibertia aff. edulis*)
11a.	Exine 3 µm thick; muri undulating; pores 6 to 7 µm diameter	Theaceae (*Archytaea aff. multiflora*)
11b.	Exine 1 µm thick; muri flat; pores 2 µm diameter	Rubiaceae (*Duroia*)
12a.	Grains scabrate	13
12b.	Grains echinate	19
13a.	Grains >30 µm	14
13b.	Grains <30 µm	16
14a.	Exine >2 µm thick	15
14b.	Exine <2 µm thick	Ulmaceae (*Celtis*)
15a.	Grains oblate 30 to 32 µm; pores 1 µm diameter; ambitus triangular	Sapindaceae (*Paullinia*)
15b.	Grains spheroidal 47 µm; pores 9 µm diameter; ambitus circular	Apocynaceae (*Odontadenia*)
16a.	Grains >20 µm; pores >3 µm diameter; annulus present	17
16b.	Grains <20 µm; pores <3 µm diameter; annulus absent	Ulmaceae (*Trema*)
17a.	Grains >20 µm	18
17b.	Grains <20 µm	Magnoliopsidae 1
18a.	Pores 3 to 5 µm diameter, vestibulate, protruding	19
18b.	Pores 7 µm diameter, aspidate	Rubiaceae (*Faramea*)

19a. Pores 3 μm diameter; annulus 1 μm thick . . . . Apocynaceae (*Malouetia aff. guatemalensis*)
19b. Pores 5 μm diameter; annulus 2 μm thick . . . . Verbenaceae (*Priva*)

20a. Grains <25 μm; spines ≤1 μm height, acute 21
20b. Grains >25 μm; spine 3 to 7 μm height, conical, rounded 23

21a. Exine <1 to 1 μm thick; pores <4 μm diameter; annulus absent 22
21b. Exine 3 μm thick; pores 5 μm diameter; annulus present . . . . Icacinaceae (*Discophora*)

22a. Grains <20 μm; pores >2.5 μm diameter . . . . Guttiferae (Clusiaceae) (*Tovomitopsis* form 2)
22b. Grains >20 μm; pores <2.5 μm diameter . . . . Euphorbiaceae (*Podocalyx aff. loranthoides*)

23a. Grains >60 μm; exine >3 μm thick (ornamentation excluded) 24
23b. Grains <50 μm; pores <3 μm thick (ornamentation excluded) 25

24a. Grains 115 to 120 μm; spine conical; pores12 μm diameter . . . . Cucurbitaceae (*Cayaponia*)
24b. Grains 60 to 64 μm; spine bottle-shaped; pores 8 μm diameter . . . . Malvaceae (*Abutilon*)

25a. Exine tectate; pores conspicuous 5 to 8 μm diameter, annulate . . . . Bombacaceae (*Scleronema*)
25b. Exine intectate; pores inconspicuous, annuli absent . . . . Theaceae (*Gordonia* [*Laplacea*])

STEPHANOPORATE

1a. Grains psilate, 5-porate, occasionally 4-porate; annuli present 1.5 to 3 μm wide; grains displaying conspicuous arcus 2 to 5 μm wide; ambitus pentagonal . . . . Betulaceae (*Alnus*)
1b. Grains scabrate, 4-porate, occasionally 5-porate; annuli and arcus absent; ambitus circular to square . . . . Euphorbiaceae (*Acalypha*)

PERIPORATE

1a. Grains with 10 or more pores 2
1b. Grains with less than 10 pores 5

2a. Pores more than 20, pores ≤2 μm diameter; grains <35 μm, displaying ornamentation 3
2b. Pores 10 connected by irregular pseudocolpi; pores 5 μm diameter; grains psilate >35 μm . . . . Malpighiaceae (*Stigmaphyllon*)

3a. Grains scabrate or granulate 23 to 28 μm; pores more than 40; exine ≤2 μm thick 4
3b. Grains echinate 29 to 31 μm; pores less than 40; exine >2 μm thick Malvaceae (*Sida*)

4a. Grains granulate with more than 80 pores; pores slightly annulate . . . . . Chenopodiaceae (*Chenopodium*)
4b. Grains scabrate with less than 80 pores; annulus absent . . . . Amaranthaceae (*Amaranthus*)

5a. Grains >25 µm, ornamented; pores 4 to 6 scattered on surface; ambitus circular or trapezoid … 6
5b. Grains psilate <25 µm; pores 6 in series of three at each face of grain resembling demicolpate condition; ambitus triangular … Loranthaceae (Viscaceae) (*Dendropemon*)

6a. Grains echinate … 7
6b. Grains verrucate, scabrate or granulate … 8

7a. Grains 26 to 45 µm; spines conical; pores inconspicuous; exine 1 to 2 µm thick … Alismataceae (*Sagittaria*)
7b. Grains 48 µm; spine rounded resembling short bacula; pores 4 µm diameter; exine 2.5 to 3 µm thick … Alismataceae (*Echinodorus*)

8a. Grains 6-porate, verrucate; exine 2 to 3 µm thick … 9
8b. Grains 5-porate, scabrate or granulate; exine ≤2 µm thick … 10

9a. Grains <30 µm; pores 3 µm diameter; verrucae 1.5 µm height … Malpighiaceae (*Dicella*)
9b. Grains >50 µm; pores 5 µm diameter; verrucae 4 µm height … Malpighiaceae (*Mascagnia*)

10a. Grains <45 µm; pores appearing as elongated ruptures (pseudopores) 4 to 15 µm length; ambitus trapezoid irregular … 11
10b. Grains >60 µm; pores circular 10 µm diameter; ambitus circular … Apocynaceae (*Prestonia*)

11a. Grains <30 µm; pseudopores <5 µm length … Cyperaceae (*Scleria*)
11b. Grains >40 µm; pseudopores 8 to 15 µm length … Cyperaceae (*Cyperus*)

FENESTRATE

Grains baculate, 18 µm, with 16 pentagonal lacunae; lacunae *ca.* 5 µm diameter; lophae simplibaculate; grains spheroidal … Amaranthaceae (*Gomphrena*)

MONOCOLPATE (MONOSULCATE)

1a. Grains <100 µm, with reticulate, baculate or scabrate pattern … 2
1b. Grains >100 µm, displaying verrucate pattern … Arecaceae (Palmae) (*Scheelea*)

2a. Grains reticulate … 3
2b. Grains baculate or scabrate … 5

3a. Grains >35 µm; sulcus as long as grain … 4
3b. Grains <35 µm; sulcus ⅔ length of grain … Arecaceae (Palmae) (*Cryosophila*)

4a. Exine 3 µm thick; brochi 1 µm wide … Bromeliaceae
4b. Exine 1.5 to 2 µm thick; brochi <1 µm wide … Arecaceae (Palmae) (*Oenocarpus*)

5a. Grains baculate … 6
5b. Grains scabrate … 7

6a. Sexine tectate, 1 µm thick; bacula 1 to 2.5 µm length, poorly packed … Arecaceae (Palmae) (*Iriartea aff. deltoidea*)
6b. Sexine intectate, 1.5 µm thick; bacula 1 µm length, densely packed … Arecaceae (Palmae) (*Geonoma* form 2)

7a. Grains >40 μm	8
7b. Grains <40 μm	9
8a. Sulcus wide, deep	Araceae (*Philodendron*)
8b. Sulcus narrow, straight	Arecaceae (Palmae) (*Syagrus*)
9a. Grains >30 μm; sulcus narrow	10
9b. Grains <30 μm; sulcus wide	Arecaceae (Palmae) (*Geonoma* form 1)
10a. Grains finely scabrate; sulcus as long as grain, margo absent	11
10b. Scabrae conspicuous; sulcus ⅔ length of grain, slightly surrounded by thin margo	Arecaceae (Palmae) (*Euterpe*)
11a. Grains oblate, 30 to 32 μm; exine up to 1.5 μm thick	Arecaceae (Palmae) (*Cheylocarpus*)
11b. Grains suboblate >32 μm; exine 1 μm thick	Arecaceae (Palmae) (*Desmoncus*)

DICOLPATE

Grains reticulate, homobrochate, brochi fine; exine variable 4 μm thick at apocolpium and 5 μm thick at mesocolpium; colpi narrow, as long as grain; grains prolate 43 to 58 × 30 μm	Acanthaceae (*Aphelandra*)

TRICOLPATE

1a. Grains ornamented, structurally simple	2
1b. Grains psilate, structurally complex, with a triradial bifurcated thickness at apocolpium resembling 3 circular annulate aspides	Loranthaceae (Viscaceae) (*Orycthanthus aff. alveolatus*)
2a. Grains with negative sculpture (reticulum)	3
2b. Grains with positive elements (scabrae, granules, clavae, spines, gemmae, bacula)	9
3a. Reticulum homobrochate	4
3b. Reticulum heterobrochate	7
4a. Grains >45 μm; exine 2 to 3 μm thick	5
4b. Grains <40 μm; exine 1 to 1.5 μm thick	6
5a. Grains prolate-spheroidal 60 to 62 μm; colpi as long as grain; ambitus circular	Bignoniaceae (*Lundia*)
5b. Grains oblate 46 to 50 μm; colpi short (brevicolpate); ambitus inter-subangular	Bombacaceae (*Quararibea*)
6a. Grains subprolate; colpi as long as grain, with equatorial constriction; ambitus circular	Bignoniaceae (*Tabebuia* form 2)
6b. Grains oblate; colpi short (brevicolpate) without equatorial constriction; ambitus inter-subangular	Bombacaceae (*Spirotheca*)
7a. Grains >25 μm; colpi short (brevicolpate); ambitus inter-subangular	8
7b. Grains <25 μm; colpi ⅓ to ½ length of grain; ambitus circular	Menispermaceae (*Cissampelos*)

8a. Grains tectate 26 μm; exine uniform 1.5 μm thick . . . . Bombacaceae (*Pseudobombax*)
8b. Grains semitectate 38 μm; exine variable 1.5 to 2.5 μm thick . . . . Bombacaceae (*Pachira aff. aquatica*)

9a. Sculpture scabrate 10
9b. Sculpture granulate, echinate, clavate, gemmate or baculate 12

10a. Grains <35 μm; exine 1.5 to 2 μm thick 11
10b. Grains >60 μm; exine 5 μm thick . . . . Bignoniaceae (*Jacaranda* form 1)

11a. Grains oblate-spheroidal <25 μm; ambitus triangular . . . . Olacaceae (*Minquartia aff. guianensis*)
11b. Grains prolate-spheroidal to subprolate >25 μm; ambitus circular trilobulate . . . . Bignoniaceae (*Arrabidaea* forms 2&3)

12a. Sculpture granulate 13
12b. Sculpture echinate, clavate, gemmate or baculate 16

13a. Ambitus triangular; colpi thin, narrow 14
13b. Ambitus circular; colpi wide, deep . . . . Bignoniaceae (*Jacaranda* form 2)

14a. Grains <35 μm; polar area 5 to 10 μm 15
14b. Grains >35 μm; polar area 13 to 18 μm . . . . Loranthaceae (Viscaceae) (*Tristerix* [*Macrosolen*])

15a. Grains 28 to 31 μm; exine 1 μm thick; polar area 10 μm . . . . Loranthaceae (Viscaceae) (*Psittacanthus*)
15b. Grains 15 to 17 μm; exine 1 to 2 μm thick; polar area 5 μm . . . . Loranthaceae (Viscaceae) (*Phoradendron*)

16a. Sculpture gemmate 17
16b. Sculpture echinate, clavate or baculate 18

17a. Grains 80 to 85 μm; exine 4 to 5 μm thick . . . . Bombacaceae (*Matisia aff. bracteolosa*)
17b. Grains 46 to 50 μm; exine 1 to 2 μm thick . . . . Bombacaceae (*Phragmotheca aff. mammosa*)

18a. Sculpture clavate or echinate 19
18b. Sculpture baculate . . . . Rubiaceae (*Psychotria* form 2)

19a. Sexine clavate, intectate 20
19b. Sexine echinate, tectate 21

20a. Grains <45 μm; exine >4 μm thick; clavae 3.5 to 5 μm height . . . . Convolvulaceae (*Maripa*)
20b. Grains >45 μm; exine 2.5 μm thick; clavae 2 μm height . . . . Convolvulaceae (*Dicranostyles aff. scandens*)

21a. Grains <70 μm; exine 1.5 to 2 μm thick 22
21b. Grains >150 μm; exine 8 to 9 μm thick . . . . Cucurbitaceae 1

22a. Grains 65 μm; ambitus inter-subangular . . . . Bombacaceae (*Rhodognaphalopsis aff. duckei*)
22b. Grains 31 to 44 μm; ambitus circular . . . . Verbenaceae (*Aegiphila*)

SYNCOLPATE

1a. Grains psilate to scabrate <25 μm, tectate . . . 2
1b. Grains pilate >45 μm, intectate; pila resembling small baculi . . . Loranthaceae (Viscaceae) 1

2a. Ambitus triangular convex; colpi displaying parasyncolpate condition . . . Loranthaceae (Viscaceae) (*Struthanthus* sp.)
2b. Ambitus triangular concave; colpi united not as above . . . Loranthaceae (Viscaceae) (*Struthanthus aff. pyrifolius*)

STEPHANOCOLPATE

1a. Grains with six or more colpi . . . 2
1b. Grains with four colpi . . . 4

2a. Grains with 6 colpi; exine 1.5 to 2 μm thick; sexine baculate or scabrate . . . 3
2b. Grains with 14 colpi; exine 2.5 to 3 μm thick; sexine reticulate . . . Acanthaceae (*Pachystachys aff. lutea*)

3a. Grains oblate 50 μm; sexine baculate; apertures short, inconspicuous, appearing as pseudoapertures; grains possibly with more than 6 colpi . . . Convolvulaceae (*Aniseia*)
3b. Grains prolate-spheroidal 22 μm; sexine psilate to scabrate; colpi well defined as long as grain . . . Labiatae (Lamiaceae) (*Hyptis*)

4a. Grains <40 μm; sexine psilate to scabrate . . . 5
4b. Grains >40 μm; sexine verrucate or reticulate . . . 6

5a. Grains suboblate 18 × 21 μm; exine 1.5 μm thick; colpi longitudinally straight . . . Magnoliopsidae 3
5b. Grains prolate-spheroidal 36 × 34 μm; exine 2.5 μm thick; opposite colpi appearing as crossed lines . . . Myrsinaceae (*Myrsine* [*Rapanea*])

6a. Sexine reticulate; exine 4 to 5 μm thick; colpi equatorially equidistant . . . 7
6b. Grains verrucate; exine 2 μm thick; colpi displaying complex position appearing as crossed . . . Convolvulaceae 1

7a. Reticulum homobrochate; muri simplibaculate; grains 42 μm . . . Bombacaceae (*Chorisia* / *Ceiba* form 1)
7b. Reticulum heterobrochate; muri duplibaculate; grains 76 to 78 μm . . . Bombacaceae (*Chorisia* / *Ceiba* form 2)

HETEROCOLPATE

1a. Grains displaying 3 colpori alternating with 3 pseudocolpi . . . 2
1b. Grains displaying 4 colpori alternating with 4 pseudocolpi . . . Magnoliopsidae 11

2a. Grains <20 μm . . . 3
2b. Grains >20 μm . . . 5

3a. Grains psilate to scabrate . . . 4
3b. Grains verrucate . . . Melastomataceae

4a. Pseudocolpi depressed; pores rectangular . . . Melastomataceae (*Clidemia*)
4b. Pseudocolpi at same level than colpi; pores lalongate . . . Melastomataceae (*Miconia*)

5a. Grains 25 µm, subprolate; polar area 5 µm 6
5b. Grains 33 µm, prolate-spheroidal; polar area 8 µm . . . . Boraginaceae (*Heliotropium*)

6a. Exine 2.5 µm thick; pores elliptic . . . . Combretaceae (*Combretum*)
6b. Exine 1 µm thick; pores circular . . . . Combretaceae (*Terminalia*)

DICOLPORATE

1a. Grains <30 µm 2
1b. Grains >30 µm 3

2a. Exine 2 to 3 µm thick; pores becoming to be colpus transversalis . . . . . . Guttiferae (Clusiaceae) (*Clusiella*)
2b. Exine 1 to 2 µm thick; pores circular or transversally parallel . . . . Apocynaceae (*Macoubea*)

3a. Sexine scabrate; exine 2 to 3 µm thick; grains structurally complex . . . . . Magnoliopsidae 4
3b. Sexine psilate; exine 1 to 2 µm thick; grains structurally simple . . . . Apocynaceae (*Secondatia*)

TRICOLPORATE

1a. Grains without sculpture (psilate) 2
1b. Grains sculptured 24

2a. Grains less than 25 µm 3
2b. Grains more than 25 µm 12

3a. Grains suboblate to oblate-spheroidal 4
3b. Grains spheroidal to subprolate 6

4a. Colpi short (brevicolpate); grains 13 to 25 µm 5
4b. Colpi ⅔ length of grain; grains less than 12 µm . . . . Elaeocarpaceae

5a. Pores lalongate; colpi 5 µm length; ambitus circular . . . . Sapotaceae (*Sarcaulus*)
5b. Pores circular; colpi 8 µm lengths; ambitus semitriangular . . . . Apocynaceae (*Macoubea aff. guianensis*)

6a. Grains prolate-spheroidal to spheroidal; polar area less than 6 µm 7
6b. Grains subprolate; polar area 8 µm . . . . Leguminosae – Papilionoideae (*Dalbergia* form 1)

7a. Pores circular 8
7b. Pores lalongate or transversally parallel 9

8a. Grains 18 × 16 µm . . . . Magnoliopsidae 9
8b. Grains 12 × 11 µm . . . . Lacistemataceae (*Lacistema*)

9a. Pores forming continuous equatorial ring (colpus equatorialis) 10
9b. Pores lalongate 6 to 8 µm wide 11

10a. Grains 20 × 18 µm . . . . Solanaceae (*Solanum* form 2)
10b. Grains 16 µm . . . . Solanaceae (*Capsicum*)

11a. Grains 21 µm; ambitus triangular . . . . Leguminosae – Caesalpinioideae (*Copaifera aff. cordifolia*)
11b. Grains 15 µm; ambitus circular . . . . Malpighiaceae (*Byrsonima*)

12a. Grains oblate to oblate-spheroidal 13
12b. Grains spheroidal to prolate 17

13a. Pores lalongate 14
13b. Pores circular, protruding . . . . . Rhamnaceae (*Gouania* form 1)

14a. Exine >4 μm thick; ambitus circular 15
14b. Exine 1.5 μm thick; ambitus triangular 16

15a. Pores transversally parallel 1 × 10 μm; margo present; grains 32 μm . . . . Humiriaceae (*Humiria aff. balsamifera*)
15b. Pores lalongate 6 × 12 μm; margo absent; grains 35 μm . . . . . Marcgraviaceae (*Souroubea*)

16a. Grains 25 to 28 μm . . . . . Sapindaceae (*Talisia*)
16b. Grains 42 × 50 μm . . . . . Magnoliopsidae 8

17a. Grains spheroidal to prolate-spheroidal 18
17b. Grains subprolate to perprolate 19

18a. Pores forming continuous equatorial ring (colpus equatorialis); grains 30 μm . . . . . Solanaceae (*Solanum* form 3)
18b. Pores lalongate 4 × 12 μm; grains 40 μm . . . . . Leguminosae – Papilionoideae (*Bowdichia aff. nitida*)

19a. Pores circular 20
19b. Pores lalongate 22

20a. Pores <2.5 μm diameter; colpi displaying equatorial constriction 21
20b. Pores 3 μm diameter; colpi common type . . . . . Sapotaceae (*Chrysophyllum*) (*Pouteria* form 3)

21a. Grains >30 μm; exine variable 1 to 2 μm thick . . . . . Leguminosae – Caesalpinioideae (*Bauhinia aff. reflexa*)
21b. Grains <30 μm; exine 1 μm thick . . . . . Flacourtiaceae (*Xylosma*)

22a. Grains <35 μm 23
22b. Grains >35 μm . . . . . Burseraceae (*Protium*)

23a. Pores forming continuous equatorial ring (colpus equatorialis); grains prolate . . . . . Solanaceae (*Solanum* form 1)
23b. Pores lalongate; grains perprolate . . . . . Umbelliferae (Apiaceae) (*Hydrocotyle*)

24a. Grains with positive sculpture 25
24b. Grains with negative sculpture 82

25a. Grains scabrate 26
25b. Grains not as above 60

26a. Grains spheroidal 27
26b. Grains oblate-spheroidal to oblate or prolate-spheroidal to prolate 29

27a. Grains <20 μm; exine 1 to 2 μm thick; margo absent 28
27b. Grains 28 to 32 μm; exine 2 to 2.5 μm thick; colpi with margo . . . . . Ochnaceae (*Ouratea*)

28a. Colpi with thin and free opercula; grains 16 μm . . . . . Euphorbiaceae (*Alchornea* form 6)
28b. Opercula absent; grains 19 to 20 μm . . . . . Leguminosae – Papilionoideae (*Machaerium*)

29a. Grains prolate-spheroidal to perprolate 30
29b. Grains oblate-spheroidal to oblate 44

30a. Grains prolate to perprolate 31

30b. Grains subprolate to prolate-spheroidal 34

31a. Grains <45 µm 32

31b. Grains >45 µm . . . . . Umbelliferae (Apiaceae) 1&2

32a. Grains <35 µm; exine 2 µm thick; costa endocolpi 2.5 to 3 µm thick; margo absent 33

32b. Grains >35 µm; exine 2.5 to 3 µm thick; costa endocolpi 4 µm; margo present . . . . . Araliaceae (*Dendropanax*)

33a. Grains perprolate 35 × 15 µm; pores circular 5 µm diameter . . . . . Euphorbiaceae (*Hyeronima*)

33b. Grains prolate 26 × 18 µm; pores elliptic 3 × 6 µm . . . . . Burseraceae (*Trattinickia*)

34a. Grains >28 µm 35

34b. Grains <28 µm 38

35a. Grains <50 µm; exine <3 µm thick 36

35b. Grains >50 µm; exine 4 to 5 µm thick . . . . . Leguminosae – Papilionoideae 1

36a. Colpi narrow, ⅔ length of grain, borders uniform 37

36b. Colpi wide, as long as grain, displaying equatorial constriction . . . . . Leguminosae – Caesalpinioideae (*Cassia* form 2)

37a. Exine 2 µm thick; pores lalongate 3 × 8 µm . . . . . Flacourtiaceae (*Casearia* form 3)

37b. Exine 1 µm thick; pores lalongate 4 × 12 µm . . . . . Burseraceae (*Crepidospermum*)

38a. Pores lalongate or rectangular 39

38b. Pores circular 40

39a. Colpi displaying slightly an equatorial constriction; ambitus circular . . . . . Bignonaceae (*Arrabidaea* form 1)

39b. Colpi narrow, without equatorial constriction; ambitus triangular . . . . . Burseraceae (*Dacryodes aff. chimatensis*)

40a. Exine variable 1 to 2.5 µm thick; polar area 10 to 11 µm 41

40b. Exine uniform 1 µm thick; polar area 7 µm 42

41a. Pores 5 µm diameter; colpi short . . . . . Sapotaceae (*Pouteria* form 2)

41b. Pores 3 µm diameter; colpi ⅔ length of grain . . . . . Flacourtiaceae (*Casearia* form 2)

42a. Pores 1 to 3 µm diameter 43

42b. Pores 7 µm diameter . . . . . Leguminosae – Papilionoideae (*Dalbergia* form 2)

43a. Colpi narrow, short, pores 1 µm diameter . . . . . Sapotaceae (*Pouteria aff. cladantha*)

43b. Colpi ⅔ length of grain; pores 3 µm diameter . . . . . Malpighiaceae (*Blepharanda*)

44a. Grains more than 30 µm 45

44b. Grains less than 30 µm 50

45a. Exine >2 µm thick 46

45b. Exine ≤2 µm thick 48

46a. Grains 32 to 38 µm; colpi as long as grai	47
46b. Grains 45 µm; colpi ⅓ length of grain	Humiriaceae (*Humiria* form 2)
47a. Grains 32 µm; exine 2 µm thick	Apocynaceae (*Aspidosperma aff. album*)
47b. Grains 38 µm; exine 3 µm thick	Apocynaceae (*Aspidosperma aff. excelsum*)
48a. Grains <32 µm; opercula absent	49
48b. Grains 32 µm; colpi with long and free conspicuous opercula	Euphorbiaceae (*Alchornea* form 3)
49a. Colpi with margo and equatorial constriction; pores lalongate	Vochysiaceae (*Vochysia*)
49b. Colpi not as above; pores transversally parallel, vestibulate	Apocynaceae (*Parahancornia* form 1)
50a. Exine 1 µm thick	51
50b. Exine 1.5 to 4 µm thick	52
51a. Grains suboblate; pores lalongate; polar area 5 to 6 µm	Chrysobalanaceae (*Licania*)
51b. Grains oblate-spheroidal; pores lolongate; polar area 12 µm	Guttiferae (Clusiaceae) (*Clusia aff. odorata*)
52a. Ambitus triangular	53
52b. Ambitus circular	55
53a. Polar area 6 to 7 µm; exine variable; margo absent	54
53b. Polar area 3 µm; exine uniform; colpi with margo	Rhamnaceae (*Gouania* form 2)
54a. Grains suboblate; pores conspicuous, protruding	Chrysobalanaceae (*Couepia*)
54b. Grains oblate-spheroidal; pores inconspicuous	Olacaceae (*Minquartia aff. guianensis*)
55a. Pores lalongate	56
55b. Pores circular	Leguminosae – Papilionoideae (*Dipterix*)
56a. Exine uniform; opercula absent	57
56b. Exine variable; colpi with opercula	Euphorbiaceae (*Alchornea* form 1)
57a. Colpi short	58
57b. Colpi ⅔ to as long as grain	59
58a. Pores 3 × 16 µm, common type	Magnoliopsidae 5
58b. Pores 2 × 6 µm, protruding, with costa endopori	Apocynaceae (*Parahancornia* form 2)
59a. Colpi displaying slightly equatorial constriction	Marcgraviaceae (*Norantea aff. guianensis*)
59b. Colpi not as above	Meliaceae (*Cabralea*)
60a. Sexine granulate	61
60b. Sexine not as above	64
61a. Grains less than 50 µm	62
61b. Grains more than 50 µm	Bignoniaceae (*Jacaranda* form 2)

62a. Grains 30 to 32 µm; colpi as long as grain 63
62b. Grains 26 µm; colpi short (brevicolpate) . . . . . . Meliaceae (*Carapa aff. guianensis*)

63a. Colpi with robust and free opercula; margo absent; polar area 5 µm . . . . Chrysobalanaceae (*Hirtella aff. racemosa*)
63b. Opercula absent; colpi with margo; polar area 10 µm . . . . . . Euphorbiaceae (*Cleidion* form 2)

64a. Sexine baculate 65
64b. Sexine not as above 69

65a. Grains oblate-spheroidal to suboblate; exine <2 µm thick 66
65b. Grains prolate-spheroidal to prolate; exine >2 µm thick 68

66a. Grains <35 µm; annulus absent 67
66b. Grains 44 µm; pores annulate . . . . . . Rubiaceae (*Coccocypselum*)

67a. Grains 30 × 33 µm; pores 5 to 6 µm diameter, depressed . . . . . . Euphorbiaceae (*Sebastiana*)
67b. Grains 18 × 21 µm; pores 2 µm diameter, common type . . . . . . Menispermaceae (*Anomospermum*)

68a. Grains 40 µm; pores circular . . . . . . Magnoliopsidae 10
68b. Grains 38 to 70 µm; pores lalongate . . . . . . Euphorbiaceae (*Mabea / Sapium*)

69a. Sexine echinate, verrucate, pilate or striate 70
69b. Sexine clavate . . . . . . Aquifoliaceae (*Ilex*)

70a. Sexine echinate 71
70b. Sexine not as above 74

71a. Spine acute, pin shaped, <1 to 3 µm length × 1 µm wide; columellae inconspicuous 72
71b. Spine conical, sharp, broad based, 1.5 to 6 µm length × 1 to 10 µm wide; columella conspicuous . . . . . . Compositae (Asteraceae)

72a. Grains <25 µm; pores circular; grains spheroidal 73
72b. Grains 48 µm; pores lalongate; grains suboblate . . . . . . Boraginaceae (*Cordia aff. nodosa*)

73a. Grains 24 µm; colpi as long as grain; surface between spine smooth . . . . . Sapotaceae (*Diploon aff. cuspidatum*)
73b. Grains 15 to 17 µm; colpi ⅔ length of grain; surface between spine scabrate . . . . . . Guttiferae (Clusiaceae) (*Tovomitopsis* form 1)

74a. Sexine verrucate or striate; grains <45 µm; exine tectate 75
74b. Sexine pilate; grains >45 µm; exine intectate . . . . . . Leguminosae – Papilionoideae 2

75a. Grains verrucate >20 µm; exine >2 µm thick 76
75b. Grains striate <20 µm; exine <1 µm thick 81

76a. Colpi with opercula 77
76b. Opercula absent 79

77a. Pores lalongate; grains 25 to 28 µm; exine uniform 2 µm thick 78
77b. Pores circular; grains 32 × 40 µm; exine variable 2 to 5 µm thick . . . . . . Euphorbiaceae (*Alchornea* form 4)

78a. Grains 25 µm . . . . . . Euphorbiaceae (*Cleidion* form 1)
78b. Grains 28 µm . . . . . . Euphorbiaceae (*Cleidion aff. praealtum*)

79a. Pores lalongate; exine variable 4 to 5 µm thick; colpi as long as grain . . . 80
79b. Pores circular; exine uniform 2 µm thick; colpi short (brevicolpate) . . . . Symplocaceae (*Symplocos*)

80a. Grains suboblate; pores vestibulate . . . . . . Chrysobalanaceae (*Chrysobalanus aff. icaco*)
80b. Grains prolate-spheroidal; pores appearing as colpus equatorialis . . . . . Leguminosae – Caesalpinioideae (*Elizabetha aff. paraensis*)

81a. Grains 10 × 5 µm . . . . . . Begoniaceae (*Begonia*)
81b. Grains 19 × 15 µm . . . . . . Anacardiaceae (*Campnosperma*)

82a. Grains reticulate or striato-reticulate 83
82b. Grains fossulate . . . . . . Rutaceae (*Adiscanthus aff. fusciflorus*)

83a. Grains striato-reticulate 84
83b. Grains reticulate 88

84a. Grains <50 µm; exine <2.5 µm thick 85
84b. Grains >50 µm; exine >2.5 µm thick . . . . . . Leguminosae – Caesalpinioideae (*Macrolobium*)

85a. Grains 38 to 39 µm 86
85b. Grains 15 to 27 µm 87

86a. Grains prolate; exine semitectate . . . . . . Leguminosae – Caesalpinioideae (*Crudia aff. amazonica*)
86b. Grains subprolate; exine tectate . . . . . . Anacardiaceae (*Tapirira aff. guianensis*)

87a. Grains 15 × 12 µm; pores lalongate . . . . . . Rutaceae (*Zanthoxylum*)
87b. Grains 27 × 23 µm; pores transversally parallel . . . . . . Anacardiaceae (*Astronium aff. graveolens*)

88a. Grains >48 µm 89
88b. Grains <48 µm 100

89a. Reticulum heterobrochate 90
89b. Reticulum homobrochate 93

90a. Muri undulating (per-reticulate condition); grains spheroidal to suboblate 91
90b. Muri uniform; grains prolate . . . . . . Flacourtiaceae (*Laetia aff. coriacea*)

91a. Grains suboblate, 50 to 58 µm; exine 6 to 7 µm thick; pores lalongate 92
91b. Grains spheroidal, 70 µm; exine 3 µm thick; pores appearing as costa equatorialis . . . . . . Euphorbiaceae (*Dalechampia aff. tiliifolia*)

92a. Grains 45 × 50 µm; exine 6 µm thick; baculi 1.5 µm wide × 3 µm height . . . . . . Euphorbiaceae (*Amanoa aff. congesta*)

92b. Grains 55 × 58 µm; exine 7 µm thick; baculi 2 µm wide × 4 µm height . . . . . Euphorbiaceae (*Amanoa* form 2)

93a. Pores lalongate to rectangular; annulus and margo absent 94
93b. Pores circular displaying "H" form, annulate; colpi with thick margo . . . Humiriaceae (*Vantanea aff. guianensis*)

94a. Grains prolate to perprolate 95
94b. Grains spheroidal to suboblate 97

95a. Grains 48 to 58 µm; colpi displaying exitus digitatus 96
95b. Grains 60 µm; colpi common type . . . . . Tiliaceae (*Heliocarpus*)

96a. Colpi common type . . . . . Umbelliferae (Apiaceae) 3
96b. Colpi with equatorial constriction and forming bridge . . . . . Vitaceae (*Cissus*)

97a. Grains 48 to 50 µm; exine 2 to 4 µm thick; pores lalongate 98
97b. Grains 63 to 65 µm; exine 5 to 7 µm thick; pores rectangular . . . . . Tiliaceae 1

98a. Grains suboblate; exine 2 µm thick; polar area 25 to 28 µm 99
98b. Grains subprolate; exine 3 to 4 µm thick; polar area 12 µm . . . . . Leguminosae – Caesalpinioideae (*Cassia* form 4)

99a. Brochi <1 µm wide; polar area 28 µm . . . . . Tiliaceae (*Apeiba* form 2)
99b. Brochi 1 µm wide; polar area 25 µm . . . . . Rubiaceae (*Psychotria* form 1)

100a. Reticulum heterobrochate 101
100b. Reticulum homobrochate 106

101a. Grains subprolate 102
101b. Grains suboblate to spheroidal 104

102a. Grains >40 µm; exine 1.5 to 2 µm thick; ambitus circular 103
102b. Grains <25 µm; exine 3 µm thick; ambitus triangular . . . . . Staphylleaceae (*Turpinia*)

103a. Exine tectate, 1.5 µm thick; pores displaying elongated apices ("H" form) . . . . . Magnoliopsidae 7
103b. Exine semitectate, 2 µm thick; pores lalongate . . . . . Rutaceae (*Citrus*)

104a. Grains spheroidal, 31 to 32 µm; exine 1 to 2.5 µm thick 105
104b. Grains suboblate, 36 µm; exine 5 µm thick . . . . . Euphorbiaceae (*Amanoa aff. nanayensis*)

105a. Grains 32 µm; exine 2.5 µm thick; pores circular, annulate . . . . . Rubiaceae (*Ferdinandusa*)
105b. Grains 21 µm; exine 1 µm thick; pores lalongate; annulus absent . . . . . Menispermaceae (*Cissampelos* form 1)

106a. Grains spheroidal 107
106b. Grains oblate-spheroidal to suboblate or prolate-spheroidal to prolate . . 109

107a. Grains 21 to 22 µm; exine 1.5 µm thick; pores lolongate 108
107b. Grains 40 µm; exine 4 µm thick; pores lalongate displaying "H" form . . . Humiriaceae (*Endopleura aff. uchi*)

108a. Colpi thin, ⅔ length of grain . . . . . Marcgraviaceae (*Marcgravia*)
108b. Colpi narrow, as long as grain . . . . . Leguminosae – Caesalpinioideae (*Schizolobium*)

109a. Grains oblate-spheroidal to suboblate 110
109b. Grains prolate-spheroidal to prolate 117

110a.	Pores without opercula	111
110b.	Pores with thin opercula	Euphorbiaceae (*Alchornea* form 2)
111a.	Exine 2 to 5 µm thick	112
111b.	Exine 1 to 2 µm thick	115
112a.	Pores circular	113
112b.	Pores lalongate or transversally parallel	114
113a.	Pores conspicuous; grains 34 × 36 µm; exine 4 µm thick	Leguminosae – Caesalpinioideae (*Martiodendron*)
113b.	Pores inconspicuous; grains 30 × 32 µm; exine 1 to 3 µm thick	Rubiaceae (*Pagamea*)
114a.	Grains <32 µm; ambitus circular; pores inconspicuous	Celastraceae (Hippocrateaceae) (*Peritassa / Plenckia*)
114b.	Grains >32 µm; ambitus triangular; pores well defined	Araliaceae (*Didymopanax*)
115a.	Grains 26 to 31 µm; pores inconspicuous apparently lalongate	116
115b.	Grains 17 µm; pores circular, well defined	Guttiferae (Clusiaceae) 1
116a.	Pores slightly protruding; polar area 16 µm	Rubiaceae (*Psychotria* forms 3&4)
116b.	Pores not as above; polar area 20 µm	Tiliaceae (*Apeiba* form 1)
117a.	Pores lolongate or circular	118
117b.	Pores lalongate or transversally parallel	124
118a.	Grains <38 µm; pores circular	119
118b.	Grains >38 µm; pores lolongate	Polygonaceae (*Symmeria aff. paniculata*)
119a.	Grains 26 to 36 µm	120
119b.	Grains 12 to 20 µm	121
120a.	Grains prolate to subprolate; pores 3 to 6 mm diameter, slightly elliptic	Erythroxylaceae (*Erythroxylum* forms 1&2)
120b.	Grains prolate-spheroidal to spheroidal; pores 10 µm diameter	Cucurbitaceae (*Posadaea*)
121a.	Grains 12 to 15 µm; exine 1 µm thick	122
121b.	Grains 19 µm; exine 1.5 to 2 µm thick	123
122a.	Grains prolate-spheroidal; colpi as long as grain	Cunoniaceae (*Weinmannia*)
122b.	Grains subprolate; colpi ⅔ length of grain	Flacourtiaceae (*Banara*)
123a.	Pores annulate	Rubiaceae (*Genipa aff. americana*)
123b.	Annuli absent	Dilleniaceae (*Tetracera*)
124a.	Grains 30 to 44 µm	125
124b.	Grains 17 to 29 µm	135
125a.	Grains subprolate to prolate	126
125b.	Grains perprolate	Umbelliferae (Apiaceae) (*Eryngium*)
126a.	Colpi displaying equatorial constriction	127
126b.	Colpi not as above	130
127a.	Colpi with exitus digitatus; grains 34 to 35 µm	128
127b.	Colpi common type; grains 40 to 44 µm	129

128a. Colpi 3.5 × 8 µm; costa endocolpi 1.5 µm thick . . . . Erythroxylaceae (*Erythroxylum* form 3)
128b. Colpi 3 × 12 µm; costa endocolpi 2 µm thick . . . . Euphorbiaceae (*Phyllanthus* form 1)

129a. Exine 2 µm thick; pores elliptic . . . . Tiliaceae (*Lueheopsis*)
129b. Exine 2.5 µm thick; pores lalongate . . . . Bignoniaceae (*Tabebuia* form 1)

130a. Exine 3 to 4 µm thick 131
130b. Exine 2 to 2.5 µm thick 133

131a. Colpi thin, narrow; exine uniform 132
131b. Colpi wide, deep; exine variable . . . . Leguminosae – Caesalpinioideae (*Cassia* form 3)

132a. Columellae (bacula) 2 µm height; pores inconspicuous . . . . Euphorbiaceae (*Amanoa* form 1)
132b. Columellae (bacula) 1 µm height; pores well defined . . . . Dilleniaceae (*Doliocarpus*)

133a. Grains 30 to 32 µm 134
133b. Grains 44 µm . . . . Euphorbiaceae (*Drypetes*)

134a. Grains subprolate . . . . Leguminosae – Papilionoideae (*Amburana aff. cearensis*)
134b. Grains prolate . . . . Tiliaceae (*Luehea*)

135a. Grains subprolate to prolate 136
135b. Grains prolate-spheroidal 139

136a. Exine 1.5 µm thick 137
136b. Exine 1 µm thick 138

137a. Pores transversally parallel 1 × 4 µm; colpi displaying equatorial constriction . . . . Araliaceae (*Schefflera aff. bangui*)
137b. Pores rectangular 7 × 9 µm; colpi not as above . . . . Leguminosae – Papilionoideae (*Centrolobium*)

138a. Pores lalongate 2 × 10 µm; colpi as long as grain . . . . Euphorbiaceae (*Phyllanthus* form 2)
138b. Pores elliptic 3 × 6 µm; colpi ⅔ length of grain . . . . Umbelliferae (Apiaceae) (*Apium*)

139a. Colpi as long as grain; brochi well defined 140
139b. Colpi ⅔ length of grain; brochi resembling baculate pattern . . . . Menispermaceae (*Abuta*)

140a. Grains <20 µm; pores inconspicuous 141
140b. Grains >20 µm; pores well defined . . . . Magnoliopsidae 6

141a. Grains 19 × 16 µm; pores depressed . . . . Leguminosae – Papilionoideae (*Myrocarpus aff. frondosus*)
141b. Grains 17 × 15 µm; pores not as above . . . . Leguminosae – Caesalpinioideae (*Cercidium*)

PERICOLPORATE

1a.	Grains 31 µm, with 6 colpori; sexine scabrate to granulate; colpi thin	Malpighiaceae (*Diacidia*)
1b.	Grains 30 to 37 µm, with 4 colpori; sexine reticulate, heterobrochate; colpi wide	Malpighiaceae (*Tetrapteryx*)

SYNCOLPORATE

1a.	Grains displaying reticulate pattern	2
1b.	Grains displaying psilate, scabrate or verrucate patterns	4
2a.	Grains >30 µm; exine 3 to 5 µm thick; colpi junction straight; ambitus circular	3
2b.	Grains <30 µm; exine 2 µm thick; colpi junction forming a small triangle (parasyncolporate condition)	Sapindaceae (*Cupania* form 2)
3a.	Reticulum homobrochate; pores 16 µm diameter; grains 48 to 52 µm	Bignoniaceae (*Mansoa*)
3b.	Reticulum heterobrochate; pores 5 µm diameter; grains 38 µm	Leguminosae – Caesalpinioideae (*Cenostigma / Caesalpinia*)
4a.	Grains ≥ 16 µm; sexine scabrate	5
4b.	Grains <16 µm; sexine psilate	Myrtaceae (*Eugenia* form 1)
5a.	Grains spheroidal to prolate	6
5b.	Grains suboblate to oblate	7
6a.	Grains >35 µm; exine variable 1.5 to 5 µm thick; pores lalongate, vestibulate; colpi junction forming a small triangle	Caryocaraceae (*Caryocar*)
6b.	Grains <30 µm; exine uniform 2 µm thick; pores circular, annulate; colpi junction straight	Leguminosae – Mimosoideae (*Entada aff. monostachia*)
7a.	Colpi junction forming a small triangle (parasyncolporate condition)	8
7b.	Colpi junction straight	11
8a.	Grains >30 µm; exine >2 µm thick	9
8b.	Grains <30 µm; exine 1 µm thick	Myrtaceae (*Syzygium*)
9a.	Grains >40 µm; exine 2.5 µm thick	10
9b.	Grains 30 to 35 µm; exine 2 µm thick	Sapindaceae (*Matayba*)
10a.	Grains verrucate	Magnoliopsidae 13
10b.	Grains scabrate	Sapindaceae (*Cupania* form 1)
11a.	Grains <30 µm; scabrae uniform	12
11b.	Grains >30 µm; scabrae becoming flat wart-like structures at mesocolpium	Leguminosae – Caesalpinioideae (*Cassia* form 1)
12a.	Exine tectate	13
12b.	Exine semitectate, displaying free scabrate process at mesocolpium	Myrtaceae 2
13a.	Pores circular	14
13b.	Pores lalongate	Myrtaceae 1

14a. Pores 1 µm diameter; grains 16 µm . . . . Myrtaceae (*Eugenia* form 2)
14b. Pores 3 to 4 µm diameter; grains 24 µm . . . . Myrtaceae (*Myrcia*)

STEPHANOCOLPORATE

1a. Grains with 5 or more colpori 2
1b. Grains with 4 colpori 8

2a. Grains with 6 to 8 colpori 3
2b. Grains with 5 colpori 5

3a. Grains <25 µm; exine 2 to 3 µm thick; pores lalongate 4
3b. Grains 40 µm; exine 1 to 1.5 µm thick; pores circular . . . . Polygalaceae (*Diclidanthera*)

4a. Grains subprolate 21 × 18 µm; sexine psilate to scabrate; colpi as long as grain; pores lalongate forming an equatorial ring . . . . . . Polygalaceae (*Polygala aff. paniculata*)
4b. Grains oblate-spheroidal 20 × 22 µm; sexine reticulate, homobrochate; colpi short (brevicolpate); pores slightly lalongate . . . . Rubiaceae (*Borreria aff. densiflora*)

5a. Grains spheroidal to subprolate <50 µm; exine <3 µm thick 6
5b. Grains oblate >55 µm; exine 3 to 5 µm thick . . . . Apocynaceae (*Molongum*)

6a. Pores elliptic; colpi ⅔ length of grain, common type 7
6b. Pores circular; colpi short (brevicolpate) displaying costa endocolpi . . . . Mendonciaceae (*Mendoncia aff. glabra*)

7a. Grains spheroidal 36 µm; pores 4 × 7 µm; exine 2 µm thick . . . . Sapotaceae (*Manilkara*)
7b. Grains subprolate 46 × 34 µm; pores 5 × 10 µm; exine 2.5 to 3 µm thick . . . . Meliaceae (*Cedrela* form 1)

8a. Sexine reticulate, homobrochate 9
8b. Sexine psilate to scabrate 12

9a. Grains suboblate 10
9b. Grains subprolate to prolate 11

10a. Exine 1.5 to 2.5 µm thick; pores transversally parallel 16 µm wide; colpi longitudinally straight . . . . Cucurbitaceae (*Cyclanthera*)
10b. Exine 3 to 4 µm thick; pores circular 8 µm diameter; opposite colpi crossed line . . . . Sapotaceae (*Elaeoluma* sp.)

11a. Pores circular 7 µm diameter, with long bifurcated and conspicuous opercula . . . Euphorbiaceae (*Alchornea* form 5)
11b. Pores lalongate 3 × 7 µm; opercula absent . . . . Sapotaceae (*Elaeoluma aff. nuda*)

12a. Grains >48 µm; exine 4 to 7 µm thick 13
12b. Grains <48 µm; exine 1 to 3 µm thick 15

13a. Pores lalongate; surface smooth 14
13b. Pores circular; surface resembling rugulate pattern . . . . Meliaceae 1

14a. Grains subprolate to prolate-spheroidal 58 µm; exine 6 to 7 µm thick; costa endocolpi 5 µm thick . . . . Magnoliopsidae 12
14b. Grains spheroidal 48 µm; exine 4 to 5 µm thick; costa endocolpi 3 µm thick . . . . Meliaceae (*Cedrela* form 2)

15a.	Grains spheroidal to oblate	16
15b.	Grains prolate-spheroidal to prolate	19
16a.	Pores lalongate or lolongate	17
16b.	Pores circular	18
17a.	Grains oblate; exine 1 µm thick; colpi with conspicuous margo; pores lolongate	Apocynaceae (*Aspidosperma aff. megalocarpum*)
17b.	Grains spheroidal; exine 2 µm thick; margo absent; pores lalongate	Sapotaceae (*Manilkara*)
18a.	Pores 3 to 6 µm diameter; annulus present; grains >35 µm	Meliaceae (*Guarea*)
18b.	Pores 1.5 to 3 µm diameter; annulus absent; grains <35 µm	Euphorbiaceae (*Acalypha*)
19a.	Pores circular	20
19b.	Pores lalongate	Meliaceae (*Trichilia*)
20a.	Pores 1.5 to 2 µm diameter; grains <25 µm; exine 1 to 1.5 µm thick	21
20b.	Pores 4 µm diameter; grains 25 to 30 µm; exine 2 to 3 µm thick	Flacourtiaceae (*Casearia* form 2)
21a.	Grains <19 µm	Sapotaceae (*Pouteria aff. elegans*)
21b.	Grains >19 µm	Sapotaceae (*Pouteria* forms 1&4)

POLLEN AND SPORE DESCRIPTIONS

PTERIDOPHYTA – LYCOPODIOPSIDA

Lycopodiaceae

Lycopodium clavatum L. Plate 1:1

Spores trilete, heteropolar, radially symmetric; sclerine 3 to 4 µm thick, per-reticulate, heterobrochate; brochi variable 1 to 3 µm wide; muri with conspicuous baculi *ca.* 3 µm height; laesurae narrow, straight, extending up to border of proximal face; amb subtriangular-convex (proximal view); spores (equatorial dimension) 35 to 40 µm.

Note: Used as exotic marker.

Lycopodium aff. linifolium L. Plate 1:2

Spores trilete, heteropolar, radially symmetric; sclerine thin 1.5 µm thick, displaying spinulose pattern; spinule short <1 µm long, base wide *ca.* 1 to 1.5 µm, regularly scattered on the surface; laesurae thin, narrow, carinate, extending up to border of proximal face; amb subtriangular-convex (proximal view); spores (equatorial dimension) 32 to 33 µm.

Lycopodium aff. subulatum Desv. Plate 1:3

Spores trilete, preserving the dyads stage from the original tetrad stage, heteropolar, radially symmetric; sclerine 2 µm thick, displaying foveolate pattern; foveolae circular, homogeneous *ca.* 1 µm diameter; regularly scattered on the surface; laesurae very thin, narrow, almost reaching up to border of proximal face; amb subtriangular-convex (proximal view) and plano-convex (lateral view); spores (equatorial dimension) 34 × 43 µm; dyads 72 to 75 µm wide.

Lycopodium aff. trianae Hieron. Plate 1:4

Spores trilete, heteropolar, radially symmetric; sclerine 1.5 to 4 µm thick; sculpture reticulate, resembling a rugulate pattern; laesurae thin, commisure bordered by broad margo 2 to 5 µm; "Y" split mark extending up to ⅔ of proximal face; amb subtriangular-convex (proximal view); spores (equatorial dimension) 30 µm.

Selaginellaceae

Selaginella aff. arthritica Alston (form 1) Plate 2:8
Spores trilete, heteropolar, radially symmetric; sclerine 3.5 µm thick (excluding spinule); sculpture spinulose, spinule acute, long, wide, *ca.* 8 µm long × 4 µm wide; densely scattered on the surface; laesurae as long as the proximal face, displaying a narrow margines; amb angular to circular irregular (proximal view); spores (equatorial dimension) 58 to 67 µm (including sculptural elements).

Selaginella aff. arthritica Alston (form 2) Plate 2:9
Spores trilete, heteropolar, radially symmetric; sclerine 11 µm thick (including sculptural elements), displaying a conspicuous spinulose pattern; spinule irregular, acute, sometimes becoming to have bi- and trifurcate ends; base wide, apparently forming a continuous ring around equatorial border; distal face with a well-developed of irregular size verrucae, *ca.* from 1 × 1 µm to 5 × 8 µm; proximal face displaying small verrucae; laesurae narrow, sinuous, extending up to border of proximal face; commisure bordered by broad margo *ca.* 2.5 µm; border of margines irregular; amb subtriangular (proximal view); spores (equatorial dimension) 43 to 52 µm.

Selaginella aff. asperula Spring (form 1) Plate 1:5
Spores trilete, heteropolar, radially symmetric; sclerine 6 to 7 µm thick, displaying an irregular spinulose pattern; spinule short and scarce *ca.* <1 to 1 µm long; laesurae thin, sinuous, extending near to equator; commisure bordered by thin margo, *ca.* 1 µm wide; amb subtriangular irregular (proximal view); spores (equatorial dimension) 44 µm.

Selaginella aff. asperula Spring (form 2) Plate 1:6
Spores trilete, heteropolar, radially symmetric; sclerine 1.5 to 2 µm thick, with verrucae-gemmae elements densely arranged, 1 × 1 µm to 2 × 4 µm wide; laesurae thin, sinuous, extending up to ¾ of proximal face; commisure bordered by thin margo, *ca.* 1 µm wide; amb subtriangular-convex (proximal view); spores variable in size *ca.* 26 to 30 µm (equatorial dimension).

Selaginella aff. asperula Spring (form 3) Plate 1:7
Spores trilete, heteropolar, radially symmetric; sclerine irregular and variable in size, from 1 to 5 µm thick, with spinulose-like pattern; spinule scarce and irregularly scattered on the surface; laesurae narrow, undulate, extending up to border of proximal face; commisure bordered by thin margo, *ca.* 1 µm wide; amb subtriangular-convex (proximal view); spores (equatorial dimension) 28 × 36 µm.

Selaginella aff. asperula Spring (form 4) Plate 2:10
Spores trilete, heteropolar, radially symmetric; sclerine 3 to 7 µm thick, verrucate, with a laminar pattern resembling columellate condition as occur in angiosperms; laesurae inconspicuous, irregular; amb subtriangular to circular (proximal view); spores (equatorial dimension) 47 µm.

Selaginella aff. asperula Spring (form 5) Plate 2:11
Spores trilete, heteropolar, radially symmetric; sclerine 4 to 5 µm thick, pattern apparently psilate; laesurae inconspicuous, undulating, with ends not well defined; commisure bordered by margines; amb angular to circular (proximal view); spores (equatorial dimension) 37 µm.

Selaginella aff. exaltata (Kuntze) Spring in Mart. Plate 2:12
Spores trilete, heteropolar, radially symmetric; sclerine 4 µm thick (including ornamentation); with dense and well developed gemmate pattern; laesurae narrow, extending almost to equator; commisure bordered by broad margo 1.5 µm wide; amb subtriangular-semiconcave (proximal view); spores (equatorial dimension) 36 µm.

Selaginella sp. Pal. (form 1) Plate 2:13
Spores trilete, heteropolar, radially symmetric; sclerine 1 to 1.5 µm thick with slightly irregular border; psilate pattern; laesurae narrow, as long as spore; amb subtriangular (proximal view); spores (equatorial dimension) 18 µm.

Selaginella sp. Pal. (form 2) Plate 2:14
Spores trilete, heteropolar, radially symmetric; sclerine 3 µm thick, with spinulose pattern; spinule scarce, irregularly scattered on the surface; laesurae narrow, undulated, extending up to border of proximal face; amb subtriangular-convex (proximal view); spores (equatorial dimension) 32 µm.

PTERIDOPHYTA – FILICOPSIDA

Aspleniaceae (Dryopteridaceae)

Asplenium sp. L. (form 1) Plate 3:15
Spores monolete, heteropolar, bilaterally symmetric; sclerine 4 µm thick, apparently granulate; granules irregular *ca.* 1 µm height; densely scattered on the surface; perine present with low undulating membraneous ridges; laesura straight, narrow, as long as spore, *ca.* 56 µm; amb plano-convex (lateral view) and elliptic-biconvex (proximal view); spores 70 to 72 µm × 46 to 48 µm (lateral view).

Asplenium sp. L. (form 2) Plate 3:16
Spores monolete, heteropolar, bilaterally symmetric; sclerine 2.5 to 3 µm thick, apparently psilate; perine present spinulose-granulate with low undulating membraneous ridges; spinule irregular *ca.* 3 µm height; laesura straight, narrow, ⅓ of the spore, *ca.* 16 µm long; thick margo present, acute, *ca.* 2.5 to 3 µm wide; amb elliptic to circular (proximal view); spores 48 to 50 µm × 46 to 48 µm (proximal view).

Asplenium sp. L. (form 3) Plate 3:17
Spores monolete, preserving its dyad stage, heteropolar, bilaterally symmetric; sclerine inconspicuous *ca.* 3 µm thick; perine present, membraneous, spinulose, fimbriate; spinule, variable in size *ca.* 9 to 20 µm; laesura not visible, apparently margo present; amb concave-convex (lateral view); spores 144 × 99 µm (lateral view); dyads 205 µm.

Asplenium sp. L. (form 4) Plate 3:18
Spores monolete, heteropolar, bilaterally symmetric; sclerine *ca.* 1 µm thick, psilate; perine present slightly scabrate with thin and irregular undulating membrane; projections of perine appearing as a conical spinule *ca.* 2 to 5 µm long; laesura ½ length of spore; thick margo present, *ca.* 2 µm wide; amb plano-convex (lateral view); spores 38 × 28 µm (lateral view).

Blechnaceae

Blechnum sp. L. Plate 3:19
Spores monolete, heteropolar, bilaterally symmetric; sclerine 1 µm thick, gemmate; gemmae variable in size *ca.* <1 to 2.5 µm height; densely scattered on the surface; laesura ½ length of spore, *ca.* 22 µm borders well defined, acute edges; margo thin, *ca.*1 µm wide; amb elliptic-biconvex (proximal view); spores 44 to 45 µm × 34 to 35 µm (proximal view).

Cyatheaceae

Alsophila sp. R. Br. Plate 3:20
Spores trilete, heteropolar, radially symmetric; sclerine 1 µm thick, psilate; laesurae straight, narrow, as long as proximal face; "Y" split mark conspicuous with thick margo, *ca.* 2.5 µm wide; amb subtriangular-semiconcave (proximal view); spores 26 µm.

Cyathea sp. Sm. (form 1) Plate 4:21
Spores trilete, heteropolar, radially symmetric; sclerine 1 µm thick, perforate-verrucate, flat verrucae densely scattered on both faces; laesurae wide, covering ¾ of the proximal face; "Y" split mark inconspicuous; amb subtriangular (proximal view); spores 28 µm.

Cyathea sp. Sm. (form 2) Plate 4:22
Spores trilete, heteropolar, radially symmetric; sclerine 1 µm thick (excluding ornamentation), densely verrucate; verrucae *ca.* 3 to 4 µm height × 2 to 5 µm wide, crowded; "Y" split mark inconspicuous, masked by ornamentation; amb subtriangular-semiconcave (proximal view); spores 48 to 54 µm.

Cyathea sp. Sm. (form 3) Plate 4:23
Spores trilete, heteropolar, radially symmetric; sclerine 9 to 10 µm thick, psilate; laesurae narrow, straight, as long as radius; margo ample, thick covering ¾ of surface of spore, *ca.* 10 to 15 µm wide; amb subtriangular-concave (proximal view); spores 90 to 110 µm.

Cyathea sp. Sm. (form 4) Plate 4:24
Spores trilete, heteropolar, radially symmetric; sclerine <1 to 1 µm thick, psilate; laesurae narrow, sinuous, becoming to be as long as radius; apparently thick margo present, *ca.* 7 µm wide, covering ⅔ of the proximal face; amb subtriangular-concave (proximal view); spores 40 to 45 µm.

Cyathea sp. Sm. (form 5) Plate 4:25
Spores trilete, heteropolar, radially symmetric; sclerine 1 µm thick (excluding ornamentation), verrucate; flat verrucae *ca.* 4 µm height × 8 to 10 µm wide; laesurae covering ⅔ of the proximal face; "Y" split mark triangular-concave, opened; with acute ends; margo variable *ca.* 1 to 5 µm; amb subtriangular (proximal view); spores *ca.* 70 µm.

Dicksoniaceae

Dicksonia sp. L'Hérit (form 1) Plate 4:26
Spores trilete, heteropolar, radially symmetric; sclerine 1 µm thick, thickening at angles *ca.* 3 µm as separate "tectum-like structure"; spores slightly scabrate; "Y" split mark conspicuous, laesurae straight, as long as radius, 1 µm wide; margo not well defined *ca.* 4 µm wide; amb subtriangular-semiconcave (proximal view); spores 40 to 45 µm.

Dicksonia sp. L'Hérit (form 2) Plate 5:27
Spores trilete, heteropolar, radially symmetric; sclerine 3 µm thick, thickening at angles *ca.* 4 to 5 µm; spores psilate to slightly scabrate; "Y" split mark conspicuous, laesurae straight, as long as radius, *ca.* 2 µm wide; margo undulating thicker at center of each radius *ca.* 3 µm wide; amb subtriangular (proximal view); spores 53 to 55 µm.

Dicksonia sp. L'Hérit (form 3) Plate 5:28
Spores trilete, heteropolar, radially symmetric; sclerine 2 to 3 µm thick, thickening at angles *ca.* 6 µm; spores slightly scabrate; laesurae acute, as long as radius, 1 to 3 µm wide; margo inconspicuous, thicker at center of each "Y"split mark radius, *ca.* 2.5 to 3 µm wide; amb subtriangular (proximal view); spores 70 to 75 µm.

Gleicheniaceae

Dicranopteris sp. Scop. Plate 5:29

Spores monolete, heteropolar, bilaterally symmetric; sclerine 1 μm thick, psilate; laesura short ⅓ length of spore; margo inconspicuous *ca.* 4 μm wide; amb plano-convex (lateral view); spores 49 × 31 μm (lateral view).

Grammitidaceae

Grammitis sp. Sw. Plate 5:30

Spores trilete, heteropolar, radially symmetric; sclerine 1 to 1.5 μm thick, apparently verrucate; flat verrucae irregular, variable in size *ca.* 2 μm in height × 3 to 4 μm wide, crowded at angles; "Y" split mark inconspicuous; laesurae and margo not visible; amb subtriangular-convex (proximal view); spores *ca.* 60 μm.

Hymenophyllaceae

Hymenophyllum aff. ciliatum Sw. Plate 6:31

Spores trilete, heteropolar, radially symmetric; sclerine 2 μm thick, densely and finely baculate, resembling thin pila, <1 μm height; laesurae narrow, straight, as long as radius, *ca.* 1 μm wide; margo *ca.* 4 μm wide; amb subtriangular-convex (proximal view); spores 80 μm.

Trichomanes aff. elegans Rich. Plate 6:32

Spores trilete, heteropolar, radially symmetric; sclerine 1 μm thick, ornamentation variable finely baculate, scabrate to spinulose; spinule 1 to 2 μm height; "Y" split mark conspicuous, laesurae straight, ¾ of radius, *ca.* 1 μm wide; thin margo present *ca.* 2 μm wide; amb irregular subtriangular-convex to rounded (proximal view); spores 48 to 50 μm.

Polypodiaceae

Polypodium sp. L. (form 1) Plate 6:33

Spores monolete, heteropolar, bilaterally symmetric; sclerine 1 to 1.5 μm thick (ornamentation excluded), verrucate; verrucae densely aggregated, protruding, *ca.* 3 μm height × 8 μm wide; laesura narrow, ⅔ length of spore; margo thin, 1 to 1.5 μm wide; amb elliptic-biconvex (proximal view) and plano-convex (lateral view); spores 60 to 65 μm × 38 to 40 μm (proximal view).

Polypodium sp. L. (form 2) Plate 6:34

Spores monolete, heteropolar, bilaterally symmetric; sclerine 6 μm thick (including ornamentation), verrucate; verrucae irregular, protruding, *ca.* 5 to 6 μm height × 2 to 10 μm wide; laesura inconspicuous; amb plano-convex (lateral view); spores 78 × 48 μm (lateral view).

Polypodium sp. L. (form 3) Plate 6:35

Spores monolete, heteropolar, bilaterally symmetric; sclerine 2 μm thick (ornamentation excluded), verrucate; verrucae protruding, some gemmae-like, *ca.* 5 μm height × 5 to 8 μm wide; laesura narrow, ⅔ length of spore; margo thick 4 to 5 μm wide; amb concave-convex (lateral view); spores 70 × 35 μm (lateral view).

Polypodium sp. L. (form 4) Plate 7:36

Spores monolete, heteropolar, bilaterally symmetric; sclerine 2 μm thick (including ornamentation), verrucate; verrucae small and flat scarcely dispersed on the surface, *ca.* 1 μm height; laesura narrow, sinuous, ⅔ length of spore, *ca.* 25 μm; margo inconspicuous or absent; amb elliptic-biconvex (proximal view); spores 47 × 27 μm (proximal view).

Pteridaceae

Pteris sp. L. Plate 7:37
Spores trilete, heteropolar, radially symmetric; sclerine 7 μm thick (including ornamentation), verrucate with rugulate-like pattern; verrucae flat and irregular, variable in size *ca.* 3 to 4 μm height × 6 to 10 μm wide; "Y" split mark inconspicuous; laesurae as long as radius; amb subtriangular (proximal view); spores *ca.* 58 μm.

Schizaeaceae

Schizaea aff. pennula Sw. Plate 7:38
Spores monolete, heteropolar, bilaterally symmetric; sclerine 1 μm thick, foveolate; foveolae uniform *ca.* 1 μm diameter; laesura inconspicuous, apparently ⅔ length of spore; amb concave-convex (lateral view); spores 56 × 28 μm (lateral view).

Thelypteridaceae

Thelypteris sp. Schmidel (form 1) Plate 7:39
Spores monolete, heteropolar, bilaterally symmetric; sclerine 3 to 4 μm thick, psilate; laesura undulating as long as spore; perine present appressed to sclerine forming fine spinulose crest; margo, if present, *ca.* 5 μm thick; amb concave-convex (lateral view); spores 75 × 34 μm (lateral view).

Thelypteris sp. Schmidel (form 2) Plate 7:40
Spores monolete, heteropolar, bilaterally symmetric; sclerine 3 μm thick, psilate; laesura undulating as long as spore; perine present appressed to sclerine forming conspicuous spinulose crest; spinule densely distributed on the surface, *ca.* 4 μm height; apparently margo absent; amb elliptic-biconvex (proximal view); spores 58 × 34 μm (proximal view).

Other Filicopsida

Filicopsida 1. Plate 7:41
Spores trilete, heteropolar, radially symmetric; sclerine 5 to 6 μm thick, per-reticulate, heterobrochate; brochi variable in size *ca.* 2 to 14 μm wide, muri thick, undulating, 3 to 4 μm wide; laesurae inconspicuous; apparently margo absent; amb subtriangular-convex (proximal view); spores *ca.* 84 × 50 μm.
Note: *aff. Crassoretitriletes vanraadshoovenii* (Germeraad, Hopping et Muller, 1968, in Hoogiemstra, 1984).

Filicopsida 2. Plate 7:42
Spores trilete, heteropolar, radially symmetric; sclerine 4 μm thick at inter-angles and 5 to 6 μm at angles, finely scabrate; laesurae thin, as long as radius; margo absent; amb subtriangular-concave (proximal view); spores *ca.* 32 to 38 μm.
Note: *aff.* #318 (Hooghiemstra, 1984).

Filicopsida 3. Plate 8:43
Spores monolete, heteropolar, bilaterally symmetric; sclerine 3.5 μm thick, apparently foveolate; foveolae *ca.* 1 μm diameter, irregularly scattered on the surface, resembling small perforations; laesura ⅔ length of spore, wide, acute, *ca.* 2.5 × 38 μm; margo thick *ca.* 4 μm wide; amb elliptic (proximal view); spores 67 × 54 μm (proximal view).

GYMNOSPERMAE

Podocarpaceae

Podocarpus sp. L'Hérit. ex Pers. (form 1) Plate 8:44

Grains vesiculate (bisaccate), diploxylonoid; bilaterally symmetric, heteropolar; exine tectate, *ca.* 2 µm thick; sexine reticulate, heterobrochate, muri thin <1 µm, lumina 3 to 11 µm wide; corpus rounded, with homogeneous cappa and cappula, corpus *ca.* 30 × 35 µm; sacci globose, linear orientation, sacci sometimes joined, *ca.* 38 to 40 µm × 30 µm; total length of grains *ca.* 68 to 70 µm. Tree.

Podocarpus sp. L'Hérit. ex Pers. (form 2) Plate 8:45

Grains vesiculate (bisaccate), diploxylonoid; bilaterally symmetric, heteropolar; exine tectate, *ca.* 1 to 1.5 µm thick; sexine reticulate, heterobrochate, muri thin <1 µm, lumina 2 to 6 µm wide; corpus elliptic, with cappa thicker than cappula, corpus *ca.* 15 to 17 µm × 25 to 28 µm; sacci globose, linear or angular orientation, *ca.* 21 × 22 to 25 µm; total length of grain *ca.* 42 to 55 µm. Tree.

Podocarpus sp. L'Hérit. ex Pers. (form 3) Plate 8:46

Grains vesiculate (bisaccate), apparently haploxylonoid; bilaterally symmetric, heteropolar; exine tectate *ca.* 1 µm thick; sexine reticulate, heterobrochate, muri thin <1 µm, lumina 1 to 5 µm wide; corpus trapezoid *ca.* 18 × 20 to 32 µm, cappa 1 µm thick × 32 µm wide, cappula <1 µm thick × 20 µm wide; sacci globose, angular orientation, *ca.* 20 × 25 µm; total length of grain *ca.* 58 µm. Tree.

ANGIOSPERMAE – MAGNOLIOPSIDAE (Dicotyledoneae)

Acanthaceae

Aphelandra sp. R. Br. Plate 9:47

Grains monad, isopolar, bilaterally symmetric; dicolpate; exine tectate 4 µm thick at apocolpium and 5 µm thick at mesocolpium; sexine reticulate, homobrochate, brochi <1 µm wide, muri simplicolumellate, columellae densely scattered, 1 to 4 µm length × 1 µm wide; colpi as long as grain, narrow, straight; grains prolate, 43 to 58 µm × 28 to 30 µm. Shrub, herb.

Pachystachys aff. lutea Nees. Plate 9:48

Grains monad, isopolar, radially symmetric; stephanocolpate, apparently 14 colpi; exine tectate 2.5 to 3 µm thick; sexine reticulate, heterobrochate, brochi *ca.* 1 to 4 µm wide, simplicolumellate, columellae (bacula) densely scattered, *ca.* 3 µm length × <1 µm wide, muri <1 µm thick; colpi as long as grain, narrow; grains subprolate, 47 × 38 µm. Shrub.

Amaranthaceae

Amaranthus sp. L. Plate 9:49

Grains monad, apolar, asymmetric; periporate; exine tectate 1.5 to 2 µm thick; sexine finely scabrate; pores <40 per grain, circular *ca.* 2 µm diameter, mesoporium *ca.* 2.5 µm; grains spheroidal, 23 to 26 µm. Herb.

Gomphrena sp. L. Plate 9:50

Grains monad, apolar, asymmetric; lophate; exine tectate 1.5 µm thick; sexine baculate, bacula *ca.* 1 µm height; 16 lacunae per grain, lacunae pentagonal, *ca.* 5 µm wide, lophae simplibaculate; grains spheroidal, 18 µm. Herb.

Anacardiaceae

Astronium aff. graveolens Jacq. Plate 9:51
Grains monad, isopolar, radially symmetric; tricolporate; exine tectate 1.2 μm thick; sexine striato-reticulate, homobrochate; striae longitudinally oriented; muri and lumina <1 μm wide, simplicolumellate; pores transversally parallel, *ca.* 3 μm height × 6 μm wide; colpi as long as grain; polar area *ca.* 4 μm; amb circular-trilobulate; grains subprolate to prolate-spheroidal, 27 × 23 μm. Tree.

Campnosperma sp. Thwaites Plate 9:52
Grains monad, isopolar, radially symmetric; tricolporate; exine tectate <1 μm thick; sexine finely striate, striae oriented longitudinally; colpi narrow, as long as grain; pores endexinic, inconspicuous; grains subprolate, 19 × 15 μm. Tree.

Tapirira aff. guianensis Aublet Plate 9:53
Grains monad, isopolar, radially symmetric; tricolporate; exine tectate 2 μm thick; sexine striato-reticulate, homobrochate; striae fine, longitudinally oriented; muri and lumina <1 μm wide; pores endexinic almost circular *ca.* 6 to 7 μm diameter; colpi as long as grain × 1.5 μm wide, costa endocolpi 3 μm thick; polar area *ca.* 10 μm; amb circular trilobulate; grains subprolate to prolate, 39 × 28 μm. Tree.

Annonaceae

Annona sp. L. (form 1) Plate 9:54
Grains monad, apolar, asymmetric; inaperturate; exine intectate 1 μm thick; sexine baculate, bacula <1 μm height, densely scattered on surface resembling a perforate pattern; amb folded, irregular to elongate with acute edges; grains 45 to 62 μm (long axis) × 30 μm (short axis). Tree, liana.

Annona sp. L. (form 2) Plate 10:55
Grains monad, apolar, asymmetric; inaperturate; exine intectate 3 μm thick; sexine baculate, bacula 3 μm height × <1 μm wide, densely scattered on surface; amb elliptic to rounded, frequently folded; grains 69 μm (long axis) × 55 μm (short axis). Tree, liana.

Diclinanona aff. tessmanii Diels Plate 10:56
Grains monad, apolar, asymmetric; inaperturate; exine intectate 3 μm thick; sexine baculate, bacula 3 μm height × 1 to 1.5 μm wide, densely scattered on surface resembling a perforate pattern; amb rounded to almost circular, folded; grains 80 to 85 μm (long axis) × 70 μm (short axis). Tree.

Unonopsis aff. pittieri Saff. Plate 10:57
Grains monad, apolar, asymmetric; inaperturate; exine intectate <1 μm thick; sexine baculate, bacula <1 μm height, densely scattered on surface; amb elliptic with acute edges, folded; grains 32 μm (long axis) × 18 μm (short axis). Tree.

Apocynaceae

Aspidosperma aff. album (Vahl) Benoist. Plate 10:58
Grains monad, isopolar, radially symmetric; tricolporate; exine tectate 2 μm thick at mesocolpium and 3 μm thick at aperture; sexine scabrate; pores depressed, lalongate; colpi wide, as long as grain, margo present, conspicuous, *ca.* 3 μm thick, extending to center of apocolpium resembling the syncolpate condition; amb circular trilobulate; grains oblate, 32 μm. Tree.

Aspidosperma aff. excelsum Benth. Plate 10:59
Grains monad, isopolar, radially symmetric; tricolporate appearing as triporate; exine tectate 3 µm thick; sexine scabrate; aperture poorly defined due to presence of thick costa colpi; pores depressed, annulate, circular *ca.* 3 µm diameter; colpi wide, as long as grain, margo present, conspicuous, thick; amb circular irregular, folded; grains oblate 38 µm. Tree.

Aspidosperma aff. megalocarpum Müell Arg. Plate 10:60
Grains monad, isopolar, radially symmetric; stephanocolporate (4-colporate); exine tectate 1 µm thick; sexine scabrate; pores depressed, lolongate, inconspicuous; colpi wide, ⅔ length of grain, rounded ends; margo present, *ca.* 2 to 2.5 µm thick, almost joined at apocolpium; amb square tetra-lobulate; grains oblate, 31 × 40 µm. Tree.

Lacmellea sp. Karsten Plate 11:61
Grains monad, isopolar, radially symmetric; triporate; exine tectate 1.5 µm thick; sexine psilate; pores lolongate, resembling short colpi, 4 µm diameter; amb circular; grains peroblate, 12 × 27 µm. Tree.

Macoubea aff. guianensis Aublet Plate 11:62
Grains monad, isopolar, radially symmetric (triaperturate) and bilaterally symmetric (diaperturate); tricolporate and less common displaying dicolporate condition; exine tectate 1 µm thick; sexine psilate; pores atrium type, circular *ca.* 3 µm diameter; colpi thin, short (brevicolpate), not well defined, *ca.* 8 µm length, rounded ends; amb semitriangular round to elliptic; grains suboblate, 13 × 18 µm. Tree.

Macoubea aff. sprucei (Müell. Arg.) Markgraf Plate 11:63
Grains monad, isopolar, bilaterally symmetric; dicolporate; exine tectate 1.5 µm thick at mesocolpium and 2 µm thick at aperture; sexine psilate; pores atrium type, transversally parallel, *ca.* 1 µm height × 10 µm wide; colpi thin, as long as grain; amb elliptic to rounded; grains oblate, 17 × 25 µm. Tree.

Malouetia aff. guatemalensis (Müell. Arg.) Standley. Plate 11:64
Grains monad, isopolar, radially symmetric; triporate; exine tectate 1.5 µm thick; sexine scabrate; pores circular *ca.* 3 µm diameter, protruding, annulate, annuli 1 µm thick; amb circular; grains suboblate, 20 × 24 µm. Tree, shrub.

Molongum sp. Pichon (form 1) Plate 11:65
Grains monad, isopolar, radially symmetric; stephanocolporate (5-colporate); exine tectate 3 µm thick; sexine psilate; pores depressed, inconspicuous, circular; colpi wide, ⅔ length of grain, rounded ends, 12 µm length; polar area between adjacent colpi 24 µm; amb circular to pentagonal; grains oblate, *ca.* 56 µm. Tree, liana.

Molongum sp. Pichon (form 2) Plate 11:66
Grains monad, isopolar, radially symmetric; stephanocolporate (5-colporate); exine tectate 4 to 5 µm thick; sexine psilate; pores endexinic, circular, *ca.* 4 to 6 µm diameter, with slightly annuli; colpi thin, straight, as long as grain; amb circular; grains apparently suboblate, *ca.* 66 µm. Tree, liana.

Odontadenia sp. Benth. Plate 11:67
Grains monad, isopolar, radially symmetric; triporate; exine tectate 2.5 µm thick; sexine scabrate to baculate; pores circular *ca.* 9 µm diameter, annulate, annuli *ca.* 3 µm wide, defined by small and irregular bacula; amb circular; grains spheroidal, *ca.* 47 µm. Liana.

Parahancornia sp. Ducke (form 1) Plate 12:68
Grains monad, isopolar, radially symmetric; tricolporate; exine tectate 2 µm thick; sexine finely scabrate, *ca.* 1 µm thick; nexine clearly differentiable from sexine *ca.* 1 µm thick; pores vestibulate, protruding, transversally parallel, *ca.* 1 µm height × 6 µm wide, costa endopori 3 µm thick resembling annulate condition; colpi thin, short, inconspicuous; amb circular; grains suboblate, 24 × 30 µm. Tree.

Parahancornia sp. Ducke (form 2) Plate 12:69
Grains monad, isopolar, radially symmetric; tricolporate; exine tectate 2 μm thick; sexine psilate to finely scabrate; nexine clearly differentiable from sexine *ca.* 1 μm thick; pores protruding, lalongate, *ca.* 2 μm height × 6 μm wide, costa endopori 2 μm thick; colpi thin, short, inconspicuous, *ca.* 8 μm length; amb circular; grains oblate-spheroidal, 23 × 25 μm. Tree.

Prestonia sp. R. Br. Plate 12:70
Grains monad, apolar, asymmetric; periporate (5-porate); exine tectate 2 μm thick; sexine scabrate; sexine clearly differentiated from nexine, *ca.* 1 μm thick ea.; pores circular-irregular, scattered, *ca.* 10 μm diameter; amb circular; grains spheroidal, *ca.* 60 to 65 μm. Liana, vine.

Secondatia sp. A. DC. (form 1) Plate 12:71
Grains monad, isopolar, bilaterally symmetric; dicolporate; exine tectate 1 to 1.5 μm thick; sexine psilate; pores endexinic, atrium type, circular, *ca.* 7 μm diameter, annulate, annuli thin <1 μm thick; colpi narrow, straight, ¾ length of grain, with costa endocolpi; amb elliptic; grains oblate, 27 × 45 μm. Liana.

Secondatia sp. A. DC. (form 2) Plate 12:72
Grains monad, isopolar, bilaterally symmetric; dicolporate; exine tectate 1.5 to 2 μm thick; sexine psilate; pores endexinic, depressed, inconspicuous; colpi narrow, short, ⅔ of grain; amb elliptic; grains oblate, 18 × 30 μm. Liana.

Aquifoliaceae

Ilex sp. L. Plate 12:73
Grains monad, isopolar, radially symmetric; tricolporate; exine tectate *ca.* 1.5 μm thick; sexine clavate; clavae long, densely packed, *ca.* 1.5 μm length; pores inconspicuous; colpi as long as grain × 2 μm wide, acute ends; amb circular; grains spheroidal, 14 μm. Tree.

Ilex aff. breviceps L. Plate 12:74
Grains monad, isopolar, radially symmetric; tricolporate; exine tectate *ca.* 1 μm thick (excluding sculptural elements); sexine clavate; clavae uniform, *ca.* 2 μm long; pores and colpi inconspicuous; amb circular to trilobulate; grains oblate-spheroidal to spheroidal, 18 × 23 μm. Tree.

Ilex aff. divaricata Reisseck Plate 12:75
Grains monad, isopolar, radially symmetric; tricolporate; exine tectate *ca.* 2 μm thick (excluding sculptural elements); sexine clavate; clavae uniform, densely packed, *ca.* 4 μm long; pores circular inconspicuous; colpi wide, deep, sharp, ¾ length of grain; polar area *ca.* 8 μm wide; amb circular; grains prolate spheroidal to spheroidal, 34 to 36 μm × 26 to 30 μm. Tree.

Ilex aff. dumosa L. Plate 12:76
Grains monad, isopolar, radially symmetric; tricolporate; exine tectate *ca.* 1 μm thick (excluding sculptural elements); sexine clavate; clavae uniform, short and broad, *ca.* 1 μm length × 1 μm wide; pores endexinic, circular, 1 μm diameter; colpi as long as grain, borders not well defined; amb circular to trilobulate; grains subprolate, 25 × 18 μm. Tree.

Araliaceae

Dendropanax sp. Decne & Planchon Plate 13:77
Grains monad, isopolar, radially symmetric; tricolporate; exine tectate *ca.* 2.5 to 3 µm thick; sexine psilate to finely scabrate; pores lalongate, 3 µm height × 8 µm wide; colpi narrow, as long as grain, with a thick costa endocolpi, *ca.* 4 µm; margo 2 µm wide almost rounded at apocolpium; amb circular trilobulate; grains prolate, 44 × 35 µm. Tree.

Didymopanax aff. acuminata Marchal Plate 13:78
Grains monad, isopolar, radially symmetric; tricolporate; exine tectate 3 µm thick; sexine reticulate, homobrochate, brochi <1 µm wide, muri simplibaculate, bacula 1.5 to 2 µm height × <1 µm wide; pores lalongate 3 µm height × 9 µm wide; colpi as long as grain with rounded ends; polar area 12 µm wide; amb circular to semitriangular; grains suboblate, 29 × 34 µm. Tree.

Didymopanax. aff. morototoni (Aublet) Decne & Planchon Plate 13:79
Grains monad, isopolar, radially symmetric; tricolporate; exine tectate *ca.* 5 µm thick; sexine reticulate, homobrochate, brochi 1 to 2 µm wide, muri simplibaculate, bacula 2 to 2.5 µm height × <1 µm wide; pores lalongate, *ca.* 5 µm height × 16 µm wide; colpi narrow, straight, acute ends, as long as grain; amb triangular; grains suboblate, 40 × 45 µm. Tree.

Schefflera aff. bangui Harms Plate 13:80
Grains monad, isopolar, radially symmetric; tricolporate; exine tectate *ca.* 1.5 µm thick; sexine reticulate, homobrochate, brochi *ca.* 1 µm wide, muri simplibaculate, bacula <1 µm height; pores transversally parallel, *ca.* 1 µm height × 4 µm wide; colpi 2 µm wide × as long as grain; amb circular trilobulate; grains subprolate, 28 × 24 µm. Tree.
Note: This genus is known to occur in Amazonia. This form is very similar to that of *S. bangui*, for which there is no information on the type locality.

Begoniaceae

Begonia sp. L. Plate 13:81
Grains monad, isopolar, radially symmetric; tricolporate; exine tectate <1 µm thick; sexine finely striate, appearing as psilate; striae longitudinally oriented; pores inconspicuous; colpi thin, as long as grain, <1 µm; amb circular trilobulate; grains subprolate, 10 × 5 µm. Herb.

Betulaceae

Alnus sp. Miller (form 1) Plate 14:82
Grains monad, isopolar, radially symmetric; stephanoporate (5-porate); exine tectate *ca.* 2 µm thick; sexine psilate to slightly scabrate; pores circular, protruding, resembling atrium type, *ca.* 2.5 µm diameter, costa ectopori *ca.* 3 µm thick, annuli 3 µm wide connecting adjacent pores forming thickened and curved bands (arcus) *ca.* 3 to 5 µm wide; amb circular irregular to pentagonal; grains apparently oblate, 26 µm. Tree.

Alnus sp. Miller (form 2) Plate 14:83
Grains monad, isopolar, radially symmetric; stephanoporate (5-porate); exine tectate *ca.* 1 µm thick; sexine psilate to slightly scabrate; pores circular, protruding, resembling atrium type, *ca.* 1.5 µm diameter, costa ectopori *ca.* 2 µm thick, annuli 1.5 µm wide connecting adjacent pores forming thickened and curved bands (arcus) *ca.* 2 µm wide; amb pentagonal irregular; grains apparently oblate, 21 µm. Tree.

Alnus sp. Miller (form 3) Plate 14:84

Grains monad, isopolar, radially symmetric; stephanoporate (4-porate); exine tectate *ca.* 2 µm thick; sexine scabrate; pores circular, protruding, atrium type, *ca.* 2 µm diameter, costa ectopori *ca.* 3 µm thick, annuli 2 µm wide connecting adjacent pores forming thickened bands (arcus) *ca.* 4 µm wide; amb square rounded; grains oblate, 32 µm. Tree.

Bignoniaceae

Arrabidaea sp. DC. (form 1) Plate 14:85

Grains monad, isopolar, radially symmetric; tricolporate; exine tectate 1.5 µm thick, densely columellate; sexine scabrate to granulate; pores lalongate, endexinic, depressed, 3 µm height × 6 µm wide; colpi narrow, as long as grain, displaying slightly an equatorial constriction; polar area *ca.* 7 µm; amb circular; grains subprolate, 24 µm × 16 to 17 µm. Liana.

Arrabidaea sp. DC. (form 2) Plate 14:86

Grains monad, isopolar, radially symmetric; tricolpate; exine tectate 2 µm thick, densely columellate; sexine scabrate, columellae appearing as short bacula; pores apparently absent; colpi wide, as long as grain, deep; polar area *ca.* 6 µm; amb circular trilobulate; grains subprolate, 32 × 22 µm. Liana.

Arrabidaea sp. DC. (form 3) Plate 14:87

Grains monad, isopolar, radially symmetric; tricolpate; exine tectate 1.5 µm thick, densely columellate; sexine scabrate; pores apparently absent; colpi narrow, as long as grain, deep, with conspicuous and thick costa (margines) *ca.* 2 to 2.5 µm thick; polar area *ca.* 5 µm; amb circular trilobulate; grains prolate spheroidal, 27 µm × 24 to 26 µm. Liana.

Jacaranda sp. Juss. (form 1) Plate 14:88

Grains monad, isopolar, radially symmetric; tricolpate; exine tectate 5 µm thick, densely columellate, columellae resembling small pila; sexine scabrate; pores apparently absent; colpi wide, as long as grain, deep; polar area *ca.* 22 µm; amb circular; grains apparently oblate, 68 to 72 µm. Tree.

Jacaranda sp. Juss. (form 2) Plate 15:89

Grains monad, isopolar, radially symmetric; tricolpate (apparently tricolporate); exine tectate 2 µm thick, columellate; sexine granulate resembling reticulate pattern; pores if present, lalongate, seen as lineal shade; colpi wide, as long as grain, deep; polar area *ca.* 18 µm; amb circular; grains apparently oblate, 58 to 60 µm.

Lundia sp. DC. Plate 15:90

Grains monad, isopolar, radially symmetric; tricolpate appearing as tricolporate; exine tectate 3 µm thick, densely columellate; sexine finely reticulate, homobrochate, brochi <1 µm wide, muri simplicolumellate, columellae resembling small bacula; pores if present, inconspicuous; colpi wide, as long as grain, deep, with fine costa endocolpi; polar area *ca.* 16 µm; amb circular; grains prolate spheroidal, 60 to 62 µm × 50 to 55 µm. Liana.

Mansoa sp. DC. Plate 16:93

Grains monad, isopolar, radially symmetric; syncolporate; exine tectate 3 µm thick, densely columellate; sexine reticulate, homobrochate, brochi *ca.* 1 µm wide, muri thin, simplibaculate; pores appearing as circular, *ca.* 16 µm diameter; colpi wide, deep, as long as grain, almost united at apocolpium; amb circular; grains oblate, 48 to 52 µm. Liana.

Memora aff. pedunculata (Vell.) Miers Plate 15:91
Grains monad, apolar, asymmetric; inaperturate; exine semitectate 4 μm thick, densely columellate; sexine coarse per-reticulate, heterobrochate, forming <20 small isolated patches (areolae) sourrounded by continuous free nexine, resembling as syncolpate condition; muri simplibaculate ca 1 μm wide, slightly undulating, lumina variable in size 3 to 5 μm wide, with small free bacula; amb circular irregular; grains spheroidal, 35 to 40 μm. Liana.

Memora sp. Miers Plate 15:92
Grains monad, apolar, asymmetric; inaperturate; exine semitectate 3 μm thick, densely columellate; sexine reticulate, apparently homobrochate, forming isolated crowded patches (areolae) not well defined; muri simplibaculate, baculi crowded, *ca.* 2 μm height; lumina uniform; amb circular irregular; grains spheroidal, 40 μm. Liana.

Tabebuia sp. Gomes ex DC. (form 1) Plate 16:94
Grains monad, isopolar, radially symmetric; tricolporate appearing as tricolpate; exine tectate 2.5 μm thick; sexine reticulate, homobrochate, brochi less than 1 μm wide, muri simplicolumellate; pores lalongate, endexinic, inconspicuous; colpi as long as grain, wide, displaying equatorial constriction; polar area 6 μm; amb circular trilobulate; grains subprolate, 40 to 43 μm × 30 to 34 μm. Tree.

Tabebuia sp. Gomes ex DC. (form 2) Plate 16:95
Grains monad, isopolar, radially symmetric; tricolpate; exine tectate 1 to 1.5 μm thick, columellate; sexine finely reticulate, homobrochate, brochi <1 μm wide, muri simplibaculate; pores if present, inconspicuous; colpi narrow, as long as grain, with equatorial constriction; amb apparently circular; grains subprolate, 36 × 25 μm. Tree.

Bombacaceae

Chorisia sp. Kunth / *Ceiba* sp. Miller (form 1) Plate 16:96
Grains monad, isopolar, radially symmetric; stephanocolpate (4-colpate); exine tectate 4 μm thick, densely columellate; sexine reticulate, homobrochate, brochi *ca.* 1 μm wide, muri simplibaculate, bacula thin *ca.* 3 μm length; pores if present, inconspicuous; colpi narrow, short, acute ends; polar area between adjacent colpi 20 μm; amb circular to square; grains oblate, 42 μm. Tree.
Note: Grains similar to those of *Ceiba*, but differs from it by the presence of four apertures instead of three.

Chorisia sp. Kunth / *Ceiba* sp. Miller (form 2) Plate 17:101
Grains monad, isopolar, radially symmetric; stephanocolpate (4-colpate); exine tectate 4 to 5 μm thick, densely columellate; sexine reticulate, heterobrochate, brochi irregular 1 to 3 μm wide, muri duplibaculate, bacula thin *ca.* 2.5 μm long; pores if present, inconspicuous; colpi wide, deep, short, rounded ends; polar area between adjacent colpi *ca.* 35 μm; amb square tetra-lobulate; grains oblate, 76 to 78 μm. Tree.
See previous note.

Matisia aff. bracteolosa Ducke Plate 16:97
Grains monad, isopolar, radially symmetric; brevi-tricolpate; exine semitectate *ca.* 4 to 5 μm thick; sexine densely gemmate-baculate, gemmae variable in size, rounded, *ca.* 2 to 5 μm, bacula thin short, *ca.* 1.5 μm length; colpi masked by gemmae; amb inter-subangular; grains oblate, 80 to 85 μm. Tree.

Pachira aff. aquatica Aublet Plate 16:98
Grains monad, isopolar, radially symmetric; brevi-tricolpate; exine semitectate *ca.* 1.5 μm thick at angles and 2.5 μm thick at aperture; sexine reticulate, heterobrochate, tectum displaying free bacula bordering aperture; brochi <1 to 1.5 μm wide, muri simplibaculate, bacula <1 to 2 μm length; colpi short, acute ends; amb inter-subangular; grains oblate, 38 μm. Tree.

Phragmotheca aff. mammosa Alverson Plate 16:99
Grains monad, isopolar, radially symmetric; brevi-tricolpate; exine semitectate *ca.* 1 to 2 μm thick; sexine gemmate-baculate, gemmae variable, rounded, 2 to 5 μm height, bacula thin short, *ca.* 1 μm length; pores if present, inconspicuous; colpi short, wide, acute ends; grains folded, amb inter-subangular; grains oblate, 46 to 50 μm. Tree.

Pseudobombax sp. Dugand Plate 16:100
Grains monad, isopolar, radially symmetric; brevi-tricolpate, rarely tetracolpate; exine tectate *ca.* 1.5 μm thick, densely columellate; sexine reticulate, heterobrochate, fine brochi at mesocolpium area becoming to be irregular at apocolpium, *ca.* <1 to 2 μm wide, muri simplibaculate, bacula <1 μm length; colpi short, acute ends; polar area between adjacent colpi 14 μm wide; amb inter-subangular rounded; grains oblate, 26 μm. Tree.

Quararibea sp. Aublet Plate 17:102
Grains monad, isopolar, radially symmetric; brevi-tricolpate; exine tectate *ca.* 2 μm thick, densely columellate; sexine reticulate, homobrochate, brochi *ca.* 2 μm wide, muri simplibaculate, bacula *ca.* 1.5 μm length × 1 μm wide; colpi short, wide, acute ends, borders not well defined; amb inter-subangular, irregularly rounded; grains oblate, 46 to 50 μm. Tree.

Rhodognaphalopsis aff. duckei Robyns Plate 18:103
Grains monad, isopolar, radially symmetric; brevi-tricolpate; exine tectate *ca.* 2 μm thick, densely columellate; sexine echinate-scabrate; scabrae <1 μm height, spine very thin, straight, *ca.* 1 μm length, scarcely scattered on the surface; pores if present, inconspicuous; colpi short, acute ends; polar area between adjacent colpi 40 μm; amb inter-subangular rounded; grains oblate, 65 μm. Tree, shrub.

Scleronema aff. micranthum Ducke Plate 18:104
Grains monad, isopolar, radially symmetric; triporate; exine tectate *ca.* 2 μm thick, moderately columellate; sexine echinate-scabrate; scabrae <1 μm height, spine conical, short, *ca.* 2 μm length × 1.5 μm wide, scarcely scattered on the surface; pores apparently circular, *ca.* 8 μm diameter, annulate, annuli thin, *ca.* 1 μm wide; amb circular; grains oblate, 45 μm. Tree.

Scleronema sp. Benth. Plate 18:105
Grains monad, isopolar, radially symmetric; triporate; exine tectate *ca.* 2 to 2.5 μm thick, moderately columellate; sexine echinate-scabrate; scabrae <1 μm height, spine conical, short, *ca.* 3 μm length × 1.5 μm wide, densely scattered on the surface; pores circular, *ca.* 5 μm diameter, annulate, annuli thin, *ca.* 2 μm wide, borders defined by small scabrae; amb circular; grains suboblate to oblate-spheroidal, 36 to 38 μm × 40 to 42 μm. Tree.

Spirotheca sp. Ulbr. Plate 18:106
Grains monad, isopolar, radially symmetric; brevi-tricolpate; exine tectate *ca.* 1.5 μm thick, densely columellate; sexine reticulate, homobrochate, brochi *ca.* 1 μm wide, muri simplibaculate, bacula <1 μm length; colpi short, acute; polar area between adjacent colpi 23 μm wide; amb inter-subangular rounded; grains oblate, 38 to 40 μm. Tree.

Bombacaceae 1. Plate 18:107

Grains monad, isopolar, radially symmetric; triporate; exine tectate *ca.* 1.5 μm thick, densely columellate; sexine reticulate, homobrochate, brochi *ca.* 1 μm wide, muri simplibaculate, bacula <1 μm length; pores appearing slightly lolongate, *ca.* 8 μm diameter, costa endopori 4 μm thick, annuli *ca.* 3 μm wide defined by small bacula; amb inter-subangular rounded; grains oblate, 40 μm.

Boraginaceae

Cordia aff. nodosa Lam. Plate 19:108

Grains monad, isopolar, radially symmetric; tricolporate; exine tectate 2 μm thick, densely columellate; sexine echinate-scabrate; pores lalongate, endexinic, inconspicuous; colpi, narrow, acute ends, as long as grain; scabrae <1 μm length, spine acute, *ca.* 3 μm height × 1 μm wide; polar area *ca.* 25 μm; amb circular; grains suboblate, 48 μm. Tree, shrub.

Heliotropium sp. L. Plate 19:109

Grains monad, isopolar, radially symmetric; heterocolpate; exine tectate 3 μm thick; sexine psilate to slightly scabrate; 3 colpori alternating with 3 μm pseudocolpi; pores lalongate, 3 μm height × 8 μm wide; colpi narrow, straight, as long as grain, free ends at apocolpium; polar area between adjacent colpi *ca.* 8 μm; amb circular hexa-lobulate; grains prolate-spheroidal, 33 × 31 μm. Herb.

Tournefortia sp. L. Plate 19:110

Grains monad, apolar, asymmetric; apparently inaperturate or aperture masked by ornamentation; exine tectate 3 to 4 μm thick (including sculptural elements); sexine strongly gemmate, gemmae rounded, variable in size 1 to 4 μm length; amb circular; grains spheroidal, 22 μm. Liana.

Burseraceae

Crepidospermum sp. Hook. f. Plate 19:111

Grains monad, isopolar, radially symmetric; tricolporate; exine tectate 1.5 μm thick; sexine psilate to slightly scabrate; pores lalongate, *ca.* 4 μm height × 12 μm wide; colpi, narrow, ⅔ length of grain, with costa endocolpi; amb circular; grains subprolate, 34 × 25 μm. Tree.

Dacryodes aff. chimatensis Cuatrec. Plate 19:112

Grains monad, isopolar, radially symmetric; tricolporate; exine tectate 1.5 μm thick; sexine scabrate to granulate; pores rectangular, *ca.* 3 μm height × 6 μm wide; colpi, narrow, ⅔ length of grain; costa endocolpi 1.5 μm thick; polar area broad; amb triangular rounded; grains subprolate, 23 to 25 μm × 18 to 20 μm. Tree.

Protium sp. Burm. f. (form 1) Plate 19:113

Grains monad, isopolar, radially symmetric; tricolporate; exine tectate 1.5 μm thick; sexine psilate; pores lalongate, *ca.* 3 μm height × 10 μm wide; colpi, narrow, straight, ⅔ length of grain, with costa endocolpi, *ca.* 2 μm thick, clearly separated from sexine; amb circular; grains prolate, 37 × 24 μm. Tree.

Protium sp. Burm. f. (form 2) Plate 19:114

Grains monad, isopolar, radially symmetric; tricolporate; exine tectate 2 μm thick; sexine psilate; pores lalongate, *ca.* 4 μm height × 10 μm wide; colpi, narrow, as long as grain, with costa endocolpi, *ca.* 3.5 μm thick; amb circular; grains prolate, 40 × 25 μm. Tree.

Trattinickia sp. Willd. Plate 19:115

Grains monad, isopolar, radially symmetric; tricolporate; exine tectate 2 µm thick; sexine psilate to slightly scabrate; pores elliptic, *ca.* 3 µm height × 6 µm wide; colpi, narrow, ⅔ length of grain, with costa endocolpi, *ca.* 3 µm thick; amb circular; grains prolate, 26 × 18 µm. Tree.

Caryocaraceae

Caryocar sp. L. (form 1) Plate 19:116

Grains monad, isopolar, radially symmetric; syncolporate (parasyncolporate condition); exine tectate 3 µm thick at mesocolpium and 4 µm thick at apocolpium; sexine slightly scabrate; pores elliptic *ca.* 7 µm height × 16 µm wide, protuberant resembling vestibulate type; colpi, narrow, straight, with costa endocolpi *ca.* 7 µm thick displaying exitus digitatus; colpi as long as grain to join at apocolpium forming a conspicuous triangle (apocolpial field) *ca.* 9 µm wide, seen in equatorial view as globose protuberance; amb triangular; grains prolate-spheroidal, 42 × 40 µm. Tree.

Caryocar sp. L. (form 2) Plate 20:117

Grains monad, isopolar, radially symmetric; syncolporate (displaying parasyncolporate condition) (also appearing as 4-colporate); exine tectate 3.5 µm thick at mesocolpium and 5 µm thick at apocolpium; sexine scabrate; pores apparently lalongate, vestibulate, not well defined; colpi, narrow, straight, with costa endocolpi, as long as grain becoming to join at pole forming a triangle (apocolpial field) *ca.* 16 µm wide, appearing as a protuberance; amb angular-square; grains subprolate, 54 × 50 µm. Tree.

Caryocar sp. L. (form 3) Plate 20:118

Grains monad, isopolar, radially symmetric; syncolporate (displaying parasyncolporate condition); exine tectate 1.5 µm thick at mesocolpium and 5 µm thick at apocolpium; sexine psilate to slightly scabrate; pores lalongate, *ca.* 3 µm height × 8 µm wide; colpi, narrow, straight, with costa endocolpi *ca.* 4 µm thick, as long as grain; becoming to join at apocolpium forming a globose protuberance; grains prolate, 38 × 24 µm. Tree.

Cecropiaceae

Cecropia sp. Loefl. (form 1) Plate 20:119

Grains monad, isopolar, bilaterally symmetric; diporate; exine tectate *ca.* 1 µm thick; sexine scabrate becoming to granulate; pores circular 1.5 to 2 µm diameter, labrum-like; amb circular; grains subprolate, 14 × 8 µm. Tree, shrub.

Cecropia sp. Loefl. (form 2) Plate 20:120

Grains monad, isopolar, bilaterally symmetric; diporate; exine tectate <1 µm thick; sexine psilate; pores circular <1 µm diameter, common type; amb circular; grains subprolate, 9 × 6 µm. Tree, shrub.

Cecropia sp. Loefl. (form 3) Plate 20:121

Grains monad, isopolar, bilaterally symmetric; diporate; exine tectate *ca.* 1 µm thick; sexine scabrate; pores circular *ca.* 1 µm diameter, labrum type; amb circular; grains prolate-spheroidal, 14 × 12 µm. Tree, shrub.

Celastraceae (Hippocrateaceae)

Hippocratea aff. volubilis L. Plate 20:122

Polyad with 16 grains, 46 to 50 µm, grouped as 4 tetrahedral tetrads, 26 × 21 µm ea. Individual grains heteropolar, bilaterally symmetric; triporate; exine tectate 1.5 µm thick; sexine reticulate, heterobrochate, brochi 1 to 2 µm wide, muri simplibaculate, bacula 2 µm height × 1 to 2 µm wide; pores circular, *ca.* 1 µm diameter; amb circular; grains suboblate, 13 × 16 µm. Liana.

Peritassa aff. laevigata (Hoffm.) A.C. Sm. Plate 20:123

Grains monad, isopolar, radially symmetric; tricolporate appearing as tricolpate; exine tectate 4 μm thick at mesocolpium and 1.5 μm thick at aperture area, densely columellate; sexine reticulate, homobrochate, brochi less than 1 μm wide, muri simplicolumellate; pores if present, inconspicuous; colpi as long as grain, wide, ends rounded; polar area 5 μm; amb circular trilobulate; grains oblate spheroidal, 30 × 32 μm. Liana.

Plenckia sp. Reisseck Plate 20:124

Grains monad, isopolar, radially symmetric; tricolporate; exine tectate 4 μm thick, densely columellate; sexine reticulate, homobrochate, brochi *ca.* 1 μm wide, muri simplibaculate, bacula crowded, thin, 3.5 μm length × less than 1 μm wide; pores transversally parallel, inconspicuous; colpi ⅔ length of grain, very thin, straight; polar area 12 μm; amb circular; grains oblate, 17 × 27 μm. Shrub.

Chenopodiaceae

Chenopodium sp. L. Plate 21:125

Grains monad, apolar, asymmetric; periporate; exine tectate 1 μm thick; sexine granulate; pores ≥80 per grain, circular *ca.* 1.5 μm diameter, annulate, annuli <1 μm wide, length between adjacent pores *ca.* 2 μm; amb circular; grains spheroidal, 27 to 28 μm. Herb.

Chloranthaceae

Hedyosmum sp. Sw. (form 1) Plate 21:126

Grains monad, apolar, asymmetric; inaperturate; exine intectate 4 μm thick; sexine clavate, clavae *ca.* 3 μm height, densely scattered on the surface; amb circular; grains spheroidal, 38 μm. Tree.

Hedyosmum sp. Sw. (form 2) Plate 21:127

Grains monad, apolar, asymmetric; inaperturate; exine intectate 3 μm thick; sexine clavate, clavae *ca.* 2 μm height, densely scattered on the surface; amb circular; grains spheroidal, 48 μm. Tree.

Hedyosmum sp. Sw. (form 3) Plate 21:128

Grains monad, apolar, asymmetric; inaperturate; exine intectate 1.5 μm thick; sexine clavate resembling fine reticulum; clavae 1.5 μm height, densely scattered on the surface; amb circular; grains spheroidal, 30 to 32 μm. Tree.

Chrysobalanaceae

Chrysobalanus aff. icaco L. Plate 21:129

Grains monad, isopolar, radially symmetric; tricolporate; exine tectate 4 μm thick at mesocolpium and 5 μm thick at aperture area, densely columellate; sexine verrucate, flat verrucae *ca.* 2 μm height × 3 μm wide, poorly scattered on surface; pores lalongate *ca.* 6 μm height × 14 μm wide, endexinic, forming conspicuous vestibulum; colpi with costa endocolpi, colpi as long as grain with borders limited by verrucae; polar area 15 μm; amb circular; grains suboblate, 35 × 45 μm. Shrub.

Couepia sp. Aublet Plate 21:130

Grains monad, isopolar, radially symmetric; tricolporate; exine tectate 2.5 μm thick at mesocolpium and 1.5 μm thick at aperture level; sexine psilate to finely scabrate; pores apparently lalongate, slightly protruding; colpi inconspicuous ⅔ length of grain; polar area 7 μm; amb triangular; grains oblate, 15 × 22 μm. Tree.

Hirtella aff. racemosa Lam. Plate 21:131
Grains monad, isopolar, radially symmetric; tricolporate; exine tectate 2 µm thick, densely columellate; sexine granulate; pores transversally parallel *ca.* 2 to 3 µm height × 16 µm wide, endexinic; colpi with irregular margines; colpi as long as grain; polar area 5 µm; amb triangular irregular with undulating sides; grains suboblate, 20 × 32 µm. Tree.

Licania sp. Aublet Plate 21:132
Grains monad, isopolar, radially symmetric; tricolporate; exine tectate 1 µm thick; sexine psilate to slightly scabrate; pores poorly defined, apparently lalongate; colpi as long as grain, almost joined at apex; polar area 5 to 6 µm; amb triangular; grains oblate, 18 × 25 to 26 µm. Tree.

Combretaceae

Combretum sp. Loefl. Plate 22:133
Grains monad, isopolar, radially symmetric; heterocolpate (3 colpori alternating with 3 pseudocolpi); exine tectate 2.5 µm thick, slightly columellate; sexine granulate; pores elliptic, 3 µm height × 8 µm wide; colpi narrow, straight, as long as grain; polar area between adjacent colpi *ca.* 5 µm; amb circular hexa-lobulate; grains subprolate, 25 × 20 µm. Liana.

Terminalia sp. L. Plate 22:134
Grains monad, isopolar, radially symmetric; heterocolpate (3 colpori alternating with 3 pseudocolpi); exine tectate 1 µm thick; sexine scabrate; pores endexinic, inconspicuous, apparently rounded-like *ca.* 5 µm diameter; pseudocolpi longest than colpori; colpi narrow, displaying slightly an equatorial constriction; polar area between adjacent colpi *ca.* 5 µm; amb circular hexa-lobulate; grains subprolate, 25 × 18 µm. Tree.

Compositae (Asteraceae)

Group *Baccharis* L. Plate 22:135
Grains monad, isopolar, radially symmetric; tricolporate; exine tectate <1 µm thick (ornamentation excluded); sexine echinate; spine thin, sharp, conical, *ca.* 3 µm height × 1 µm wide; aperture inconspicuous, masked by sculptural elements; amb circular; grains spheroidal, 18 µm (including ornamentation). Shrub, tree, liana.

Group *Eupatorium* L. Plate 22:136
Grains monad, isopolar, radially symmetric; tricolporate; exine tectate 3 to 4 µm thick (ornamentation excluded), columellate; sexine clearly separated from nexine at mesocolpium, echinate; spine scarce, short, sharp, conical with ample base scabrate and undulating, *ca.* 1.5 µm height × 4 to 5 µm wide; pores circular 3 µm diameter; colpi narrow, as long as grain; polar area *ca.* 8 µm; amb circular; grains spheroidal, 26 to 28 µm (including ornamentation). Herb.

Group *Senecio* L. (form 1) Plate 22:137
Grains monad, isopolar, radially symmetric; tricolporate; exine tectate 4 µm thick (ornamentation excluded); sexine joined to nexine by scarce columellae, displaying a conspicuous layer separation; sexine echinate, spine long, sharp, conical, *ca.* 6 µm height × 10 µm wide, with scabrate process between them; pores apparently circular, inconspicuous; colpi as long as grain; amb circular; grains spheroidal, 40 µm (including ornamentation). Herb, shrub, small tree and liana.

Group *Senecio* L. (form 2) Plate 22:138
Grains monad, isopolar, radially symmetric; tricolporate; exine tectate 7 μm thick (ornamentation excluded); sexine strongly separated from nexine at mesocolpium area; sexine echinate, spine, short, sharp, conical, wide base, *ca.* 4 μm height × 8 μm wide, with scabrate process between them; pores circular 3 μm diameter, common type; colpi as long as grain; amb circular; grains spheroidal, 38 to 40 μm (including ornamentation). Herb, shrub, small tree and liana.

Convolvulaceae

Aniseia sp. Choisy Plate 22:139
Grains monad, isopolar, radially symmetric; stephanocolpate (6-colpate); grains structurally complex; exine intectate 2 μm thick; sexine strongly baculate, bacula uniform *ca.* 2 μm height × 1 μm wide; aperture not well defined, appearing as pseudocolpi; grains could have more than 6 apertures, becoming to show pericolpate condition; amb circular hexa-lobulate; grains oblate, 50 μm. Liana.

Dicranostyles aff. scandens Benth. Plate 22:140
Grains monad, isopolar, radially symmetric; tricolpate; exine intectate 2.5 μm thick; sexine clavate, clavae *ca.* 2 μm height × <1 to 1 μm wide; colpi wide, deep, as long as grain; amb circular trilobulate; grains spheroidal, 40 μm. Liana.

Maripa sp. Aublet (form 1) Plate 23:141
Grains monad, isopolar, radially symmetric; tricolpate; exine intectate 5 to 6 μm thick; sexine strongly clavate, clavae resembling baculate pattern; clavae *ca.* 5 μm height × 1 to 2 μm wide; colpi broad, deep, ⅔ length of grain, borders surrounded by irregular clavae; polar area 22 μm; amb triangular; grains oblate, 55 to 60 μm. Liana.

Maripa sp. Aublet (form 2) Plate 23:142
Grains monad, isopolar, radially symmetric; tricolpate; exine intectate 4 μm thick; sexine strongly clavate, clavae resembling baculate pattern; clavae *ca.* 3.5 μm height × 1 μm wide; colpi broad, deep, as long as grain; amb circular trilobulate; grains spheroidal, 48 to 50 μm. Liana.

Convolvulaceae 1. Plate 23:143
Grains monad, isopolar, radially symmetric; stephanocolpate (4-colpate); exine tectate 2 μm thick; sexine finely verrucate, appearing as irregular flat verucae, *ca.* <1 μm height × 1 to 3 μm wide; colpi displaying complex view, broad, deep, acute ends, as long as grain; length between adjacent colpi 5 μm and between opposite colpi 20 μm; amb square tetra-lobulate; grains apparently oblate, 50 μm.

Cucurbitaceae

Cayaponia sp. A. Silva Manso Plate 23:144
Grains monad, apolar, asymmetric; triporate, grains appearing folded apparently with more than 3 pores; exine tectate 3 to 4 μm thick (ornamentation excluded); sexine echinate-scabrate; spine conical, rounded ends, *ca.* 6 to 7 μm height × 3 to 4 μm wide; surface between spine with small scabrae; pores circular *ca.* 12 μm diameter, annulate, annuli 4 μm thick; amb irregularly rounded; 115 to 120 μm. Liana.

Cyclanthera sp. Schrader Plate 23:145
Grains monad, isopolar, radially symmetric; stephanocolporate (4-colporate); exine tectate 1.5 μm thick at apocolpium and 2.5 μm thick at mesocolpium; sexine finely reticulate, homobrochate, brochi 1 μm wide, muri simplibaculate; pores apparently transversally parallel, *ca.* 16 μm wide; colpi narrow, ⅘ length of grain, acute ends; length between adjacent colpi 18 μm and between opposite colpi 30 μm; amb square rounded; grains oblate, 45 μm. Vine.

Posadaea sp. Cogn. Plate 24:146

Grains monad, isopolar, radially symmetric; tricolporate; exine tectate 2 µm thick, densely columellate; sexine reticulate, homobrochate, brochi 1 µm wide, muri simplibaculate, bacula thin, 1 µm length × less than 1 µm wide; pores endexinic, inconspicuous, apparently circular, *ca.* 10 µm diameter; colpi ¾ length of grain, thin, straight; amb circular; grains prolate-spheroidal to spheroidal, 36 × 34 µm. Liana.

Cucurbitaceae 1. Plate 24:147

Grains monad, apolar, asymmetric; tricolpate, grains appearing broken with apparently more than 3 colpi; exine tectate 8 to 9 µm thick (ornamentation excluded), strongly columellate, with a baculate-like pattern; sexine echinate; spine short, sharp, conical, *ca.* 2 µm height × 2 µm wide, surface between spine undulating; bacula straight *ca.* 6 to 7 µm height × 2 µm wide; amb irregular rounded, 150 to 160 µm (including sculptural elements).

Cunoniaceae

Weinmannia sp. L. Plate 24:148

Grains monad, isopolar, radially symmetric; tricolporate; exine tectate 1 µm thick, densely columellate; sexine reticulate, homobrochate, brochi <1 µm wide, muri simplibaculate; pores endexinic, inconspicuous, apparently circular; colpi as long as grain, thin, straight; amb circular; grains prolate-spheroidal to spheroidal, 14 to 15 µm × 11 to 12 µm. Tree.

Dilleniaceae

Doliocarpus sp. Rolander Plate 24:149

Grains monad, isopolar, radially symmetric; tricolporate; exine tectate 3 µm thick, densely columellate; sexine reticulate, homobrochate, brochi 1.5 µm wide, muri simplibaculate, bacula thin, 1 µm length × 1 µm wide; pores endexinic, lalongate, sharp, *ca.* 5 µm height × 16 µm wide; colpi ¾ length of grain, thin, straight; polar area 9 to 10 µm; amb circular; grains subprolate, 35 × 26 µm. Liana.

Tetracera sp. L. Plate 24:150

Grains monad, isopolar, radially symmetric; tricolporate; exine tectate 2 µm thick, densely columellate; sexine reticulate, homobrochate, brochi <1 µm wide, muri simplibaculate, bacula thin, <1 µm size; pores endexinic, inconspicuous, apparently circular; colpi as long as grain, wide; polar area 5 µm; amb circular; grains prolate-spheroidal, 19 × 16 µm. Liana.

Elaeocarpaceae

Elaeocarpaceae 1. Plate 24:151

Grains monad, isopolar, radially symmetric; tricolporate; exine tectate 1 µm thick; sexine slightly psilate appearing as rugulate; aperture not well defined, bordered by small scabrae; pores endexinic, inconspicuous, apparently circular; colpi ⅔ length of grain, straight, wide; amb circular; grains oblate-spheroidal to spheroidal, 11 × 12 µm.

Ericaceae

Agarista sp. D. Don ex G. Don f. / *Gaylussacia* sp. Kunth (form 1) Plate 24:152

Multiplanar tetrahedral tetrad, 42 to 44 µm length. Individual grains isopolar, radially symmetric; tricolporate; exine tectate 2 µm thick; sexine scabrate to granulate; pores lalongate to transversally parallel running at same line of grain junction, inconspicuous; colpi straight, ¾ length of grain, 1 µm wide; polar area 8 µm; amb circular; grains suboblate, 24 × 34 µm. Shrub.

Note: Our reference collection suggests that *Agarista* and *Gaylussacia* have the same pollen type. *Agarista* has a disjunct distribution (Andes and southeastern Brazil). *Gaylussacia* is mostly found in southern Brazil. This pollen type occurred in glacial sediments of lake Pata together with *Podocarpus, Ilex* and *Hedyosmun*, thus suggesting cooler and humid conditions.

Agarista sp. D. Don ex G. Don f. / *Gaylussacia* sp. Kunth (form 2) Plate 25:153
Multiplanar tetrahedral tetrad, 54 to 58 µm length. Individual grains isopolar, radially symmetric; tricolporate; exine tectate 1 to 1.5 µm thick, densely columellate; sexine scabrate; pores transversally parallel, inconspicuous masked by the junction area of grains; colpi narrow, ⅔ length of grain, with costa endocolpi, 3.5 µm thick, margo 2 µm wide, displaying apparently a slightly equatorial constriction; polar area 10 µm; amb circular; grains suboblate, 34 × 41 µm. Shrub.
See previous note.

Agarista sp. D. Don ex G. Don f. / *Gaylussacia* sp. Kunth (form 3) Plate 25:154
Multiplanar tetrahedral tetrad, 60 to 62 µm length. Individual grains isopolar, radially symmetric; tricolporate; exine tectate 1 to 1.5 µm thick; sexine scabrate to finely rugulate; pores transversally parallel, inconspicuous masked by the junction area of grains; colpi short, narrow, *ca.* 24 µm length × 1.5 to 2 µm wide, with a thick, margo undulate, *ca.* 2.5 µm wide, apparently interrupted at the apertural area; polar area 18 to 20 µm; amb circular; grains suboblate, 38 × 46 µm. Shrub.
See previous note.

Gaultheria sp. Kalman ex L. Plate 25:155
Multiplanar tetrahedral tetrad, 68 to 72 µm length. Individual grains isopolar, radially symmetric; tricolporate; exine tectate 3 µm thick, densely columellate; sexine granulate resembling fine baculi; aperture inconspicuous; pores transversally parallel, extending along the junction area of grains, *ca.* 25 µm; colpi short, wide, acute ends, *ca.* 5 µm wide × ½ length of grain; polar area 20 µm; amb circular; grains oblate-spheroidal, 40 × 52 µm. Shrub.

Erythroxylaceae

Erythroxylum sp. P. Browne (form 1) Plate 26:156
Grains monad, isopolar, radially symmetric; tricolporate; exine tectate 2 µm thick, densely columellate; sexine reticulate, homobrochate, brochi <1 µm wide, muri simplibaculate, bacula thin, <1 µm size; pores endexinic, elliptic to circular, 3 µm diameter; colpi narrow, as long as grain, with costa endocolpi *ca.* 1.5 µm thick, displaying exitus digitatus at pores area; amb circular; grains prolate, 28 × 18 µm. Tree, shrub.

Erythroxylum sp. P. Browne (form 2) Plate 26:157
Grains monad, isopolar, radially symmetric; tricolporate; exine tectate 1 µm thick, columellate; sexine reticulate, homobrochate, brochi <1 µm wide, muri simplibaculate, bacula <1 µm size; pores endexinic, apparently lalongate 3.5 µm height × 8 µm wide; colpi narrow, as long as grain, with costa endocolpi 1.5 µm thick, forming conspicuous exitus digitatus and displaying equatorial constriction becoming to bridge; amb circular; grains prolate, 34 × 24 µm. Tree, shrub.

Erythroxylum sp. P. Browne (form 3) Plate 26:158
Grains monad, isopolar, radially symmetric; tricolporate; exine tectate 1 to 1.5 µm thick, columellate; sexine reticulate, homobrochate, brochi <1 µm wide, muri simplibaculate, bacula <1 µm size; pores endexinic, elliptic to circular, 6 µm diameter; colpi narrow, as long as grain, with costa endocolpi *ca.* 3 µm thick, forming exitus digitatus and displaying equatorial constriction; amb circular; grains prolate, 26 × 18 µm. Tree, shrub.

Erythroxylum sp. P. Browne (form 4) Plate 26:159
Grains monad, isopolar, radially symmetric; tricolporate; exine tectate 1.5 µm thick, columellate; sexine reticulate, homobrochate, brochi 1 µm wide, muri simplibaculate, bacula <1 µm size; pores endexinic, apparently circular 6 µm diameter; colpi narrow, as long as grain, displaying an equatorial constriction, with costa endocolpi 1.5 µm thick, forming conspicuous exitus digitatus; amb circular; grains subprolate, 27 × 19 µm. Tree, shrub.

Euphorbiaceae

Acalypha sp. L. (form 1) Plate 26:160
Grains monad, isopolar, radially symmetric; stephanocolporate (4-colporate appearing as porate grains; occasionally 5-colporate); exine tectate 1 µm thick; sexine scabrate; pores circular *ca.* 1.5 µm diameter, aspidate type; colpi apparently absent, if present, small and inconspicuous; amb circular to tetra-angular; grains oblate-spheroidal, 14 × 18 to 20 µm. Herb.

Acalypha sp. L. (form 2) Plate 26:161
Grains monad, isopolar, radially symmetric; stephanocolporate (4-colporate appearing as porate grains; occasionally 5-colporate); exine tectate 1.5 to 2 µm thick; sexine scabrate; pores circular *ca.* 3 µm diameter, aspidate type; colpi apparently absent, if present, small and inconspicuous; amb circular to pentagonal; grains suboblate, 25 × 33 µm. Herb.

Acalypha sp. L. (form 3) Plate 26:162
Grains monad, isopolar, radially symmetric; stephanocolporate (4-colporate appearing as porate grains); exine tectate 1.5 µm thick; sexine scabrate; pores circular, protruding, 2.5 µm diameter, aspidate; colpi apparently absent, if present, small, inconspicuous; amb circular to square; grains oblate, 20 µm. Herb.

Alchornea sp. Sw. (form 1) Plate 26:163
Grains monad, isopolar, radially symmetric; tricolporate; exine tectate 2 µm thick at mesocolpium to 4 µm thick at aperture; sexine scabrate, densely columellate; pores lalongate to transversally parallel, endexinic, *ca.* 2 × 6 to 8 µm; colpi as long as grain, with a conspicuous thin opercula, *ca.* 2 µm thick; amb circular; grains suboblate, 20 × 24 to 26 µm. Tree.
Note: This morphological type includes *Aparisthmium* and *Alchorneopsis*. Both genera have been reported in "terra firme" forest in the Brazilian Amazon.

Alchornea sp. Sw. (form 2) Plate 26:164
Grains monad, isopolar, radially symmetric; tricolporate; exine tectate 1 to 1.5 µm thick, densely columellate; sexine reticulate, homobrochate, brochi <1 µm wide, muri simplibaculate; pores endexinic, inconspicuous; colpi 5 µm wide, as long as grain, with a conspicuous thin opercula, *ca.* 1.5 µm wide; polar area 8 µm; amb circular; grains oblate-spheroidal to spheroidal, 20 to 21 µm × 21 to 24 µm. Tree.
See previous note.

Alchornea sp. Sw. (form 3) Plate 26:165
Grains monad, isopolar, radially symmetric; tricolporate; exine tectate 1 to 1.5 µm thick, slightly columellate; sexine scabrate; pores apparently endexinic; colpi, deep, 8 µm wide, as long as grain, with a conspicuous long and free opercula; opercula straight, *ca.* 14 µm length × 2 µm thick; polar area 4 µm; amb circular; grains oblate, 32 to 35 µm. Tree.
See previous note.

Alchornea sp. Sw. (form 4) Plate 26:166
Grains monad, isopolar, radially symmetric; tricolporate; exine tectate 2 μm thick at mesocolpium to 5 μm thick at aperture area, densely columellate; sexine finely scabrate-verrucate, flat verrucae <1 μm height; pores endexinic, apparently circular, *ca.* 6 μm diameter; colpi ⅔ length of grain with well developed opercula, *ca.* 3.5 μm. wide; amb circular; grains suboblate, 32 × 40 μm. Tree.
See previous note.

Alchornea sp. Sw. (form 5) Plate 27:167
Grains monad, isopolar, radially symmetric; stephanocolporate (4-colporate); exine tectate 2.5 to 3 μm thick, densely columellate; sexine finely reticulate, homobrochate, brochi <1 μm wide, muri simplibaculate; pores circular, endexinic, *ca.* 10 μm diameter; colpi 5 μm wide, as long as grain, with a conspicuous opercula; opercula long, appearing as bifurcate, *ca.* 10 μm length × 2 μm wide; polar area between adjacent colpi 8 μm; amb circular to square; grains subprolate, 35 × 28 μm. Tree.
See previous note.

Alchornea sp. Sw. (form 6) Plate 27:168
Grains monad, isopolar, radially symmetric; tricolporate; exine tectate 1 μm thick, slightly columellate; sexine finely scabrate; pores endexinic, inconspicuous; colpi ⅔ length of grain with a conspicuous thin and free opercula, *ca.* 5 μm wide, apparently bifurcate; amb circular; grains spheroidal, 16 μm. Tree.
See previous note.

Amanoa aff. congesta Hayden Plate 27:171
Grains monad, isopolar, radially symmetric; tricolporate; exine tectate 6 μm thick, densely columellate; sexine per-reticulate, heterobrochate, brochi 3 to 8 μm wide, lumina irregular, muri undulating, simplibaculate, bacula conspicuous *ca.* 1.5 μm wide × 3 μm height; pores endexinic, apparently elliptic to lalongate, *ca.* 5 μm height × 10 μm wide; colpi as long as grain, displaying costa endocolpi *ca.* 4 μm thick; polar area 7 μm; amb circular; grains suboblate, 45 × 50 μm. Tree.

Amanoa aff. nanayensis Hayden Plate 27:170
Grains monad, isopolar, radially symmetric; tricolporate; exine tectate 5 μm thick, columellate; sexine per-reticulate, heterobrochate, brochi 3 to 6 μm wide, lumina irregular, simplibaculate, bacula conspicuous *ca.* 1 μm wide × 2.5 μm height; colpi wide, as long as grain; polar area 12 μm; amb circular; grains suboblate, 36 μm. Tree.

Amanoa sp. Aublet (form 1) Plate 27:169
Grains monad, isopolar, radially symmetric; tricolporate; exine tectate 3 μm thick, columellate; sexine reticulate, homobrochate, brochi 1 to 1.5 μm wide, muri simplibaculate, bacula conspicuous *ca.* 1 μm wide × 2 μm height; pores endexinic, apparently lalongate; colpi ⅔ length of grain; amb circular; grains prolate, 38 × 28 μm. Tree.

Amanoa sp. Aublet (form 2) Plate 27:172
Grains monad, isopolar, radially symmetric; tricolporate; exine semitectate 7 μm thick, densely columellate; sexine per-reticulate, heterobrochate, brochi 5 to 12 μm wide, lumina irregular occasionally with free baculate columellae; muri simpli to duplobaculate, bacula *ca.* 2 μm wide × 4 μm height; pores endexinic, lalongate, *ca.* 5 μm height; colpi as long as grain, displaying costa endocolpi *ca.* 5 μm thick; amb circular; grains oblate-spheroidal, 55 × 58 μm. Tree.

Cleidion aff. praealtum Croizat Plate 28:173
Grains monad, isopolar, radially symmetric; tricolporate; exine tectate 2.5 μm thick; sexine verrucate, resembling rugulate pattern, slightly columellate; flat verrucae *ca.* 1 to 2 μm wide × 1 μm height; pores inconspicuous, endexinic, apparently lalongate; colpi wide, ¾ length of grain, with robust and free opercula; polar area 9 μm; amb circular; grains suboblate, 25 × 28 μm. Tree.

Cleidion sp. Blume (form 1) Plate 28:174
Grains monad, isopolar, radially symmetric; tricolporate; exine tectate 1.5 to 2 μm thick; sexine granulate, slightly columellate; pores inconspicuous, endexinic; colpi wide, as long as grain, with a conspicuous, long, robust and free opercula; operculum occasionally appearing bifurcate; polar area 10 μm; amb circular; grains oblate, 15 × 28 to 30 μm. Tree.

Cleidion sp. Blume (form 2) Plate 28:175
Grains monad, isopolar, radially symmetric; tricolporate; exine tectate 2 μm thick; sexine verrucate, columellate; verrucae flat, crowded; pores inconspicuous; colpi wide, as long as grain, with a conspicuous, long, robust and free opercula; polar area 4 μm; amb circular; grains oblate, 25 μm. Tree.

Croton sp. L. (form 1) Plate 28:176
Grains monad, apolar, asymmetric; inaperturate; exine intectate 6 μm thick; sexine clavate, clavae densely distributed, displaying the typical "croton pattern," forming "rosets" of five elements; occasionally sculptural elements with baculate pattern; clavae isodiametric *ca.* 4 μm length × 2 μm wide, triangular ends; amb circular; grains spheroidal, 80 μm. Tree, shrub.

Croton sp. L. (form 2) Plate 28:177
Grains monad, apolar, asymmetric; inaperturate; exine intectate 4 to 5 μm thick; sexine clavate, clavae densely distributed displaying the typical "croton pattern" arranged in "rosets" of five elements; clavae isodiametric *ca.* 4 μm length × 3 μm wide, triangular ends; amb circular; grains spheroidal, 58 to 60 μm. Tree, shrub.

Dalechampia aff. tiliifolia Lam. Plate 28:178
Grains monad, isopolar, radially symmetric; tricolporate; exine semitectate 3 μm thick, columellate; sexine per-reticulate, heterobrochate, brochi 3 to 10 μm wide, lumina irregular occasionally with free bacula; muri simplibaculate, bacula *ca.* <1 μm wide × 2.5 μm height; pores transversally parallel, with costa endopori becoming to be costa transversalis, appearing as an equatorial thickness around grain; colpi thin <1 μm wide, straight, ½ length of grain; amb circular; grains spheroidal, 70 μm. Vine.

Drypetes sp. Vahl Plate 29:179
Grains monad, isopolar, radially symmetric; tricolporate; exine tectate 2 μm thick; sexine reticulate, homobrochate, brochi 1 μm wide, muri simplibaculate, bacula *ca.* <1 μm wide × 1 μm height; pores lalongate *ca.* 4 μm height × 12 μm wide; colpi 2 μm wide, ⅔ length of grain, displaying costa endocolpi *ca.* 3 μm thick; amb circular; grains subprolate, 44 × 34 μm. Tree.

Hyeronima sp. Allemão Plate 29:180
Grains monad, isopolar, radially symmetric; tricolporate; exine tectate 2 μm thick; sexine psilate to slightly scabrate; pores elliptic to circular *ca.* 5 μm diameter; colpi thin, as long as grain, displaying costa endocolpi *ca.* 2.5 μm thick, forming exitus digitatus at pores area; amb circular; grains perprolate, 35 × 15 μm. Tree.

Mabea sp. Aublet (form 1) Plate 29:181
Grains monad, isopolar, radially symmetric; tricolporate; exine intectate 2 μm thick; sexine baculate, resembling verrucate pattern, bacula 1 μm height × 1 μm wide; pores, endexinic, lalongate *ca.* 8 μm height × 22 μm wide; colpi narrow, as long as grain, with costa endocolpi *ca.* 2 μm thick; polar area 13 μm; amb circular; grains subprolate, 56 × 47 μm. Tree, shrub.

Mabea sp. Aublet (form 2) Plate 29:182
Grains monad, isopolar, radially symmetric; tricolporate; exine intectate 3 μm thick; sexine baculate, bacula, thin, 1 μm height; pores, endexinic, lalongate *ca.* 4 μm height × 15 to 20 μm wide, almost united with each other, resembling a continuous equatorial ring; colpi 3 μm wide, as long as grain, with costa endocolpi *ca.* 2 μm thick; amb circular; grains prolate, 50 × 33 μm. Tree, shrub.

Mabea sp. Aublet (form 3) Plate 30:184
Grains monad, isopolar, radially symmetric; tricolporate; exine intectate 3 μm thick; sexine baculate, bacula 1 μm height × <1 μm wide; pores, endexinic, lalongate *ca.* 6 μm height × 18 μm wide; colpi 4 μm wide to very thin at apocolpium, as long as grain, with costa endocolpi *ca.* 3 μm thick, displaying equatorial constriction appearing as a bridge; polar area 14 μm; amb circular; grains subprolate to prolate-spheoridal, 56 × 50 μm. Tree, shrub.

Pera aff. glabrata (Schott) Baillon Plate 29:183
Grains monad, isopolar, radially symmetric; tricolporate; exine tectate 1 μm thick; sexine psilate to slightly scabrate; pores circular slightly lolongate, 2 μm diameter; colpi thin, straight, short, ⅓ length of grain, with thin costa endocolpi <1 μm thick; amb circular; grains oblate-spheroidal to spheroidal, 24 × 26 μm. Tree.

Phyllanthus sp. L. (form 1) Plate 30:185
Grains monad, isopolar, radially symmetric; tricolporate; exine tectate 2 μm thick, densely columellate; sexine reticulate, homobrochate, brochi <1 μm wide, muri simplibaculate, bacula *ca.* 1.5 μm height; pores, endexinic, lalongate *ca.* 3 μm height × 12 μm wide; colpi narrow, straight, as long as grain, with costa endocolpi *ca.* 2 μm thick, displaying a delicate exitus digitatus; amb circular; grains subprolate, 35 × 25 μm. Herb, tree, shrub.

Phyllanthus sp. L. (form 2) Plate 30:186
Grains monad, isopolar, radially symmetric; tricolporate; exine tectate 1 μm thick, densely columellate; sexine reticulate, homobrochate, brochi <1 μm wide, muri simplibaculate, bacula *ca.* 1 μm height; pores endexinic, lalongate *ca.* 2 μm height × 10 μm wide, almost joining between them; colpi thin, straight, as long as grain, with costa endocolpi *ca.* 2 μm thick; amb circular; grains subprolate, 29 × 22 μm. Herb, tree, shrub.

Podocalyx aff. loranthoides Klotzsch. Plate 30:187
Grains monad, isopolar, radially symmetric; triporate; exine tectate 1 μm thick (ornamentation excluded), finely columellate; sexine echinate, spine short *ca.* 1 μm height, acute, broad base; pores, circular, *ca.* 2 to 2.5 μm diameter, occasionally displaying thin covering membrane; amb circular; grains oblate-spheroidal, 21 × 23 μm. Tree.

Sapium sp. P. Browne (form 1) Plate 30:188
Grains monad, isopolar, radially symmetric; tricolporate; exine intectate 3 μm thick; sexine baculate, bacula 2 μm height; pores, endexinic, lalongate *ca.* 6 μm height × 20 μm wide; colpi narrow, 2 μm wide, as long as grain, with costa endocolpi *ca.* 5 μm thick, forming a conspicuous exitus digitatus; amb circular; grains subprolate, 70 × 59 μm. Tree.

Sapium sp. P. Browne (form 2) Plate 31:189
Grains monad, isopolar, radially symmetric; tricolporate; exine intectate 2 μm thick; sexine baculate, bacula thin, crowded, 1.5 μm height; pores, endexinic, lalongate *ca.* 3.5 μm height × 10 μm wide; colpi thin, straight, as long as grain, with costa endocolpi *ca.* 2 μm thick, forming exitus digitatus; amb circular trilobulate; subprolate, 38 × 28 μm. Tree.

Sapium sp. P. Browne (form 3) Plate 31:190
Grains monad, isopolar, radially symmetric; tricolporate; exine intectate 3.5 μm thick; sexine baculate, bacula 2.5 μm height × 1 μm wide; pores, endexinic, lalongate, not well defined; colpi narrow, 1 μm wide, ⅔ length of grain, with costa endocolpi *ca.* 4 μm thick, displaying equatorial constriction; amb circular; grains subprolate, 68 × 52 μm. Tree.

Sebastiana sp. Sprengel Plate 31:191
Grains monad, isopolar, radially symmetric; tricolporate; exine intectate 1.5 to 2 μm thick; sexine baculate appearing as reticulate pattern; bacula thin, 1.5 μm height; pores, endexinic, depressed, elliptic appearing as circular, *ca.* 5 μm height × 6 μm wide; colpi narrow, as long as grain, displaying slightly an equatorial constriction; amb circular trilobulate; grains oblate-spheroidal, 30 × 33 μm. Shrub.

Flacourtiaceae

Banara sp. Aublet Plate 31:192
Grains monad, isopolar, radially symmetric; tricolporate; exine tectate 1 μm thick; sexine finely reticulate, homobrochate, brochi <1 μm wide, muri simplibaculate; pores circular to slightly lalongate, inconspicuous; colpi straight, thin, ⅔ length of grain; amb circular; grains subprolate, 13 × 10 μm. Tree.

Casearia sp. Jacq. (form 1) Plate 31:193
Grains monad, isopolar, radially symmetric; tricolporate; exine tectate 2 to 2.5 μm thick, columellate; sexine psilate to slightly scabrate; pores circular, *ca.* 3 μm diameter; colpi thin, straight, ⅔ length of grain, with fine costa endocolpi *ca.* 1 μm thick; polar area 10 μm; amb circular; grains prolate-spheroidal, 22 × 20 μm. Tree.

Casearia sp. Jacq. (form 2) Plate 31:194
Grains monad, isopolar, radially symmetric; stephanocolporate (4-colporate); exine tectate 2.5 to 3 μm thick, moderately columellate; sexine psilate to slightly scabrate; pores appearing as circular, *ca.* 4 μm diameter; colpi narrow, ¾ length of grain, with costa endocolpi *ca.* 2 μm thick, displaying conspicuous exitus digitatus; polar area 10 μm; amb circular to square; grains subprolate, 30 × 24 μm. Tree.

Casearia sp. Jacq. (form 3) Plate 31:195
Grains monad, isopolar, radially symmetric; tricolporate; exine tectate 2 μm thick; sexine psilate to slightly scabrate; pores lalongate 3 μm height × 8 μm wide; colpi narrow, 22 μm length × 1 μm wide, with costa endocolpi *ca.* 2 μm thick; amb circular; grains subprolate, 31 × 26 μm. Tree.

Laetia aff. coriacea Spruce ex Benth. Plate 32:196
Grains monad, isopolar, radially symmetric; tricolporate; exine semitectate 2 μm thick at apocolpium and 5 to 7 μm thick at mesocolpium, strongly columellate; sexine reticulate, heterobrochate, brochi *ca.* 1 μm wide, diminishing toward aperture, muri simplibaculate, bacula thin, 1.5 μm height at apocolpium to 7 μm height at mesocolpium, becoming to be free at equator; pores transversally parallel 2 μm height × 10 μm wide, acute ends; colpi narrow, straight, ¾ length of grain, with costa endocolpi *ca.* 4 μm thick; amb circular; grains prolate, 55 × 33 μm. Tree.
Note: This type is very similar to *Aphelandra* (Acanthaceae), but separated from it by the presence of three colpori.

Xylosma sp. Foster f. Plate 32:197

Grains monad, isopolar, radially symmetric; tricolporate; exine tectate 1 µm thick; sexine psilate; pores inconspicuous; colpi thin, as long as grain, displaying thin margo interrupted at equator by equatorial constriction, appearing as small bridge; polar area 7 µm; amb circular; grains subprolate, 28 × 23 µm. Tree, shrub.

Gentianaceae

Schultesia sp. Mart. Plate 32:198

Multiplanar tetrahedral tetrad, 65 to 75 µm length. Individual grains isopolar, radially symmetric; tricolporate; exine intectate 4 µm thick; sexine densely baculate, bacula *ca.* 3 µm length × 1 µm wide; aperture inconspicuous, pores apparently lalongate; colpi short, narrow; amb circular; grains oblate-spheroidal, *ca.* 40 × 50 µm. Herb.

Guttiferae (Clusiaceae)

Clusia aff. odorata Seemann Plate 32:199

Grains monad, isopolar, radially symmetric; tricolporate; exine tectate 1 µm thick, slightly columellate; sexine scabrate; pores circular appearing slightly lolongate, *ca.* 6 µm diameter; colpi ⅔ length of grain, narrow at poles and wider at equator; polar area 12 µm; amb circular; grains oblate-spheroidal, 18 × 20 µm. Tree, shrub.

Clusiella sp. Planchon & Triana Plate 32:200

Grains monad, isopolar, bilaterally symmetric; dicolporate; exine tectate variable in size, 2 to 3 µm thick at apocolpium and 3 µm thick at equator, slightly columellate; sexine psilate; pores inconspicuous, apparently lalongate to colpus transversalis; colpi short, thin, with costa endocolpi ca 2 µm. thick; amb apparently elliptic ; grains suboblate, 22 × 26 µm. Tree, shrub.

Kielmeyera sp. Mart. Plate 32:201

Multiplanar tetrahedral tetrad, 44 to 48 µm length. Individual grains isopolar, radially symmetric; tricolporate; exine tectate 2 to 2.5 µm thick, densely columellate; sexine finely baculate, bacula thin 1.5 µm height × <1 µm wide; pores transversally parallel, inconspicuous masked by the junction area of grains; colpi as long as grain, with thick margo 4 to 5 µm wide; amb circular with mesocolpium borders concave as triangular outline; grains suboblate, 25 × 32 µm. Shrub.

Tovomitopsis sp. Planchon & Triana (form 1) Plate 32:202

Grains monad, isopolar, radially symmetric; tricolporate; exine tectate <1 to 1 µm thick; sexine echinate-scabrate, spine acute, scarcely scattered, <1 µm height; surface between spine slightly scabrate; pores circular, *ca.* 2 µm diameter; colpi ⅔ length of grain, inconspicuous; polar area 5 µm; amb circular; grains spheroidal, 15 to 17 µm. Tree.

Tovomitopsis sp. Planchon & Triana (form 2) Plate 32:203

Grains monad, isopolar, radially symmetric; triporate; exine tectate <1 to 1 µm thick; sexine echinate, spine acute, scarcely scattered, <1 µm height; pores circular, *ca.* 3 to 3.5 µm diameter, apparently covered by thin ectexinic membrana; amb circular; grains spheroidal, 19 to 20 µm. Tree.

Guttiferae 1. Plate 32:204

Grains monad, isopolar, radially symmetric; tricolporate appearing as triporate; exine tectate 1 µm thick; sexine reticulate, homobrochate, brochi <1 µm wide; pores circular 2 to 2.5 µm diameter; colpi thin, as long as grain; amb circular; grains suboblate, 14 × 17 µm.

Humiriaceae

Endopleura aff. uchi (Huber) Cuatrec. Plate 33:205

Grains monad, isopolar, radially symmetric; tricolporate; exine tectate 4 μm thick, columellate; sexine reticulate, homobrochate, brochi *ca.* 1 μm wide, muri simplibaculate, bacula <1 μm height; pores lalongate, *ca.* 2 μm height × 8 μm wide, with longitudinally elongations at apices ("H" form); colpi narrow, depressed, as long as grain, with thick margo *ca.* 7 to 8 μm wide, appearing as aperture protuberant; polar area 12 μm; amb semitriangular concave; grains spheroidal, 40 μm. Tree.

Humiria aff. balsamifera (Aublet) A. St-Hil. Plate 33:206

Grains monad, isopolar, radially symmetric; tricolporate; exine tectate 4 μm thick, columellate, displaying a conspicuous stratification: nexine 1 μm thick, columella 1.5 μm height, sexine 1.5 μm thick; sexine psilate; pores thin, transversally parallel, *ca.* 1 μm height × 10 μm wide; colpi short, narrow, ¼ length of grain (brevicolpate), with thick margo *ca.* 6 μm thick × 4 μm wide; polar area as wide as equatorial diameter; amb circular; grains suboblate, 27 × 32 μm. Tree.

Humiria sp. Aublet Plate 33:207

Grains monad, isopolar, radially symmetric; tricolporate; exine tectate 3 μm thick, columellate, displaying conspicuous stratification: nexine 1 μm thick, columella 1 μm height, sexine 1 μm thick; sexine scabrate; pores circular 7 μm diameter; colpi short, wide, elongate, ⅓ length of grain (brevicolpate), with thick margo 4 μm thick × 5 μm wide, appearing undulating and broken; polar area 22 μm; amb circular; grains oblate-spheroidal, 43 × 45 μm. Tree.

Vantanea aff. guianensis Aublet Plate 34:208

Grains monad, isopolar, radially symmetric; tricolporate; exine tectate 3 μm thick, densely columellate; sexine reticulate, homobrochate, brochi *ca.* <1 μm wide, muri simplibaculate, bacula <1 μm height; pores circular, annulate, slightly protuberant, *ca.* 14 μm diameter with longitudinally elongations at apices ("H" form); colpi narrow, as long as grain, displaying margo *ca.* 4 μm thick; amb circular to triangular; grains subprolate, 62 × 50 μm. Tree.

Icacinaceae

Discophora sp. Miers Plate 34:209

Grains monad, isopolar, radially symmetric; triporate; exine tectate 3 μm thick (ornamentation included), slightly columellate, stratification conspicuous; sexine echinate, spine short, acute, moderately scattered, <1 μm height; pores circular, 5 μm diameter; thin annulus present; amb circular; grains oblate-spheroidal to spheroidal, 17 × 19 μm. Tree.

Labiatae (Lamiaceae)

Hyptis sp. Jacq. Plate 34:210

Grains monad, isopolar, radially symmetric; stephanocolpate (6-colpate); exine tectate 1.5 μm thick, columellate; sexine psilate to slightly scabrate; colpi narrow, as long as grain; polar area between adjacent colpi 4 μm; amb circular hexa-lobulate; grains prolate-spheroidal to spheroidal, 22 × 20 μm. Herb.

Lacistemataceae

Lacistema sp. Sw. Plate 34:211

Grains monad, isopolar, radially symmetric; tricolporate; exine tectate 1 μm thick, columellate; sexine psilate; aperture inconspicuous, apparently pores circular and colpi short; polar area 5 μm; amb circular; grains prolate-spheroidal to spheroidal, 12 × 11 μm. Tree, shrub.

Leguminosae – Caesalpinioideae

Bauhinia aff. reflexa Schery Plate 34:212

Grains monad, isopolar, radially symmetric; tricolporate; exine tectate 2 µm thick, slightly columellate; sexine psilate; pores lolongate, 2.5 µm height × 2 µm wide; colpi wide, ¾ length of grain, *ca.* 26 µm, displaying moderate equatorial constriction; amb circular; grains prolate to subprolate, 32 × 24 µm. Liana.

Cassia sp. L. (form 1) Plate 34:213

Grains monad, isopolar, radially symmetric; syncolporate; exine tectate 1 µm thick at mesocolpium and 2 µm thick at aperture, columellate; sexine scabrate to slightly verrucate; verrucae flat, wide; pores lalongate; colpi wide, joined at apocolpium, with thin costa endocolpi; amb circular to slightly triangular; grains suboblate, 32 to 36 µm. Tree.

Cassia sp. L. (form 2) Plate 34:214

Grains monad, isopolar, radially symmetric; tricolporate; exine tectate 2 to 2.5 µm thick, columellate; sexine scabrate; pores lalongate 3 µm height × 14 µm wide; colpi 3 to 4 µm wide, depressed, as long as grain, with thin costa endocolpi, displaying equatorial constriction; amb circular; polar area 4 µm; grains prolate-spheroidal, 40 × 38 µm. Tree.

Cassia sp. L. (form 3) Plate 35:215

Grains monad, isopolar, radially symmetric; tricolporate; exine tectate 3 µm thick at mesocolpium and 4 µm thick at aperture, densely columellate; sexine reticulate, homobrochate, brochi *ca.* 1 µm wide, muri simplibaculate, bacula thin *ca.* 1 µm height; pores lalongate appearing as rectangular, *ca.* 6 µm height × 16 µm wide, protruding; colpi as long as grain × 6 µm wide, with costa endocolpi *ca.* 2 µm thick; polar area 8 µm; amb circular; grains subprolate, 40 × 36 µm. Tree.

Cassia sp. L. (form 4) Plate 35:216

Grains monad, isopolar, radially symmetric; tricolporate; exine tectate 3 to 4 µm thick, columellate; sexine reticulate, homobrochate, brochi *ca.* 1 µm wide, muri simplibaculate, bacula thin *ca.* 2 µm height; pores apparently lalongate, *ca.* 10 µm height; colpi as long as grain, displaying slightly an equatorial constriction; polar area 12 µm; amb circular to triangular; grains spheroidal, 50 µm. Tree.

Cenostigma sp. Tul. / *Caesalpinia* sp. L. Plate 35:217

Grains monad, isopolar, radially symmetric; syncolporate; exine tectate 5 µm thick at mesocolpium diminishing to 2 to 3 µm thick at aperture, densely columellate, displaying a conspicuous stratification, nexine <1 µm thick, sexine 2 to 4 µm thick; sexine reticulate, heterobrochate, brochi 1 µm wide diminishing toward aperture, muri simplibaculate, bacula thin *ca.* 2.5 µm height; pores apparently circular *ca.* 5 µm diameter; colpi wide, with costa ectocolpi 2 µm thick; amb circular; grains suboblate, 38 µm. Tree, shrub. Note: These two taxa show similar morphological characters.

Cercidium sp. Tul. Plate 35:218

Grains monad, isopolar, radially symmetric; tricolporate becoming to syncolporate; exine tectate 1.5 µm thick, columellate; sexine reticulate, homobrochate, brochi <1 µm wide, muri simplibaculate, bacula thin, short; aperture inconspicuous, pores apparently lalongate; colpi narrow, as long as grain; amb circular; grains prolate-spheroidal, 17 × 15 µm. Tree.

Copaifera aff. cordifolia Hayne Plate 35:219

Grains monad, isopolar, radially symmetric; tricolporate; exine tectate 1 µm thick; sexine psilate; pores transversally parallel, *ca.* 1 µm height × 8 µm wide; colpi as long as grain, displaying slightly an equatorial constriction; polar area 5 µm; amb triangular; grains spheroidal, 21 µm. Tree.

Crudia aff. amazonica Sprengel ex Benth. Plate 35:220
Grains monad, isopolar, radially symmetric; tricolporate; exine semitectate 1 µm thick; sexine striato-reticulate, homobrochate, brochi <1 µm wide, muri simplibaculate; striae thin, crowded, <1 µm wide, longitudinally oriented; aperture inconspicuous masked by ornamentation; amb circular; grains prolate, 38 × 16 µm. Tree.

Dinizia aff. excelsa Ducke Plate 36:221
Multiplanar decussate tetrad, 60 to 62 µm length. Individual grains isopolar, radially symmetric; tricolporate; exine intectate 4 to 5 µm thick; sexine baculate to clavate with irregular and rounded elements diminishing toward aperture and apocolpium, *ca.* 1 to 3 µm height × 1 to 4 µm wide; pores circular 5 µm diameter, annulate, annulus 1.5 µm thick; colpi wide, as long as grain, with rounded apices; polar area 4 µm; amb circular; grains suboblate, 24 × 35 µm. Tree.

Dinizia sp. Ducke Plate 36:222
Multiplanar decussate tetrad, 62 to 66 µm length. Individual grains isopolar, radially symmetric; tricolporate; exine intectate 5 to 6 µm thick; sexine baculate to clavate with irregular and rounded elements diminishing toward aperture and apocolpium, *ca.* 1 to 4 µm height × 2 to 8 µm wide; pores circular 5 µm diameter, annulate, annulus 2 µm thick; colpi wide, as long as grain, with rounded apices; polar area 12 µm; amb circular; grains suboblate, 40 × 52 µm. Tree.

Elizabetha aff. paraensis Ducke Plate 36:223
Grains monad, isopolar, radially symmetric; tricolporate; exine tectate 2 to 3 µm thick (ornamentation included); sexine verrucate, flat verrucae 1 to 2 µm height × 2 to 3 µm wide; pores apparently lalongate, 9 µm height appearing as colpus equatorialis; colpi thin, ⅔ length of grain; amb circular; grains prolate-spheroidal, 32 × 30 µm. Tree.

Macrolobium aff. acaciaefolium (Benth.) Benth. Plate 37:225
Grains monad, isopolar, radially symmetric; tricolporate; exine semitectate 4 µm thick, columellate; sexine striato-reticulate, homobrochate, brochi 1 µm wide, muri simplibaculate; striae 1 µm wide, longitudinally oriented; pores endexinic, circular to slightly lalongate, 6 µm diameter; colpi 6 µm wide, as long as grain; amb circular; grains prolate, 65 × 45 µm. Tree, shrub.

Macrolobium aff. angustifolium (Benth.) Cowan Plate 37:226
Grains monad, isopolar, radially symmetric; tricolporate; exine semitectate 3 µm thick, columellate; sexine striato-reticulate, homobrochate, brochi <1 µm wide appearing as a perforate pattern, muri simplibaculate; striae densely distributed, thick 1 to 1.5 µm wide, longitudinally oriented; pores endexinic, circular to slightly lolongate, 7 to 8 µm diameter; colpi 6 µm wide, as long as grain; amb circular; grains prolate, 54 × 34 µm. Tree.

Martiodendron sp. Gleason Plate 36:224
Grains monad, isopolar, radially symmetric; tricolporate; exine tectate 4 µm thick, densely columellate; sexine reticulate, homobrochate, brochi 1 µm wide, muri simplibaculate; pores circular appearing as hexagonal, 12 µm diameter; colpi 6 µm wide, as long as grain; amb circular; grains oblate-spheroidal, 34 × 36 µm. Tree.

Schizolobium sp. Vogel. Plate 37:227
Grains monad, isopolar, radially symmetric; tricolporate; exine tectate 1.5 µm thick, densely columellate; sexine reticulate, homobrochate, brochi 1 µm wide, muri thin, simplibaculate; pores inconspicuous, apparently lolongate; colpi narrow, as long as grain; polar area 5 µm; amb circular; grains spheroidal, 21 µm. Tree.

Leguminosae – Mimosoideae

Acacia sp. Miller — Plate 37:228

Polyad, elliptic, with 16 grains symmetrically arrangement, 32 × 25 µm. Individual grains apolar, asymmetric, apparently periporate; exine tectate <1 µm thick; sexine psilate; pores inconspicuous, at angles of contact between grains; grains displaying a rounded-trapezoid shape, *ca.* 10 µm. Tree, liana.

Entada aff. monostachia DC. — Plate 38:229

Grains monad, isopolar, radially symmetric; syncolporate; exine tectate 2 µm thick, scarcely columellate; sexine psilate to slightly scabrate; pores circular 5 µm diameter, annulate, annulus 1.5 µm thick; colpi 1.5 µm wide; grains spheroidal, 27 µm. Liana.

Inga sp. Miller — Plate 38:230

Polyad, elliptic to circular, with 16 grains symmetrically arrangement, 37 × 31 µm. Individual grains apolar, asymmetric, periporate, 4 to 5 pores; exine tectate 1 µm thick; sexine scabrate; pores small, circular, <1 µm diameter, at angles of contact between grains; amb appearing as pentagonal; grains suboblate, displaying a rounded-trapezoid, *ca.* 10 µm. Tree.

Mimosa aff. bimucronata (DC.) Okem — Plate 38:231

Polyad, circular, with 8 grains, 13 × 11 µm. Individual grains apolar, asymmetric; apparently periporate; exine tectate <1 µm thick; sexine psilate; pores inconspicuous; grains suboblate, displaying a rounded irregular shape, 5 × 7 µm. Tree, shrub.

Mimosa aff. extensa Benth. — Plate 38:232

Polyad, elliptic to circular, 8 grains symmetrically arrangement, *ca.* 17 × 13 µm. Individual grains apolar, asymmetric; periporate, 4 to 5 pores; exine tectate <1 µm thick; sexine psilate; pores inconspicuous, <0.5 µm; grains suboblate, displaying a rounded-trapezoid shape, *ca.* 7 × 10 µm. Herb.

Mimosa aff. micropteris Benth. — Plate 38:233

Multiplanar tetrahedral tetrad, 15 µm. Individual grains apolar, asymmetric; periporate; exine tectate <15 µm thick; sexine psilate; pores inconspicuous; amb circular; grains suboblate, displaying a trapezoidal shape, *ca.* 12 µm. Shrub.

Mimosa aff. pteridifolia Benth. — Plate 38:234

Uniplanar tetragonal tetrad, 18 × 20 µm. Individual grains apolar, asymmetric; triporate; exine tectate 1 µm thick; sexine psilate; pores inconspicuous, at angles of contact between grains; amb circular, grains oblate-spheroidal, 10 × 11 µm. Shrub.

Mimosa aff. punctulata Spruce ex Benth. — Plate 38:235

Polyad, circular hexa-lobulate, with 6 grains, 13 × 15 µm. Individual grains apolar, asymmetric; apparently inaperturate; exine tectate 1 µm thick, slightly columellate; sexine psilate to scabrate; grains suboblate, displaying a trapezoid shape, *ca.* 6 × 5 µm. Shrub.

Mimosa aff. rufescens Benth. — Plate 38:236

Polyad, circular, 12 grains symmetrically arrangement, *ca.* 28 µm. Individual grains apolar, asymmetric; periporate, 4 to 5 pores; exine tectate 1 µm thick; sexine psilate; pores inconspicuous, at angles of contact between grains; amb tetra to pentagonal; grains suboblate, displaying a rounded-trapezoid shape, *ca.* 9 × 14 µm. Liana.

Mimosa sp. L. Plate 38:237
Polyad, circular to rectangular, 6 grains symmetrically arrangement, *ca.* 30 × 35 µm. Individual grains apolar, asymmetric; periporate, more than 4 pores; exine tectate 1 to 1.5 µm thick; sexine psilate to slightly scabrate; pores circular 2 µm diameter; amb circular; grains suboblate, displaying a rounded-trapezoid shape, 17 × 20 µm. Tree, shrub, herb, liana.
Note: This form could belong to a much longer polyad.

Parkia aff. multijuga Benth. Plate 38:238
Polyad, elliptic, 12 to 16 grains symmetrically arranged, 154 to 158 µm × 108 to 112 µm. Individual grains apolar, asymmetric; aperture if present, inconspicuous; exine tectate 3.5 µm thick; sexine per-reticulate, heterobrochate, resembling a verrucate pattern; brochi 3 to 10 µm wide, muri 2 to 5 µm thick, simplibaculate, undulating as a flat verrucae *ca.* 2 to 3 µm height × 2 to 5 µm wide; amb circular; grains suboblate, 40 × 50 µm. Tree.

Leguminosae – Papilionoideae

Amburana aff. cearensis (Allemão) A.C. Sm. Plate 38:239
Grains monad, isopolar, radially symmetric; tricolporate; exine tectate 2 to 2.5 µm thick, slightly undulating, densely columellate; sexine reticulate, homobrochate, brochi <1 µm wide, muri thin, simplibaculate; pores endexinic, lalongate, *ca.* 4 µm height × 8 µm wide; colpi narrow, with costa endocolpi *ca.* 2 µm thick, as long as grain; amb circular; grains subprolate, 30 × 25 µm. Tree.

Bowdichia aff. nitida Spruce. ex Benth. Plate 38:240
Grains monad, isopolar, radially symmetric; tricolporate; exine tectate 1.5 µm thick, slightly columellate; sexine psilate; pores endexinic, lalongate, *ca.* 4 µm height × 12 µm wide; colpi narrow, with a protruding costa ectocolpi *ca.* 2.5 µm, displaying equatorial constriction; colpi as long as grain; polar area 10 µm; amb circular; grains spheroidal, 40 µm. Tree.

Centrolobium sp. Mart. ex Benth. Plate 39:241
Grains monad, isopolar, radially symmetric; tricolporate; exine tectate 1.5 µm thick, columellate; sexine reticulate, homobrochate, brochi <1 µm wide, muri, simplibaculate; pores rectangular, *ca.* 7 µm height × 9 µm wide; colpi narrow, straight, as long as grain, displaying slightly an equatorial constriction; amb circular trilobulate; grains subprolate, 28 × 22 µm. Tree.

Dalbergia sp. L. f. (form 1) Plate 39:242
Grains monad, isopolar, radially symmetric; tricolporate; exine tectate 1 µm thick, columellate; sexine psilate; pores circular 8 µm diameter, slightly protuberant; colpi narrow, with costa ectocolpi, as long as grain; polar area 8 µm; amb circular; grains subprolate, 24 × 20 µm. Liana.

Dalbergia sp. L. f. (form 2) Plate 39:243
Grains monad, isopolar, radially symmetric; tricolporate; exine tectate 1 µm thick, columellate; sexine psilate to slightly scabrate; pores circular 7 µm diameter; colpi narrow, as long as grain; polar area 6 µm; amb circular; grains subprolate, 19 × 16 µm. Liana.

Dipteryx sp. Schreber. Plate 39:244
Grains monad, isopolar, radially symmetric; tricolporate; exine tectate 2 to 2.5 µm thick, columellate; sexine scabrate resembling a rugulate pattern; pores endexinic, circular 5 to 6 µm diameter; colpi straight, 2 µm wide, as long as grain; polar area 8 µm; amb circular; grains suboblate, 24 × 27 µm. Tree.

Machaerium sp. Pers. (form 1) Plate 39:245

Grains monad, isopolar, radially symmetric; tricolporate; exine tectate 1 µm thick, columellate; sexine psilate to slightly scabrate; pores circular 3 µm diameter; colpi thin, as long as grain; amb circular; grains spheroidal, 19 µm. Liana.

Machaerium sp. Pers. (form 2) Plate 39:246

Grains monad, isopolar, radially symmetric; tricolporate; exine tectate 1.5 to 2 µm thick, columellate; sexine psilate to slightly scabrate; pores elliptic 3 µm height × 6 µm wide; colpi thin, as long as grain, with thin costa endocolpi *ca.* 1 µm thick; amb circular to triangular; grains prolate-spheroidal to spheroidal, 20 × 18 µm. Liana.

Myrocarpus aff. frondosus Allemão Plate 39:247

Grains monad, isopolar, radially symmetric; tricolporate; exine tectate 1.5 to 2 µm thick, columellate; sexine reticulate, homobrochate, brochi <1 µm wide; pores apparently lalongate, endexinic, depressed; colpi, narrow, deep, as long as grain; polar area *ca.* 6 µm; amb circular trilobulate; grains prolate-spheroidal, 19 µm × 16 to 18 µm. Tree.

Papilionoideae 1. Plate 39:248

Grains monad, isopolar, radially symmetric; tricolporate; exine tectate 4 to 5 µm thick, densely columellate, stratification conspicuous, nexine 1 µm thick, sexine 3.5 to 4 µm thick; sexine scabrate; pores lalongate 7 µm height × 18 µm wide; colpi narrow, as long as grain, with slightly costa endocolpi; amb circular; grains subprolate, 56 × 48 µm.

Papilionoideae 2. Plate 39:249

Grains monad, isopolar, radially symmetric; tricolporate; exine intectate 4 µm thick at mesocolpium and 7 µm thick at aperture; sexine pilate resembling baculate pattern; pila thin × 3 µm height; pores rectangular rounded, 6 µm height × 10 µm wide; colpi narrow, straight, ½ length of grain, with protruding costa ectocolpi *ca.* 5 µm thick; polar area 20 µm; amb triangular; grains prolate-spheroidal, 48 × 45 µm.

Loranthaceae (Viscaceae)

Dendropemon sp. (Blume) Reichb. Plate 39:250

Grains monad, apolar, asymmetric; periporate; exine tectate 1 µm thick; sexine psilate; pores 6 in series of three at each face of grain as demicolpate condition; pores circular 3 µm diameter; amb triangular rounded; grains apparently oblate, size 23 µm. Vine.

Orycthanthus aff. alveolatus (HBK) Kuijt Plate 39:251

Grains monad, isopolar, radially symmetric; tricolpate (structurally complex); exine tectate 1 µm thick; sexine psilate, with a triradial bifurcated thickness at apocolpium, appearing as circular structures, resembling annulate aspides, 8 µm diameter; colpi inconspicuous; amb circular; grains oblate, 33 µm. Vine.

Phoradendron sp. Nutt Plate 40:252

Grains monad, isopolar, radially symmetric; tricolpate; exine tectate 2 µm thick at mesocolpium to 1 µm thick at aperture, densely columellate; sexine granulate; pores if present inconspicuous; colpi narrow, ⅔ length of grain, displaying margo 2 µm wide; polar area 5 µm; amb triangular; grains oblate-spheroidal, 15 to 16 µm × 17 µm. Vine.

Psittacanthus sp. Mart. Plate 40:253
Grains monad, isopolar, radially symmetric; tricolpate; exine tectate 1 μm thick; columellate; sexine granulate; colpi short, narrow; polar area 10 μm; amb triangular concave (inter-angular); grains peroblate, 28 to 31 μm. Vine.

Struthanthus aff. pyrifolius (HBK) G. Don. Plate 40:254
Grains monad, isopolar, radially symmetric; syncolpate; exine tectate 1.5 μm; sexine psilate; colpi narrow, rounded at apocolpium; amb triangular slightly concave; grains oblate, 20 to 25 μm. Vine.

Struthanthus sp. Mart. Plate 40:255
Grains monad, isopolar, radially symmetric; syncolpate; exine tectate 1 μm thick; sexine psilate to slightly scabrate; colpi narrow, joined at apocolpium forming small triangle (apocolpial field); amb triangular; grains oblate, 23 μm. Vine.

Tristerix sp. Mart. (*Macrosolen* sp. (Blume) Reichb.) (form 1) Plate 40:256
Grains monad, isopolar, radially symmetric; tricolpate; exine tectate 3 μm thick; columellate; sexine granulate; colpi long, narrow; with costa ectocolpi appearing as a rounded and protruding aperture; polar area 18 μm; amb triangular concave (inter-angular); grains peroblate, 50 to 68 μm. Vine.

Tristerix sp. Mart. (*Macrosolen* sp. (Blume) Reichb.) (form 2) Plate 40:257
Grains monad, isopolar, radially symmetric; tricolpate; exine tectate 2 μm thick at mesocolpium and 1 μm thick at aperture, columellate; sexine moderately granulate at mesocolpium becoming to psilate at aperture; colpi long, thin, displaying demicolpate condition; polar area 13 μm; amb triangular concave (inter-angular); grains peroblate, 36 to 44 μm Vine.

Loranthaceae 1. Plate 40:258
Grains monad, isopolar, radially symmetric; syncolpate; exine intectate 1.5 μm thick; sexine pilate appearing as baculate pattern; pila <1 μm height, scarcely scattered; colpi long, *ca.* 2 μm wide, joined at apocolpium in straight angle; amb triangular slightly concave (inter-angular); grains peroblate, 47 to 53 μm.

Malpighiaceae

Blepharanda sp. Griseb. Plate 41:259
Grains monad, isopolar, radially symmetric; tricolporate; exine tectate 1 μm thick, stratification conspicuous; sexine psilate to slightly scabrate; pores circular 3 μm diameter; colpi thin, ¾ length of grain; amb circular; grains prolate-spheroidal, 15 × 13 μm. Shrub.

Byrsonima sp. Rich. ex Kunth. Plate 41:260
Grains monad, isopolar, radially symmetric; tricolporate; exine tectate 1 μm thick; sexine psilate; pores lalongate, 2 μm height × 6 μm wide; colpi thin, as long as grain; amb circular; grains prolate-spheroidal, 15 × 14 μm. Tree.

Diacidia sp. Griseb. Plate 41:261
Grains monad, apolar, asymmetric; pericolporate (6-colporate); exine tectate 2 μm thick; sexine scabrate to granulate; pores circular 4 to 5 μm diameter; length of mesoporium 14 μm; colpi very thin, 20 μm length, becoming to be inconspicuous; amb circular; grains spheroidal, 31 μm. Tree, shrub.

Dicella sp. Griseb. Plate 41:262
Grains monad, apolar, asymmetric; periporate (6-porate); exine tectate 2 to 3 µm thick; sexine verrucate, flat verrucae variable in size *ca.* 1 to 1.5 µm height × 2 to 3 µm wide; pores circular 3 µm diameter; length of mesoporium 13 µm; amb circular; grains spheroidal, 27 to 29 µm. Liana.

Mascagnia sp. (DC.) Colla Plate 41:263
Grains monad, apolar, asymmetric; periporate (6-porate); exine tectate 3 µm thick; sexine densely verrucate, verrucae *ca.* 2 to 4 µm height × 3 to 5 µm wide; pores masked by ornamentation, circular 5 µm diameter; amb circular; grains spheroidal, 50 µm. Liana.

Stigmaphyllon sp. A. Juss. Plate 41:264
Grains monad, apolar, asymmetric; periporate (10-porate), with thin pseudocolpi; exine tectate 4 µm thick; sexine psilate; pores circular, 5 µm diameter; length of mesoporium 16 µm; amb circular; grains spheroidal, 38 to 40 µm. Liana.

Tetrapterys sp. Cav. Plate 41:265
Grains monad, apolar, asymmetric; pericolporate (4-colporate), aperture apparently not equidistant; exine tectate 2 µm thick, columellate; sexine reticulate, heterobrochate, resembling rugulate pattern; brochi fine <1 to 1.5 µm wide, muri simplibaculate; pores circular 4 to 5 µm diameter; colpi wide, with rounded ends, 20 µm length × 4 µm wide; amb circular irregular; grains spheroidal, 30 to 37 µm. Liana.

Malvaceae

Abutilon sp. Miller Plate 42:266
Grains monad, isopolar, radially symmetric; triporate; exine tectate 4 µm thick (ornamentation excluded), densely columellate; sexine echinate-scabrate; spine bottle-shaped, *ca.* 6 to 7 µm height × 2 µm wide, scarcely scattered; surface scabrate, scabrae <1 µm; pores circular 8 µm diameter; colpi, if present, short, inconspicuous; amb circular; grains spheroidal, 60 to 64 µm (including ornamentation). Herb, shrub.

Sida sp. L. Plate 42:267
Grains monad, apolar, asymmetric; periporate (>20 pores); exine tectate 2 to 2.5 µm thick (ornamentation excluded); sexine echinate-scabrate; spine conical-shaped, *ca.* 1.5 µm height × 1.5 µm wide; surface scabrate, scabrae <1 µm; pores circular, 2 µm diameter; amb circular; grains spheroidal, 29 to 31 µm (including ornamentation). Herb.

Marcgraviaceae

Marcgravia sp. L. Plate 42:268
Grains monad, isopolar, radially symmetric; tricolporate; exine tectate 1.5 µm thick; sexine reticulate, homobrochate, brochi <1 µm wide; pores endexinic, apparently lolongate, *ca.* 7 µm height × 1.5 µm wide; colpi thin, ⅔ length of grain; amb circular; grains spheroidal, 22 µm. Herb.

Norantea aff. guianensis Aublet Plate 42:269
Grains monad, isopolar, radially symmetric; tricolporate; exine tectate 2.5 µm thick; sexine psilate to scabrate; pores lalongate, *ca.* 3 µm height × 10 µm wide; colpi thin, as long as grain, with costa endocolpi *ca.* 3 µm thick; displaying slightly an equatorial constriction; polar area ca 14 µm; amb circular; grains oblate-spheroidal, 22 × 24 µm. Liana.

Souroubea sp. Aublet Plate 42:270

Grains monad, isopolar, radially symmetric; tricolporate; exine tectate 4 to 5 µm thick; sexine psilate; pores endexinic, lalongate, *ca.* 6 µm height × 12 µm wide; colpi thin, ½ length of grain, with costa endocolpi *ca.* 3 µm thick; polar area ca 15 µm; amb circular; grains suboblate to oblate-spheroidal, 31 × 35 µm. Liana.

Melastomataceae

Clidemia sp. D. Don Plate 42:271

Grains monad, isopolar, radially symmetric; heterocolpate; exine tectate 1 µm thick, slightly columellate; sexine psilate; 3 colpori alternating with 3 pseudocolpi; pores lalongate, appearing as rectangular, 4 µm height × 7 µm wide; colpi thin, straight, ⅔ length of grain; pseudocolpi depressed, thin; polar area between adjacent colpi *ca.* 3.5 µm; amb circular trilobulate, with true colpi at angles; grains spheroidal, 16 to 17 µm. Shrub.

Miconia sp. Ruíz & Pavón (form 1) Plate 42:272

Grains monad, isopolar, radially symmetric; heterocolpate; exine tectate 1 µm thick, slightly columellate; sexine scabrate; 3 colpori alternating with 3 pseudocolpi; pores apparently lolongate to circular, endexinic, *ca.* 2 µm diameter; colpi as long as grain, thin; pseudocolpi inconspicuous; polar area between adjacent colpi *ca.* 2 µm; amb triangular hexa-lobulate, with true colpi at angles; grains prolate-spheroidal, 16 to 18 µm × 14 to 15 µm. Tree.

Miconia sp. Ruíz & Pavón (form 2) Plate 42:273

Grains monad, isopolar, radially symmetric; heterocolpate; exine tectate <1 µm thick; sexine psilate; 3 colpori alternating with 3 pseudocolpi; aperture inconspicuous; colpi and pseudocolpi as long as grain, thin; pseudocolpi inconspicuous; amb triangular trilobulate, with the true colpi at angles; grains prolate, 15 × 8 µm. Tree.

Miconia sp. Ruíz & Pavón (form 3) Plate 42:274

Grains monad, isopolar, radially symmetric; heterocolpate; exine tectate 1 µm thick; sexine psilate; 3 colpori alternating with 3 pseudocolpi; pores apparently lalongate, endexinic, inconspicuous; colpi and pseudocolpi as long as grain, thin; polar area between adjacent colpi *ca.* 25 µm; amb circular hexa-lobulate; grains spheroidal, 16 µm. Tree.

Melastomataceae 1. Plate 42:275

Grains monad, isopolar, radially symmetric; heterocolpate; exine tectate <1 to 1 µm thick; sexine verrucate, verrucae flat, <1 µm height; 3 colpori alternating with 3 pseudocolpi; pores inconspicuous; colpi and pseudocolpi thin, as long as grain; polar area between adjacent colpi 3 µm; amb circular hexa-lobulate; grains prolate-spheroidal, 14 × 12 µm.

Meliaceae

Cabralea sp. A. Juss. Plate 42:276

Grains monad, isopolar, radially symmetric; tricolporate; exine tectate 2 µm thick, columellate; sexine psilate to slightly scabrate; pores lalongate, acute, 7 × 10 µm; colpi narrow, ⅔ length of grain, with costa endocolpi 2.5 µm thick; amb circular; grains oblate-spheroidal, 25 × 27 µm. Tree.

Carapa aff. guianensis Aublet Plate 43:277
Grains monad, isopolar, radially symmetric; tricolporate; exine tectate 1 μm thick, densely columellate; sexine granulate; pores endexinic, circular to slightly lalongate, 6 μm diameter; colpi thin, short (brevicolpate) *ca.* 4 μm height × <1 μm wide; amb circular; grains oblate, 21 to 22 μm × 26 μm. Tree.

Cedrela sp. P. Browne (form 1) Plate 43:278
Grains monad, isopolar, radially symmetric; stephanocolporate (5-colporate); exine tectate 2.5 to 3 μm thick, columellate; sexine psilate to slightly scabrate; pores elliptic, 5 μm height × 10 μm wide; colpi narrow, ⅔ length of grain; amb pentagonal; grains subprolate, 46 × 34 μm. Tree.

Cedrela sp. P. Browne (form 2) Plate 43:279
Grains monad, isopolar, radially symmetric; stephanocolporate (4-colporate); exine tectate 4 to 5 μm thick; sexine psilate to slightly scabrate; pores lalongate, acute, 5 μm height × 12 μm wide; colpi 3 μm wide × as long as grain, with costa endocolpi *ca.* 2 μm thick; amb square; grains subprolate to prolate-spheroidal, 48 × 43 μm. Tree.

Guarea aff. glabra Vahl Plate 43:280
Grains monad, isopolar, radially symmetric; stephanocolporate (4-colporate); exine tectate 1.5 μm thick, columellate; sexine psilate; pores circular 5 to 6 μm diameter, with costa endopori *ca.* 3 μm thick; annulate, annuli 1.5 μm wide; colpi thin, straight, as long as pore; amb circular appearing as square; grains suboblate, 36 × 40 μm. Tree.

Guarea sp. Allam. ex L. Plate 44:281
Grains monad, isopolar, radially symmetric; stephanocolporate (4-colporate); exine tectate 1.5 to 2 μm thick, columellate; sexine psilate; pores circular to slightly lalongate 3 μm wide; annulate, annuli thin; colpi thin, straight, as long as grain; opposite colpi cross-lined; amb circular; grains suboblate, 34 × 37 μm. Tree.

Trichilia sp. P. Browne (form 1) Plate 44:282
Grains monad, isopolar, radially symmetric; stephanocolporate (4-colporate); exine tectate 1 μm thick, columellate; sexine psilate to slightly scabrate; aperture bordering by small scabrate process; pores apparently lalongate, inconspicuous; colpi narrow, ⅔ length of grain, opposite colpi appearing as crossed line; amb circular tetra-lobulate; grains prolate-spheroidal, 23 × 20 μm. Tree.

Trichilia sp. P. Browne (form 2) Plate 44:283
Grains monad, isopolar, radially symmetric; stephanocolporate (4-colporate); exine tectate 1 μm thick at mesocolpium and 3 μm thick at aperture, columellate; sexine psilate; pores lalongate 4 μm height × 8 μm wide; colpi narrow, ⅔ length of grain, displaying costa endocolpi 1.5 μm thick; amb circular to square; grains subprolate, 26 × 22 μm. Tree.

Trichilia sp. P. Browne (form 3) Plate 44:284
Grains monad, isopolar, radially symmetric; stephanocolporate (4-colporate); exine tectate 1 to 1.5 μm thick at mesocolpium and 4 μm thick at aperture; sexine psilate; pores lalongate 3.5 μm height × 7 μm wide; colpi narrow, ⅔ length of grain, displaying costa endocolpi 3 to 4 μm thick; amb circular to square; grains subprolate, 35 × 29 μm. Tree.

Trichilia sp. P. Browne (form 4) Plate 44:285
Grains monad, isopolar, radially symmetric; stephanocolporate (4-colporate); exine tectate 2 to 3 μm thick at mesocolpium and 5 to 6 μm thick at aperture, columellate; sexine psilate; pores slightly lalongate 3 μm height × 6 μm wide, with protruding costa ectopori 2 μm thick, resembling annulate condition; colpi narrow, ⅓ length of grain, displaying costa endocolpi 3 μm thick; amb square, with aperture at inter-angles; grains prolate, 46 × 26 μm. Tree.

Trichilia sp. P. Browne (form 5) Plate 44:286
Grains monad, isopolar, radially symmetric; stephanocolporate (4-colporate); exine tectate 1.5 to 2 μm, columellate; sexine psilate to slightly scabrate; pores lalongate to linear 1 μm height × 3 μm wide; colpi narrow, ⅔ length of grain, displaying costa endocolpi; amb square, with aperture at inter-angles; grains subprolate, 28 × 20 μm. Tree.

Meliaceae 1. Plate 44:287
Grains monad, isopolar, radially symmetric; stephanocolporate (4-colporate); exine tectate 4 μm thick; sexine psilate resembling rugulate pattern; pores circular, depressed, 6 to 7 μm diameter; colpi narrow, inconspicuous, ⅓ length of grain, with costa endocolpi 3 μm thick; amb square, with aperture at angles; grains suboblate, 60 × 45 μm.

Mendonciaceae

Mendoncia aff. glabra (P. & E.) Nees Plate 44:288
Grains monad, isopolar, radially symmetric; stephanocolporate (5-colporate); exine tectate 2 μm thick, columellate; sexine scabrate; pores endexinic, circular 8 μm diameter, containing the colpi; colpi short (brevicolpate) 8 μm length × 2 μm wide, with costa endocolpi 3 μm thick; amb circular; grains spheroidal, 44 μm. Liana.

Menispermaceae

Abuta sp. Aublet Plate 44:289
Grains monad, isopolar, radially symmetric; tricolporate; exine tectate 1.5 μm thick, densely columellate; sexine reticulate, homobrochate; brochi <1 μm wide, simplibaculate; bacula thin, *ca.* 1 μm height; pores lalongate, 1.5 μm height × 3 μm wide; colpi thin, ⅔ length of grain; amb circular; grains prolate-spheroidal, 16 × 15 μm. Liana.

Anomospermum sp. Miers Plate 44:290
Grains monad, isopolar, radially symmetric; tricolporate; exine intectate 1 to 1.5 μm thick; sexine baculate; bacula scarcely scattered, thin, <1 μm height, resembling short clavae; pores apparently circular 2 μm diameter; colpi narrow, ⅓ length of grain; polar area 12 μm; amb circular; grains oblate-spheroidal, 18 × 21 μm. Liana.

Cissampelos sp. L. (form 1) Plate 44:291
Grains monad, isopolar, radially symmetric; tricolpate apparently tricolporate; exine tectate 1 μm thick; sexine reticulate, heterobrochate, brochi <1 to 3 μm wide, simplibaculate; pores apparently present, lalongate, inconspicuous; colpi thin, ⅔ length of grain; amb circular; grains spheroidal, 21 μm. Liana.

Cissampelos sp. L. (form 2) Plate 45:292
Grains monad, isopolar, radially symmetric; tricolpate; exine tectate 1 μm thick, columellate; sexine reticulate, heterobrochate, brochi <1 to 1 μm wide, simplibaculate, muri *ca.* 0.5 μm wide; pores if present, inconspicuous; colpi thin, ⅓ length of grain; amb circular; grains oblate-spheroidal, 15 × 17 μm. Liana.

Cissampelos sp. L. (form 3) Plate 45:293
Grains monad, isopolar, radially symmetric; tricolpate; exine tectate 1.5 μm thick, columellate; sexine reticulate, heterobrochate, brochi <1 to 2.5 μm wide, diminishing toward apocolpium and aperture, simplibaculate; pores if present, inconspicuous; colpi thin, ⅓ length of grain; polar area 13 μm; amb circular; grains oblate-spheroidal, 19 × 22 μm. Liana.

Moraceae

Brosimum sp. Sw. Plate 45:294
Grains monad, apolar, asymmetric; diporate; exine tectate 1 μm thick; sexine psilate; pores slightly aspidate, circular, 2 μm diameter; amb circular; grains spheroidal, 24 μm. Tree.

Moraceae 1. Plate 45:295
Grains monad, apolar, asymmetric; diporate; exine tectate 1 μm thick; sexine psilate; pores circular, 2.5 μm diameter; amb circular to elliptical; grains oblate, 12 × 15 μm.

Myrsinaceae

Myrsine sp. L. (*Rapanea* sp. Aublet) Plate 45:296
Grains monad, isopolar, radially symmetric; stephanocolporate (4-colporate) frequently appearing as 4-colpate; exine tectate *ca.* 2 μm thick, columellate; sexine slightly scabrate; pores endexinic, inconspicuous, apparently circular, *ca.* 4 μm diameter; colpi narrow, ⅔ length of grain, with thin costa endocolpi, displaying exitus digitatus; opposite colpi crossed line; polar area variable, 4 μm wide between adjacent colpi and 16 μm wide between opposite colpi; amb circular to irregularly square; grains subprolate, 28 × 20 μm. Tree.

Myrtaceae

Eugenia sp. L. (form 1) Plate 45:297
Grains monad, isopolar, radially symmetric; syncolporate; exine tectate <1 μm thick; sexine psilate; pores apparently annulate, inconspicuous, circular, *ca.* 1 μm diameter; colpi very thin, joined at apices; amb triangular rounded; grains oblate to peroblate, 9 μm × 15 to 16 μm. Tree, shrub.

Eugenia sp. L. (form 2) Plate 45:298
Grains monad, isopolar, radially symmetric; syncolporate; exine tectate <1 μm thick; sexine scabrate; pores circular, *ca.* 1 μm diameter; colpi thin, joined at apices; amb triangular; grains oblate, 16 μm. Tree, shrub.

Myrcia sp. DC. ex Guillemin Plate 45:299
Grains monad, isopolar, radially symmetric; syncolporate; exine tectate 1 μm thick; sexine psilate to slightly scabrate; pores aspidate type, circular, *ca.* 3 to 4 μm diameter; colpi very thin, joined at apices; amb triangular; grains oblate, 24 μm. Tree.

Syzygium sp. Gaertner Plate 45:300
Grains monad, isopolar, radially symmetric; syncolporate (displaying parasyncolporate condition); exine tectate 1 µm thick; sexine scabrate; pores vestibulate type, inconspicuous; colpi narrow, joined at apices forming an ample triangle (apocolpial field) *ca.* 3 µm wide; amb triangular; grains oblate, 25 µm. Tree.

Myrtaceae 1. Plate 45:301
Grains monad, isopolar, radially symmetric; syncolporate; exine tectate 1 µm thick; sexine psilate to slightly scabrate; pores aspidate type, lalongate, 3 µm height × 8 µm wide; colpi very thin, inconspicuous, joined at apices; amb triangular rounded; grains oblate, 18 × 30 µm.

Myrtaceae 2. Plate 45:302
Grains monad, isopolar, radially symmetric; syncolporate; exine semitectate 1.5 to 2 µm thick; sexine scabrate; free scabrate process at mesocolpium; pores slightly vestibulate type, apparently lalongate, inconspicuous; colpi very thin, joined at apices; amb triangular; grains oblate, 28 to 30 µm.

Ochnaceae

Ouratea sp. Aublet Plate 45:303
Grains monad, isopolar, radially symmetric; tricolporate; exine tectate 2 to 2.5 µm thick, columellate; sexine scabrate; pores endexinic, apparently circular, *ca.* 4 µm diameter; colpi short, *ca.* 18 to 20 µm length × 2 to 3 µm wide, with conspicuous margo. 2 µm thick; margo bordering by crowded scabrae; amb circular; grains spheroidal, 28 to 32 µm. Tree, shrub.

Olacaceae

Minquartia aff. guianensis Aublet Plate 45:304
Grains monad, isopolar, radially symmetric; tricolpate apparently tricolporate; exine tectate 2 µm thick at mesocolpium diminishing to 1 µm thick at aperture, columellate; sexine psilate to slightly scabrate; pores if present, endexinic, inconspicuous; colpi 1 to 2 µm wide, as long as grain, with thin costa ectocolpi; polar area 6 µm; amb triangular; grains oblate-spheroidal, 20 × 22 µm. Tree.

Onagraceae (Oenotheraceae)

Onagraceae 1. Plate 45:305
Grains monad, apolar, asymmetric; diporate; exine tectate 1 µm thick at mesoporium and 3 µm thick at aperture; sexine psilate; pores circular, *ca.* 1 µm diameter, protruding, resembling vestibulate condition, annulate, annuli 2.5 µm thick, 1.5 µm wide; amb elliptical; grains oblate, 11 × 16 µm.

Polygalaceae

Diclidanthera sp. Mart. Plate 46:306
Grains monad, isopolar, radially symmetric; stephanocolporate (6-colporate); exine tectate 1 to 1.5 µm thick; sexine slightly scabrate; pores endexinic, inconspicuous, apparently circular, *ca.* 3 µm diameter; colpi narrow, straight, as long as grain; polar area between adjacent colpi 6 µm; amb circular; grains suboblate; 40 µm. Liana.

Polygala aff. paniculata L. Plate 46:307
Grains monad, isopolar, radially symmetric; stephanocolporate (8-colporate); sexine tectate 2 µm thick at apocolpium and 3 µm thick at mesocolpium; sexine psilate to slightly scabrate; pores lalongate, 2 µm height × 5 µm wide joining at apices displaying a continuous equatorial ring; costa endopori 1 to 1.5 µm thick; colpi narrow, straight line, as long as grain; polar area between adjacent colpi 4 µm; amb circular; grains subprolate, 21 × 18 µm. Herb.

Polygonaceae

Symmeria aff. paniculata Benth. Plate 46:308
Grains monad, isopolar, radially symmetric; tricolporate; exine tectate 2 µm thick, densely columellate; sexine reticulate, homobrochate, brochi 1 µm wide, muri very thin, simplibaculate; pores endexinic, apparently lolongate; colpi straight, as long as grain; polar area 14 µm; amb circular; grains prolate-spheroidal, 40 × 28 µm. Tree.

Proteaceae

Euplassa sp. Salisb. Plate 46:309
Grains monad, isopolar, radially symmetric; triporate; exine tectate 3 µm thick, densely columellate; sexine reticulate, homobrochate, brochi 1 µm wide, muri thin, simplibaculate, bacula 1.5 µm height; pores circular 3 µm diameter, annulate, annuli 2.5 to 3 µm wide; amb triangular concave, with aperture at angles; grains oblate, 30 × 52 µm. Tree.

Roupala sp. Aublet (form 1) Plate 46:310
Grains monad, isopolar, radially symmetric; triporate; exine tectate 2 µm thick, columellate; sexine reticulate, homobrochate, brochi <1 µm wide; pores circular 8 µm; amb triangular, with aperture at angles; grains oblate, 36 µm. Tree.

Roupala sp. Aublet (form 2) Plate 46:311
Grains monad, isopolar, radially symmetric; triporate; exine tectate 1.5 µm thick, columellate; sexine apparently psilate to slightly scabrate; pores circular 6 µm; amb triangular slightly concave, with aperture at angles; grains oblate, 26 µm. Tree.

Rhamnaceae

Gouania sp. Jacq. (form 1) Plate 47:312
Grains monad, isopolar, radially symmetric; tricolporate; exine tectate 2 µm thick, slightly columellate, displaying conspicuous stratification; sexine psilate; pores protruding, circular 5 µm diameter; colpi wide, 8 µm wide × as long as grain, with protuberant margo *ca.* 3.5 to 4 µm thick; polar area 6 µm; amb circular to triangular; grains suboblate, 24 × 27 µm. Liana.

Gouania sp. Jacq. (form 2) Plate 47:313
Grains monad, isopolar, radially symmetric; tricolporate; exine tectate 2 µm thick, slightly columellate, displaying conspicuous stratification; sexine scabrate; pores endexinic, protruding, resembling a vestibulate type, inconspicuous, circular; colpi thin, straight, as long as grain, with margo *ca.* 4 to 5 µm thick; polar area 3 µm; amb triangular convex; grains oblate-spheroidal, 20 × 22 µm. Liana.

Rubiaceae

Alibertia aff. edulis (A. Rich.) A. Rich. Plate 47:314
Grains monad, isopolar, radially symmetric; triporate; exine semitectate 1 to 1.5 μm thick, columellate; sexine reticulate, heterobrochate, brochi <1 to 2 μm wide, muri thin, simplibaculate, bacula 1 μm height, occasionally free bacula; pores circular 3 μm diameter, annulate, annuli 2 to 2.5 μm wide; amb triangular rounded; grains oblate, 30 to 32 μm. Tree.

Borreria aff. densiflora DC. Plate 47:315
Grains monad, isopolar, radially symmetric; stephanocolporate (6 to 7-colporate); exine tectate 2.5 μm thick; sexine reticulate, homobrochate, resembling baculate pattern; brochi <1 μm wide, muri simplibaculate, bacula thin, 1 μm height; pores endexinic, inconspicuous, apparently slightly lalongate; colpi short (brevicolpate), very thin, 5 μm length; polar area as wide as grain; amb circular; grains oblate-spheroidal, 20 × 22 μm. Herb.

Coccocypselum sp. P. Browne Plate 47:316
Grains monad, isopolar, radially symmetric; apparently tricolporate; exine intectate 1.5 μm thick; sexine baculate, bacula 1 μm height, densely scattered; pores depressed, apparently circular, annulate; colpi short, 10 μm wide × ⅓ length of grain; polar area 23 μm; amb circular; grains oblate, 44 μm. Herb.

Coussarea aff. hydrangeifolia (Benth.) Benth. & Hook. Plate 47:317
Grains monad, apolar, asymmetric; diporate; exine intectate 1 thick; sexine scabrate to granulate; pores circular, *ca.* 8 to 10 μm diameter, annulate, annuli very distinctive appearing as rings at apices, 5 to 6 μm thick, densely scabrate; amb elliptical; grains spheroidal, 22 to 24 μm. Tree, shrub.

Duroia sp. L. f. Plate 47:319
Grains monad, isopolar, radially symmetric; triporate; exine tectate 1 μm thick; sexine reticulate, hetereobrochate, brochi 1 to 2.5 μm wide, muri thin, simplibaculate, bacula 1 μm height; pores circular, 2 μm diameter, annulate, annuli 2 to 3 μm thick; amb circular; grains oblate, 43 to 45 μm. Tree.

Faramea sp. Aublet Plate 47:318
Grains monad, isopolar, radially symmetric; triporate; exine tectate 1 μm thick; sexine scabrate to slightly baculate; pores circular, *ca.* 7 μm diameter, aspidate, annulate, annuli 2 μm thick, bordering by densely micro-baculate process; amb triangular with aperture at angles; grains oblate, 24 to 26 μm. Tree, shrub.

Ferdinandusa sp. Pohl Plate 48:322
Grains monad, isopolar, radially symmetric; tricolporate; exine tectate 2.5 μm thick, densely columellate; sexine reticulate, heterobrochate, brochi 1 to 2.5 μm wide, muri 1 μm thick, simplibaculate, bacula 1.5 μm height × 1 μm wide; pores circular, 6 μm diameter, apparently annulate, annuli thin; colpi 3 μm wide, as long as grain; polar area 7 μm; amb circular; grains spheroidal, 32 μm. Tree.

Genipa aff. americana L. Plate 47:320
Grains monad, isopolar, radially symmetric; tricolporate; exine tectate 1.5 μm thick, densely columellate; sexine reticulate, homobrochate, brochi 1 μm wide, muri simplibaculate; pores circular, 4 to 5 μm diameter, annulate, annuli thin, 1 μm thick; colpi 3 μm wide, as long as grain; polar area 5 μm; amb circular; grains prolate-spheroidal, 19 × 18 μm. Tree.

Pagamea sp. Aublet — Plate 47:321

Grains monad, isopolar, radially symmetric; tricolporate; exine tectate 1.5 µm thick at apocolpium to 3 µm thick at aperture, columellate; sexine reticulate, homobrochate, brochi 1 µm wide, muri thin, simplibaculate, bacula 1 µm height; pores apparently circular, inconspicuous; colpi thin; amb circular; grains oblate-spheroidal, 30 × 32 µm. Tree, shrub.

Psychotria sp. L. (form 1) — Plate 48:323

Grains monad, isopolar, radially symmetric; tricolporate; sexine tectate 2 µm thick, densely columellate; sexine reticulate, homobrochate, brochi 1 µm wide, muri simplibaculate, bacula 1 µm height; pores apparently lalongate, inconspicuous; colpi short, narrow, ⅓ length of grain; polar area 25 µm; amb circular; grains suboblate, 48 to 50 µm. Tree, shrub.

Psychotria sp. L. (form 2) — Plate 48:324

Grains monad, isopolar, radially symmetric; tricolpate; exine intectate 2 µm thick; sexine baculate resembling clavate pattern; bacula crowded, 1 µm height × 0.5 to 1 µm wide; colpi wide, deep, ⅔ length of grain; polar area 18 µm; amb circular; grains suboblate, 33 µm. Tree, shrub.

Psychotria sp. L. (form 3) — Plate 48:325

Grains monad, isopolar, radially symmetric; tricolporate; exine tectate 2 µm thick, densely columellate; sexine reticulate, homobrochate, brochi 1 µm wide, simplibaculate; aperture inconspicuous, grains folded; amb circular; grains suboblate, 30 µm. Tree, shrub.

Psychotria sp. L. (form 4) — Plate 48:326

Grains monad, isopolar, radially symmetric; tricolporate; exine tectate 1 µm thick, densely columellate; sexine reticulate, homobrochate, brochi 1 µm wide, muri simplibaculate, bacula 1 µm height; pores circular, 1.5 µm diameter, slightly protruding; colpi straight, ⅓ length of grain; polar area 16 µm; amb circular; grains suboblate, 26 to 28 µm. Tree, shrub.

Rutaceae

Adiscanthus aff. fusciflorus Ducke — Plate 49:328

Grains monad, isopolar, radially symmetric; tricolporate; exine semitectate 2 to 2.5 µm thick, columellate; sexine complex, fossulate-verrucate resembling per-reticulate pattern; fossulae elongated 2 to 5 µm length × 1 to 2.5 µm wide, displaying a longitudinally orientation, muri simplibaculate, bacula 1 µm height, occasionally free; flat verrucae on the surface; pores lalongate, 3 µm height × 10 µm wide; colpi thin, straight, as long as grains; amb circular; grains prolate, 49 × 29 µm. Tree.

Citrus sp. L. — Plate 48:327

Grains monad, isopolar, radially symmetric; tricolporate; exine semitectate 2 µm thick, densely columellate; sexine reticulate, heterobrochate, brochi 1 to 3 µm wide, muri 1 µm thick, simplibaculate, bacula 2 µm height × 1 µm wide, occasionally free bacula into lumina; pores endexinic, inconspicuous, apparently lalongate; colpi wide, as long as grain; polar area 14 µm; amb circular; grains prolate, 40 × 34 µm. Tree, shrub.

Zanthoxylum sp. L. — Plate 49:329

Grains monad, isopolar, radially symmetric; tricolporate; sexine tectate 1 µm thick, columellate; sexine striato-reticulate, homobrochate, brochi longitudinally oriented, 1 µm wide, muri simplibaculate; pores inconspicuous apparently lalongate; colpi thin, as long as grain, with slightly costa endocolpi; amb circular; grains subprolate, 15 × 12 µm. Tree, shrub.

Sapindaceae

Cupania sp. L. (form 1) Plate 49:330
Grains monad, isopolar, radially symmetric; syncolporate (displaying parasyncolporate condition); exine tectate 2.5 μm thick, densely columellate; sexine scabrate; pores endexinic, depressed, circular 5 μm diameter; colpi 2 μm wide, joined at apices displaying small triangle (apocolpial field); amb triangular-concave; grains oblate, 42 to 45 μm. Tree.

Cupania sp. L. (form 2) Plate 49:331
Grains monad, isopolar, radially symmetric; syncolporate (displaying parasyncolporate condition); exine tectate 2 μm thick, densely columellate; sexine reticulate, homobrochate, brochi <1 μm wide, muri simplibaculate; pores endexinic, depressed, circular, inconspicuous; colpi 1 μm wide, joined at apices displaying small triangle (apocolpial field); amb triangular convex; grains oblate, 20 × 27 μm. Tree.

Matayba sp. Aublet Plate 49:332
Grains monad, isopolar, radially symmetric; syncolporate (displaying parasyncolporate condition); exine tectate 2 μm thick, densely columellate; sexine scabrate; pores endexinic, depressed, apparently circular; colpi narrow, joined at apices displaying small triangle (apocolpial field), *ca.* 5 μm wide; amb triangular; grains oblate, 30 to 35 μm. Tree.

Paullinia sp. L. Plate 49:333
Grains monad, isopolar, radially symmetric; triporate; exine semitectate 1 μm thick at mesoporium and 3 μm thick at aperture, slightly columellate; sexine scabrate occasionally displaying fine perforations; pores circular *ca.* 1 μm diameter, endexinic, drop type appearing as protruding; costa ectopori 3 μm thick; amb triangular irregular; grains oblate, 30 to 32 μm. Liana.

Talisia sp. Aublet Plate 49:334
Grains monad, isopolar, radially symmetric; tricolporate; exine tectate 1.5 μm thick; sexine psilate; pores endexinic, depressed, lalongate *ca.* 2 μm height × 6 μm wide, forming small vestibulum; colpi thin, short, *ca.* 6 μm length; amb triangular irregularly concave-convex; grains oblate, 25 to 28 μm. Tree.

Sapotaceae

Chrysophyllum sp. L. Plate 49:335
Grains monad, isopolar, radially symmetric; tricolporate; exine tectate 1 μm thick; sexine psilate; pores circular 3 μm diameter; colpi thin, ⅔ length of grain, with slightly costa ectocolpi; amb circular; grains prolate, 30 × 22 μm. Tree.

Diploon aff. cuspidatum (Hoehne) Cronq. Plate 49:336
Grains monad, isopolar, radially symmetric; tricolporate; exine intectate 1 μm thick; sexine echinate, appearing as pilate pattern; spine 1 μm length; pores circular 2 μm diameter; colpi thin, as long as grain; amb circular; grains spheroidal, 24 μm. Tree.

Elaeoluma aff. nuda (Baehni) Aubrév. Plate 50:338
Grains monad, isopolar, radially symmetric; stephanocolporate (4-colporate); exine tectate 2 to 2.5 μm thick, columellate; sexine reticulate, homobrochate, brochi 1 to 2 μm wide; pores lalongate, *ca.* 3 μm height × 7 μm wide; colpi as long as grain, opposite colpi crossed-lined; polar area between adjacent colpi *ca.* 8 μm; amb square-like; grains subprolate, 38 to 40 μm × 32 μm. Tree, shrub.

Elaeoluma sp. Baillon Plate 50:339
Grains monad, isopolar, radially symmetric; stephanocolporate (4-colporate); exine tectate 3 to 4 μm thick, columellate; sexine reticulate, homobrochate; brochi <1 μm wide, muri simplibaculate, bacula conspicuous 3 μm height × 1 μm wide; pores endexinic, appearing as circular 8 μm diameter; colpi thin, 25 μm length, with thick costa endocolpi *ca.* 4 μm; opposite colpi crossed-lined; amb circular; grains suboblate, 44 × 50 μm. Tree, shrub.

Manilkara sp. Adams Plate 49:337
Grains monad, isopolar, radially symmetric; stephanocolporate (4-colporate occasionally with 5 colpori); exine tectate 2 μm thick; sexine psilate; pores slightly elliptic, 4 μm height × 7 μm wide; colpi narrow, ⅔ length of grain; amb circular to square ; grains spheroidal, 36 μm. Tree.

Pouteria aff. cladantha Sandw. Plate 50:340
Grains monad, isopolar, radially symmetric; tricolporate; exine tectate 1 μm thick, columellate; sexine psilate to slightly scabrate; pores circular 1 μm diameter; colpi narrow, short *ca.* 9 μm length × 1 μm wide, with very thin costa endocolpi; amb circular; grains subprolate, 16 × 13 μm. Tree.

Pouteria aff. elegans (A. DC.) Baehni Plate 50:341
Grains monad, isopolar, radially symmetric; stephanocolporate (4-colporate); exine tectate 1 μm thick at apocolpium to 1.5 μm thick at aperture; sexine psilate, slightly scabrate; pores circular 1.5 to 2 μm diameter; colpi thin, 12 μm length; polar area 5 μm between adjacent colpi; amb circular to square grains prolate to prolate-spheroidal, 17 × 15 μm. Tree.

Pouteria sp. Aublet (form 1) Plate 50:342
Grains monad, isopolar, radially symmetric; stephanocolporate (4-colporate); exine tectate *ca.* 1.5 μm thick; slightly columellate; sexine scabrate appearing as micro-reticulate pattern; pores circular, 2 μm diameter; colpi narrow, 22 μm length × 1.5 μm wide, with thin costa endocolpi; polar area 6 μm between adjacent colpi; amb square; grains prolate to subprolate, 23 to 24 μm × 15 to 17 μm. Tree.

Pouteria sp. Aublet (form 2) Plate 50:343
Grains monad, isopolar, radially symmetric; tricolporate; exine tectate 1 μm thick at apocolpium to 2.5 μm thick at aperture; sexine scabrate; pores circular 5 μm diameter; colpi short, very thin, inconspicuous, with costa endocolpi *ca.* 1.5 μm thick; polar area 11 μm; amb circular; grains subprolate, 24 × 20 μm. Tree.

Pouteria sp. Aublet (form 3) Plate 50:344
Grains monad, isopolar, radially symmetric; tricolporate; exine tectate 1 to 1.5 μm; sexine psilate; pores circular 3 μm diameter; colpi thin, ⅔ length of grain, with costa endocolpi ca 1.5 μm thick; amb circular; grains subprolate, 30 × 20 μm. Tree.

Pouteria sp. Aublet (form 4) Plate 50:345
Grains monad, isopolar, radially symmetric; stephanocolporate (4-colporate); exine tectate 1.5 μm thick; sexine psilate to slightly scabrate; pores circular 2 μm diameter; colpi thin, 14 μm length, with slightly costa endocolpi *ca.* 1 μm thick, opposite colpi crossed line; amb circular; grains subprolate, 20 × 15 μm. Tree.

Sarcaulus sp. Radlk. Plate 50:346
Grains monad, isopolar, radially symmetric; tricolporate; exine tectate 1 μm thick; sexine psilate; pores endexinic apparently lalongate, aspidate type; colpi short (brevicolpate), very thin, 5 μm length, with costa ectocolpi 1 μm thick; polar area as wide as grain; amb circular; grains suboblate, 21 × 25 μm. Tree.

Solanaceae

Capsicum sp. L. Plate 50:347

Grains monad, isopolar, radially symmetric; tricolporate; exine tectate 1 µm thick; sexine psilate; pores lalongate, becoming to join at apices displaying a continuous equatorial ring (colpus equatorialis); colpi thin, as long as grain, with costa endocolpi 1 µm thick; amb circular; grains spheroidal, 16 µm. Herb.

Solanum sp. L. (form 1) Plate 50:348

Grains monad, isopolar, radially symmetric; tricolporate; exine tectate 1 µm thick; sexine psilate; pores lalongate joining at apices displaying a continuous and protuberant equatorial ring (colpus equatorialis); colpi thin, folded, inconspicuous; amb circular; grains prolate, 30 × 22 µm. Tree, shrub, herb.

Solanum sp. L. (form 2) Plate 51:349

Grains monad, isopolar, radially symmetric; tricolporate; exine tectate 1 µm thick; sexine psilate; pores lalongate displaying a continuous and protuberant equatorial ring (colpus equatorialis), resembling aspidate type; colpi wide, as long as grain; polar area 5 µm; amb circular trilobulate; grains prolate-spheroidal, 20 × 18 µm. Tree, shrub, herb.

Solanum sp. L. (form 3) Plate 51:350

Grains monad, isopolar, radially symmetric; tricolporate; exine tectate 1.5 µm thick at apocolpium and 6 µm thick at aperture; sexine psilate; pores lalongate, almost joined at apices displaying a continuous and protuberant equatorial ring (colpus equatorialis), resembling a vestibulum; colpi narrow, as long as grain, with conspicuous costa endocolpi 6 µm thick; amb circular; grains prolate-spheroidal, 30 × 28 µm. Tree, shrub, herb.

Staphylleaceae

Turpinia sp. Vent. Plate 51:351

Grains monad, isopolar, radially symmetric; tricolporate; exine tectate 3 µm thick; sexine reticulate, heterobrochate, brochi 1 µm wide at mesocolpium diminishing to <1 µm wide at aperture, muri simplibaculate, baculi 1 µm height; pores endexinic, lalongate, inconspicuous; colpi narrow, as long as grain, with conspicuous protruding costa ectocolpi (margo) 3 µm thick, displaying slightly an equatorial constriction; amb triangular; grains subprolate, 25 × 22 µm. Tree.

Symplocaceae

Symplocos sp. Jacq. Plate 51:352

Grains monad, isopolar, radially symmetric; tricolporate; exine tectate 2 µm thick; sexine verrucate, flat verrucae <1 µm height × 1.5 to 2.5 µm wide; pores apparently circular, 3 µm diameter, drop type; colpi very thin, short (brevicolpate), ⅓ length of grain, inconspicuous; polar area as wide as grain; amb circular; grains suboblate, 30 × 34 µm. Tree, shrub.

Theaceae

Archytaea aff. multiflora Benth. Plate 51:353

Grains monad, isopolar, radially symmetric; triporate; exine tectate 3 µm thick, densely columellate; sexine per-reticulate, heterobrochate, brochi 1.5 to 4 µm wide, muri undulating, 1.5 µm thick, simplibaculate, bacula 2 µm height × 1 µm wide; pores circular, 6 to 7 µm diameter, slightly annulate; amb circular; grains oblate, 46 to 50 µm. Tree, shrub.
Note: This taxon is also classified as Guttiferae (Mabberley, 1997).

Gordonia sp. Ellis (*Laplacea* sp. Kunth) (form 1) Plate 51:354
Grains monad, isopolar, radially symmetric; triporate; exine intectate 1 μm thick (ornamentation excluded); sexine echinate-scabrate, spine conical, sharp to rounded, variable in size 3 to 6 μm height × 2 to 2.5 μm wide, scarcely dispersed; scabrae <1 μm wide densely scattered on the surface; pores apparently circular, inconspicuous; amb circular; grains oblate, 45 μm. Tree.

Gordonia sp. Ellis (*Laplacea* sp. Kunth) (form 2) Plate 51:355
Grains monad, isopolar, radially symmetric; triporate; exine intectate 2 μm thick (ornamentation excluded), columellate; sexine echinate-scabrate, spine bottle shaped, variable in size 2 to 3 μm height × 1 to 1.5 μm wide, scarcely dispersed; scabrae <1 μm wide densely scattered on the surface; pores apparently circular, inconspicuous; amb circular; grains oblate, 40 μm (ornamentation included). Tree.

Tiliaceae

Apeiba sp. Aublet (form 1) Plate 52:356
Grains monad, isopolar, radially symmetric; tricolporate; exine tectate 1 μm thick, densely columellate; sexine reticulate, homobrochate, brochi <1 μm wide, muri simplibaculate; pores apparently lalongate, inconspicuous; colpi short, 4 μm wide × ¼ length of grain; polar area 20 μm; amb circular; grains suboblate, 31 μm. Tree.

Apeiba sp. Aublet (form 2) Plate 52:357
Grains monad, isopolar, radially symmetric; tricolporate; exine tectate 2 μm thick, densely columellate; sexine finely reticulate, homobrochate, brochi <1 μm wide, appearing as pilate pattern; pores apparently lalongate, slightly protuberant, inconspicuous; colpi short (brevicolpate), 2 μm wide × ¼ length of grain; polar area 28 μm; amb circular; grains suboblate, 50 μm. Tree.

Heliocarpus sp. L. Plate 52:358
Grains monad, isopolar, radially symmetric; tricolporate; exine tectate 2.5 to 3 μm thick, densely columellate; sexine reticulate, homobrochate, brochi *ca.* 1 μm wide, muri simplibaculate, bacula conspicuous 1.5 μm height × <1 μm wide; pores elliptic, 6 μm height × 12 μm wide; colpi narrow, as long as grain, with costa endocolpi 3 μm thick; amb circular; grains prolate, 60 × 45 μm. Tree.

Luehea sp. Willd. Plate 52:359
Grains monad, isopolar, radially symmetric; tricolporate; exine tectate 2 μm thick, columellate; sexine reticulate, homobrochate, brochi <1 μm wide, muri simplibaculate; pores elliptic, 3 μm height × 6 μm wide; colpi narrow, straight, 24 μm length, with costa endocolpi 2 μm thick; polar area as wide as grain; amb circular; grains prolate, 32 × 19 μm. Tree.

Lueheopsis sp. Burret Plate 52:360
Grains monad, isopolar, radially symmetric; tricolporate; exine tectate 2 μm thick, densely columellate; sexine reticulate, homobrochate, brochi *ca.* 1 μm wide, muri simplibaculate, bacula conspicuous 1 μm height; pores endexinic, elliptic, acute, 5 μm height × 10 μm wide; colpi 2 μm wide, as long as grain, with costa endocolpi 2 μm thick, displaying slightly an equatorial constriction; amb circular trilobulate; grains prolate to subprolate, 44 × 34 μm. Tree.

Mortoniodendron sp. Standley & Steyerm Plate 53:361
Grains monad, isopolar, radially symmetric; triporate; exine tectate <1 μm thick, columellate; sexine finely reticulate, homobrochate; pores circular, vestibulate, 3 μm diameter, slightly annulate, annulus <1 μm wide; amb circular; grains spheroidal, 19 to 20 μm. Tree.

Tiliaceae 1. Plate 53:362
Grains monad, isopolar, radially symmetric; tricolporate; exine tectate 5 μm thick at mesocolpium and 7 μm thick at aperture; columellate; sexine finely reticulate, homobrochate, resembling granulate condition; brochi <1 μm wide, muri simplibaculate; pores rectangular with elongated bifurcations at apices, 7 μm height × 22 μm wide; colpi narrow, as long as grain, with costa endocolpi 4 μm thick, displaying slightly an equatorial constriction; amb triangular; grains oblate-spheroidal to spheroidal, 63 × 65 μm.

Ulmaceae

Celtis sp. L. (form 1) Plate 53:363
Grains monad, isopolar, radially symmetric; triporate, occasionally diporate; exine tectate 1.5 to 2 μm. thick, columellate; sexine scabrate; pores circular, aspidate, 3 to 4 μm diameter; amb circular; grains oblate, 20 × 34 μm. Tree, liana.

Celtis sp. L. (form 2) Plate 53:364
Grains monad, isopolar, radially symmetric; triporate, occasionally diporate; exine tectate 1.5 to 2 μm. thick, columellate; sexine scabrate; pores circular, aspidate, 3 to 4 μm diameter; amb circular; grains oblate, 20 × 34 μm. Tree, liana.

Trema sp. Lour. Plate 53:365
Grains monad, isopolar, radially symmetric; triporate; exine tectate 1 μm thick, columellate; sexine scabrate; pores circular, common type, 1.5 to 2 μm diameter; amb elliptic; grains oblate, 14 × 17 μm. Tree.

Umbelliferae (Apiaceae)

Apium sp. L. Plate 53:366
Grains monad, isopolar, radially symmetric; tricolporate; exine tectate 1 μm thick, columellate; sexine micro-reticulate, homobrochate, brochi <1 μm wide; pores endexinic, inconspicuous, apparently lalongate to elliptic, 3 μm height × 6 μm wide; colpi narrow, ⅔ as long as grain; amb circular; grains prolate, 24 × 19 μm. Herb.

Eryngium sp. L. Plate 53:367
Grains monad, isopolar, radially symmetric; tricolporate; exine tectate 1 μm thick, densely columellate; sexine micro-reticulate, homobrochate, brochi <1 μm wide, muri simplibaculate, bacula conspicuous; pores endexinic, lalongate, 2 μm height × 8 μm wide; colpi narrow, as long as grain, with costa endocolpi 3 μm thick, resembling an equatorial thickness ring; amb circular; grains prolate to perprolate, 31 × 14 μm. Herb.

Hydrocotyle L. Plate 53:368
Grains monad, isopolar, radially symmetric; tricolporate; exine tectate 1.5 μm thick; sexine psilate; pores lalongate, 3 μm height × 8 μm wide; colpi narrow, straight, as long as grain, with conspicuous costa endocolpi 2 μm thick, displaying an equatorial thickness depression; amb circular; grains prolate to perprolate, 32 × 15 μm. Herb.

Umbelliferae 1. Plate 54:369
Grains monad, isopolar, radially symmetric; tricolporate; exine tectate 1 to 2 μm thick, columellate; sexine finely scabrate; pores lalongate, 5 μm height × 14 μm wide; colpi narrow, ⅔ length of grain, with

conspicuous costa endocolpi 3 to 4 μm thick, displaying an equatorial thickness ring; amb circular; grains prolate to perprolate, 45 × 26 μm.

Umbelliferae 2. Plate 54:370
Grains monad, isopolar, radially symmetric; tricolporate; exine tectate 2 to 2.5 μm thick; sexine scabrate; pores lalongate joined at apices resembling a continuous equatorial ring (colpus equatorialis); colpi narrow, ¾ length of grain, with conspicuous costa endocolpi 4 μm thick, appearing as an equatorial thickness; amb circular; grains prolate to perprolate, 54 × 24 μm.

Umbelliferae 3. Plate 54:371
Grains monad, isopolar, radially symmetric; tricolporate; exine tectate 2 to 3 μm thick, densely columellate; sexine reticulate, homobrochate, brochi <1 μm wide; pores lalongate, acute, 6 μm height × 18 μm wide; colpi narrow, ⅔ length of grain, with conspicuous costa endocolpi 3 μm thick, appearing bifurcate as exitus digitatus; amb circular; grains prolate to perprolate, 58 × 25 μm.

Verbenaceae

Aegiphila sp. Jacq. (form 1) Plate 54:372
Grains monad, isopolar, radially symmetric; tricolpate; exine tectate 1.5 μm thick, densely columellate; sexine micro-echinate, spine short, sharp, <1 μm height, scarcely scattered; colpi short (brevicolpate); polar area as wide as grain; amb circular; grains oblate, 31 μm. Shrub, liana.

Aegiphila sp. Jacq. (form 2) Plate 54:373
Grains monad, isopolar, radially symmetric; tricolpate; exine tectate 2 to 2.5 μm thick, densely columellate; sexine echinate, spine conical, sharp, *ca.* 1 μm height, scarcely scattered; colpi inconspicuous, as long as grain, appearing as porate pattern; amb circular; grains spheroidal, 44 μm. Shrub, liana.

Priva sp. Adams Plate 54:374
Grains monad, isopolar, radially symmetric; triporate; exine tectate 1 μm thick, columellate; sexine scabrate; pores circular 5 μm diameter, protruding, annulate, annuli 2 μm thick; amb circular; grains suboblate, 16 × 21 μm. Herb.

Vitaceae

Cissus sp. L. Plate 55:376
Grains monad, isopolar, radially symmetric; tricolporate; exine tectate 2 μm thick, densely columellate; sexine micro-reticulate, homobrochate, brochi <1 μm wide, resembling scabrate pattern; pores lalongate, endexinic, masked by colpi bridge; acute, 6 μm height × 18 μm wide; colpi 2 μm wide × ¾ length of grain, with conspicuous costa endocolpi 2.5 μm thick, appearing bifurcate as exitus digitatus, displaying equatorial constriction forming a bridge-like structure; amb circular; grains prolate, 48 × 32 μm. Liana.

Vochysiaceae

Vochysia sp. Aublet Plate 54:375
Grains monad, isopolar, radially symmetric; tricolporate; exine tectate 2 μm thick, columellate; sexine scabrate; pores endexinic, lalongate, inconspicuous; colpi narrow, as long as grain, with conspicuous protruding costa ectocolpi (margo) 3 μm thick, displaying slightly an equatorial constriction; polar area 12 μm; amb triangular-concave; grains suboblate, 25 × 32 μm. Tree.

Other Magnoliopsidae

Magnoliopsidae 1. Plate 55:377

Grains monad, isopolar, radially symmetric; triporate; exine tectate 1.5 to 2 µm thick; sexine scabrate appearing as finely reticulate; scabrae <1 µm height, homogeneous; pores circular *ca.* 3 µm diameter, displaying slightly annulus; amb circular; grains oblate-spheroidal, 25 × 29 µm.

Magnoliopsidae 2. Plate 55:378

Grains monad, isopolar, radially symmetric; triporate; exine tectate 2.5 to 3 µm thick; sexine reticulate, homobrochate, brochi *ca.* 1 µm wide, muri simplibaculate, baculate crowded, homogeneous; pores inconspicuous, masked by ornamentation, apparently circular; amb circular; grains oblate-spheroidal, 23 × 25 µm.

Magnoliopsidae 3. Plate 55:379

Grains monad, isopolar, radially symmetric; stephanocolpate (4-colpate); exine tectate 1.5 µm thick, densely columellate; sexine scabrate; colpi wide, ⅔ length of grain, with thin ectexinic membrana; polar area between adjacent colpi 12 µm; amb square; grains suboblate, 18 × 21 µm.

Magnoliopsidae 4. Plate 55:380

Grains monad, heteropolar, bilaterally symmetric; dicolporate, structure complex; exine tectate 2 to 3 µm thick, densely columellate; sexine scabrate; pores endexinic, depressed, inconspicuous; colpi narrow, tortuous, as long as grain, becoming to join at apex, displaying a longitudinal ring; amb elliptic irregular; grains suboblate, 25 × 30 µm.

Magnoliopsidae 5. Plate 55:381

Grains monad, isopolar, radially symmetric; tricolporate; exine tectate 2.5 µm thick, columellate; sexine psilate to slightly scabrate; pores lalongate, acute, 3 µm height × 16 µm wide; colpi short, thin, ⅓ length of grain; amb circular; grains suboblate 22 × 26 µm.

Magnoliopsidae 6. Plate 55:382

Grains monad, isopolar, radially symmetric; tricolporate; exine tectate 1.5 µm thick; sexine finely reticulate, homobrochate, brochi <1 µm wide, muri simplibaculate; pores endexinic, transversally parallel 2 µm height × 8 µm wide; colpi thin, as long as grain, borders irregularly defined, apparently covered by thin membrana; polar area 6 µm; amb circular; grains prolate-spheroidal, 27 × 25 µm.

Magnoliopsidae 7. Plate 55:383

Grains monad, isopolar, radially symmetric; tricolporate; exine tectate 1.5 µm thick; sexine reticulate, heterobrochate, brochi 1 µm wide diminishing toward aperture, muri simplibaculate; sculpture resembling perforate pattern; pores endexinic, lalongate with longitudinally elongated apices ("H" form); colpi thin, straight, ⅔ length of grain, with conspicuous costa endocolpi 3 µm thick, interrupted at pore level; margo 3 to 4 µm wide; amb circular; grains subprolate, 44 × 36 µm.

Magnoliopsidae 8. Plate 56:384

Grains monad, isopolar, radially symmetric; tricolporate; exine tectate 1 to 2 µm thick; sexine psilate; pores lalongate, endexinic; colpi wide, as long as grain, displaying slightly costa endocolpi; polar area 8 µm; amb circular trilobutate; grains suboblate to oblate-spheroidal, 42 × 50 µm.

Magnoliopsidae 9. Plate 56:385

Grains monad, isopolar, radially symmetric; tricolporate appearing as tricolpate; exine tectate <1 µm thick; sexine psilate; aperture inconspicuous, apparently pores circular, endexinic; colpi thin, as long as grain; polar area 6 µm; amb circular; grains prolate-spheroidal to spheroidal, 18 × 16 µm.

Magnoliopsidae 10. Plate 56:386

Grains monad, isopolar, radially symmetric; tricolporate; exine intectate 2 to 2.5 μm thick; sexine baculate, bacula densely scattered, 1 μm wide × 2 μm height; pores apparently circular 6 μm diameter; colpi thin, straight, as long as grain; polar area 10 μm ; amb circular; grains prolate, 40 × 28 μm.

Magnoliopsidae 11. Plate 56:387

Grains monad, isopolar, radially symmetric; heterocolpate, grains displaying complex structure, 4 colpori alternating with 4 pseudocolpi; exine tectate <1 μm thick, slightly columellate; sexine psilate to scabrate; pores apparently lalongate, *ca.* 6 μm wide, displaying fine vestibulum; colpori narrow, as long as grain, joined at apex; pseudocolpi thin, ⅔ length of grain acute ends; amb square rounded; grains oblate, 24 μm.

Magnoliopsidae 12. Plate 56:388

Grains monad, isopolar, radially symmetric; stephanocolporate (4-colporate); exine tectate 6 to 7 μm thick, slightly columellate; nexine thicker than sexine, sexine 1.5 μm thick, nexine 5 μm thick; sexine psilate; pores endexinic, lalongate, inconspicuous; colpi narrow, ⅔ length of grain, acute ends, with costa endocolpi 5 μm thick; polar area 23 μm; amb square; grains spheroidal, 58 μm.

Magnoliopsidae 13. Plate 56:389

Grains monad, isopolar, radially symmetric; syncolporate (3-colporate); exine tectate 2 μm thick; slightly columellate; sexine finely verrucate, verrucae flat, uniform; pores endexinic, apparently lalongate, inconspicuous, vestibulate, costa ectopori 5 μm thick; colpi joined at apices forming a wide triangular parasyncolporate condition; amb triangular; grains oblate, 39 μm.

ANGIOSPERMAE – LILIOPSIDAE (Monocotyledoneae)

Alismataceae

Echinodorus sp. Rich. ex Engelm. Plate 57:390

Grains monad, apolar, asymmetric; periporate (>4 pores); exine intectate 2.5 to 3 μm thick; sexine displaying irregular echinate pattern; spine appearing as short bacula; pores apparently circular *ca.* 4 μm diameter; amb circular; grains suboblate, 48 μm. Aquatic herb.

Sagittaria sp. L. (form 1) Plate 57:391

Grains monad, apolar, asymmetric; periporate (ca. 5 pores); exine tectate 1.5 μm thick (ornamentation excluded); sexine echinate, spine conical, sharp 1 μm length; pores endexinic, masked by ornamentation and a thin membraneous process, apparently circular *ca.* 2 to 2.5 μm diameter; amb circular; grains spheroidal, 26 μm (ornamentation included). Aquatic herb.

Sagittaria sp. L. (form 2) Plate 57:392

Grains monad, apolar, asymmetric; periporate (>5 pores); exine tectate 2 μm thick (ornamentation excluded); sexine echinate, spine conical, sharp, broad based, 2.5 to 3 μm length; pores inconspicuous, masked by ornamentation, apparently circular; amb circular; grains spheroidal, 45 μm (ornamentation included). Aquatic herb.

Sagittaria sp. L. (form 3) Plate 57:393

Grains monad, apolar, asymmetric; periporate (>4 pores); exine tectate 1 to 1.5 μm thick (ornamentation excluded); sexine echinate, spine conical, thin, sharp 2 μm length; pores endexinic, inconspicuous, apparently circular; amb circular; grains spheroidal, 30 μm (ornamentation included). Aquatic herb.

Araceae

Anthurium sp. Pers. Plate 57:394

Grains monad, apolar, asymmetric; inaperturate (apparently periporate); exine tectate <1 μm thick; sexine reticulate, homobrochate, brochi fine <1 μm wide; aperture if present, inconspicuous; amb circular; grains spheroidal, 19 μm. Epiphytic herb.

Philodendron sp. Schott Plate 57:395

Grains monad, heteropolar, bilaterally symmetric; monosulcate; exine tectate 1 to 1.5 μm thick; sexine scabrate to granulate; sulcus, deep, opened, 4 to 5 μm wide, as long as grain; amb elliptic; grains oblate, 22 × 51 μm. Vine.

Arecaceae (Palmae)

Chelyocarpus sp. Dammer Plate 57:396

Grains monad, heteropolar, bilaterally symmetric; monosulcate; exine tectate 1 to 1.5 μm thick, densely columellate; sexine scabrate; sulcus narrow, as long as grain; amb elliptic, acute ends; grains suboblate, 15 to 16 μm × 30 to 32 μm. Tree.

Cryosophila sp. Blume Plate 57:397

Grains monad, heteropolar, bilaterally symmetric; monosulcate; exine tectate 1 μm thick; sexine reticulate, homobrochate, brochi 1 μm wide, muri simplibaculate, bacula conspicuous *ca.* 1 μm length; sulcus ⅔ length of grain, inconspicuous; amb elliptic; grains suboblate, 27 × 35 μm. Tree.

Desmoncus sp. Mart. Plate 37:398

Grains monad, heteropolar, bilaterally symmetric; monosulcate; exine tectate 1 μm thick, densely columellate; sexine scabrate; sulcus narrow, as long as grain; amb elliptic with acute ends; grains suboblate, 21 × 36 μm. Tree.

Euterpe sp. Mart. Plate 57:399

Grains monad, heteropolar, bilaterally symmetric; monosulcate; exine tectate 1.5 μm thick, densely columellate; sexine finely scabrate; sulcus narrow, ⅔ length of grain, *ca.* 1 μm wide × 32 μm length, with slightly margo; amb elliptic with acute ends; grains suboblate, 19 to 20 μm × 38 μm. Tree.

Geonoma sp. Willd. (form 1) Plate 58:400

Grains monad, heteropolar, bilaterally symmetric; monosulcate; exine tectate 1.5 μm thick, densely columellate; sexine scabrate; sulcus wide, as long as grain; amb elliptic; grains suboblate, 25 × 30 μm. Tree.

Geonoma sp. Willd. (form 2) Plate 58:401

Grains monad, heteropolar, bilaterally symmetric; monosulcate; exine intectate 1.5 μm thick; sexine baculate, bacula thin, short, *ca.* 1 μm height; sulcus wide, irregular, bordering by small bacula, as long as grain; amb elliptic with acute ends; grains suboblate, 22 × 35 μm. Tree.

Iriartea aff. deltoidea Ruíz & Pavón Plate 58:402

Grains monad, heteropolar, bilaterally symmetric; monosulcate; exine tectate *ca.* 1 μm thick (ornamentation excluded); sexine baculate, bacula irregular, poorly packed, 1 to 2.5 μm length; sulcus narrow, inconspicuous, as long as grain; amb circular to elliptic; displaying dimorphism; grains suboblate to spheroidal, 26 to 30 μm × 34 to 36 μm. Tree.

Mauritia aff. flexuosa L. f. (form 1) Plate 58:403

Grains monad, heteropolar, bilaterally symmetric; monoporate; exine tectate 1.5 μm thick (ornamentation excluded); sexine echinate-scabrate, slightly columellate; scarce spine appearing as bottle shaped, 3 μm length × 1.5 μm wide at base, encrusted in a small tectate depression; scabrae <1 μm, densely scattered on the grain surface; pore circular irregular, borders no well defined, 12 μm diameter; amb circular; grains spheroidal irregular, 60 to 68 μm (including ornamentation). Tree.

Mauritia aff. flexuosa L. f. (form 2) Plate 58:404

Grains monad, heteropolar, bilaterally symmetric; monoporate; exine tectate 2 μm thick (ornamentation excluded); sexine echinate-scabrate, slightly columellate; spine scarcely scattered, conical, curved, *ca.* 10 μm height × 2 μm wide at base, encrusted in a conspicuous tectate depression; scabrae *ca.* 1 μm wide, densely scattered on the grain surface; pore circular to elliptic, 15 μm diameter; amb elliptic; grains suboblate to oblate-spheroidal, 70 μm (including ornamentation). Tree.

Mauritiella aff. aculeata Kunth Plate 59:405

Grains monad, heteropolar, bilaterally symmetric; monoporate; exine tectate 3 to 4 μm thick (ornamentation excluded); sexine echinate-scabrate, slightly columellate; spine scarcely scattered, conical-rounded shaped, broad, *ca.* 6 to 7 μm length × 3 to 4 μm wide, encrusted in a small tectate depression; fine scabrae densely scattered on the grain; pore circular 8 μm diameter, with conspicuous costa ectopori, *ca.* 2 μm wide; amb circular; grains spheroidal 55 to 60 μm (including ornamentation). Tree.

Oenocarpus sp. Mart. Plate 59:406

Grains monad, heteropolar, bilaterally symmetric; monosulcate; exine tectate 1.5 to 2 μm thick; sexine reticulate, homobrochate, brochi <1 μm wide; sulcus irregular, undulating, wide, as long as grain; amb elliptic; grains suboblate, 38 to 40 μm × 58 to 60 μm. Tree.

Scheelea sp. Karsten Plate 59:407

Grains monad, heteropolar, bilaterally symmetric; monosulcate; exine tectate 3 μm thick, slightly columellate; sexine finely verrucate, verrucae flat, inconspicuous; sulcus wide, as long as grain; amb elliptic with acute ends; grains suboblate, frequently appearing as folded, 68 × 110 μm. Tree.

Syagrus sp. Mart. (form 1) Plate 59:408

Grains monad, heteropolar, bilaterally symmetric; monosulcate; exine tectate 1 μm thick; sexine scabrate, scabrae *ca.* 1 μm height; sulcus narrow, as long as grain, displaying thickness borders resembling a margo condition; amb elliptic with acute ends; grains oblate, 27 × 50 μm. Tree.

Syagrus sp. Mart. (form 2) Plate 59:409

Grains monad, heteropolar, bilaterally symmetric; monosulcate; exine tectate <1 μm thick; sexine psilate to slightly scabrate; sulcus straight, narrow, ⅔ length of grain; amb elliptic; grains suboblate, 29 to 30 μm × 42 to 44 μm. Tree.

Arecaceae 1. Plate 59:410

Grains monad, heteropolar, bilaterally symmetric; monoporate; exine tectate 1 μm thick; sexine echinate-scabrate; spine conical, sharp, short, thin *ca.* 2 μm length × 1 μm wide, encrusted in a small tectate depression; scabrae <1 μm, densely scattered on the surface; pore circular 3 μm diameter; amb circular; grains spheroidal irregular, grains frequently appearing as folded, 30 to 32 μm (including ornamentation).

Bromeliaceae

Bromeliaceae 1. Plate 59:411

Grains monad, heteropolar, bilaterally symmetric; monosulcate; exine tectate 3 µm thick; sexine reticulate, homobrochate, brochi 1 µm wide, muri simplibaculate, bacula conspicuous 2 µm height × 1 µm wide; sulcus inconspicuous, as long as grain; amb elliptic; grains oblate, 38 × 58 µm. Epiphytic herb.

Cyperaceae

Cyperus sp. L. (form 1) Plate 60:413

Grains monad, heteropolar, bilaterally symmetric; periporate (5-aperturate); exine semitectate 1.5 to 2. µm thick; sexine granulate to baculate, bacula conspicuous *ca.* 1 µm wide; pseudopores appearing as elongated ruptures of sexine, inconspicuous, irregular, *ca.* 1 µm wide × 8 µm length; 4 pores on grain sides and the fifth at basal apex; grains irregular, amb trapezoid; grains 40 × 45 µm. Herb.

Cyperus sp. L. (form 2) Plate 60:414

Grains monad, heteropolar, bilaterally symmetric; periporate (5-aperturate); exine semitectate 1 µm thick; sexine scabrate; pseudopores appearing as elongated ruptures of sexine, inconspicuous, *ca.* 2 µm wide × 15 µm length; 4 pores on grain sides and the fifth at basal apex; grains irregular, amb trapezoid; grains 40 × 42 µm. Herb.

Fimbrystylis sp. Vahl Plate 59:412

Grains monad, apolar, asymmetric; inaperturate; exine semitectate 1 µm thick; finely reticulate, homobrochate, brochi <1 µm wide, muri simplibaculate, bacula thin, short; apertures if present, inconspicuous; amb irregularly rounded; grains 34 × 40 µm. Herb.

Rhynchospora sp. Vahl Plate 60:415

Grains monad, apolar, asymmetric; apparently triporate (it could have more than 3 apertures); exine semitectate 1 µm thick; sexine reticulate, homobrochate, brochi 1 µm wide, muri simplibaculate; pores appearing as pseudoapertures, inconspicuous; grains irregular, amb circular irregular; grains 26 × 29 µm. Herb.

Scleria sp. P. Bergius Plate 60:416

Grains monad, heteropolar, bilaterally symmetric; periporate (5-aperturate); exine semitectate 1 µm thick; sexine scabrate; pseudopores appearing as elongated ruptures of sexine, frequently covered by thin sexinic membrane, irregular, *ca.* 1 µm wide × 3 to 4 µm length; 4 pores on grain sides and the fifth at basal apex; grains irregular, amb trapezoid; grains 26 × 30 µm. Herb.

Poaceae (Gramineae)

Poaceae 1. Plate 60:417

Grains monad, heteropolar, bilaterally symmetric; monoporate; exine tectate *ca.* 1.5 µm thick; sexine psilate, columellate; pore circular, drop type, 1.5 µm diameter, annulate, annulus 1 to 1.5 µm thick; amb circular; grains spheroidal irregular, 28 to 30 µm.

Poaceae 2. Plate 60:418

Grains monad, heteropolar, bilaterally symmetric; monoporate; exine tectate 1 µm thick, columellate; sexine densely scabrate; pore circular, 2 to 2.5 µm diameter, annulate, annulus thin, inconspicuous; amb circular; grains spheroidal irregular, 38 µm.

Poaceae 3. Plate 60:419

Grains monad, heteropolar, bilaterally symmetric; monoporate; exine tectate 2 μm thick; sexine psilate to slightly scabrate, columellate; pore circular, protruding, 2.5 μm diameter, annulate, annuli 4 μm thick; amb circular; grains spheroidal irregular, 35 to 37 μm.

Poaceae 4. Plate 60:420

Grains monad, heteropolar, bilaterally symmetric; monoporate; exine tectate 1 to 1.5 μm thick; sexine scabrate, densely columellate; pore circular, 5 to 6 μm diameter, annulate, annuli 4 μm wide; amb circular, frequently appearing folded; grains spheroidal irregular, 70 μm.

Poaceae 5. Plate 60:421

Grains monad, heteropolar, bilaterally symmetric; monoporate; exine tectate 1 μm thick; sexine psilate; pore circular, 4 μm diameter, annulate, annulus 3 μm wide; amb circular; grains spheroidal irregular, 47 to 50 μm.

Poaceae 6. Plate 60:422

Grains monad, heteropolar, bilaterally symmetric; monoporate; exine tectate 3 μm thick; sexine scabrate becoming to be granulate, densely columellate; pore circular, 6 to 7 μm diameter, annulate, annuli 5 μm wide; amb circular, frequently appearing folded; grains spheroidal irregular, 80 μm.

PLATES
PRANCHAS

All pollen and spores are shown on a scale of 1 millimeter = 1 micron, unless marked with an asterisk.

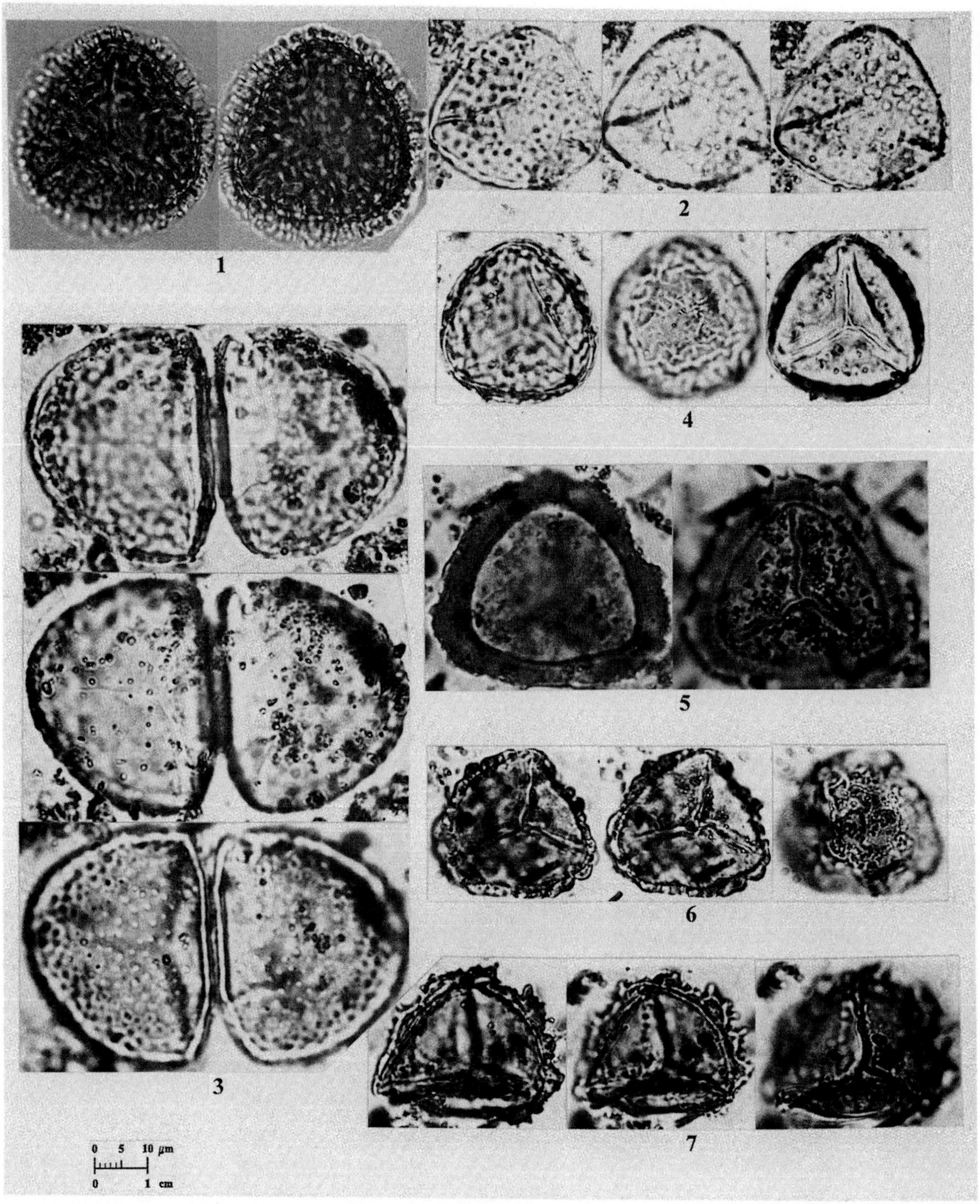

Plate/Prancha 1

PTERIDOPHYTA – LYCOPODIOPSIDA. Lycopodiaceae: *Lycopodium clavatum* (exotic marker) (1); *Lycopodium aff. linifolium* (2); *L. aff. subulatum* (3); *L. aff. trianae* (4). **Selaginellaceae:** *Selaginella aff. asperula* form 1 (5); *S. aff. asperula* form 2 (6); *S. aff. asperula* form 3 (7).

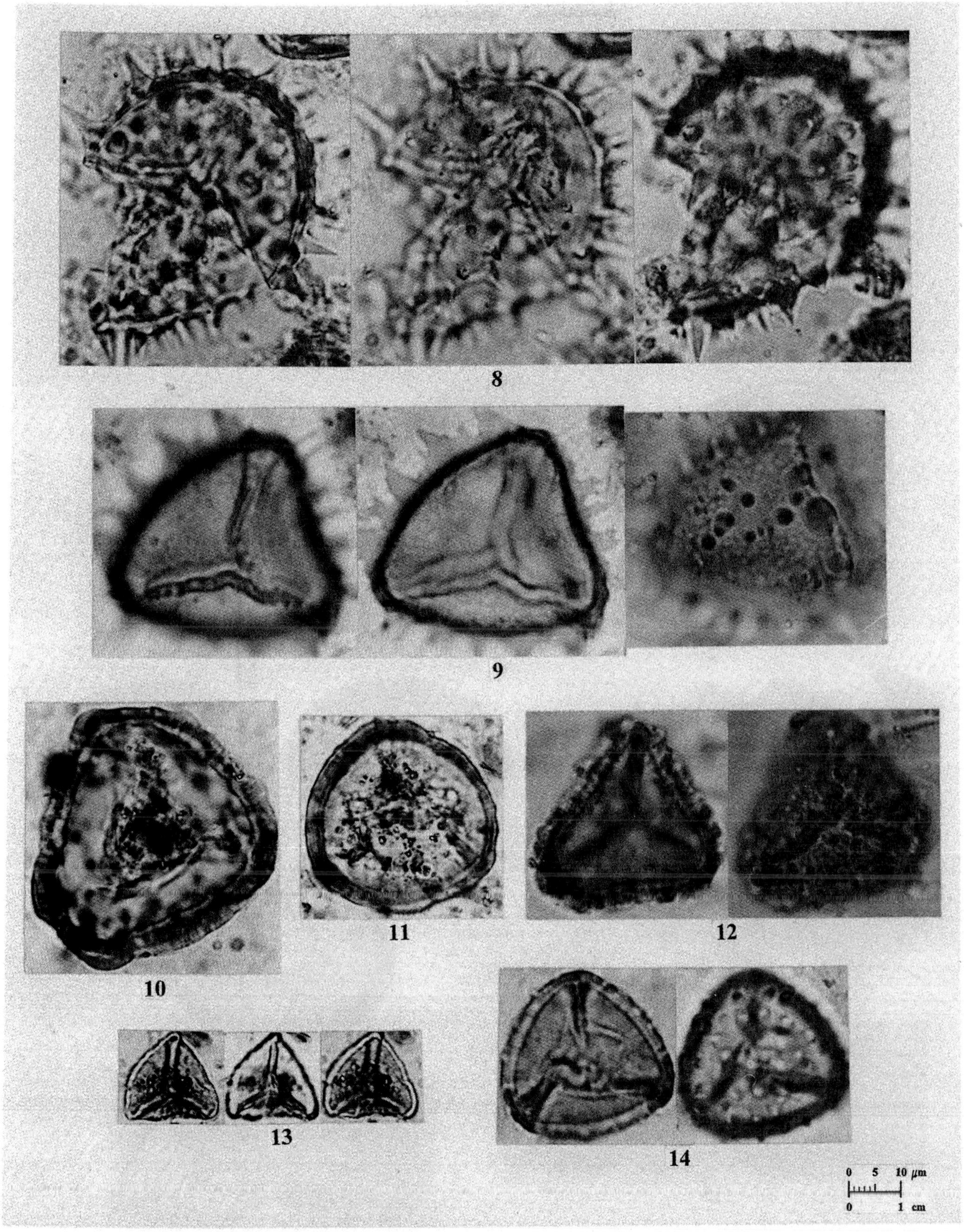

Plate/Prancha 2

Selaginellaceae: *Selaginella aff. arthritica* form 1 (8); *S. aff. arthritica* form 2 (9); *Selaginella aff. asperula* form 4 (10); *S. aff. asperula* form 5 (11); *Selaginella aff. exaltata* (12); *Selaginella* sp. form 1 (13); *Selaginella* sp. form 2 (14).

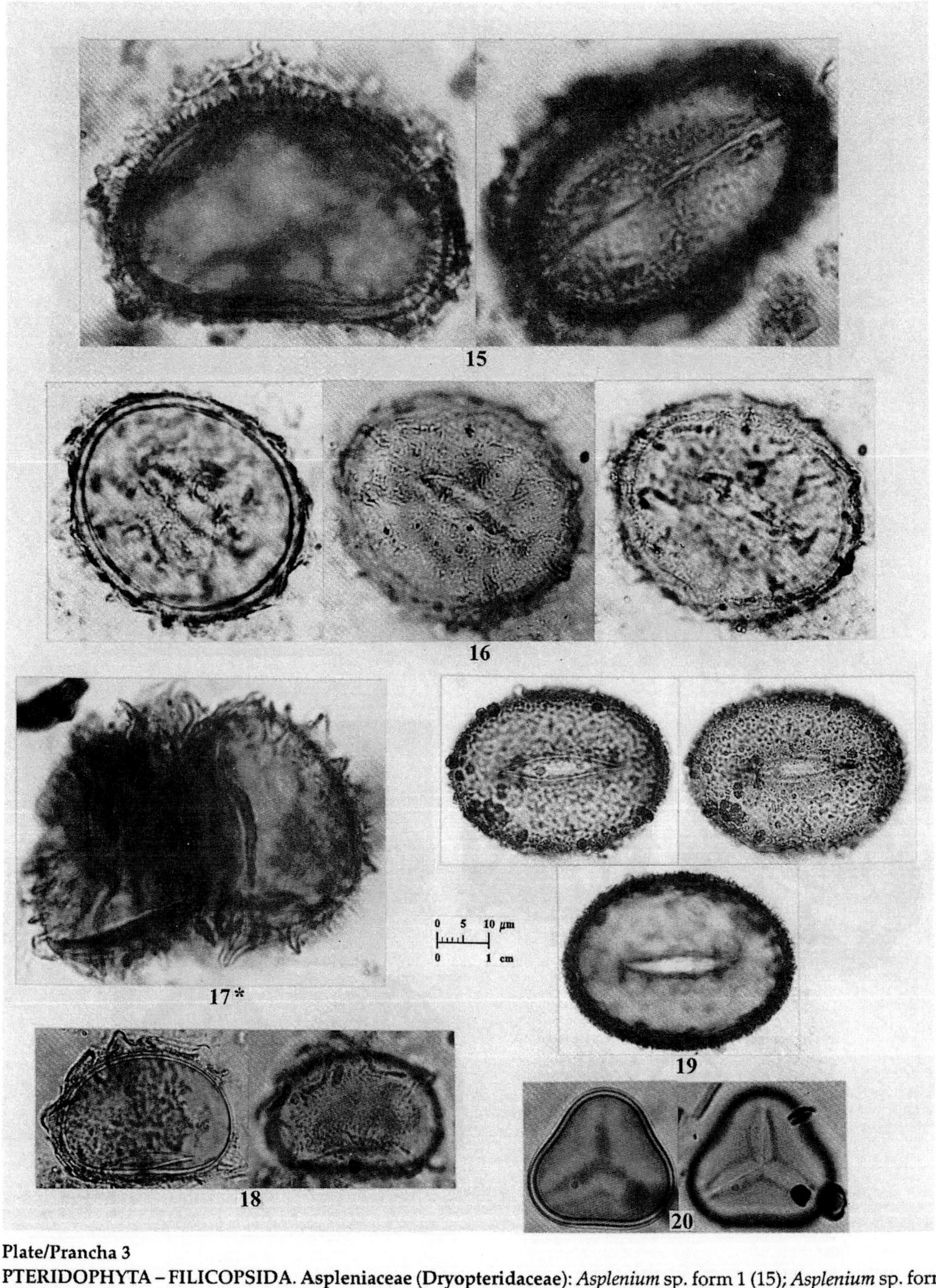

Plate/Prancha 3

PTERIDOPHYTA – FILICOPSIDA. Aspleniaceae (Dryopteridaceae): *Asplenium* sp. form 1 (15); *Asplenium* sp. form 2 (16); *Asplenium* sp. form 3 (17); *Asplenium* sp. form 4 (18). Blechnaceae: *Blechnum* sp. (19). **Cyatheaceae:** *Alsophila* sp. (20) [* #17 is reduced to ⅓ scale (1 mm = 3 µm)].

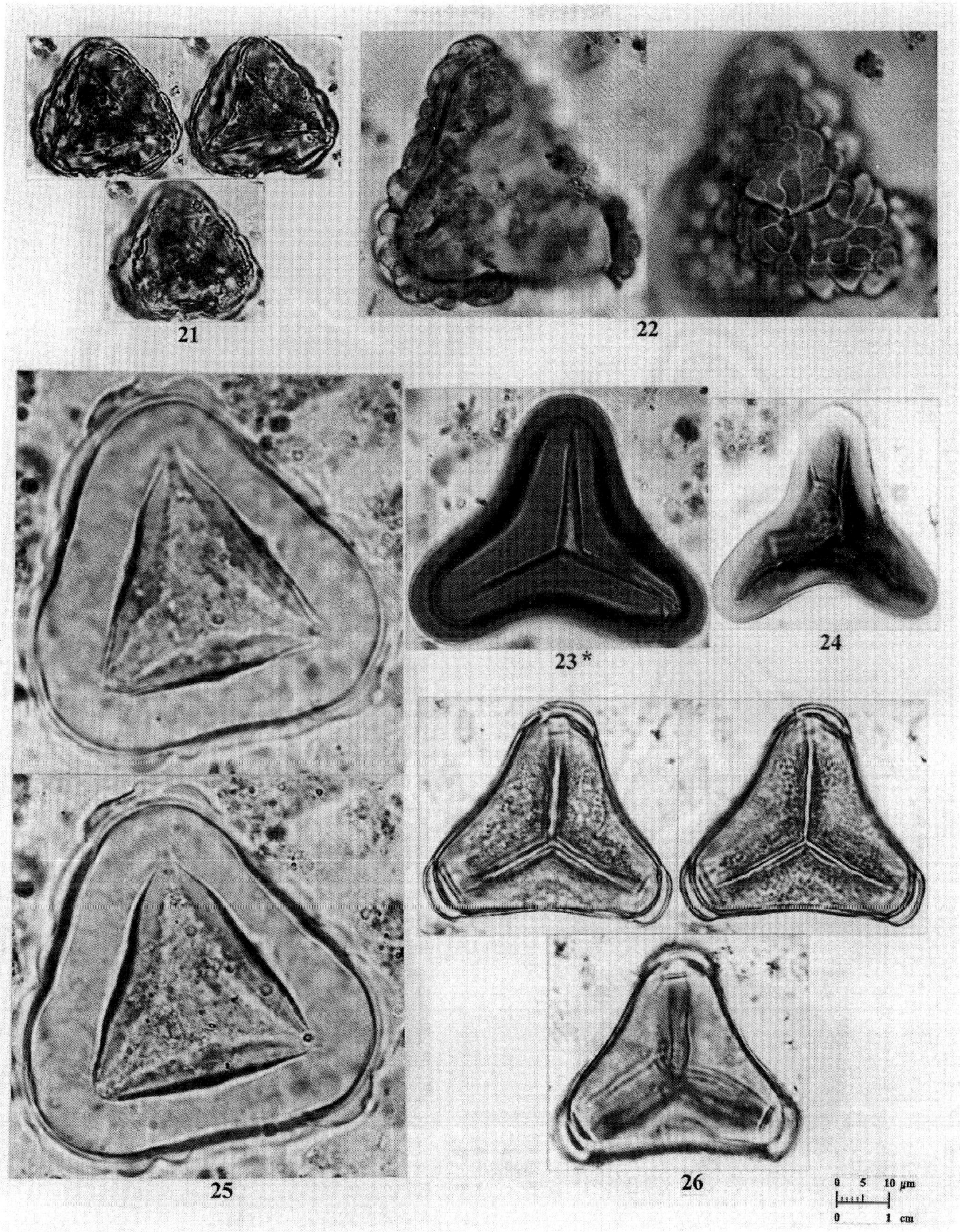

Plate/Prancha 4

Cyatheaceae: *Cyathea* sp. form 1 (21); *Cyathea* sp. form 2 (22); *Cyathea* sp. form 3 (23); *Cyathea* sp. form 4 (24); *Cyathea* sp. form 5 (25). **Dicksoniaceae**: *Dicksonia* sp. form 1 (26). [*#23 is reduced to ½ scale (1 mm = 2 μm)].

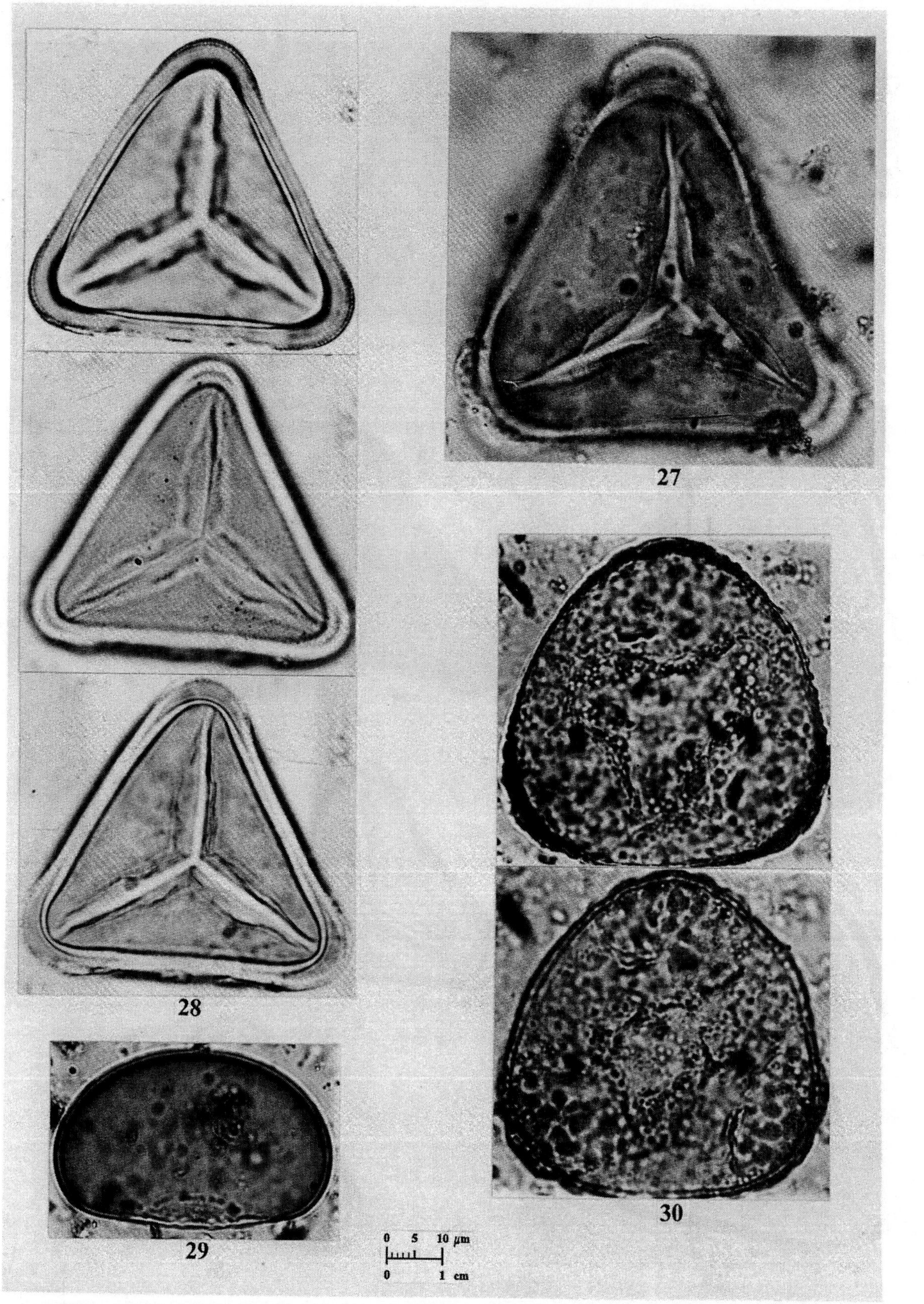

Plate/Prancha 5

Dicksoniaceae: *Dicksonia* sp. form 2 (27); *Dicksonia* sp. form 3 (28). **Gleicheniaceae**: *Dicranopteris* sp. (29). **Grammitidaceae**: *Grammitis* sp. (30).

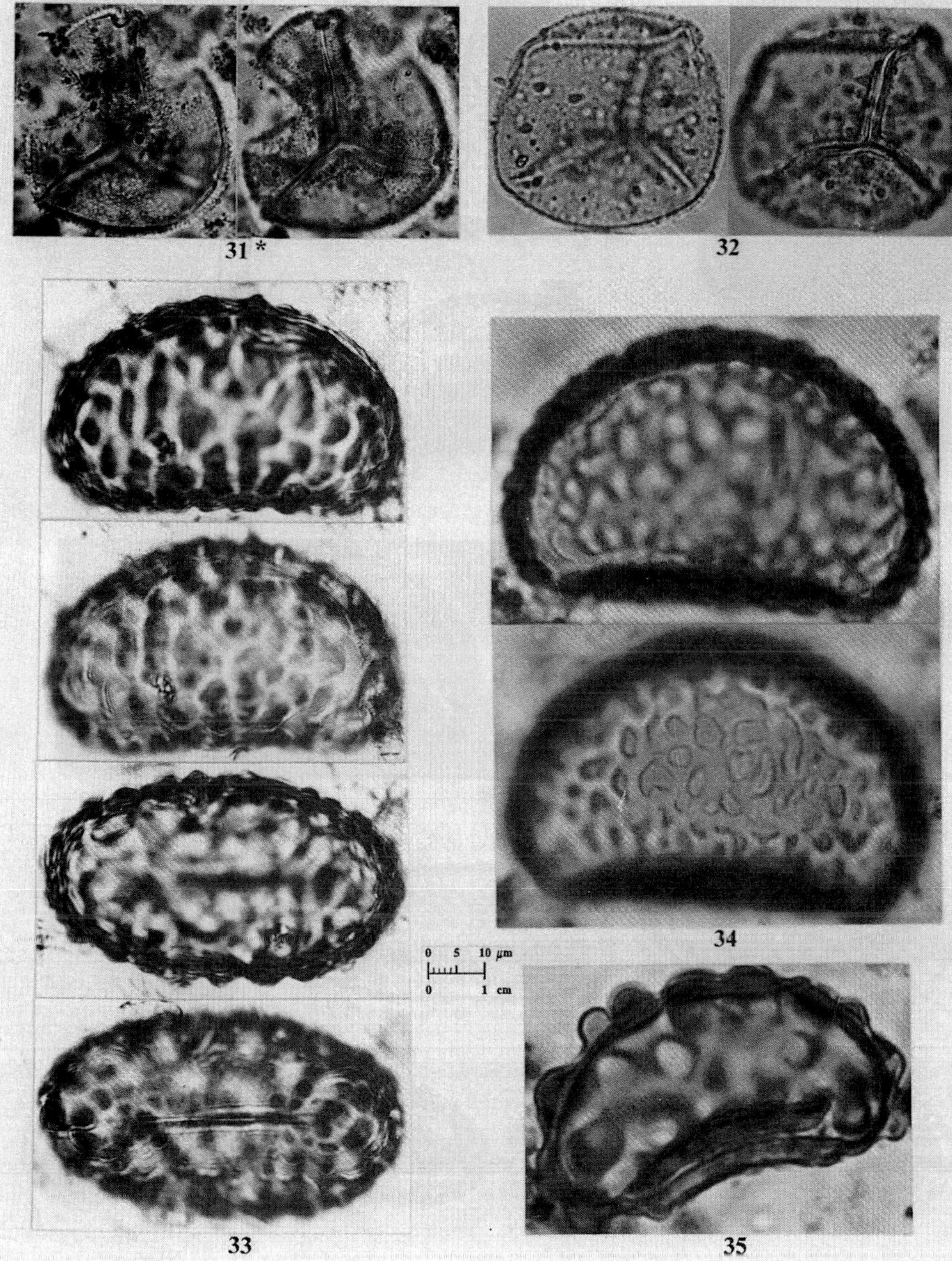

Plate/Prancha 6
Hymenophyllaceae: *Hymenophyllum aff. ciliatum* (31); *Trichomanes aff. elegans* (32). **Polypodiaceae**: *Polypodium* sp. form 1 (33); *Polypodium* sp. form 2 (34); *Polypodium* sp. form 3 (35) [*#31 is reduced to ½ scale (1 mm = 2 μm)].

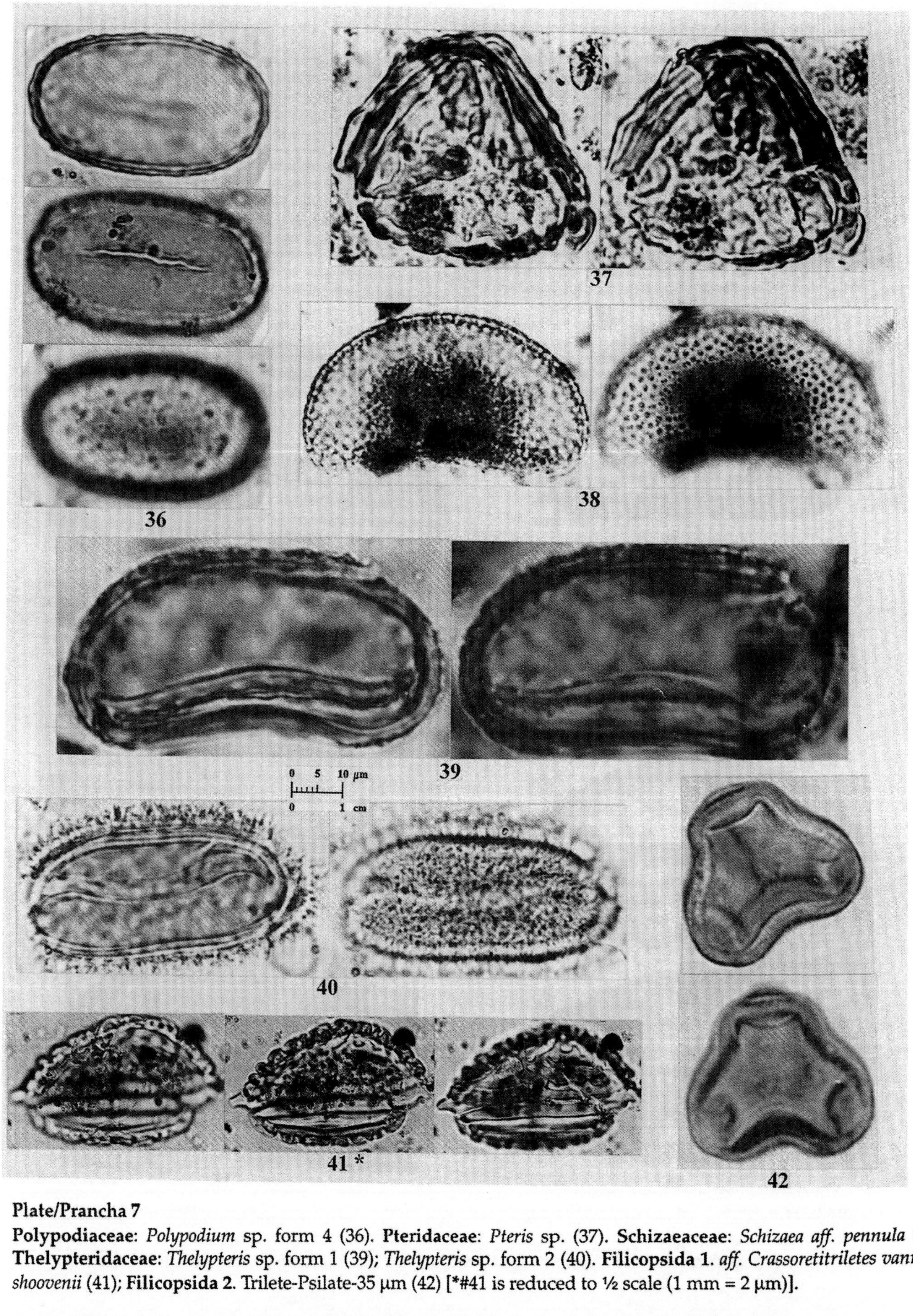

Plate/Prancha 7

Polypodiaceae: *Polypodium* sp. form 4 (36). **Pteridaceae**: *Pteris* sp. (37). **Schizaeaceae**: *Schizaea aff. pennula* (38). **Thelypteridaceae**: *Thelypteris* sp. form 1 (39); *Thelypteris* sp. form 2 (40). **Filicopsida 1.** *aff. Crassoretitriletes vanraadshoovenii* (41); **Filicopsida 2.** Trilete-Psilate-35 μm (42) [*#41 is reduced to ½ scale (1 mm = 2 μm)].

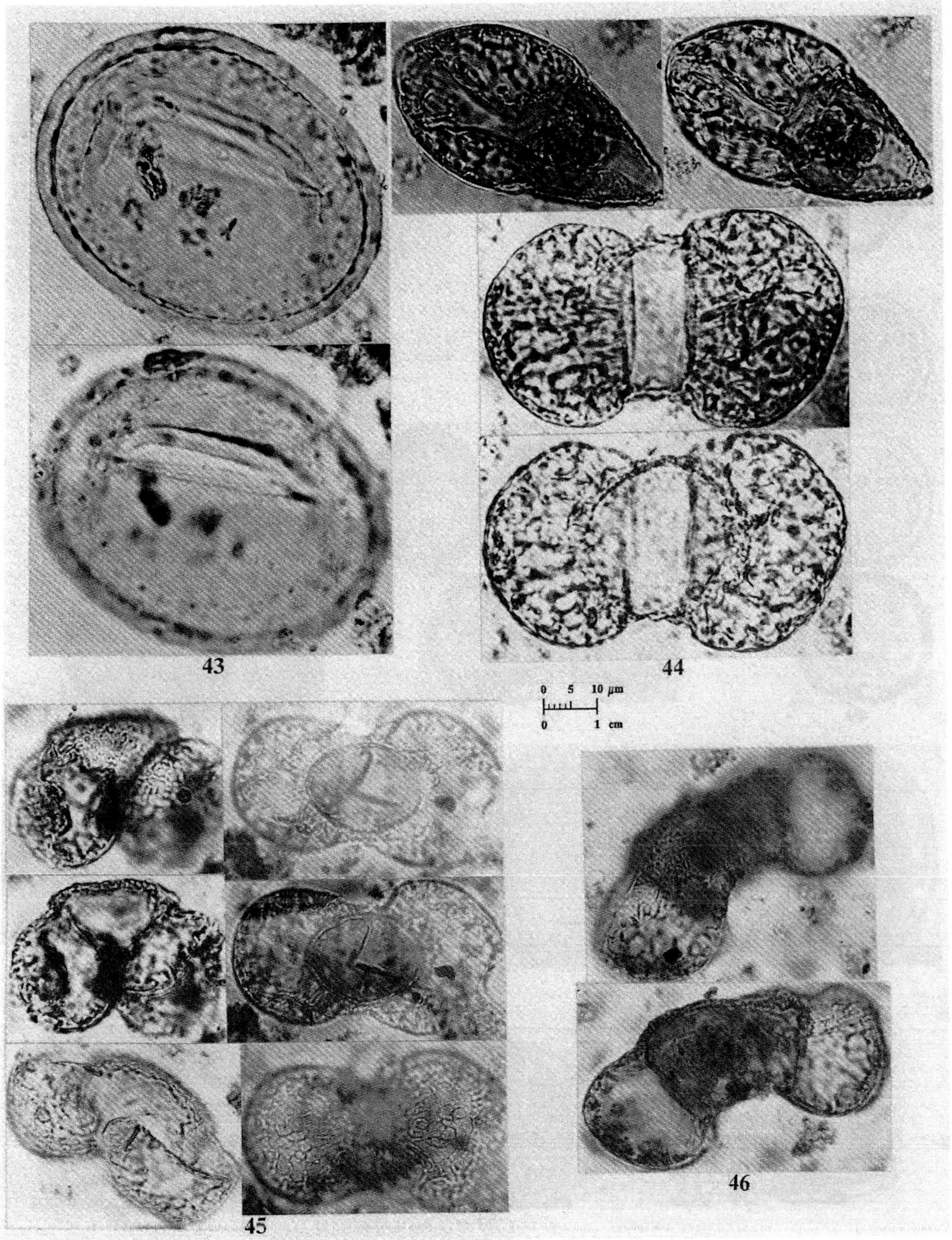

Plate/Prancha 8

Filicopsida 3. Monolete-Psilate-68 µm (43). **GYMNOSPERMAE. Podocarpaceae:** *Podocarpus* sp. form 1 (44); *Podocarpus* sp. form 2 (45); *Podocarpus* sp. form 3 (46).

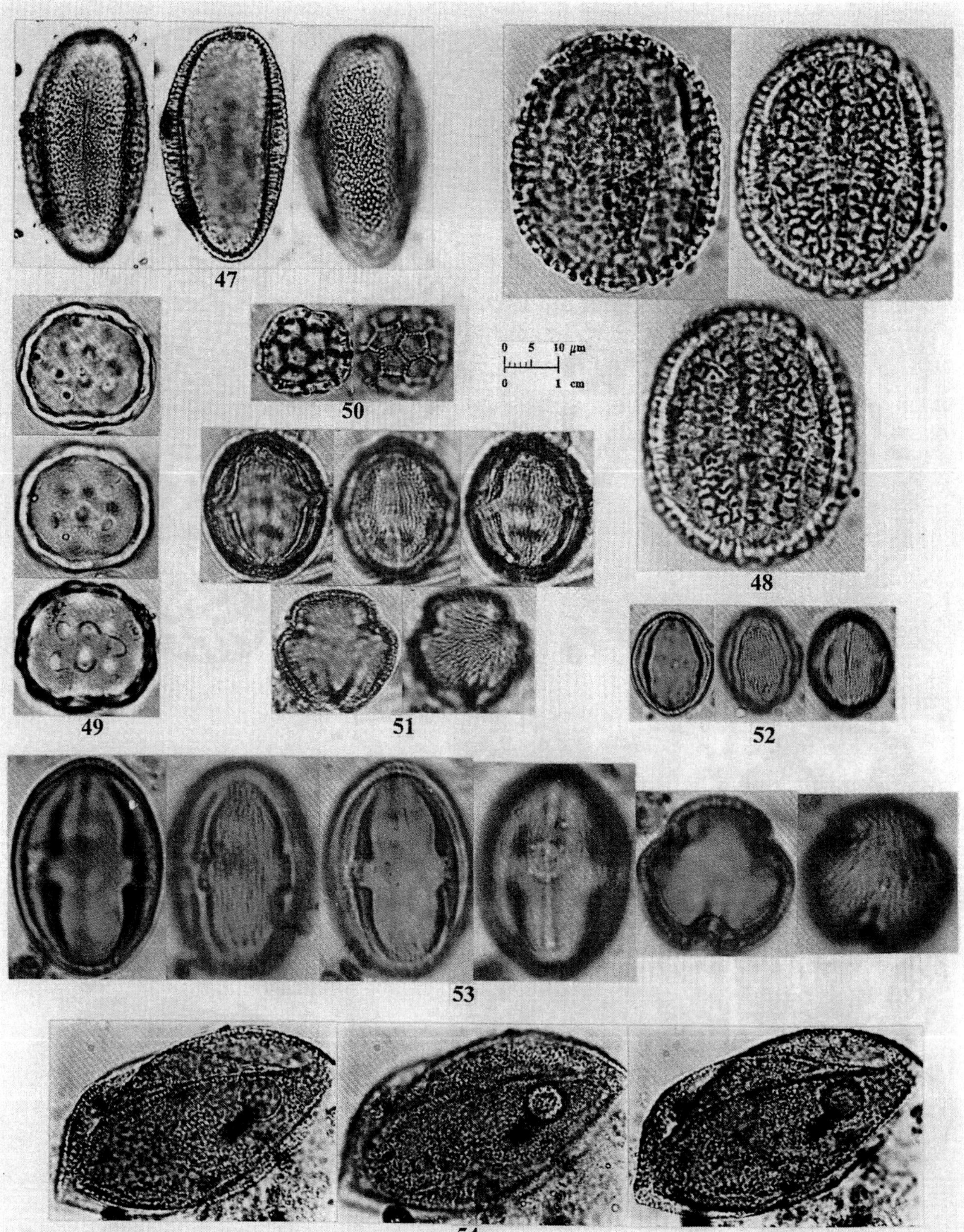

Plate/Prancha 9

ANGIOSPERMAE – MAGNOLIOPSIDAE (Dicotyledoneae). Acanthaceae: *Aphelandra* sp. (47); *Pachystachys aff. lutea* (48). **Amaranthaceae:** *Amaranthus* sp. (49); *Gomphrena* sp. (50). **Anacardiaceae:** *Astronium aff. graveolens* (51); *Campnosperma* sp. (52); *Tapirira aff. guianensis* (53). **Annonaceae:** *Annona* sp. form 1 (54).

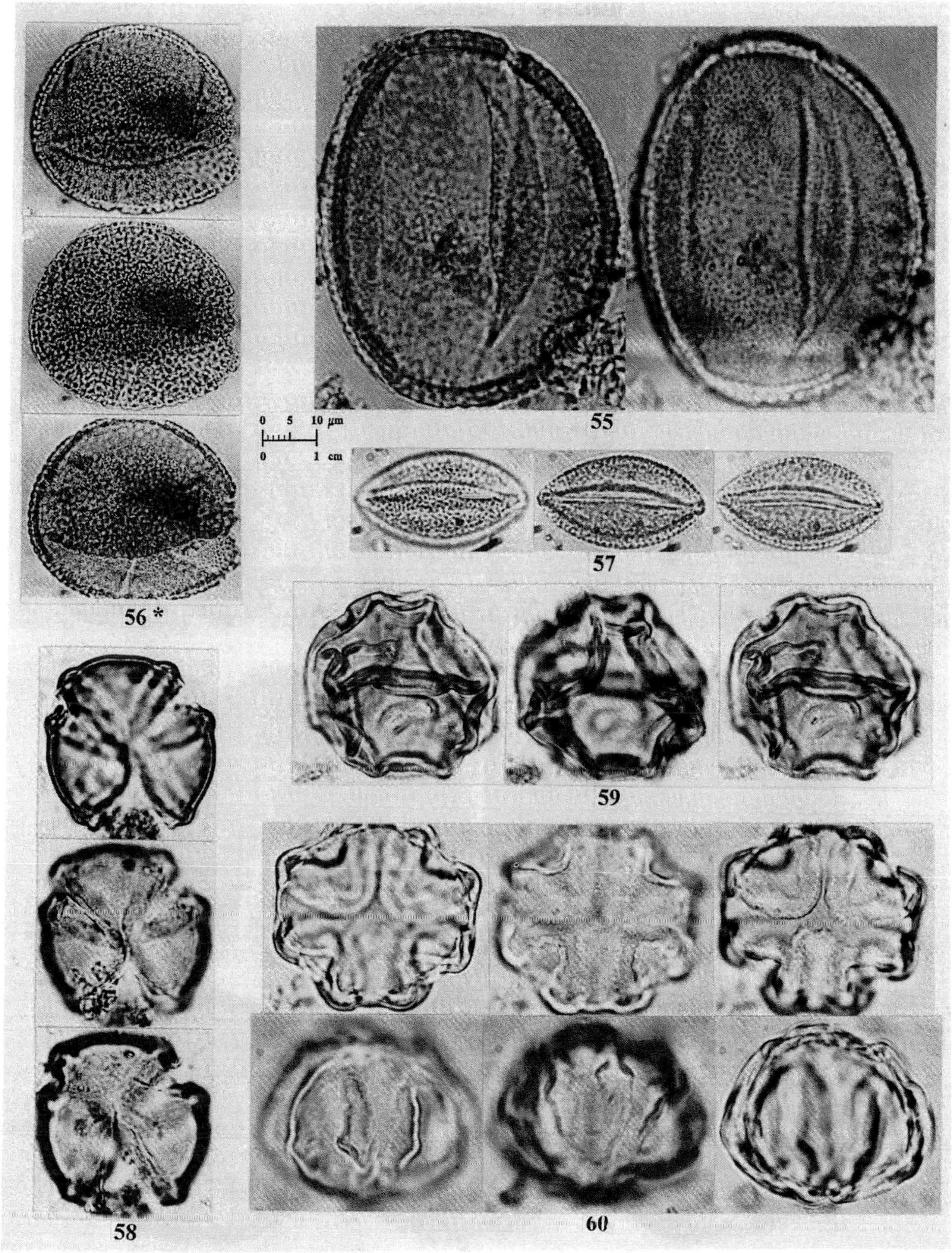

Plate/Prancha 10

Annonaceae: *Annona* sp. form 2 (55); *Diclinanona aff. tessmanii* (56); *Unonopsis aff. pittieri* (57). **Apocynaceae**: *Aspidosperma aff. album* (58); *A. aff. excelsum* (59); *A. aff. megalocarpum* (60) [*#56 is reduced to ½ scale (1 mm = 2µm)].

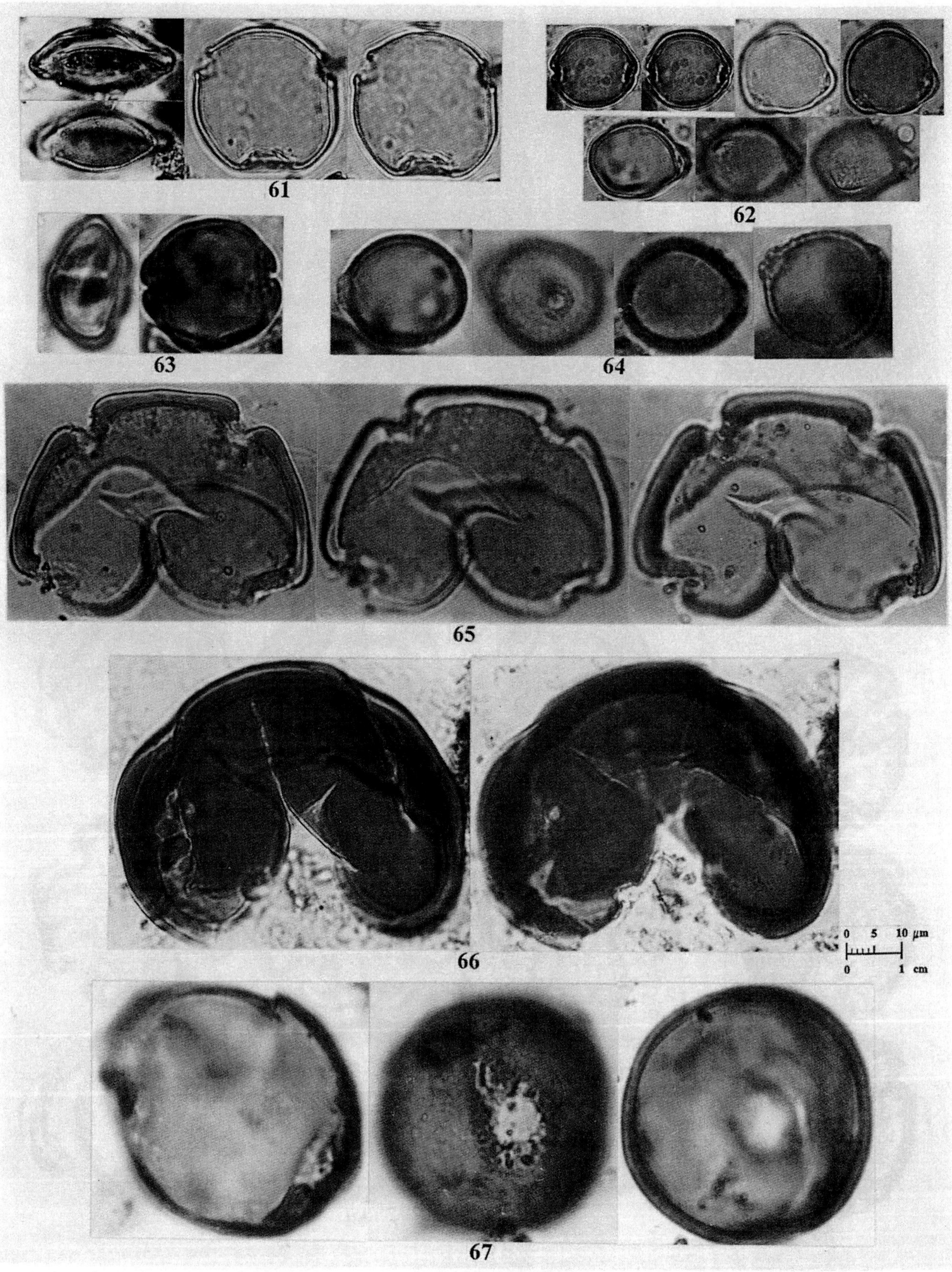

Plate/Prancha 11

Apocynaceae: *Lacmellea* sp. (61); *Macoubea aff. guianensis* (62); *M. aff. sprucei* (63); *Malouetia aff. guatemalensis* (64); *Molongum* sp. form 1 (65); *Molongum* sp. form 2 (66); *Odontadenia* sp. (67).

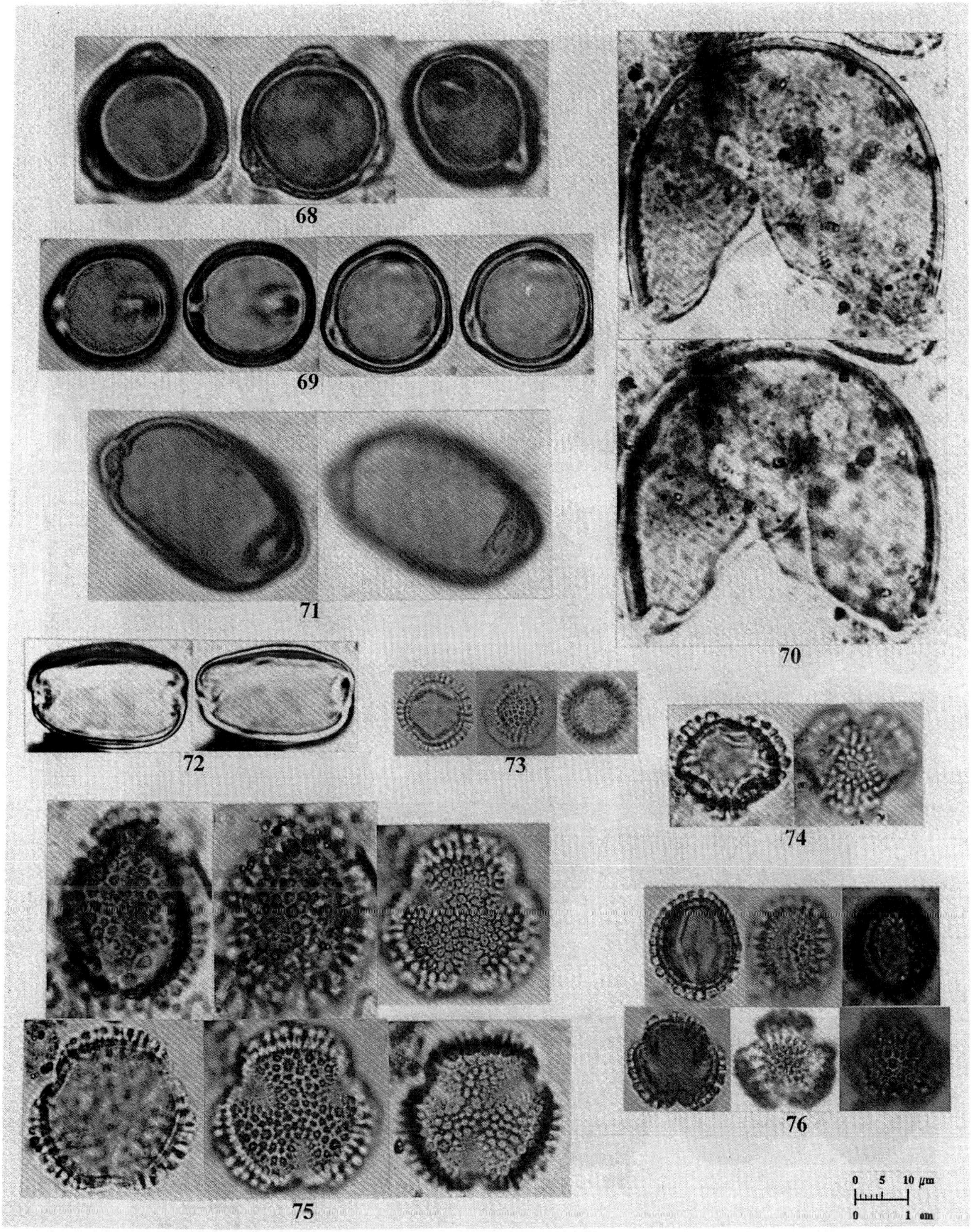

Plate/Prancha 12

Apocynaceae: *Parahancornia* sp. form 1 (68); *Parahancornia* sp. form 2 (69); *Prestonia* sp. (70); *Secondatia* sp. form 1 (71); *Secondatia* sp. form 2 (72). **Aquifoliaceae**: *Ilex* sp. (73); *I. aff. breviceps* (74); *I. aff. divaricata* (75); *I. aff. dumosa* (76).

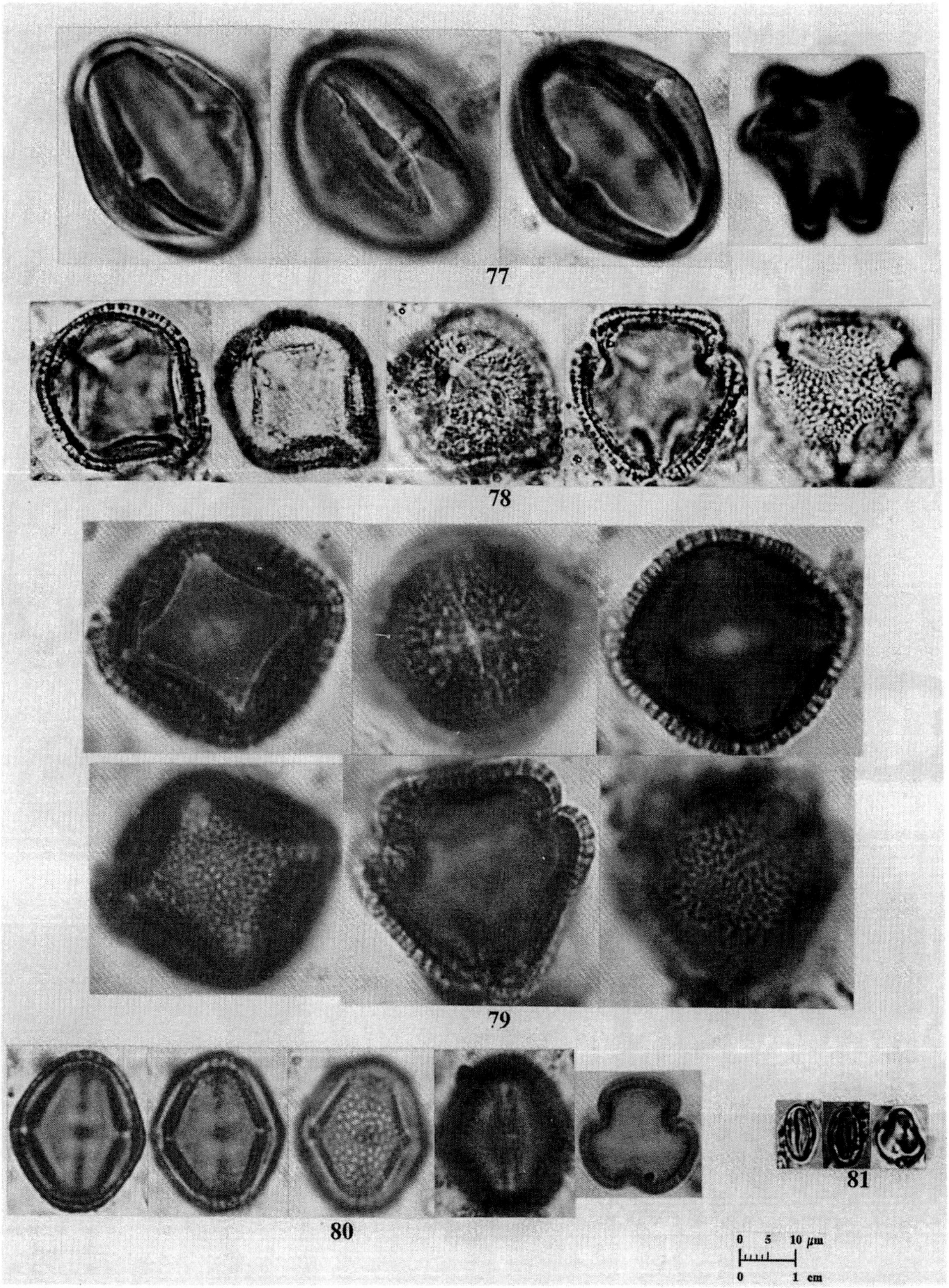

Plate/Prancha 13

Araliaceae: *Dendropanax* sp. (77); *Didymopanax aff. acuminata* (78); *D. aff. morototoni* (79); *Schefflera aff. bangui* (80). **Begoniaceae**: *Begonia* sp. (81).

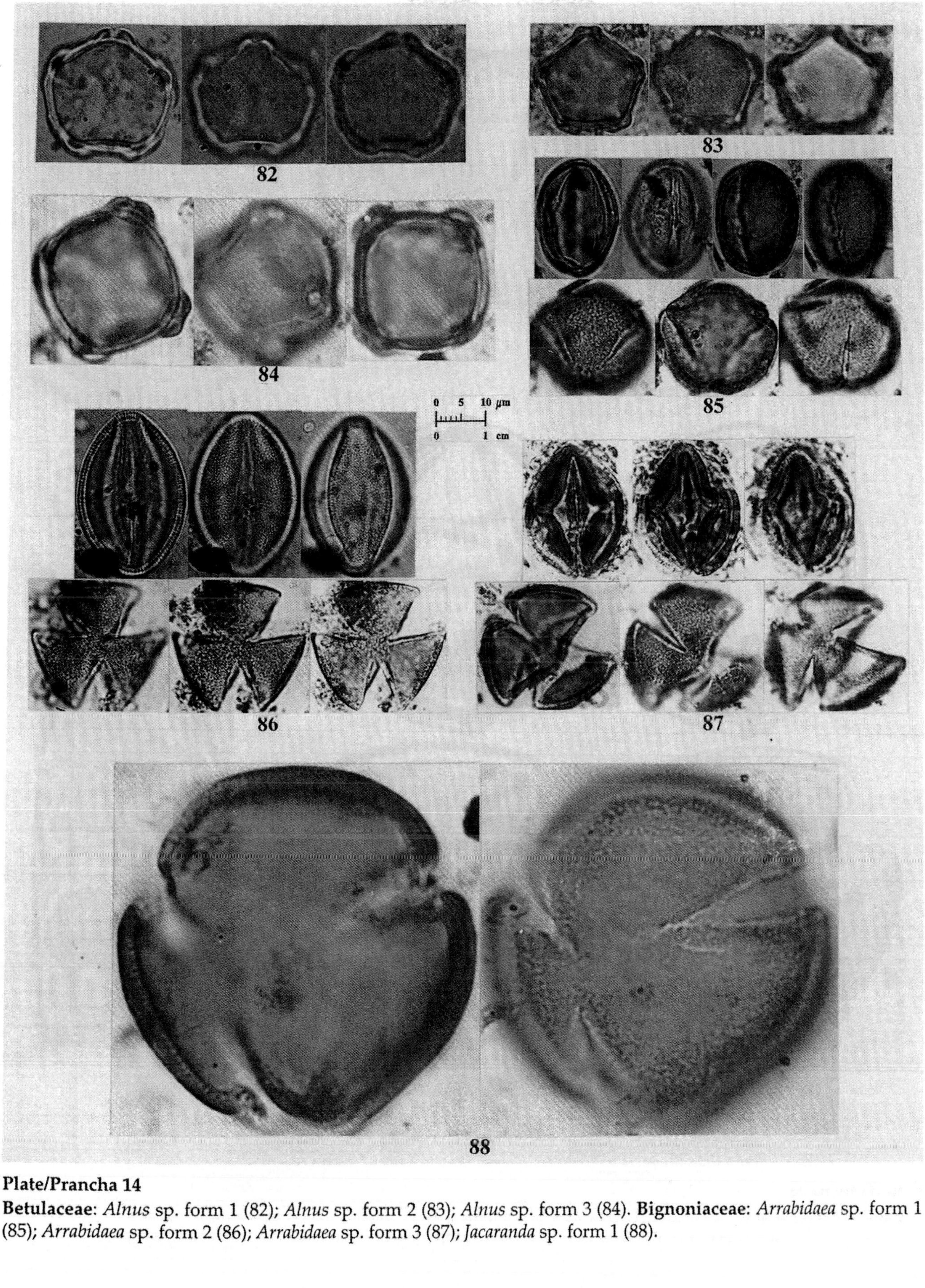

Plate/Prancha 14

Betulaceae: *Alnus* sp. form 1 (82); *Alnus* sp. form 2 (83); *Alnus* sp. form 3 (84). **Bignoniaceae**: *Arrabidaea* sp. form 1 (85); *Arrabidaea* sp. form 2 (86); *Arrabidaea* sp. form 3 (87); *Jacaranda* sp. form 1 (88).

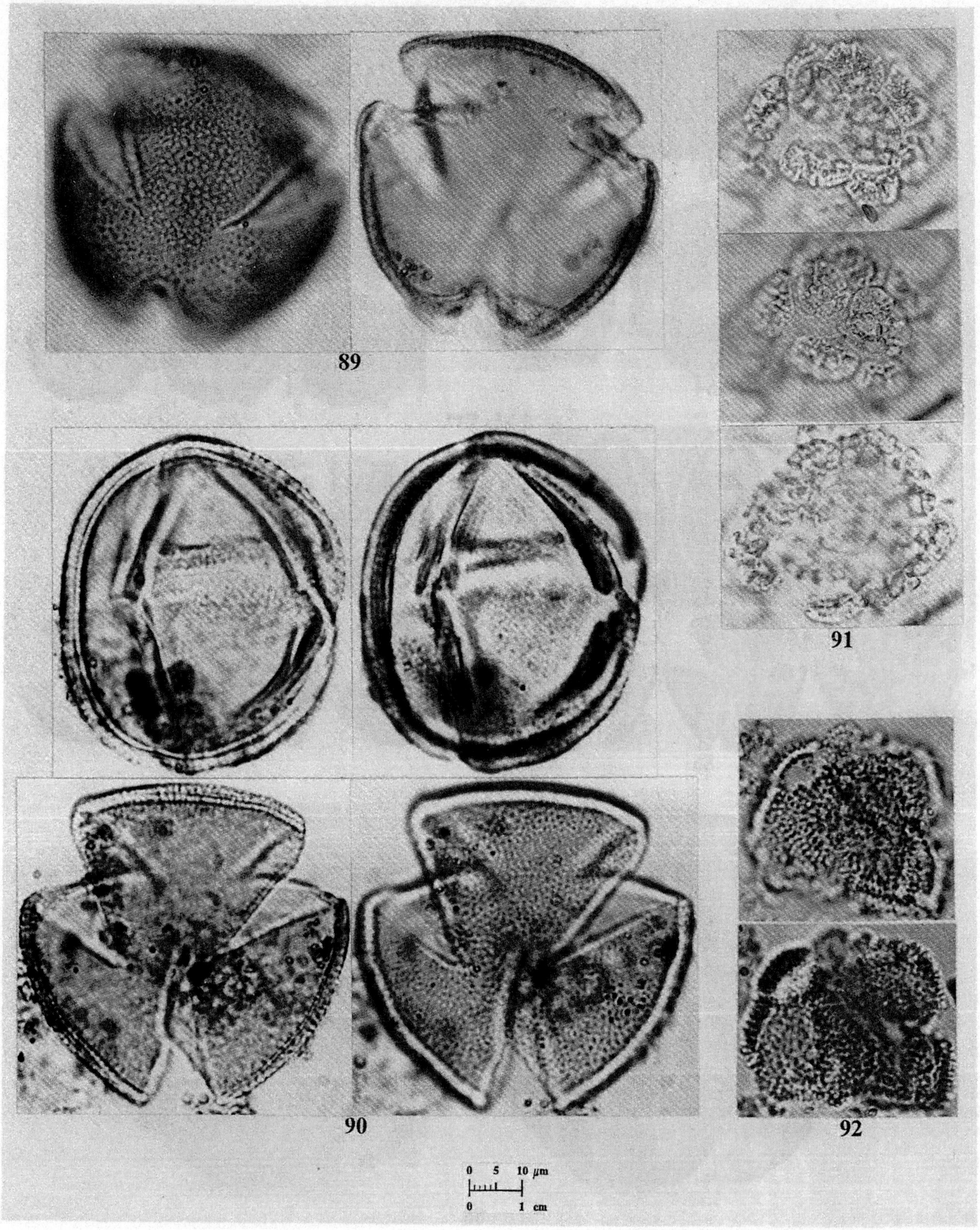

Plate/Prancha 15
Bignoniaceae: *Jacaranda* sp. form 2 (89); *Lundia* sp. (90); *Memora aff. pedunculata* (91); *Memora* sp. (92).

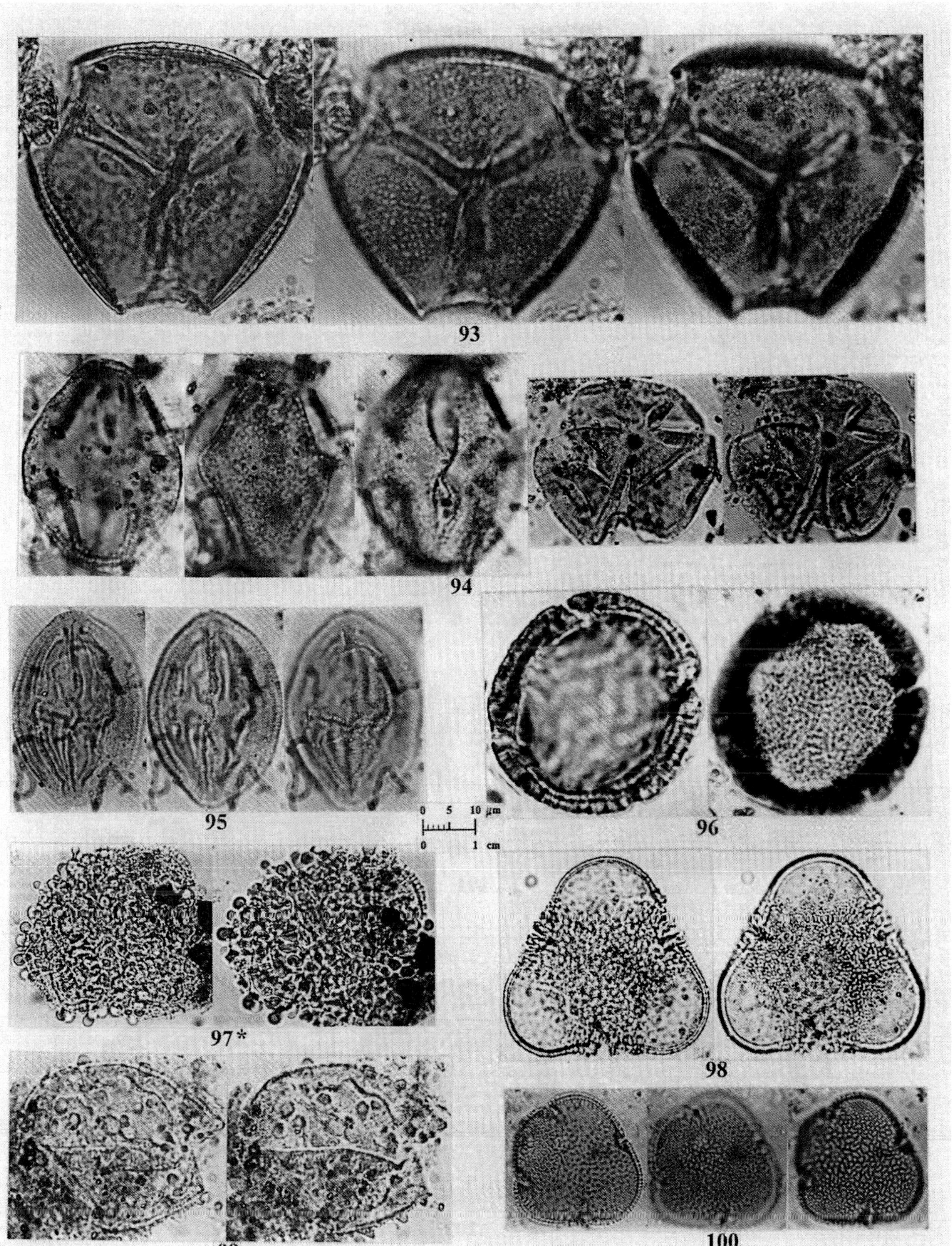

Plate/Prancha 16

Bignoniaceae: *Mansoa* sp. (93); *Tabebuia* sp. form 1 (94); *Tabebuia* sp. form 2 (95). **Bombacaceae**: *Chorisia* sp. / *Ceiba* sp. form 1 (96); *Matisia aff. bracteosa* (97); *Pachira aff. aquatica* (98); *Phragmotheca aff. mammosa* (99); *Pseudobombax* sp. (100) [*#97 is reduced to ½ scale (1 mm = 2 µm)].

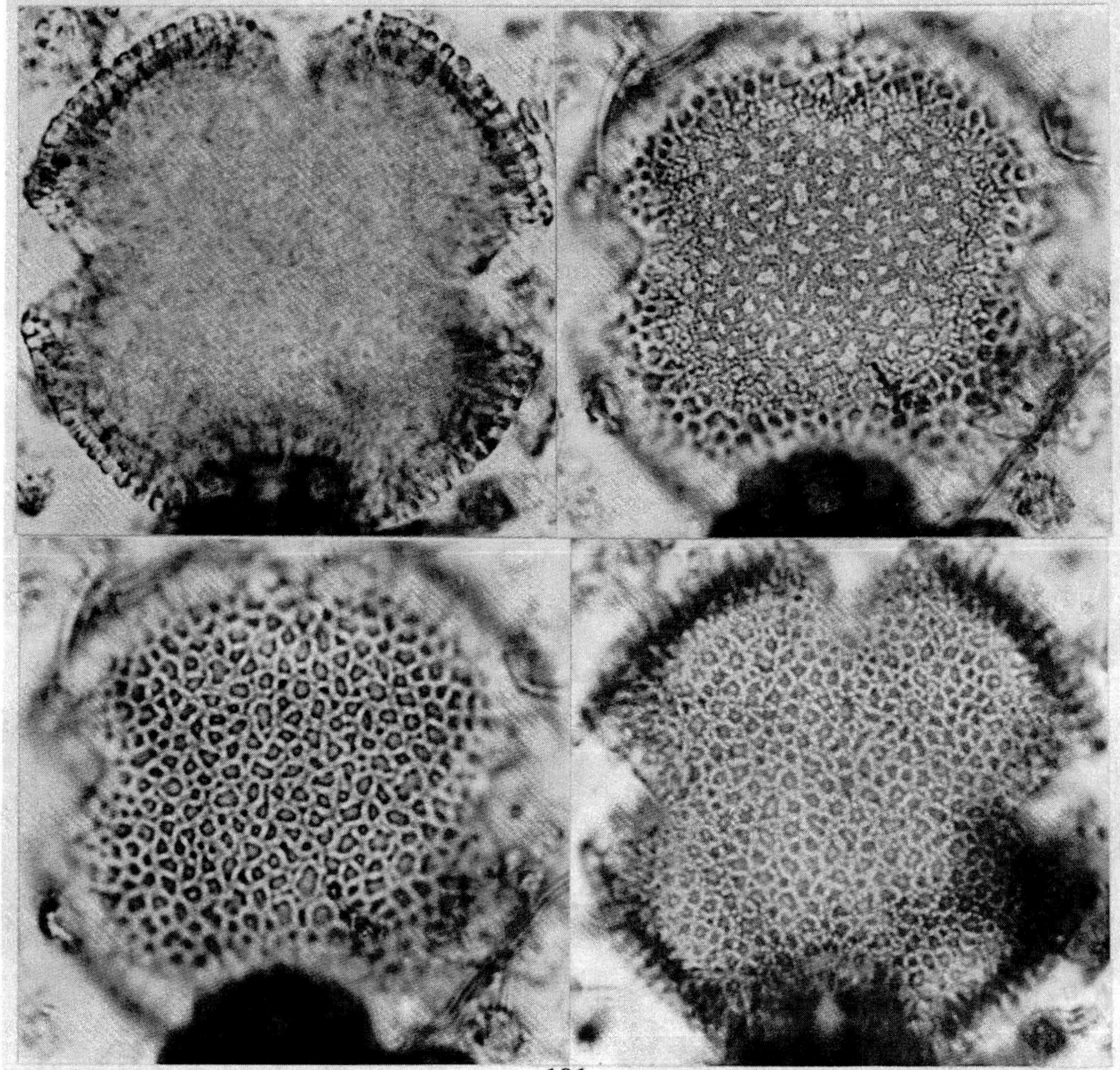

101

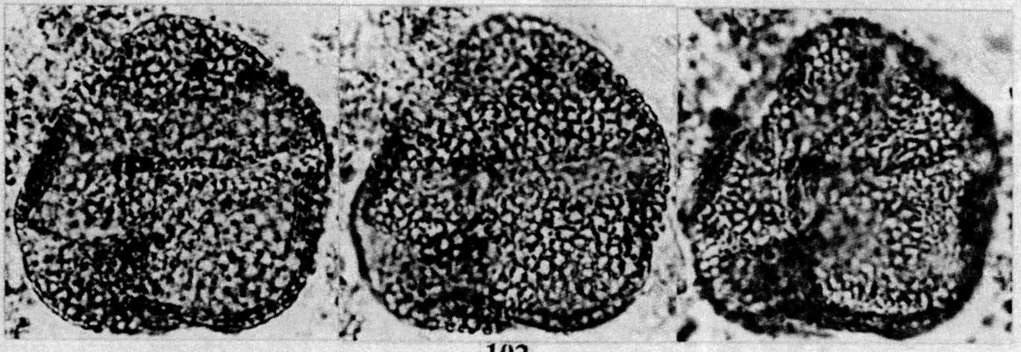

102

Plate/Prancha 17
Bombacaceae: *Chorisia* sp. / *Ceiba* sp. form 2 (101); *Quararibea* sp. (102).

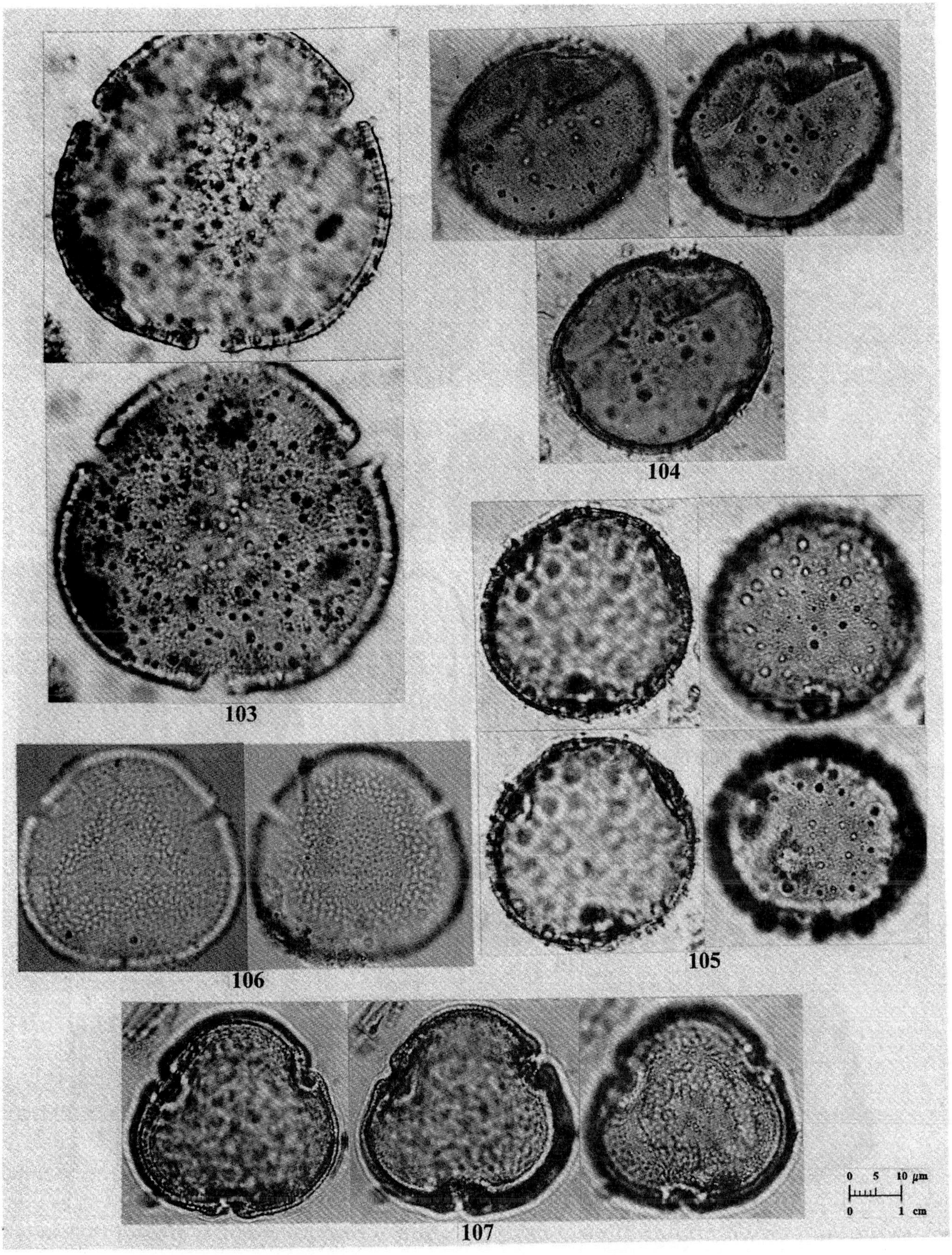

Plate/Prancha 18

Bombacaceae: *Rhodognaphalopsis aff. duckei* (103); *Scleronema aff. micranthum* (104); *Scleronema* sp. (105); *Spirotheca.* sp. (106); Bombacaceae 1 (107).

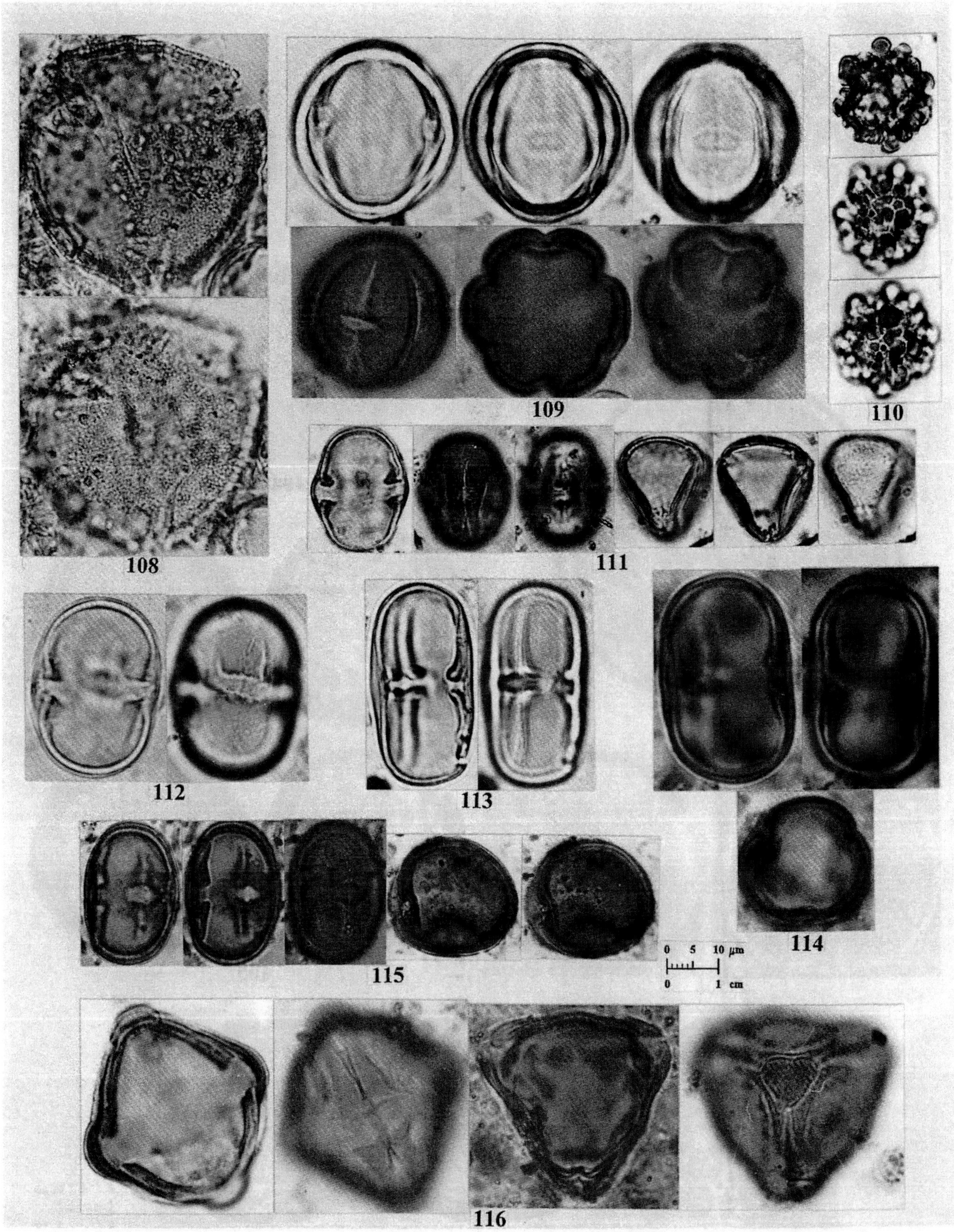

Plate/Prancha 19

Boraginaceae: *Cordia aff. nodosa* (108); *Heliotropium* sp. (109); *Tournefortia* sp. (110). **Burseraceae**: *Crepidospermum* sp. (111); *Dacryodes aff. chimatensis* (112); *Protium* sp. form 1 (113); *Protium* sp. form 2 (114); *Trattinickia* sp. (115). **Caryocaraceae**: *Caryocar* sp. form 1 (116).

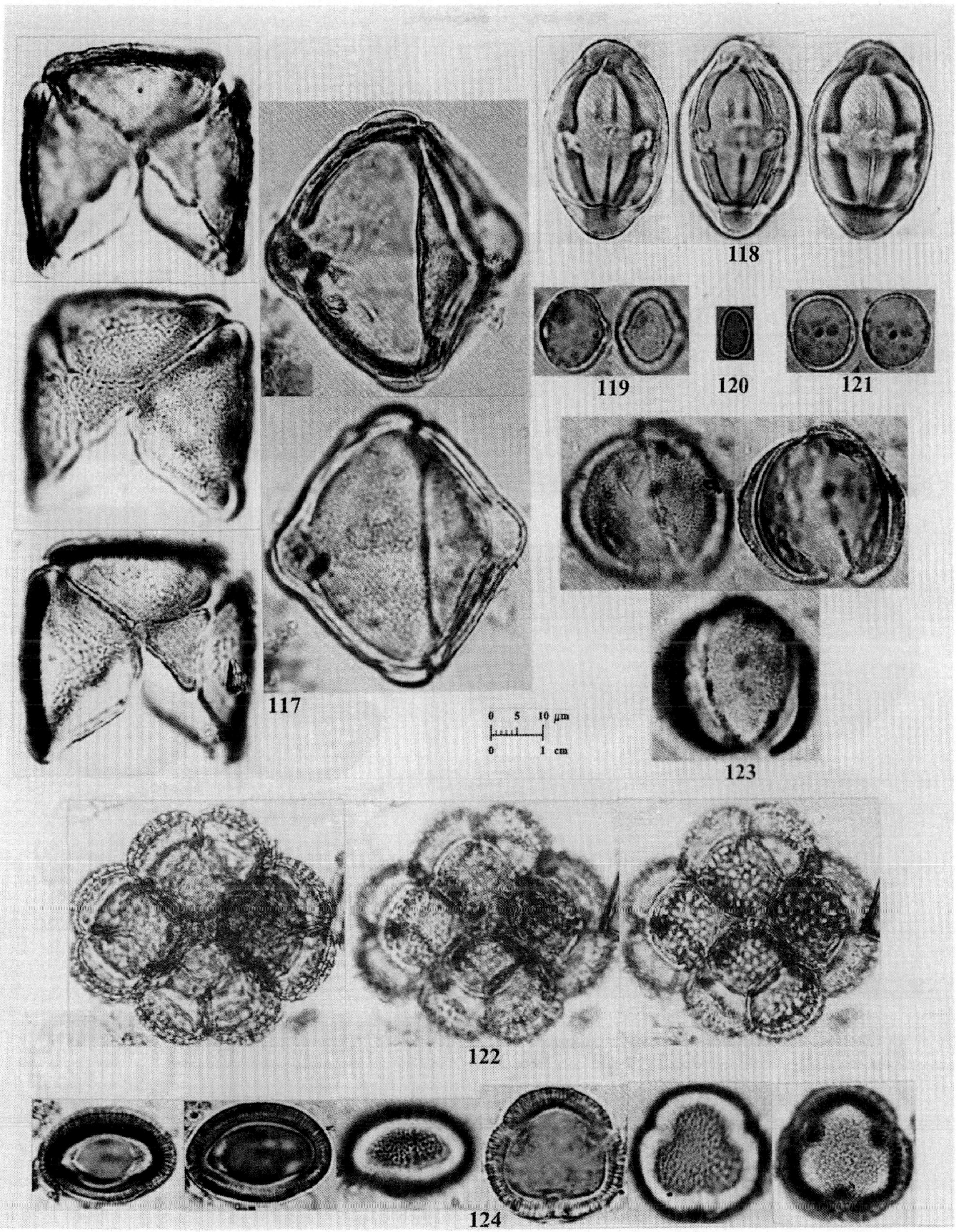

Plate/Prancha 20

Caryocaraceae: *Caryocar* sp. form 2 (117); *Caryocar* sp. form 3 (118). **Cecropiaceae**: *Cecropia* sp. form 1 (119); *Cecropia* sp. form 2 (120); *Cecropia* sp. form 3 (121). **Celastraceae (Hippocrateaceae)**: *Hippocratea aff. volubilis* (122); *Peritassa aff. laevigata* (123); *Plenckia* sp. (124).

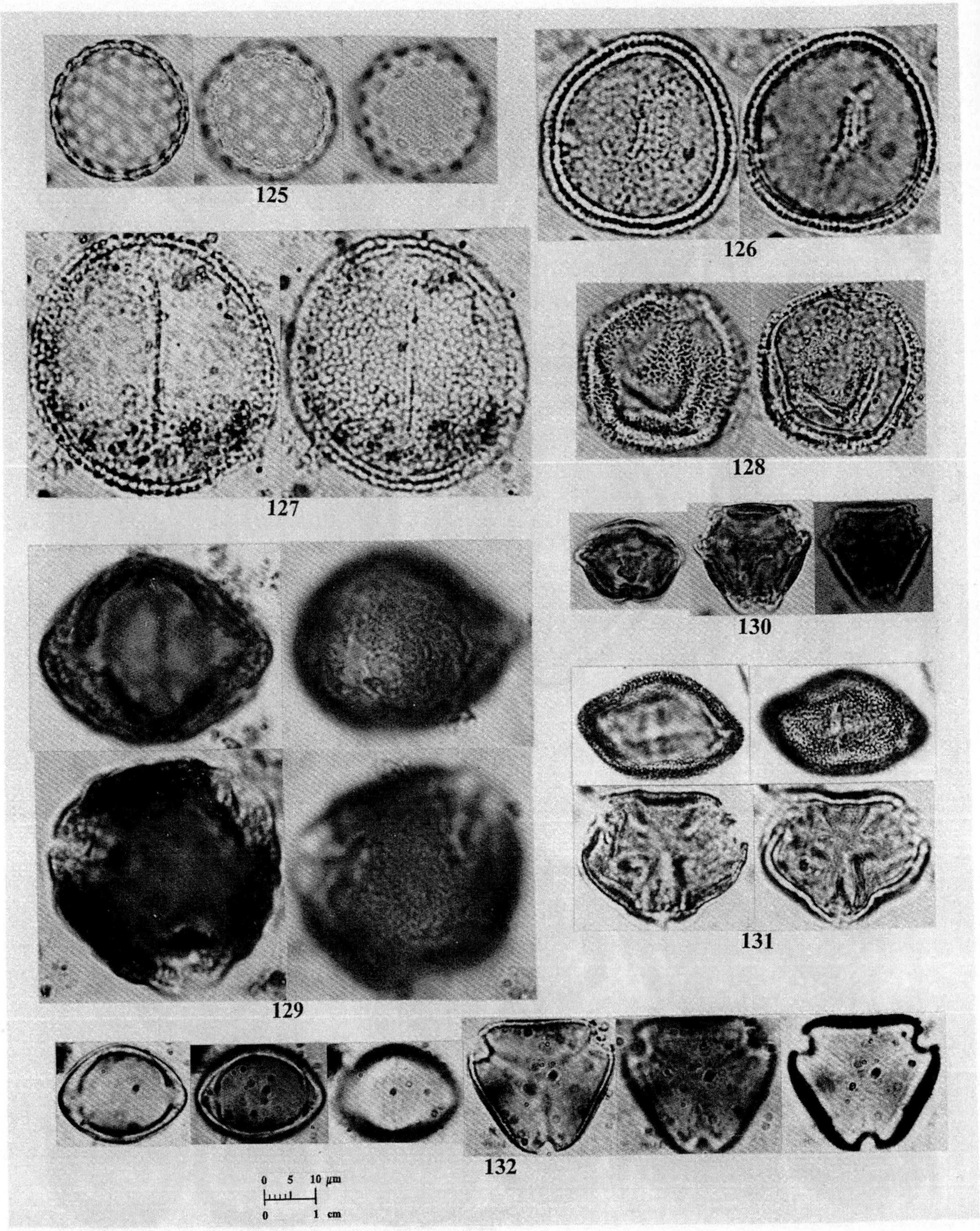

Plate/Prancha 21

Chenopodiaceae: *Chenopodium* sp. (125). **Chloranthaceae**: *Hedyosmum* sp. form 1 (126); *Hedyosmum* sp. form 2 (127); *Hedyosmum* sp. form 3 (128). **Chrysobalanaceae**: *Chrysobalanus aff. icaco* (129); *Couepia* sp. (130); *Hirtella aff. racemosa* (131); *Licania* sp. (132).

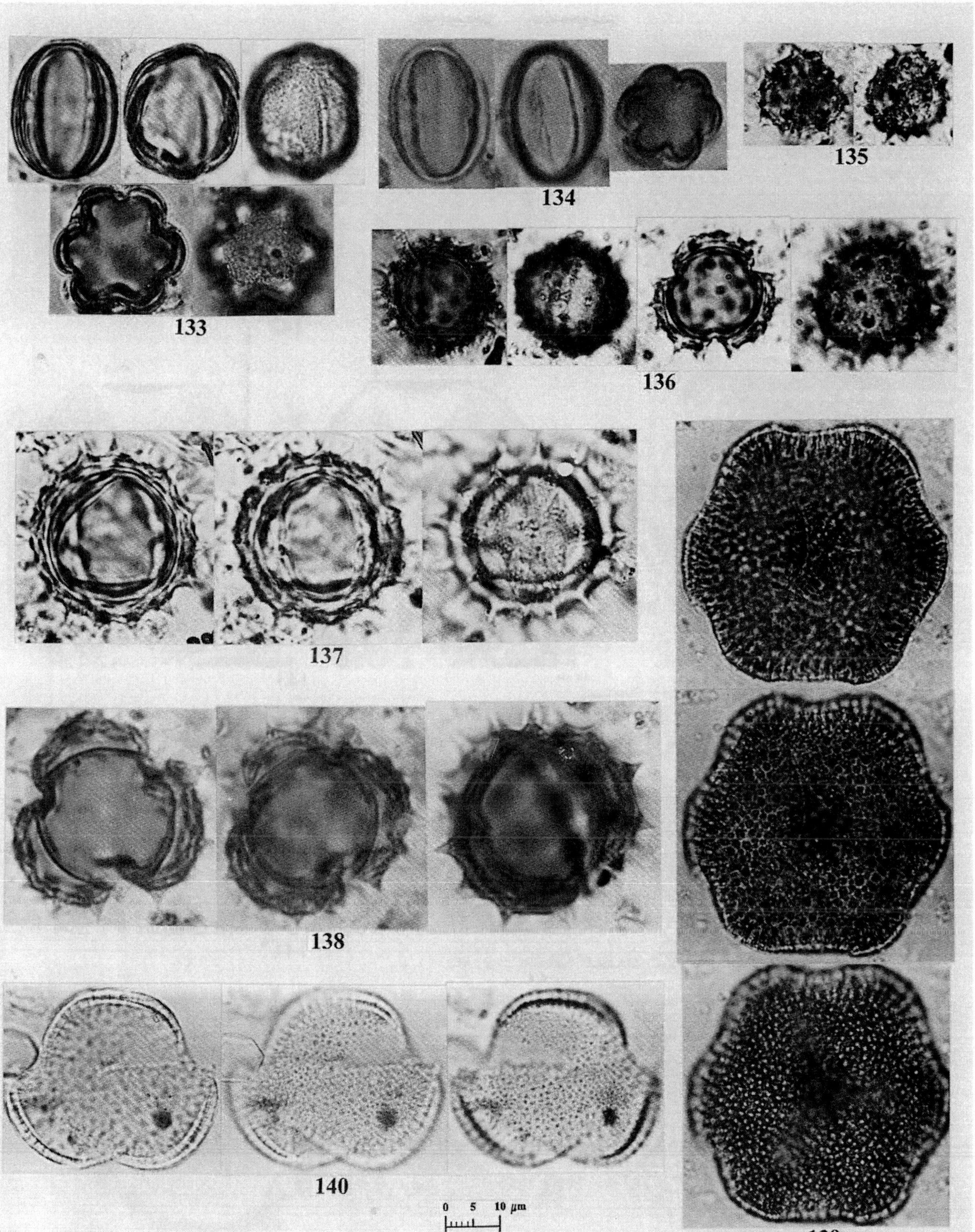

Plate/Prancha 22

Combretaceae: *Combretum* sp. (133); *Terminalia* sp. (134). **Compositae (Asteraceae)**: *Baccharis* group (135); *Eupatorium* group (136); *Senecio* group form 1 (137); *Senecio* group form 2 (138). **Convolvulaceae**: *Aniseia* sp. (139); *Dicranostyles aff. scandens* (140).

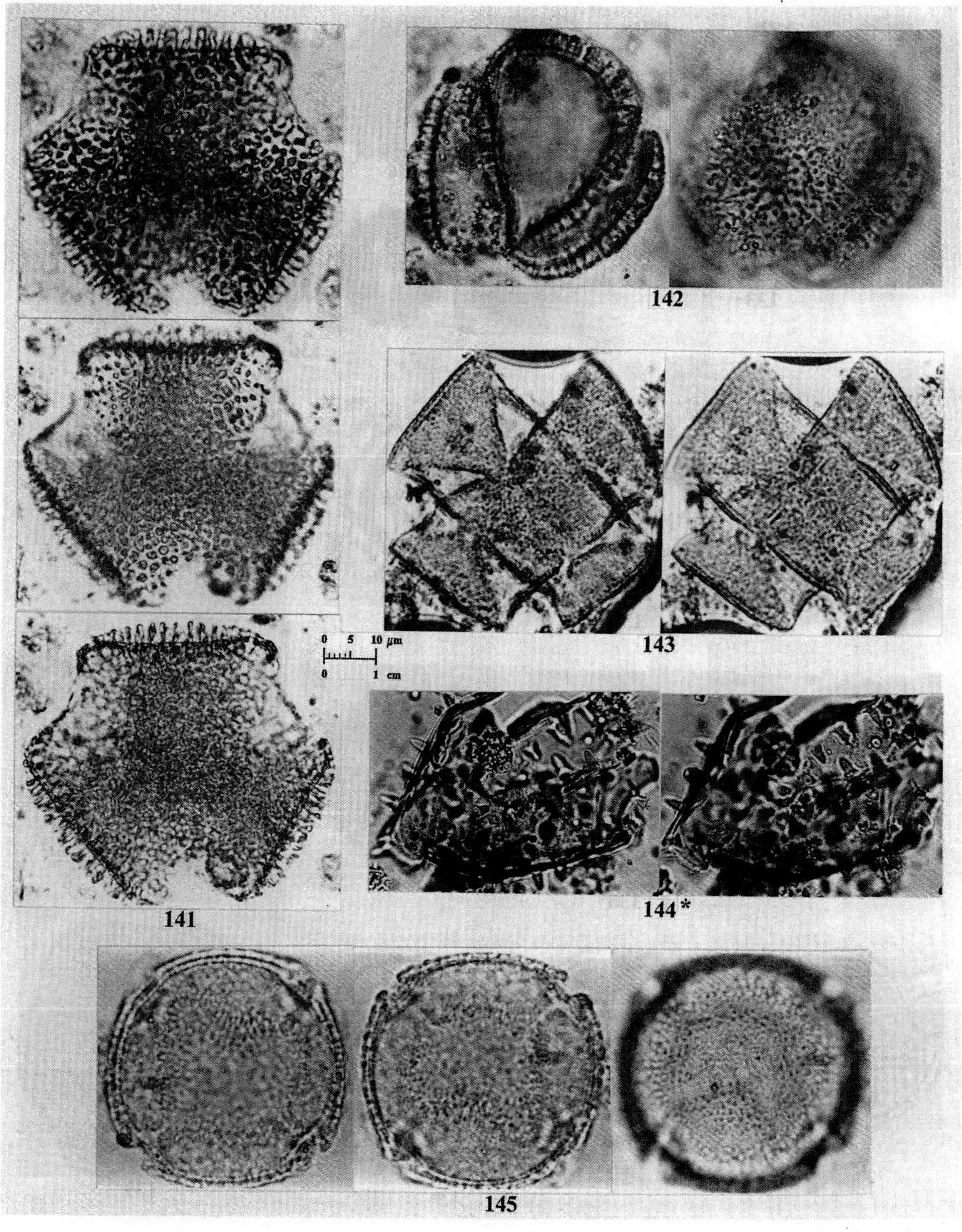

Plate/Prancha 23

Convolvulaceae: *Maripa* sp. form 1 (141); *Maripa* sp. form 2 (142); Convolvulaceae 1 (143). **Cucurbitaceae**: *Cayaponia* sp. (144); *Cyclanthera* sp. (145) [*#144 is reduced to ½ scale (1 mm = 2 μm)].

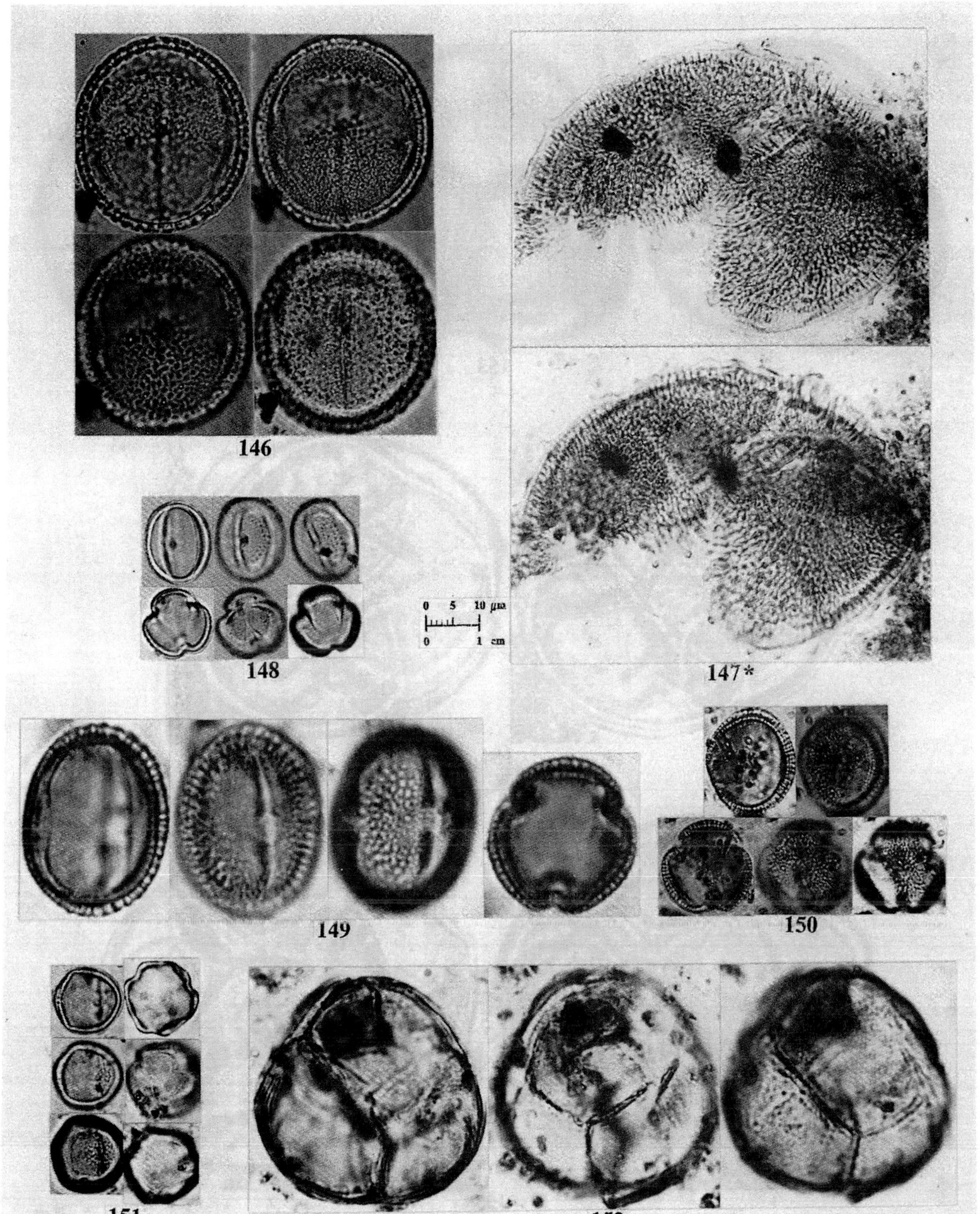

Plate/Prancha 24

Cucurbitaceae: *Posadaea* sp. (146); Cucurbitaceae 1 (147). **Cunoniaceae:** *Weinmannia* sp. (148). **Dilleniaceae:** *Doliocarpus* sp. (149); *Tetracera* sp. (150). **Elaeocarpaceae:** Elaeocarpaceae 1 (151). **Ericaceae:** *Agarista* sp. / *Gaylussacia* sp. form 1 (152) [*#147 is reduced to ½ scale (1 mm = 2 μm)].

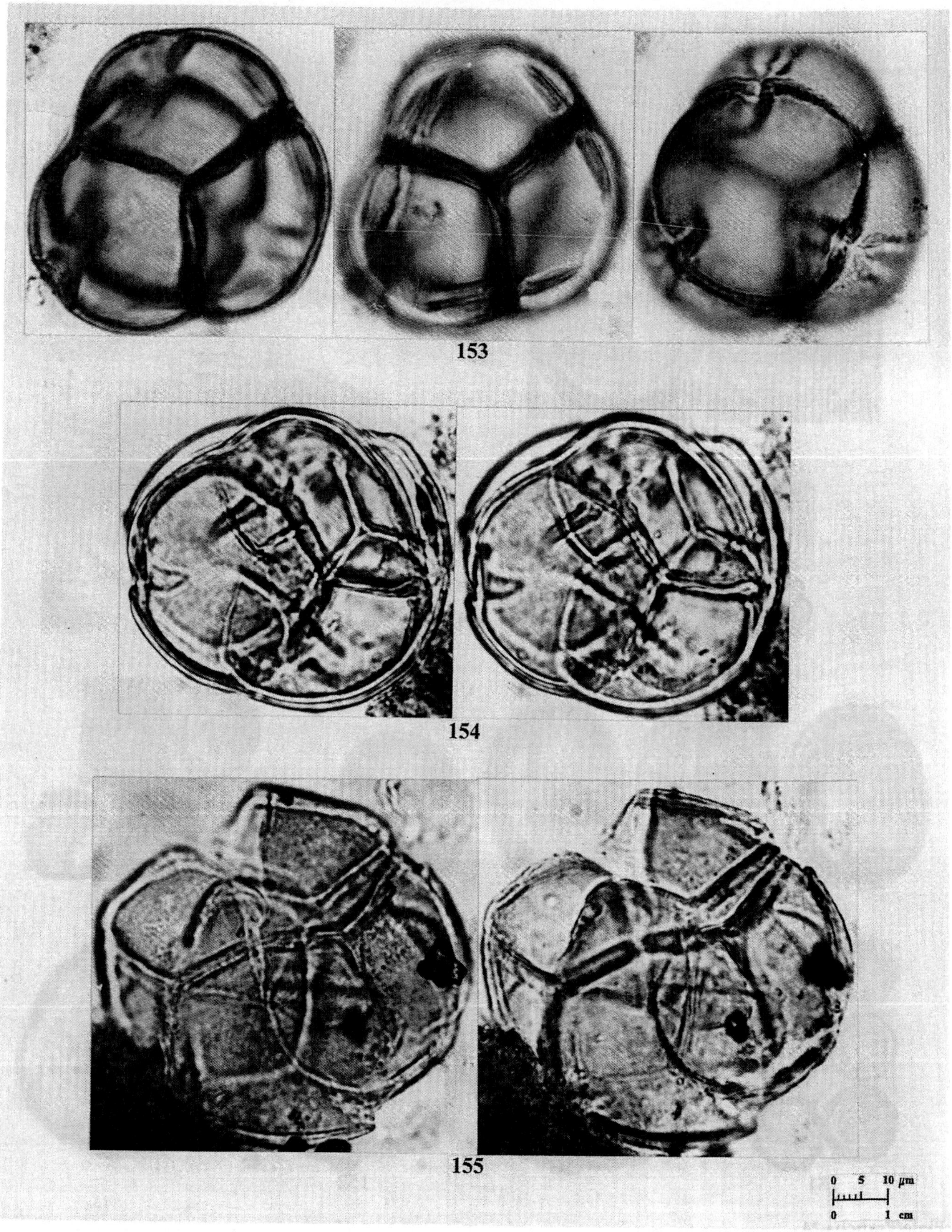

Plate/Prancha 25
Ericaceae: *Agarista* sp. / *Gaylussacia* sp. form 2 (153); *Agarista* sp. / *Gaylussacia* sp. form 3 (154); *Gaultheria* sp. (155).

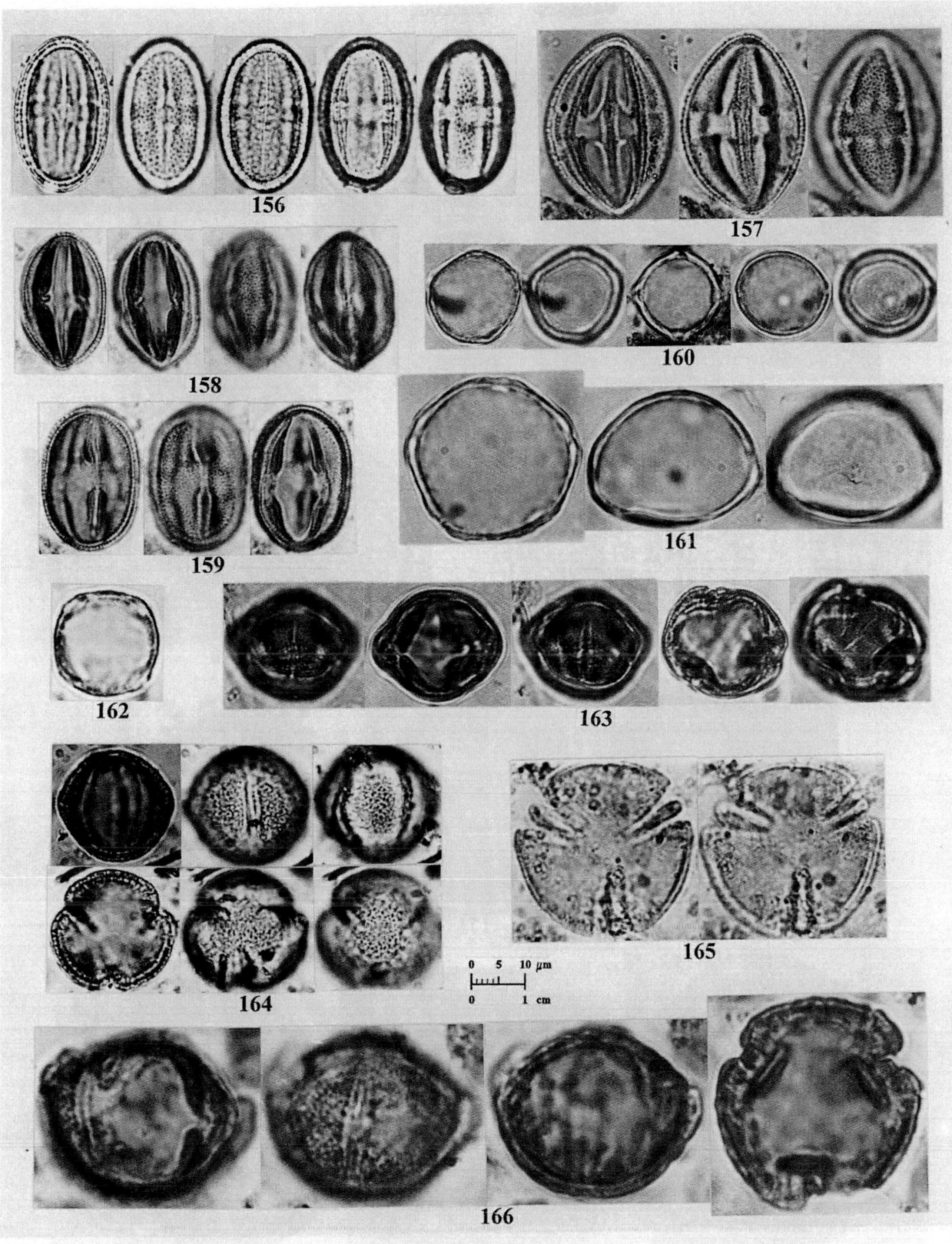

Plate/Prancha 26

Erythroxylaceae: *Erythroxylum* sp. form 1 (156); *Erythroxylum* sp. form 2 (157); *Erythroxylum* sp. form 3 (158); *Erythroxylum* sp. form 4 (159). **Euphorbiaceae**: *Acalypha* sp. form 1 (160); *Acalypha* sp. form 2 (161); *Acalypha* sp. form 3 (162); *Alchornea* sp. form 1 (163); *Alchornea* sp. form 2 (164); *Alchornea* sp. form 3 (165); *Alchornea* sp. form 4 (166).

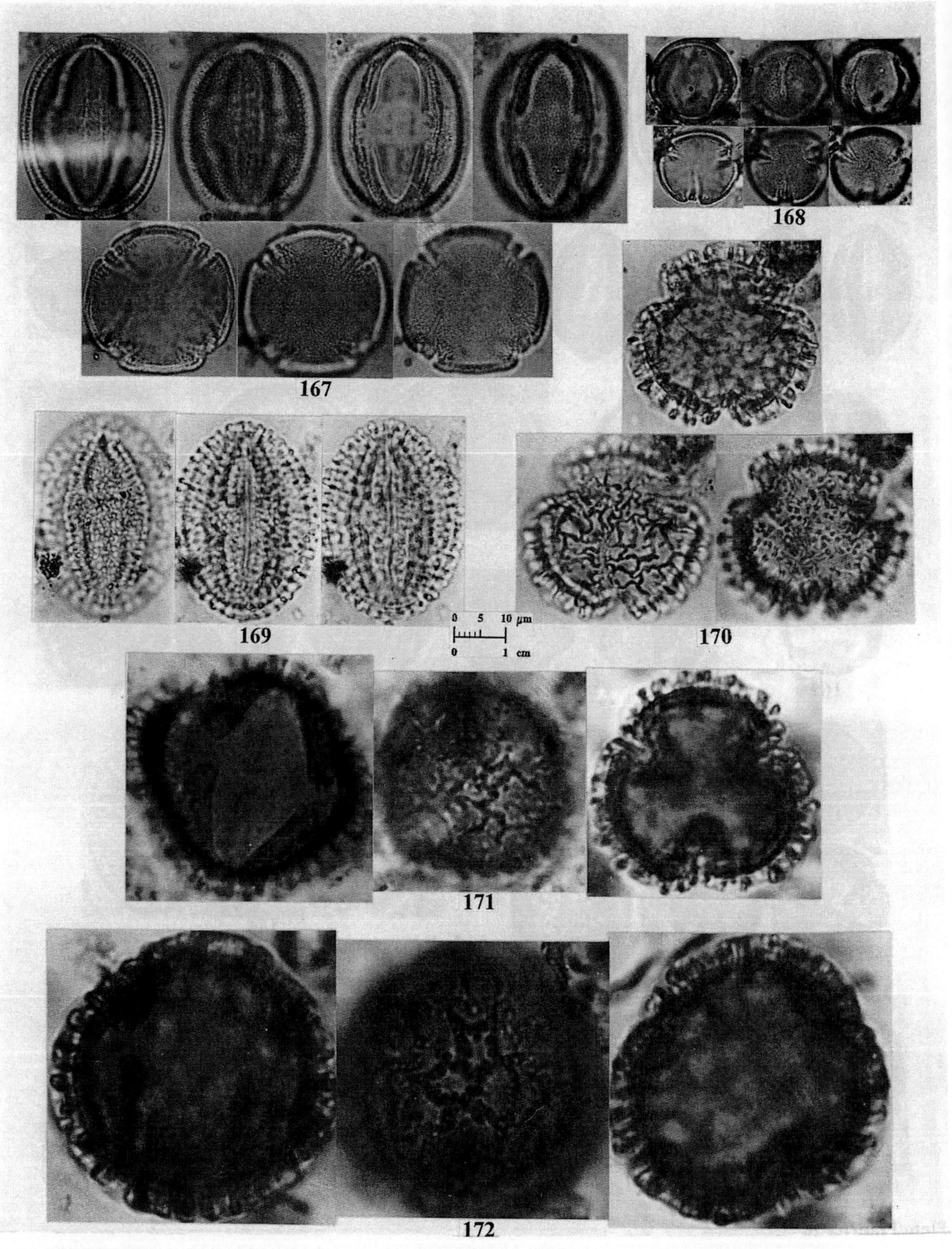

Plate/Prancha 27

Euphorbiaceae: *Alchornea* sp. form 5 (167); *Alchornea* sp. form 6 (168); *Amanoa aff. congesta* (171); *A. aff. nanayensis* (170); *Amanoa* sp. form 1 (169); *Amanoa* sp. form 2 (172).

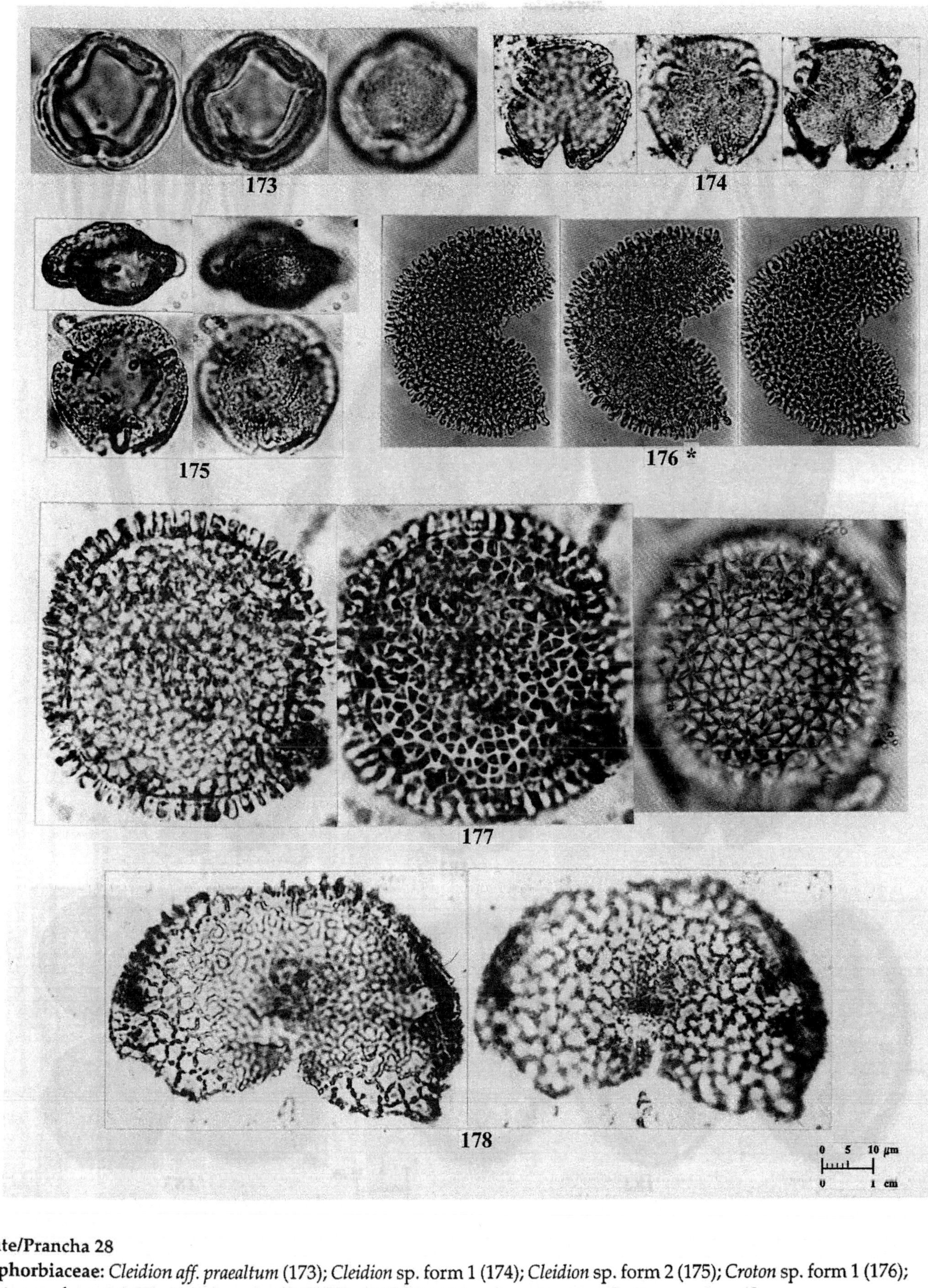

Plate/Prancha 28

Euphorbiaceae: *Cleidion aff. praealtum* (173); *Cleidion* sp. form 1 (174); *Cleidion* sp. form 2 (175); *Croton* sp. form 1 (176); *Croton* sp. form 2 (177); *Dalechampia aff. tiliifolia* (178) [*#176 is reduced to ½ scale (1 mm = 2 µm)].

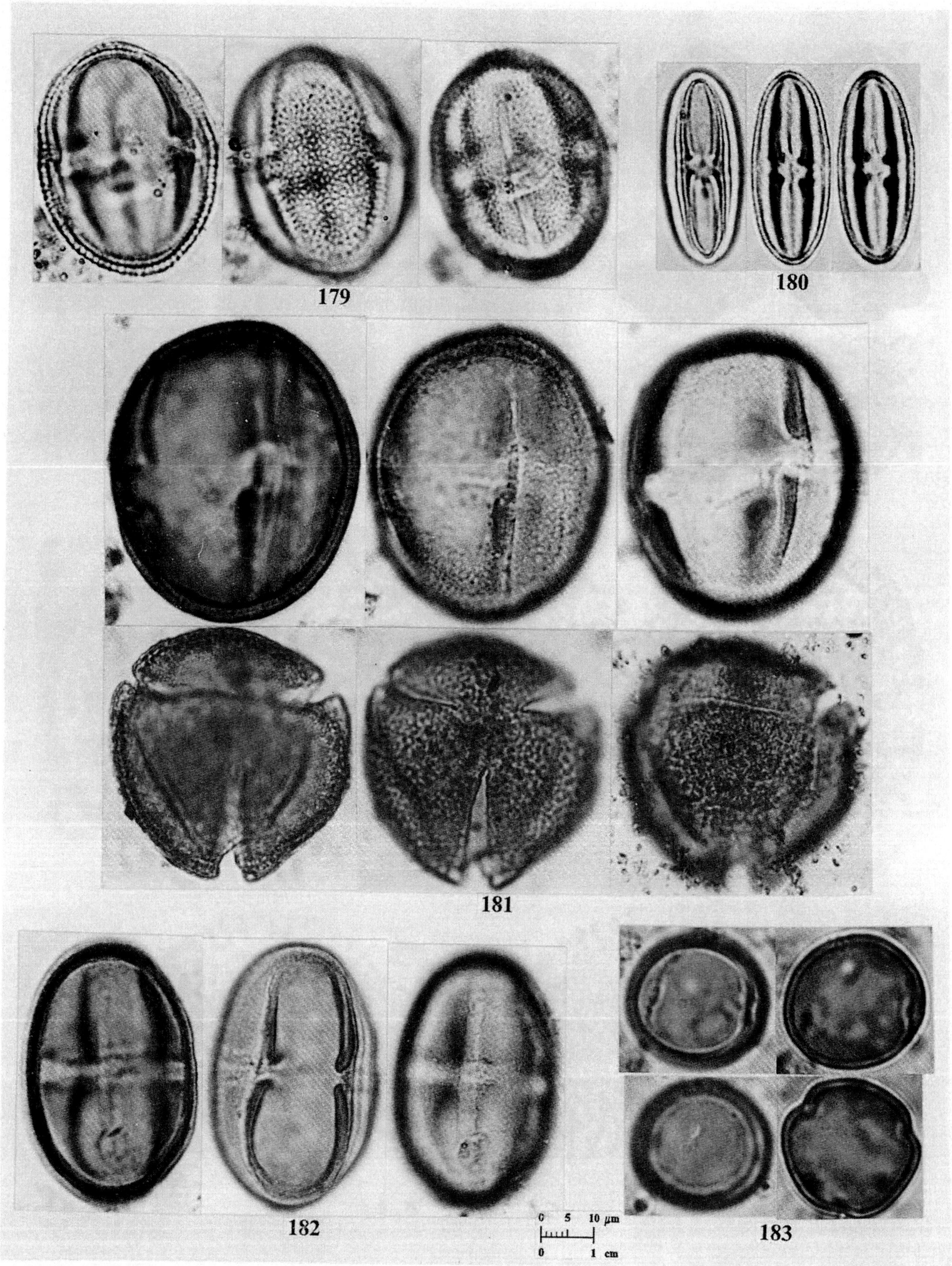

Plate/Prancha 29

Euphorbiaceae: *Drypetes* sp. (179); *Hyeronima* sp. (180); *Mabea* sp. form 1 (181); *Mabea* sp. form 2 (182); *Pera aff. glabrata* (183).

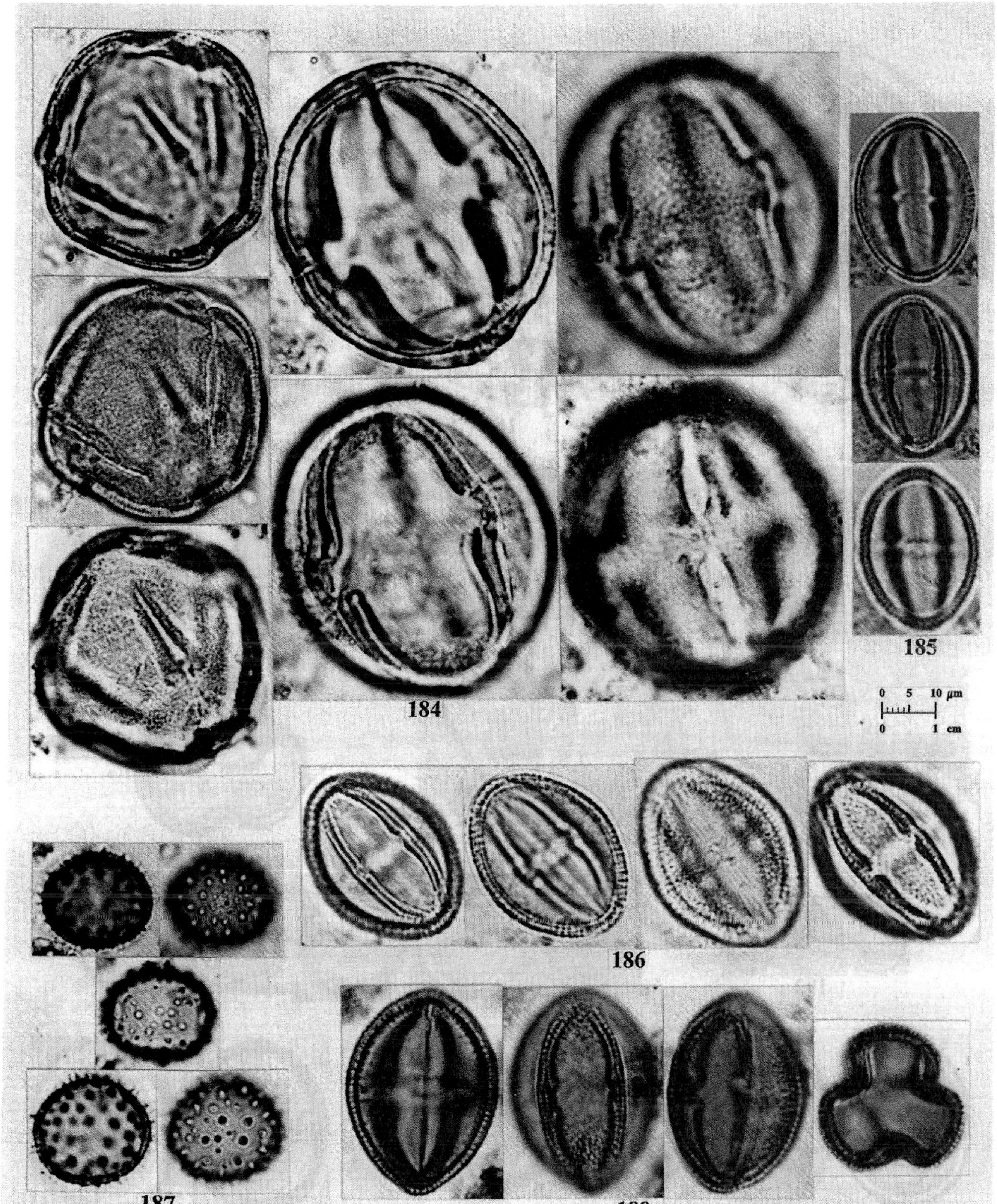

Plate/Prancha 30

Euphorbiaceae: *Mabea* sp. form 3 (184); *Phyllanthus* sp. form 1 (185); *Phyllanthus* sp. form 2 (186); *Podocalyx aff. loranthoides* (187); *Sapium* sp. form 1 (188).

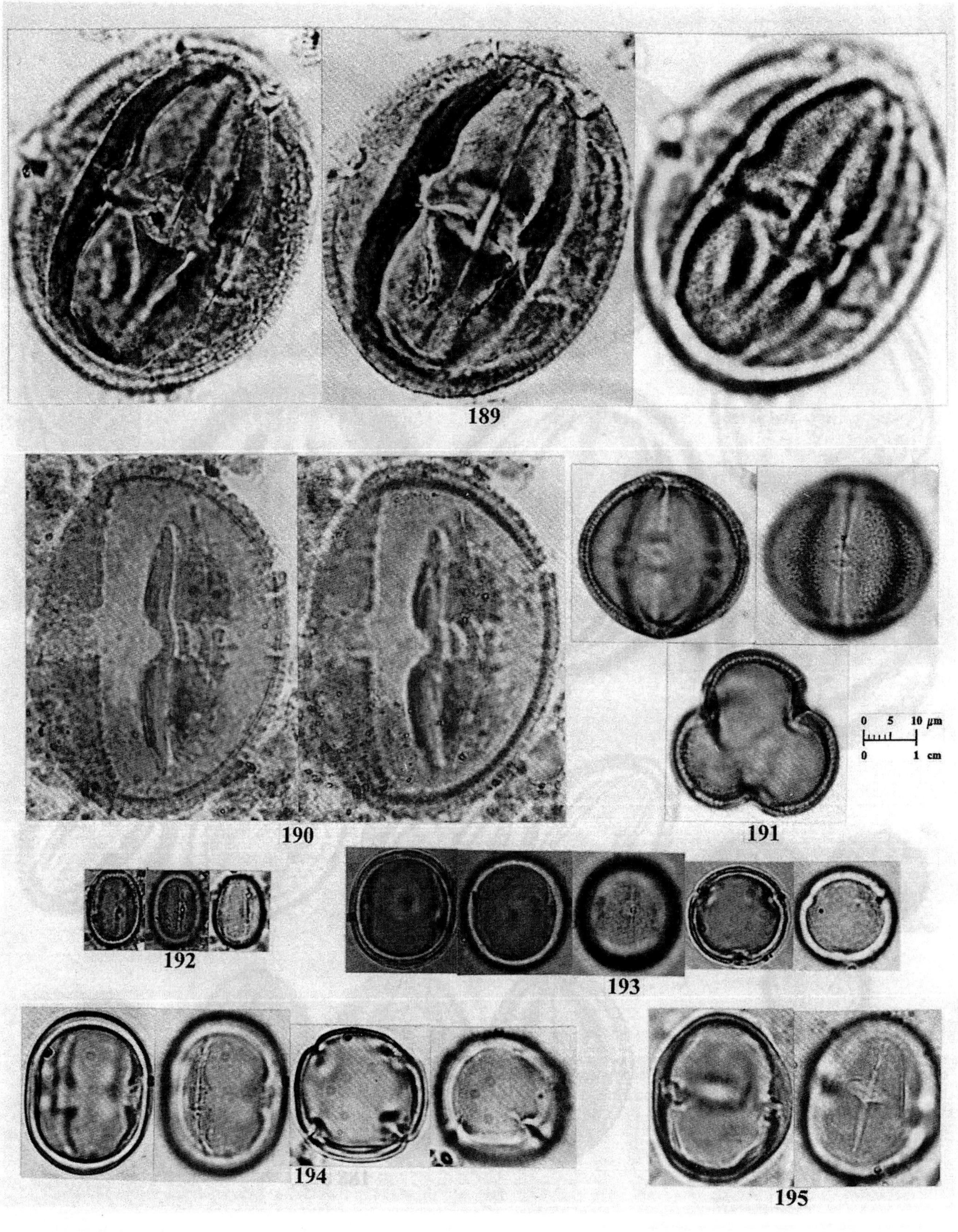

Plate/Prancha 31

Euphorbiaceae: *Sapium* sp. form 2 (189); *Sapium* sp. form 3 (190); *Sebastiana* sp. (191). **Flacourtiaceae**: *Banara* sp. (192); *Casearia* sp. form 1 (193); *Casearia* sp. form 2 (194); *Casearia* sp. form 3 (195).

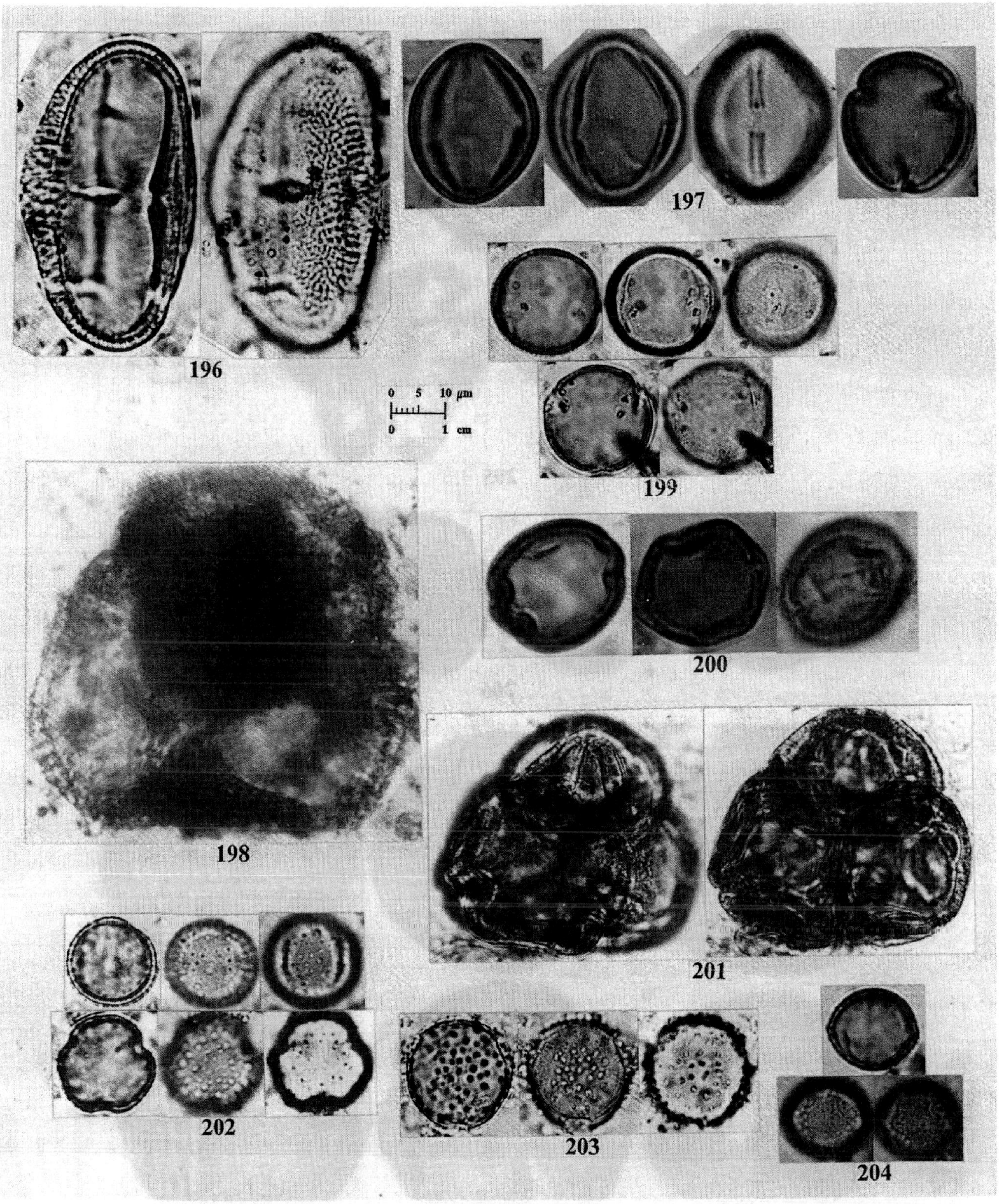

Plate/Prancha 32

Flacourtiaceae: *Laetia aff. coriacea* (196); *Xylosma* sp. (197). **Gentianaceae**: *Schultesia* sp. (198). **Guttiferae (Clusiaceae)**: *Clusia aff. odorata* (199); *Clusiella* sp. (200); *Kielmeyera* sp. (201); *Tovomitopsis* sp. form 1 (202); *Tovomitopsis* sp. form 2 (203); Guttiferae 1 (204).

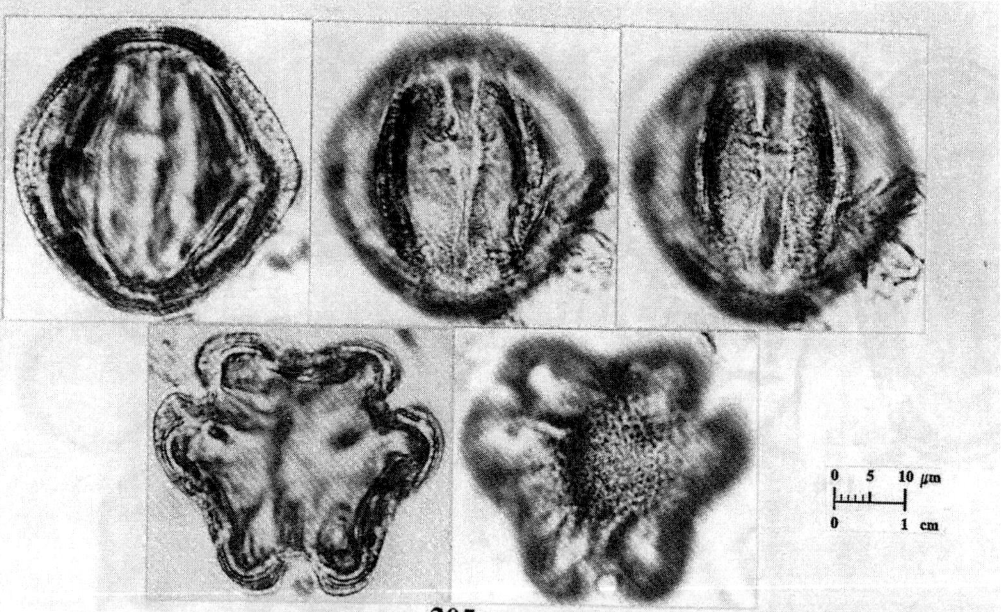

205

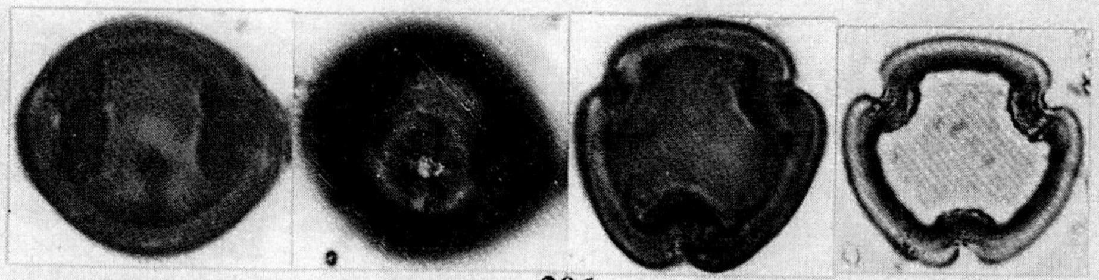

206

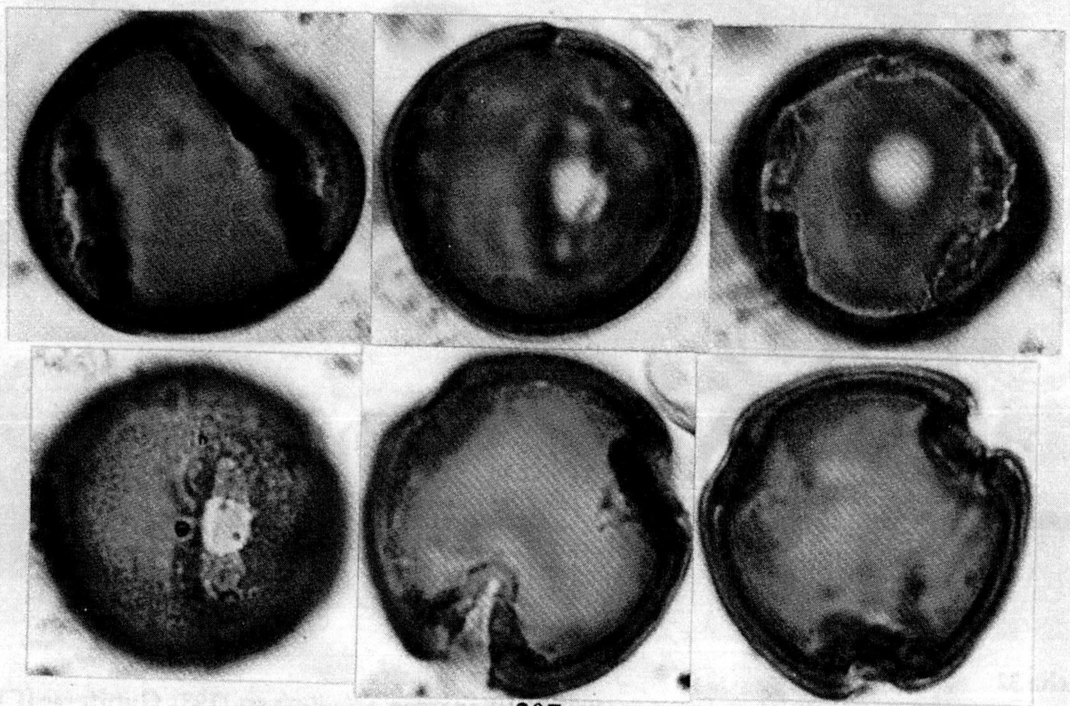

207

Plate/Prancha 33
Humiriaceae: *Endopleura aff. uchi* (205); *Humiria aff. balsamifera* (206); *Humiria* sp. (207).

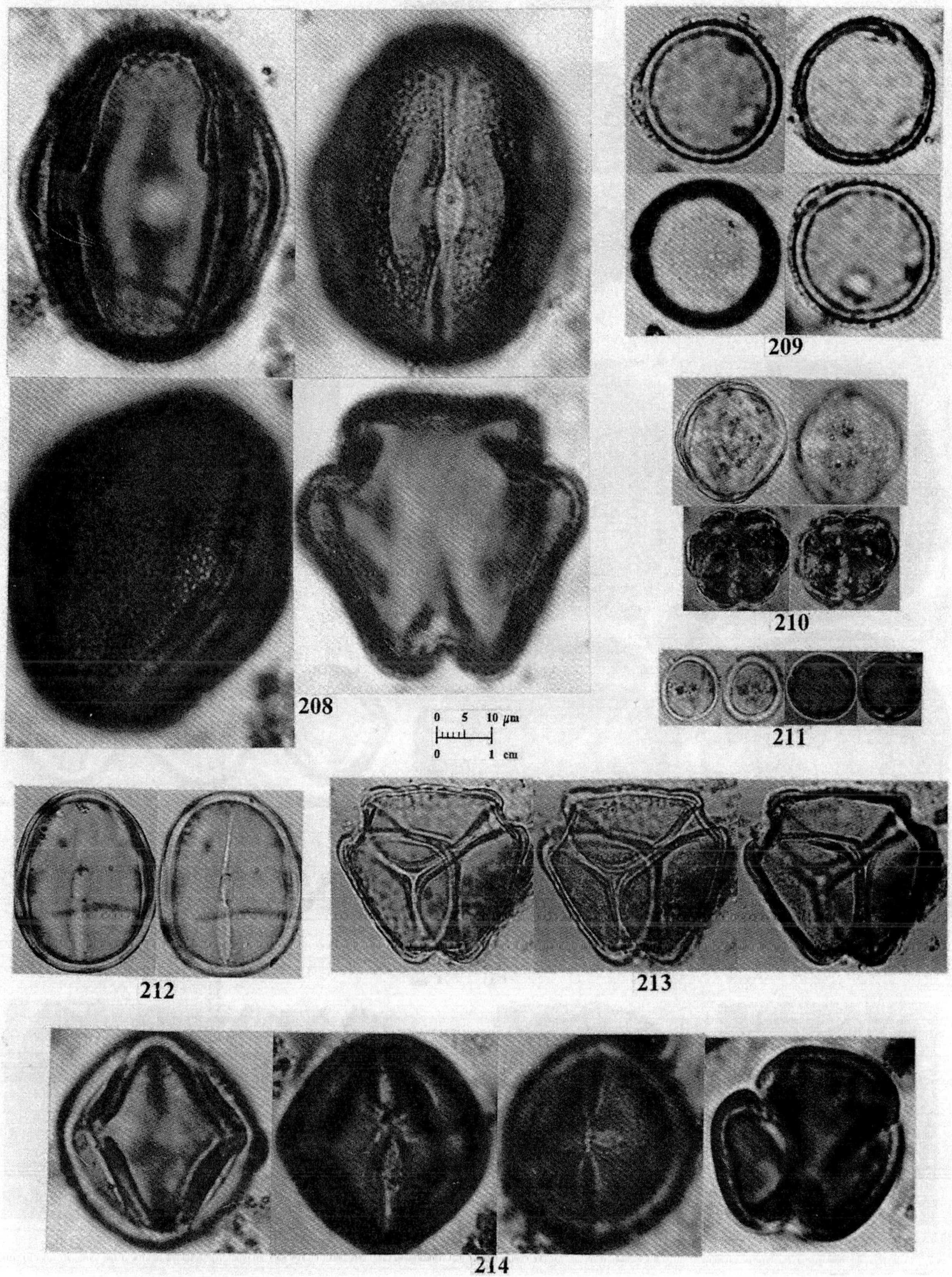

Plate/Prancha 34

Humiriaceae: *Vantanea aff. guianensis* (208). **Icacinaceae**: *Discophora* sp. (209). **Labiatae (Lamiaceae)**: *Hyptis* sp. (210). **Lacistemataceae**: *Lacistema* sp. (211). **Leguminosae – Caesalpinioideae**: *Bauhinia* sp. (212); *Cassia* sp. form 1 (213); *Cassia* sp. form 2 (214).

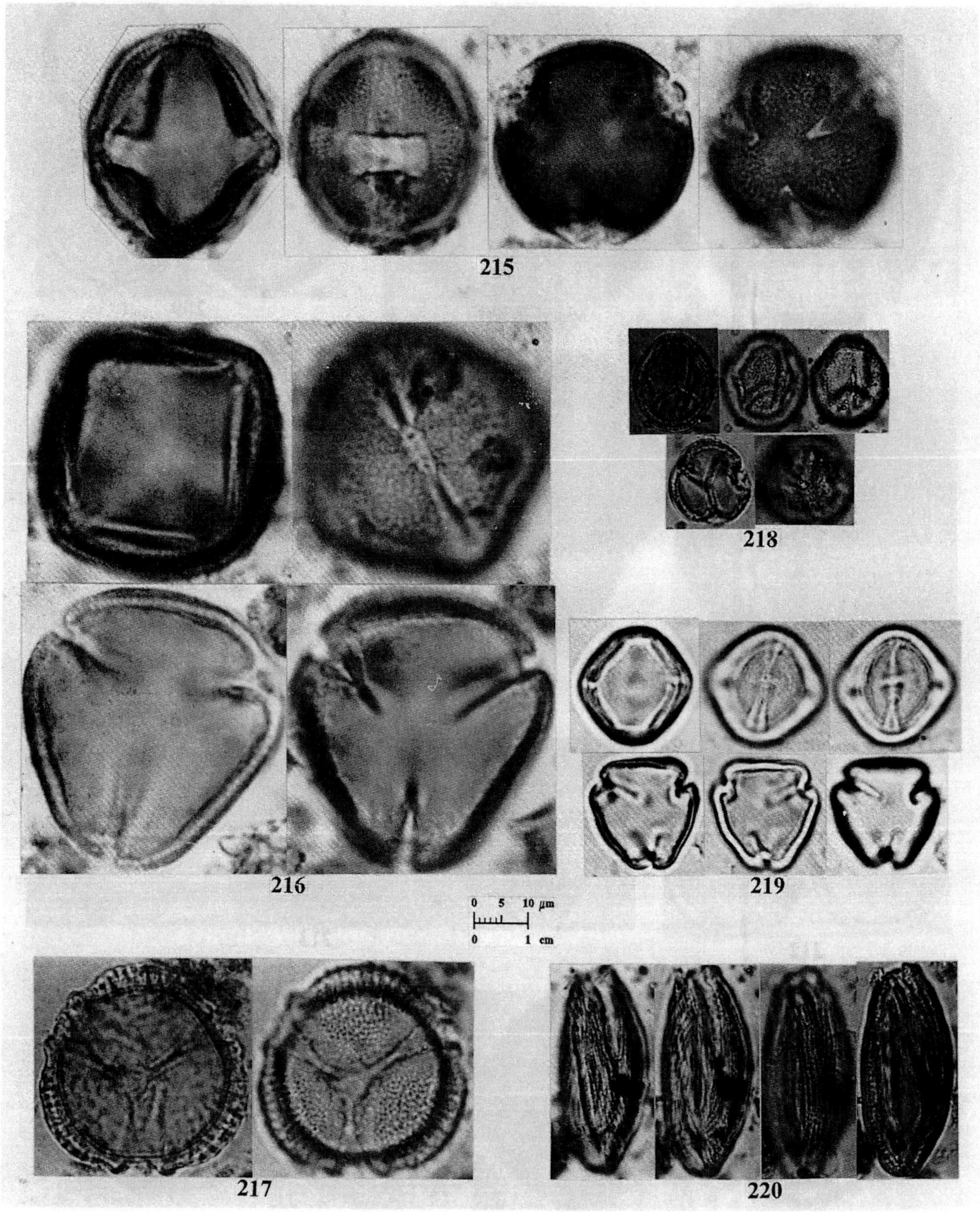

Plate/Prancha 35

Leguminosae – Caesalpinioideae: *Cassia* sp. form 3 (215); *Cassia* sp. form 4 (216); *Cenostigma* sp. / *Caesalpinia* sp. (217); *Cercidium* sp. (218); *Copaifera aff. cordifolia* (219); *Crudia aff. amazonica* (220).

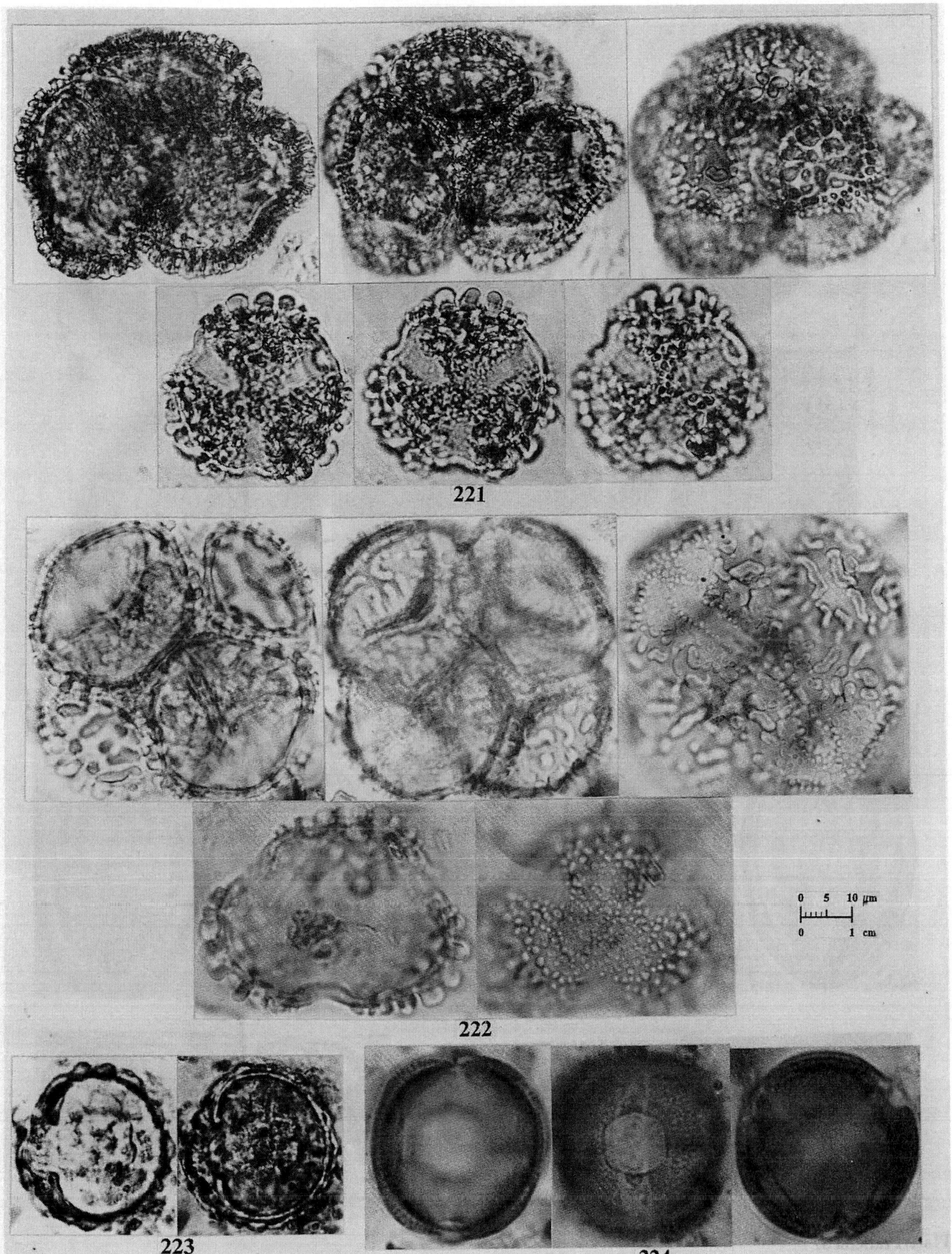

Plate/Prancha 36

Leguminosae – Caesalpinioideae: *Dinizia aff. excelsa* (221); *Dinizia* sp. (222); *Elizabetha aff. paraensis* (223); *Martiodendron* sp. (224).

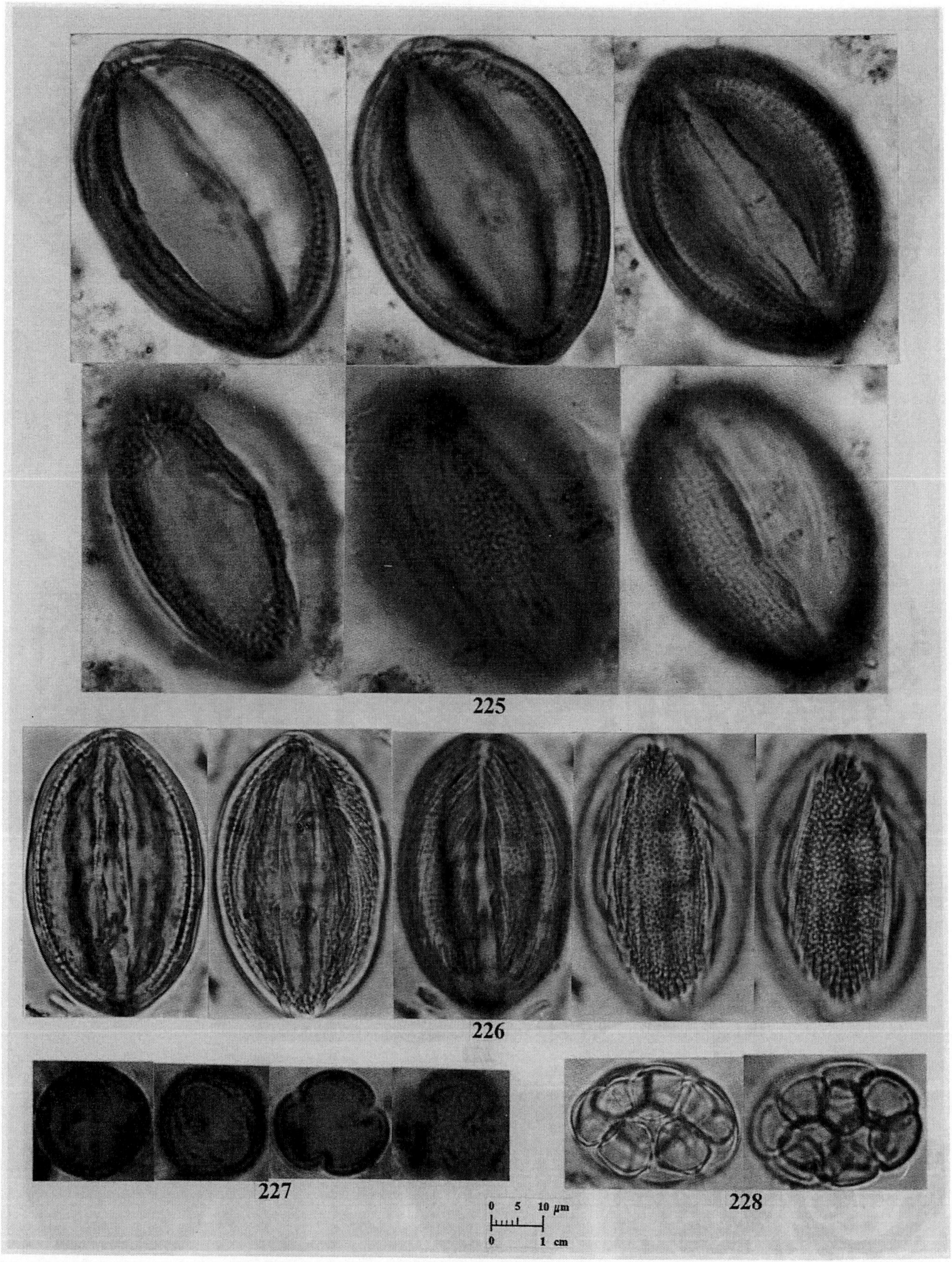

Plate/Prancha 37

Leguminosae – Caesalpinioideae: *Macrolobium aff. acaciaefolium* (225); *M. aff. angustifolium* (226); *Schizolobium* sp. (227). **Leguminosae – Mimosoideae**: *Acacia* sp. (228).

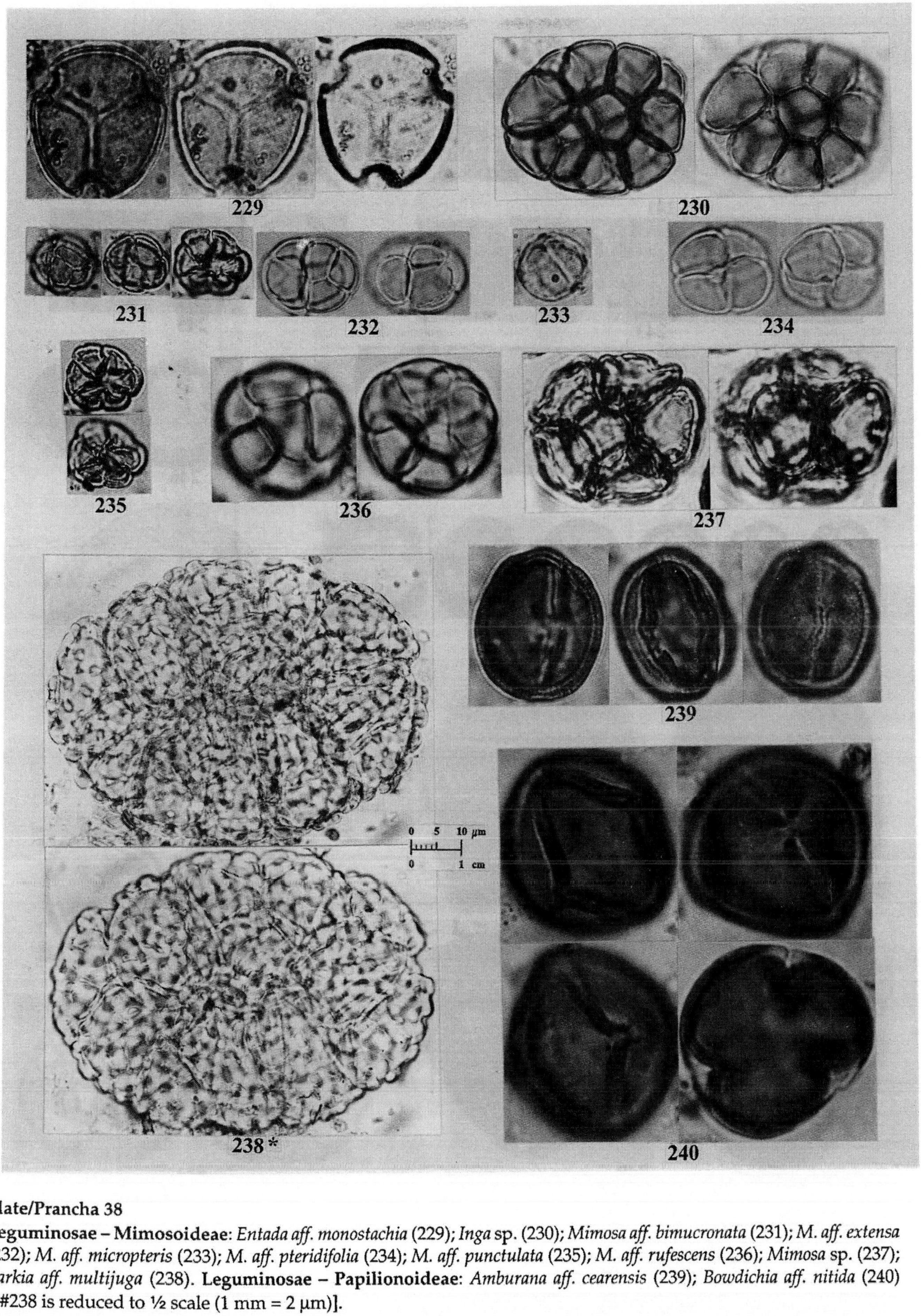

Plate/Prancha 38

Leguminosae – Mimosoideae: *Entada aff. monostachia* (229); *Inga* sp. (230); *Mimosa aff. bimucronata* (231); *M. aff. extensa* (232); *M. aff. micropteris* (233); *M. aff. pteridifolia* (234); *M. aff. punctulata* (235); *M. aff. rufescens* (236); *Mimosa* sp. (237); *Parkia aff. multijuga* (238). **Leguminosae – Papilionoideae**: *Amburana aff. cearensis* (239); *Bowdichia aff. nitida* (240) [*#238 is reduced to ½ scale (1 mm = 2 μm)].

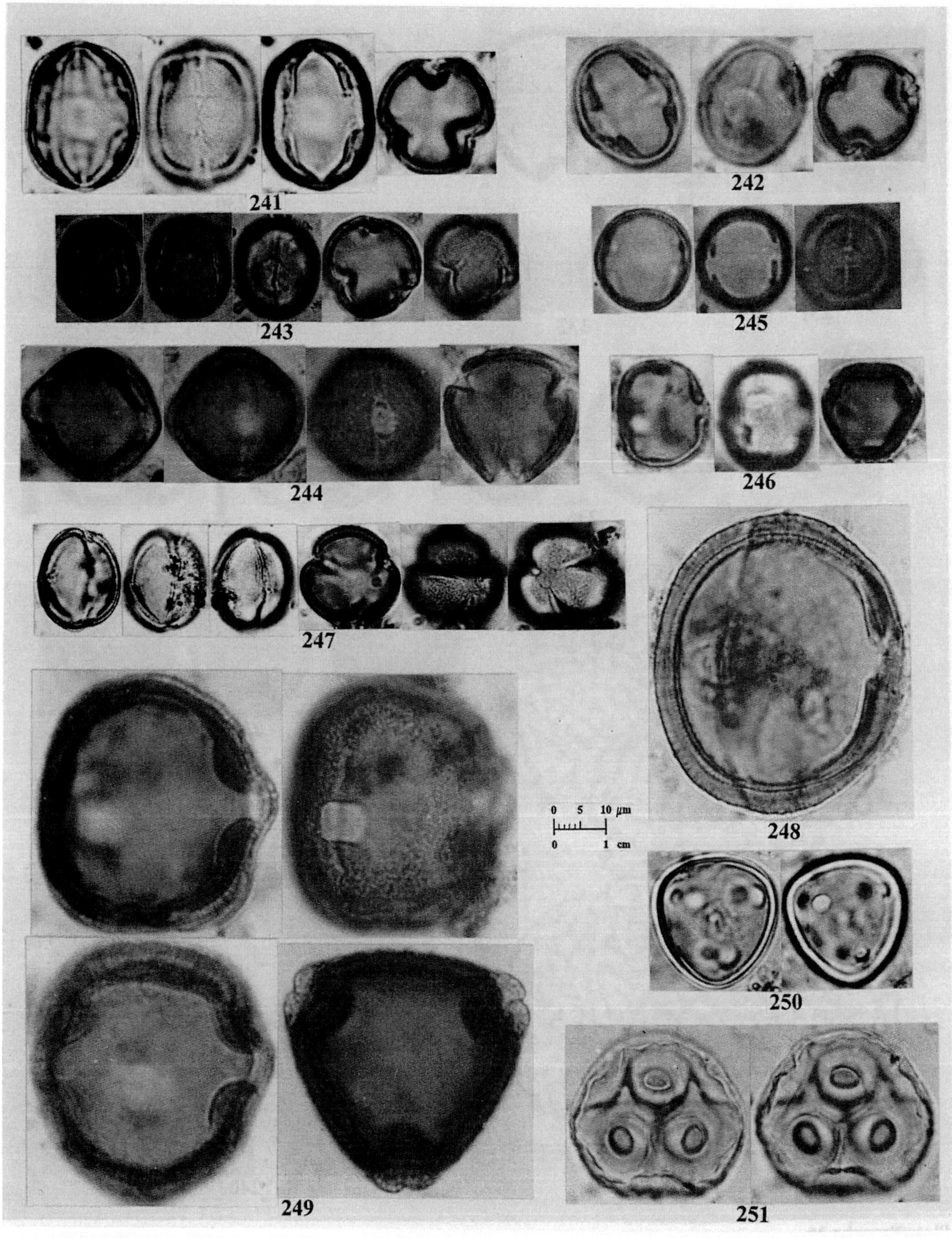

Plate/Prancha 39

Leguminosae – Papilionoideae: *Centrolobium* sp. (241); *Dalbergia* sp. form 1 (242); *Dalbergia* sp. form 2 (243); *Dipteryx* sp. (244); *Machaerium* sp. form 1 (245); *Machaerium* sp. form 2 (246); *Myrocarpus aff. frondosus* (247); Papilionoideae 1 (248); Papilionoideae 2 (249). **Loranthaceae (Viscaceae)**: *Dendropemon* sp. (250); *Orycthanthus aff. alveolatus* (251).

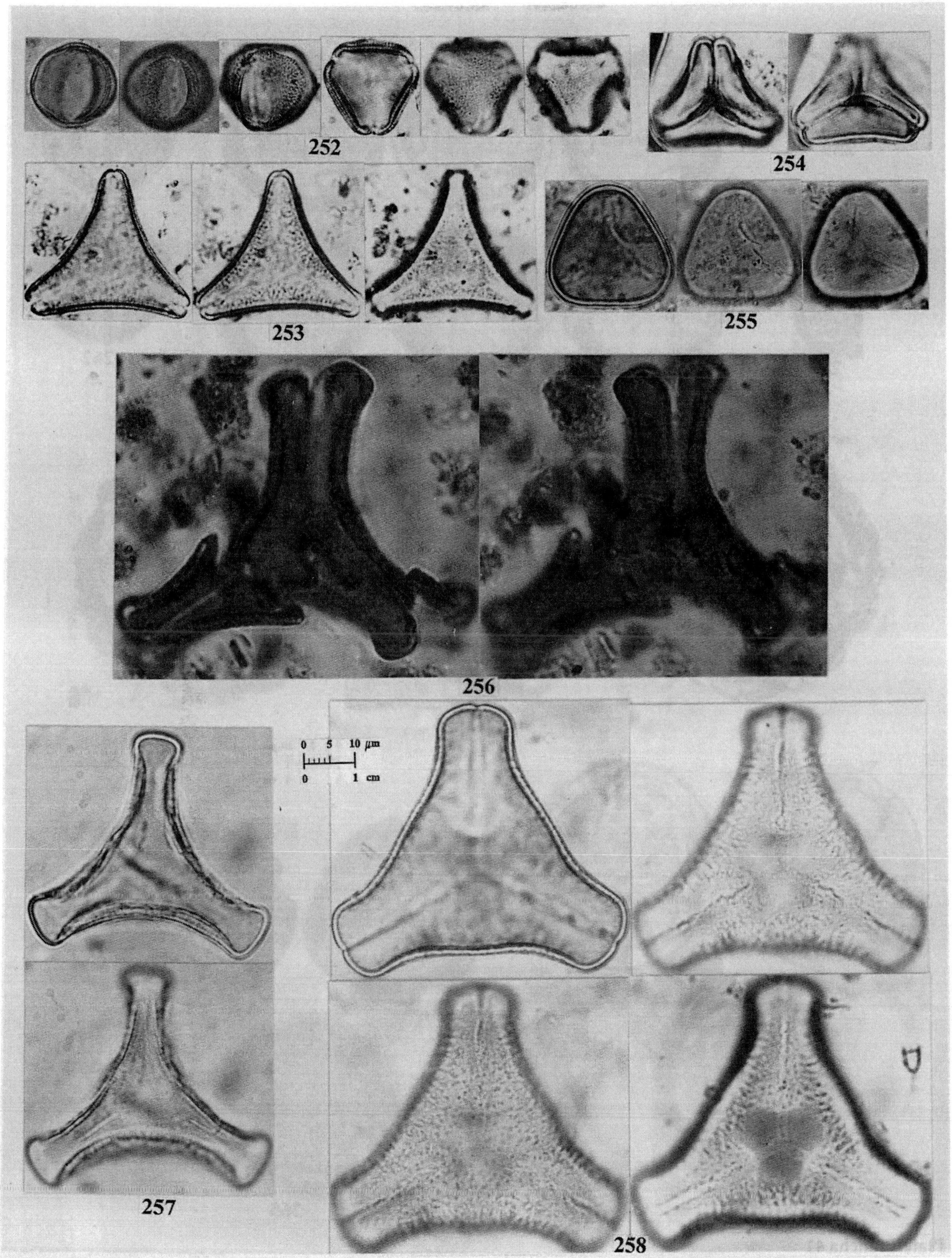

Plate/Prancha 40

Loranthaceae (Viscaceae): *Phoradendron* sp. (252); *Psittacanthus* sp. (253); *Struthanthus aff. pyrifolius* (254); *Struthanthus* sp. (255); *Tristerix* sp. (*Macrosolen* sp.) form 1 (256); *Tristerix* sp. (*Macrosolen* sp.) form 2 (257); Loranthaceae 1 (258).

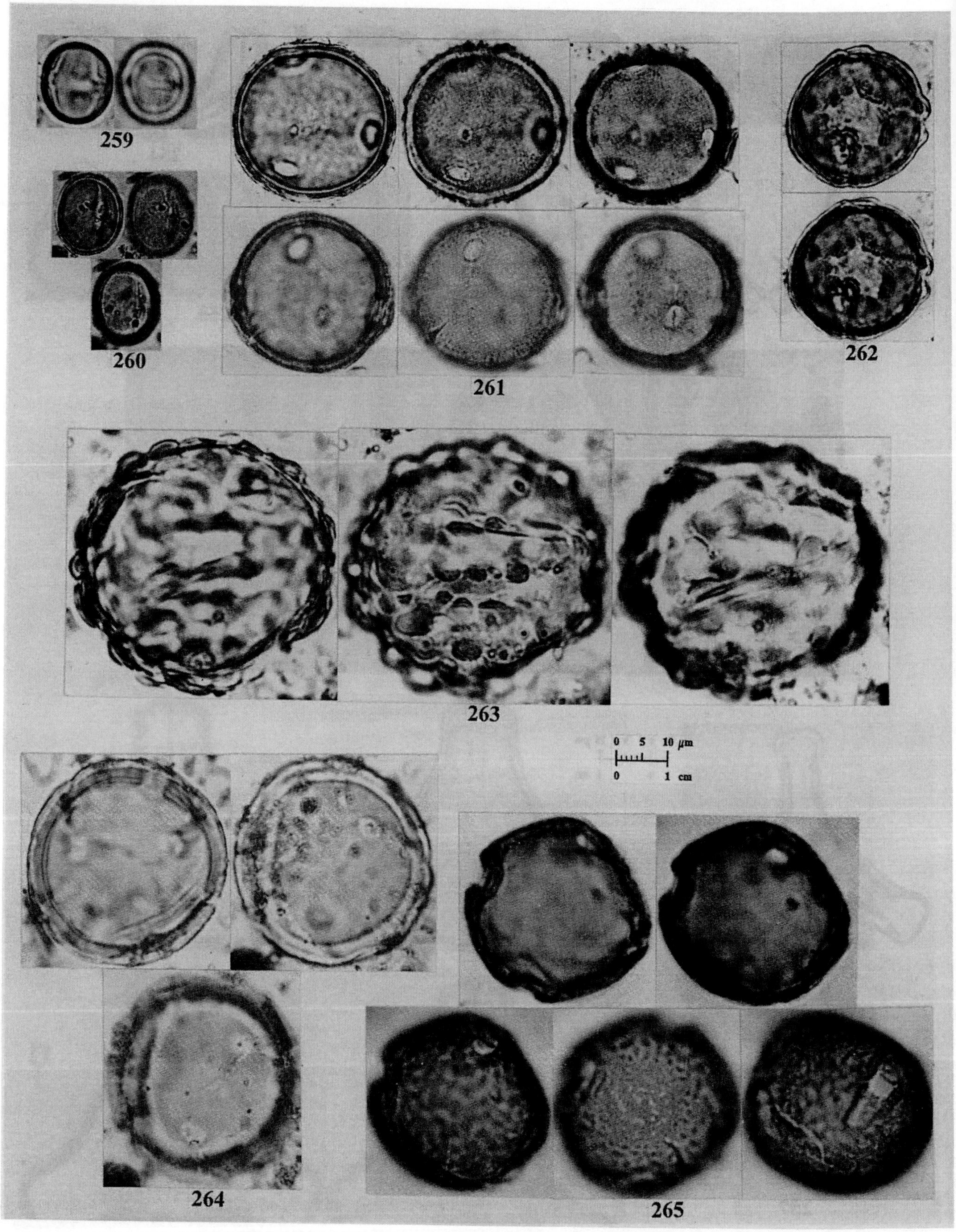

Plate/Prancha 41

Malpighiaceae: *Blepharanda* sp. (259); *Byrsonima* sp. (260); *Diacidia* sp. (261); *Dicella* sp. (262); *Mascagnia* sp. (263); *Stigmaphyllon* sp. (264); *Tetrapteryx* sp. (265).

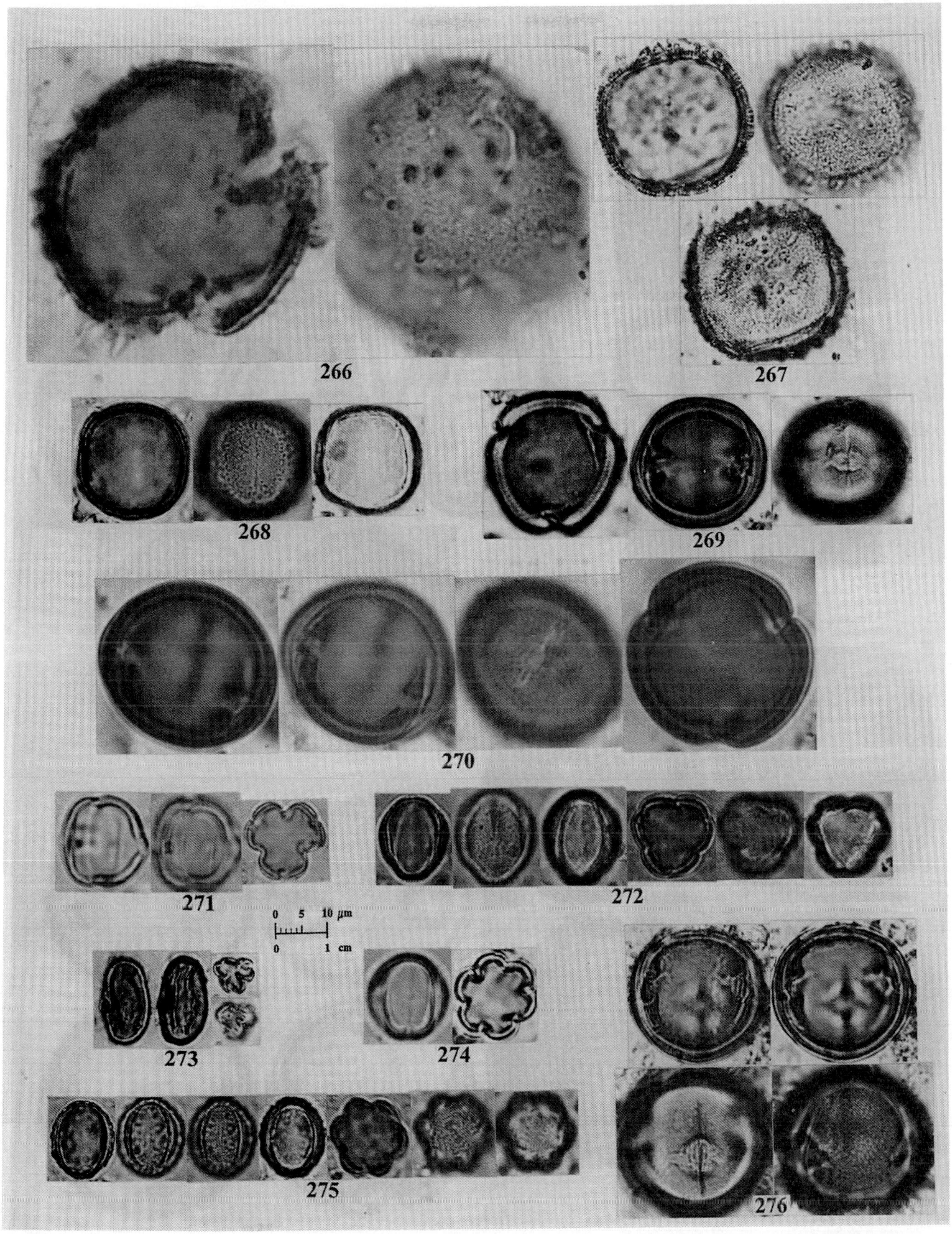

Plate/Prancha 42

Malvaceae: *Abutilon* sp. (266); *Sida* sp. (267). **Marcgraviaceae**: *Marcgravia* sp. (268); *Norantea aff. guianensis* (269); *Souroubea* sp. (270). **Melastomataceae**: *Clidemia* sp. (271); *Miconia* sp. form 1 (272); *Miconia* sp. form 2 (273); *Miconia* sp. form 3 (274); Melastomataceae 1 (275). **Meliaceae**: *Cabralea* sp. (276).

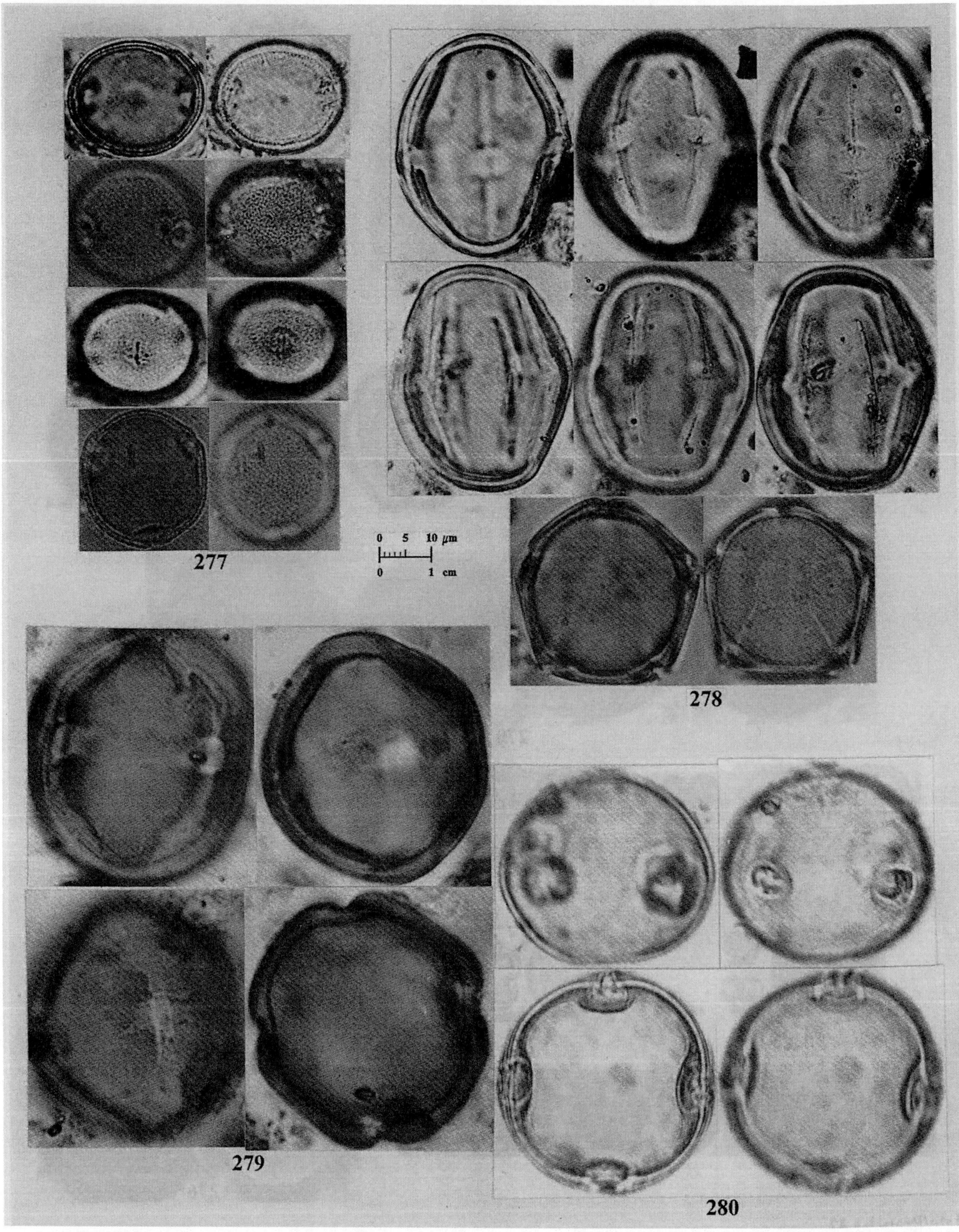

Plate/Prancha 43
Meliaceae: *Carapa aff. guianensis* (277); *Cedrela* sp. form 1 (278); *Cedrela* sp. form 2 (279); *Guarea aff. glabra* (280).

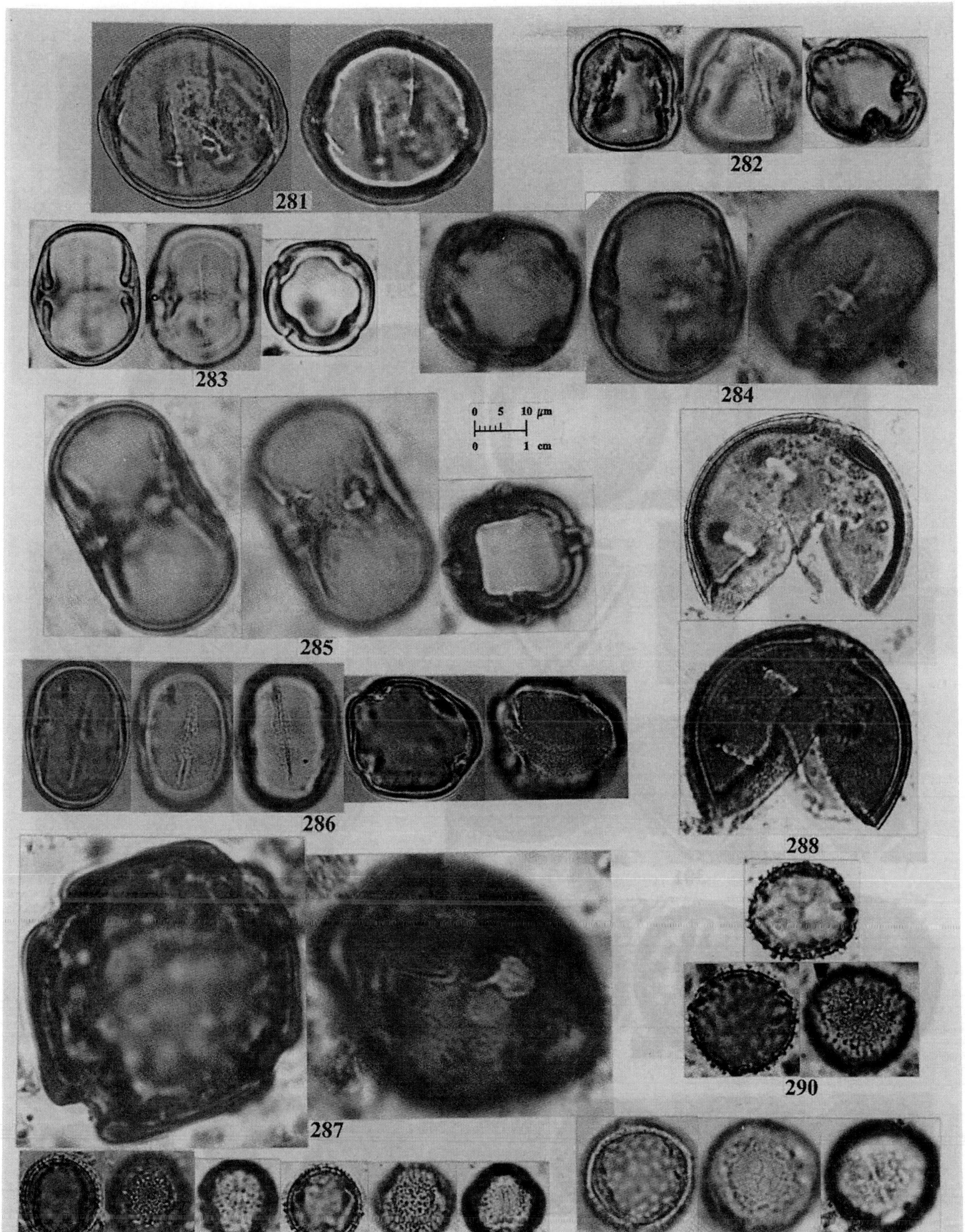

Plate/Prancha 44

Meliaceae: *Guarea* sp. (281); *Trichilia* sp. form 1 (282); *Trichilia* sp. form 2 (283); *Trichilia* sp. form 3 (284); *Trichilia* sp. form 4 (285); *Trichilia* sp. form 5 (286); Meliaceae 1 (287). **Mendonciaceae**: *Mendoncia aff. glabra* (288). **Menispermaceae**: *Abuta* sp. (289); *Anomospermum* sp. (290); *Cissampelos* sp. form 1 (291).

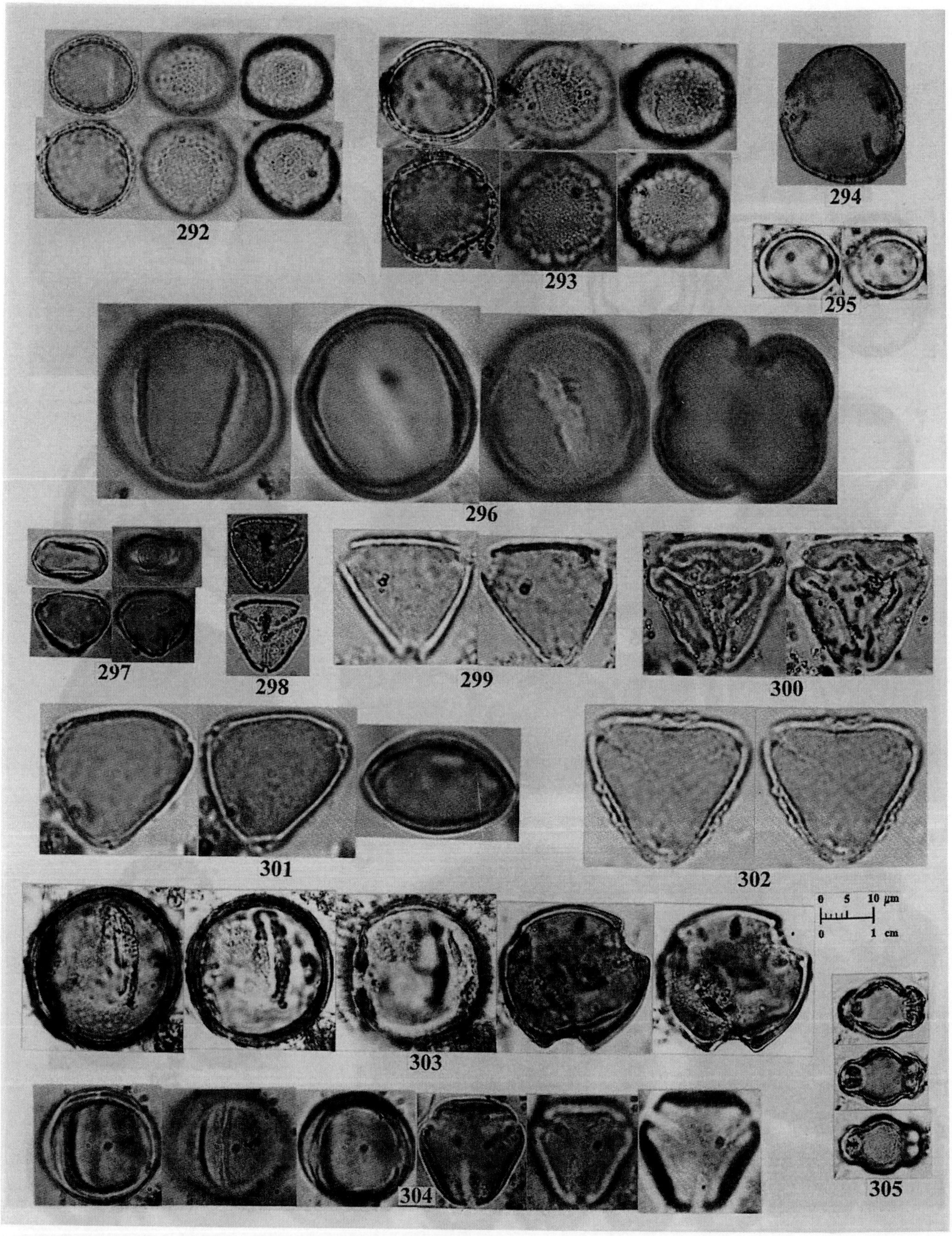

Plate/Prancha 45

Menispermaceae: *Cissampelos* sp. form 2 (292); *Cissampelos* sp. form 3 (293). **Moraceae**: *Brosimum* sp. (294); Moraceae 1 (295). **Myrsinaceae**: *Myrsine* sp. (*Rapanea* sp.) (296). **Myrtaceae**: *Eugenia* sp. form 1 (297); *Eugenia* sp. form 2 (298); *Myrcia* sp. (299); *Syzygium* sp. (300); Myrtaceae 1 (301); Myrtaceae 2 (302). **Ochnaceae**: *Ouratea* sp. (303). **Olacaceae**: *Minquartia aff. guianensis* (304). **Onagraceae** (**Oenotheraceae**): Onagraceae 1 (305).

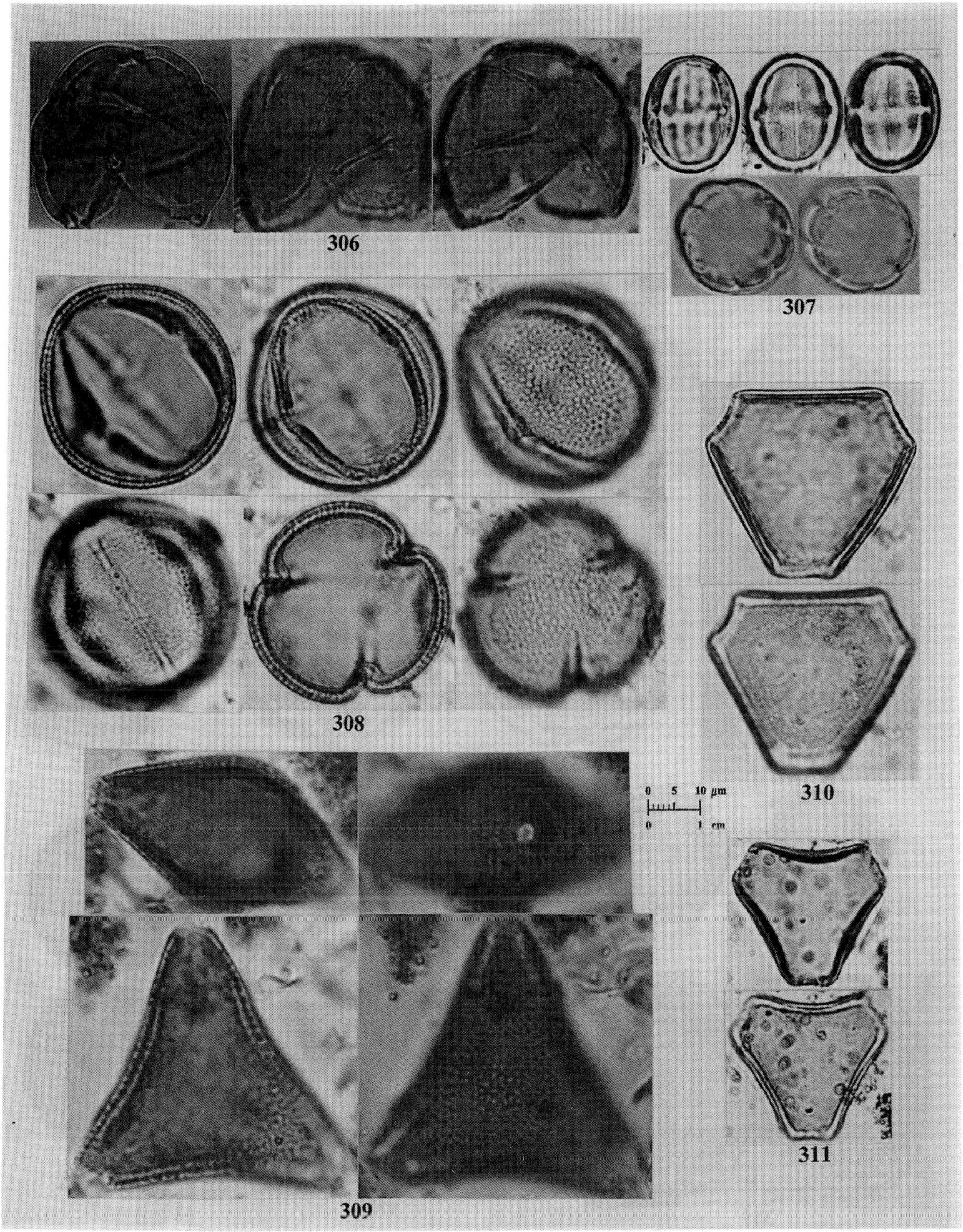

Plate/Prancha 46

Polygalaceae: *Diclidanthera* sp. (306); *Polygala aff. paniculata* (307). **Polygonaceae**: *Symmeria aff. paniculata* (308). **Proteaceae**: *Euplassa* sp. (309); *Roupala* sp. form 1 (310); *Roupala* sp. form 2 (311).

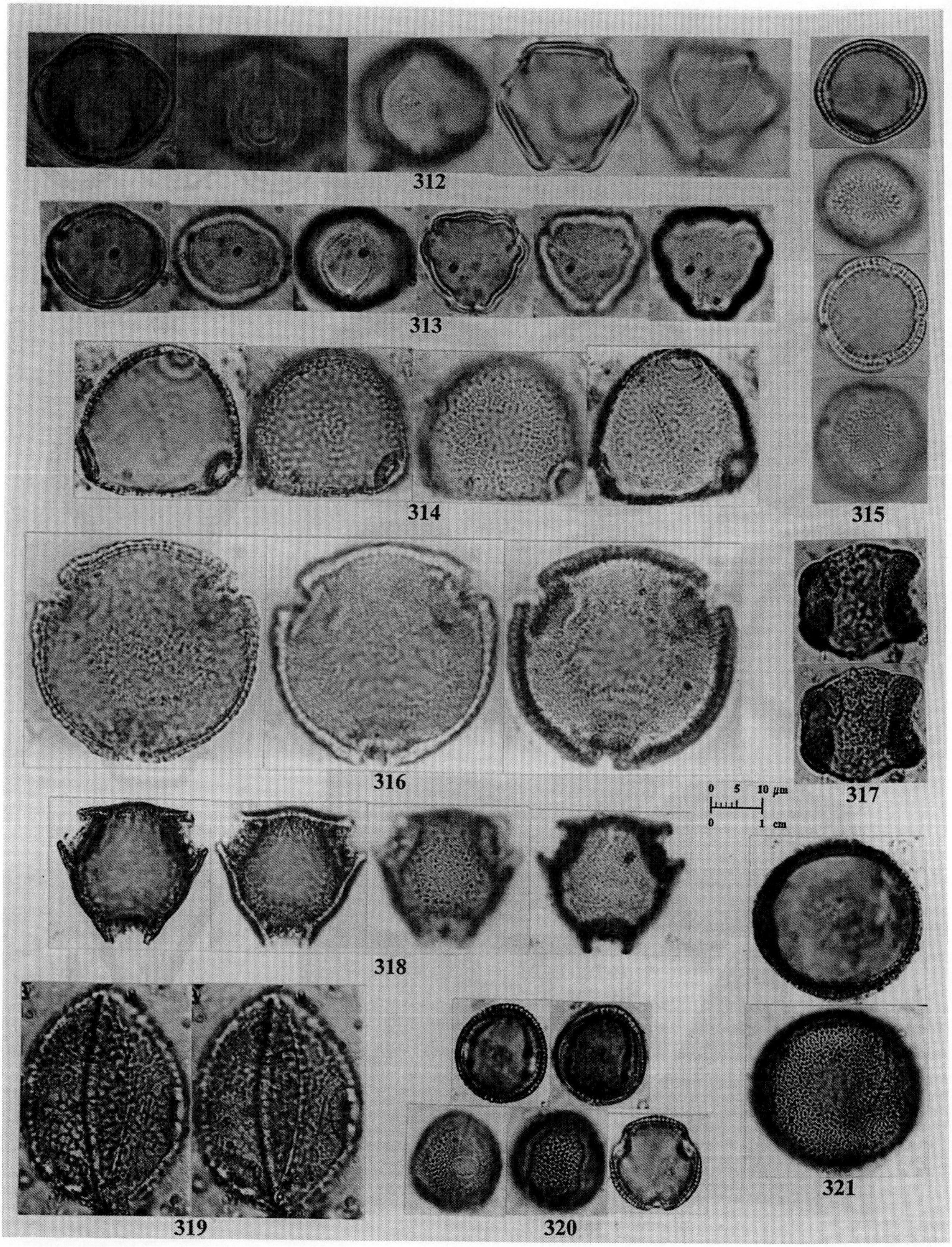

Plate/Prancha 47

Rhamnaceae: *Gouania* sp. form 1 (312); *Gouania* sp. form 2 (313). **Rubiaceae**: *Alibertia aff. edulis* (314); *Borreria aff. densiflora* (315); *Coccocypselum* sp. (316); *Coussarea aff. hydrangeifolia* (317); *Duroia* sp. (319); *Faramea* sp. (318); *Genipa aff. americana* (320); *Pagamea* sp. (321).

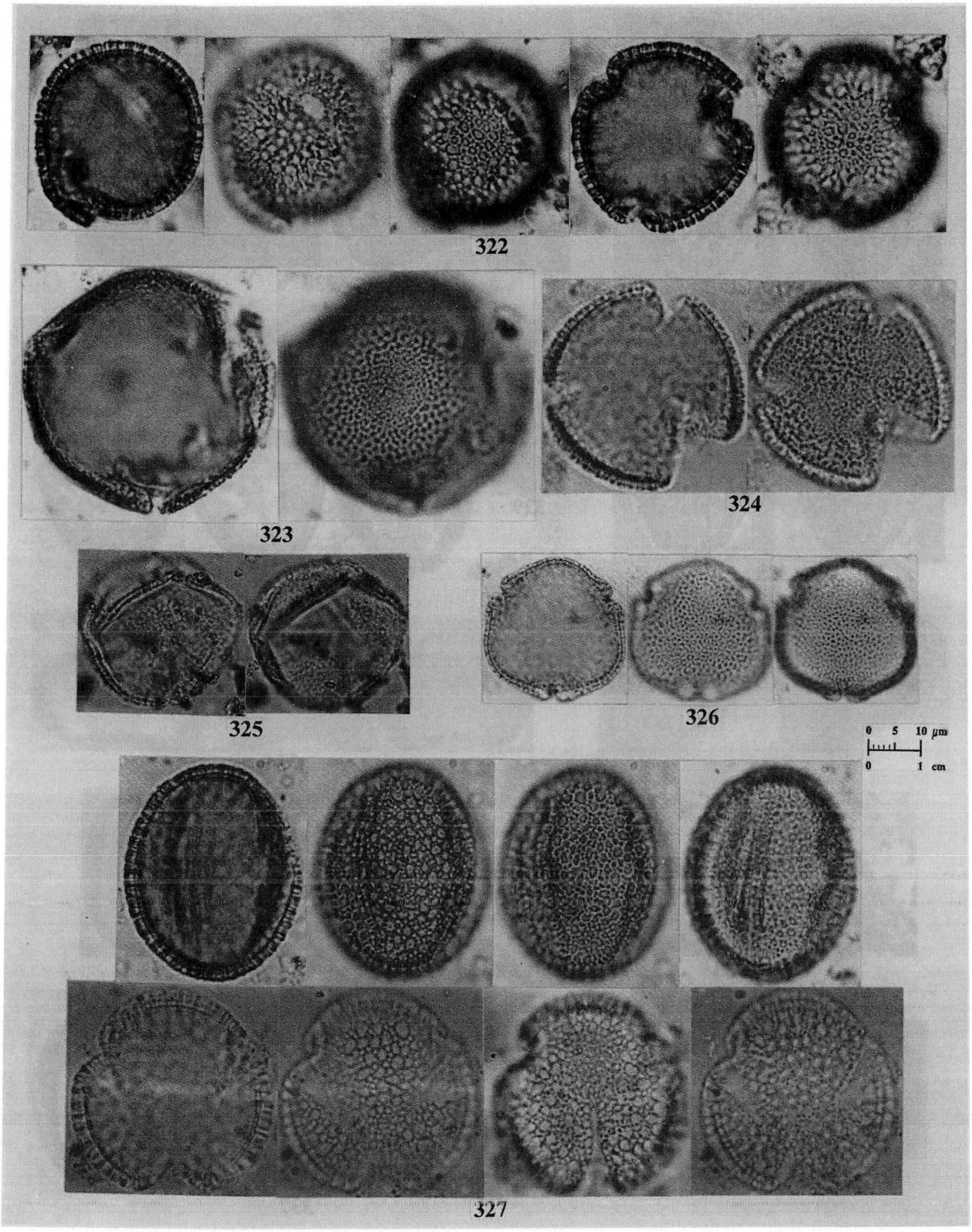

Plate/Prancha 48

Rubiaceae: *Ferdinandusa* sp. (322); *Psychotria* sp. form 1 (323); *Psychotria* sp. form 2 (324); *Psychotria* sp. form 3 (325); *Psychotria* sp. form 4 (326). **Rutaceae**: *Citrus* sp. (327).

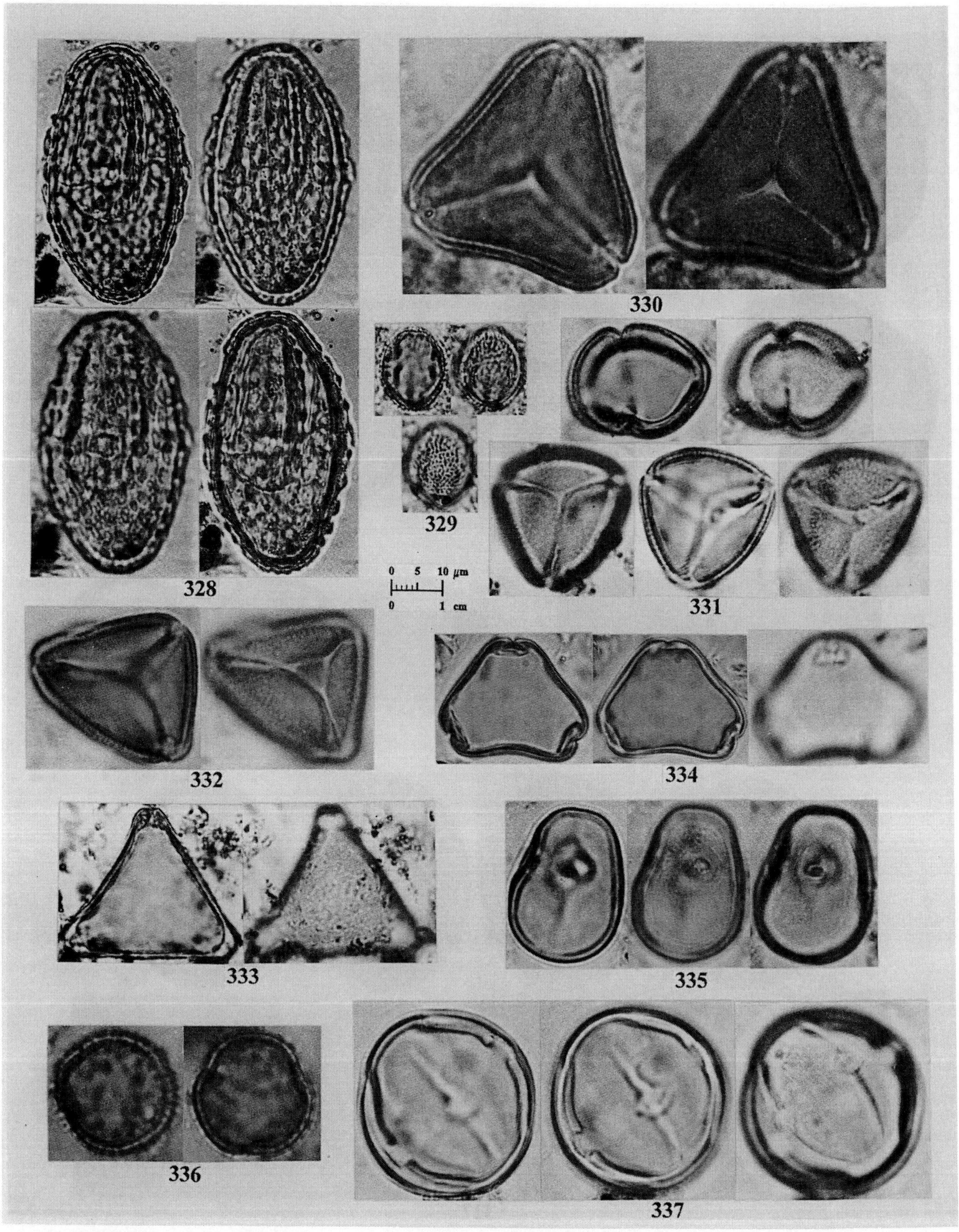

Plate/Prancha 49

Rutaceae: *Adiscanthus aff. fusciflorus* (328); *Zanthoxylum* sp. (329). **Sapindaceae**: *Cupania* sp. form 1 (330); *Cupania* sp. form 2 (331); *Matayba* sp. (332); *Paullinia* sp. (333); *Talisia* sp. (334). **Sapotaceae**: *Chrysophyllum* sp. (335); *Diploon aff. cuspidatum* (336); *Manilkara* sp. (337).

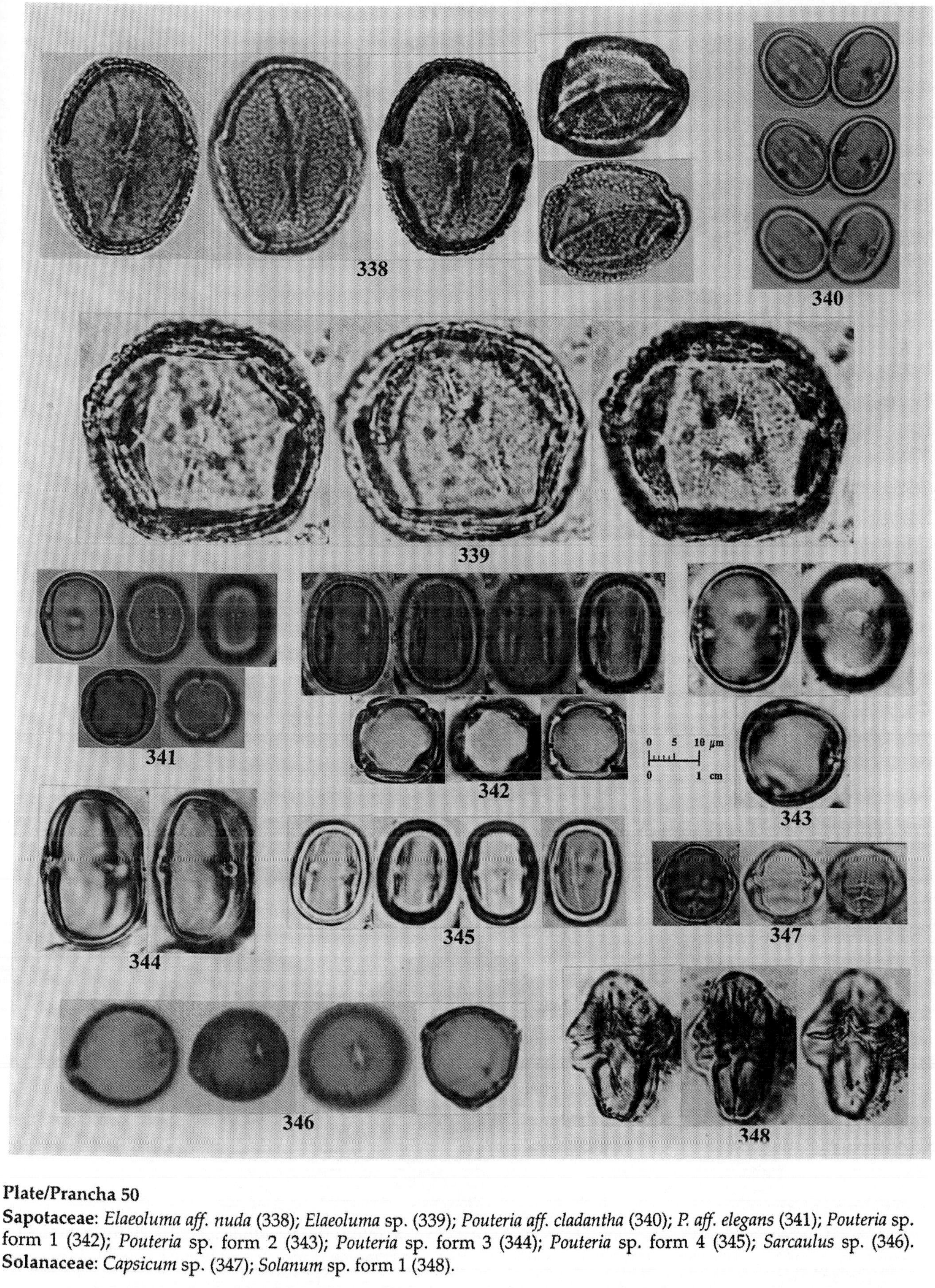

Plate/Prancha 50

Sapotaceae: *Elaeoluma aff. nuda* (338); *Elaeoluma* sp. (339); *Pouteria aff. cladantha* (340); *P. aff. elegans* (341); *Pouteria* sp. form 1 (342); *Pouteria* sp. form 2 (343); *Pouteria* sp. form 3 (344); *Pouteria* sp. form 4 (345); *Sarcaulus* sp. (346). **Solanaceae**: *Capsicum* sp. (347); *Solanum* sp. form 1 (348).

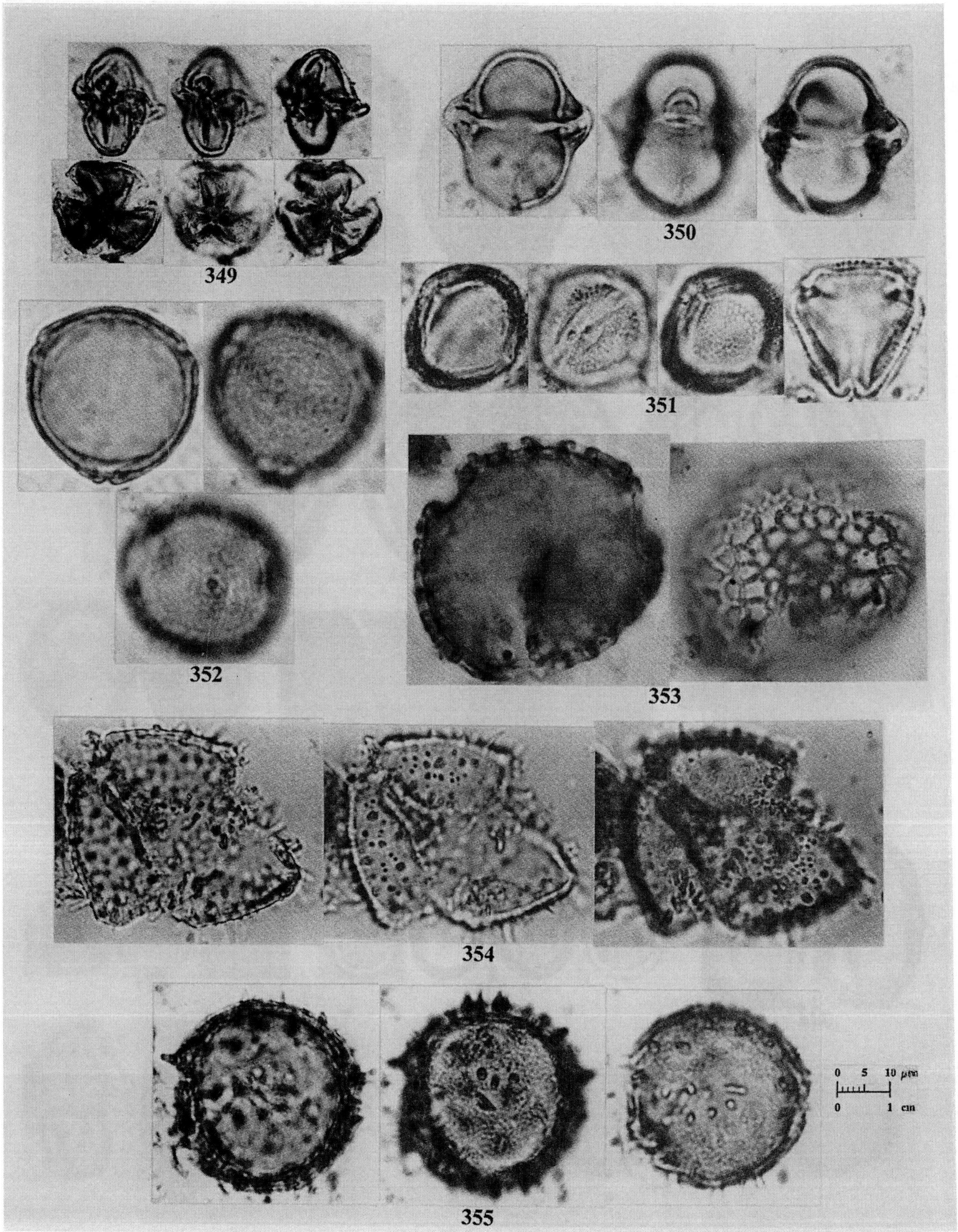

Plate/Prancha 51

Solanaceae: *Solanum* sp. form 2 (349); *Solanum* sp. form 3 (350). **Staphylleaceae**: *Turpinia* sp. (351). **Symplocaceae**: *Symplocos* sp. (352). **Theaceae**: *Archytaea aff. multiflora* (353); *Gordonia* sp. (*Laplacea* sp.) form 1 (354); *Gordonia* sp. (*Laplacea* sp.) form 2 (355).

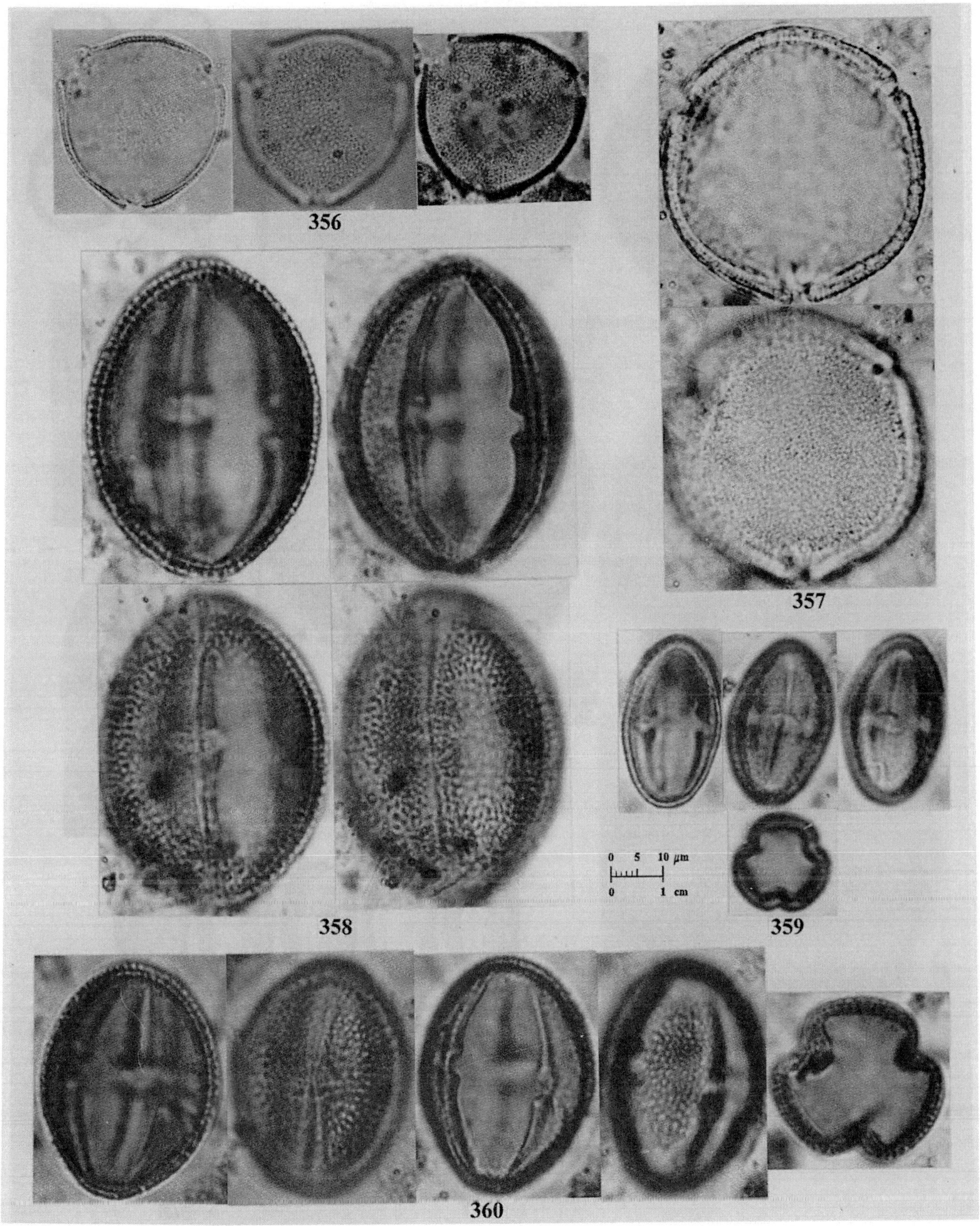

Plate/Prancha 52
Tiliaceae: *Apeiba* sp. form 1 (356); *Apeiba* sp. form 2 (357); *Heliocarpus* sp. (358); *Luehea* sp. (359); *Lueheopsis* sp. (360).

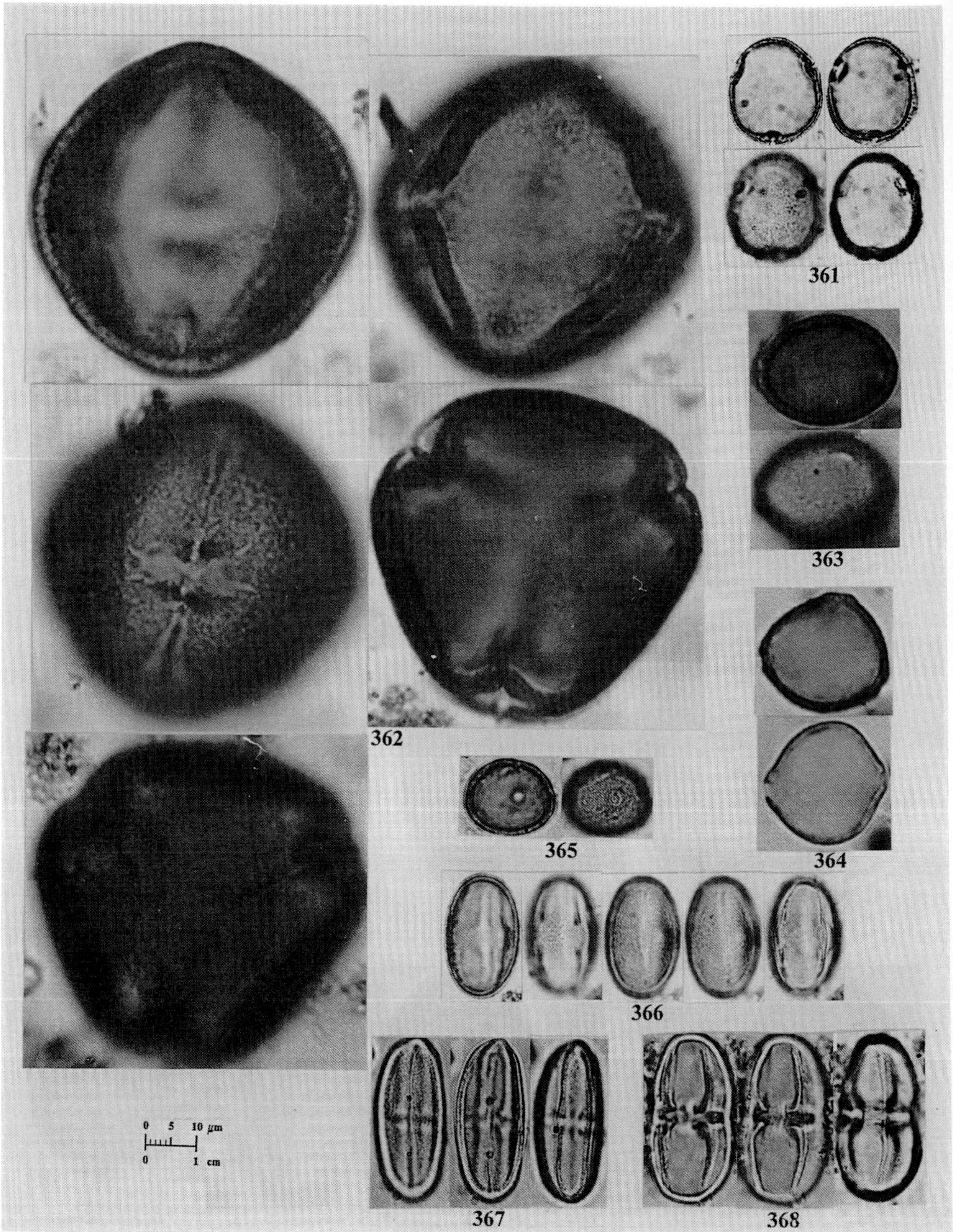

Plate/Prancha 53

Tiliaceae: *Mortoniodendron* sp. (361); Tiliaceae 1 (362). **Ulmaceae**: *Celtis* sp. form 1 (363); *Celtis* sp. form 2 (364); *Trema* sp. (365). **Umbelliferae (Apiaceae)**: *Apium* sp. (366); *Eryngium* sp. (367); *Hydrocotyle* sp. (368).

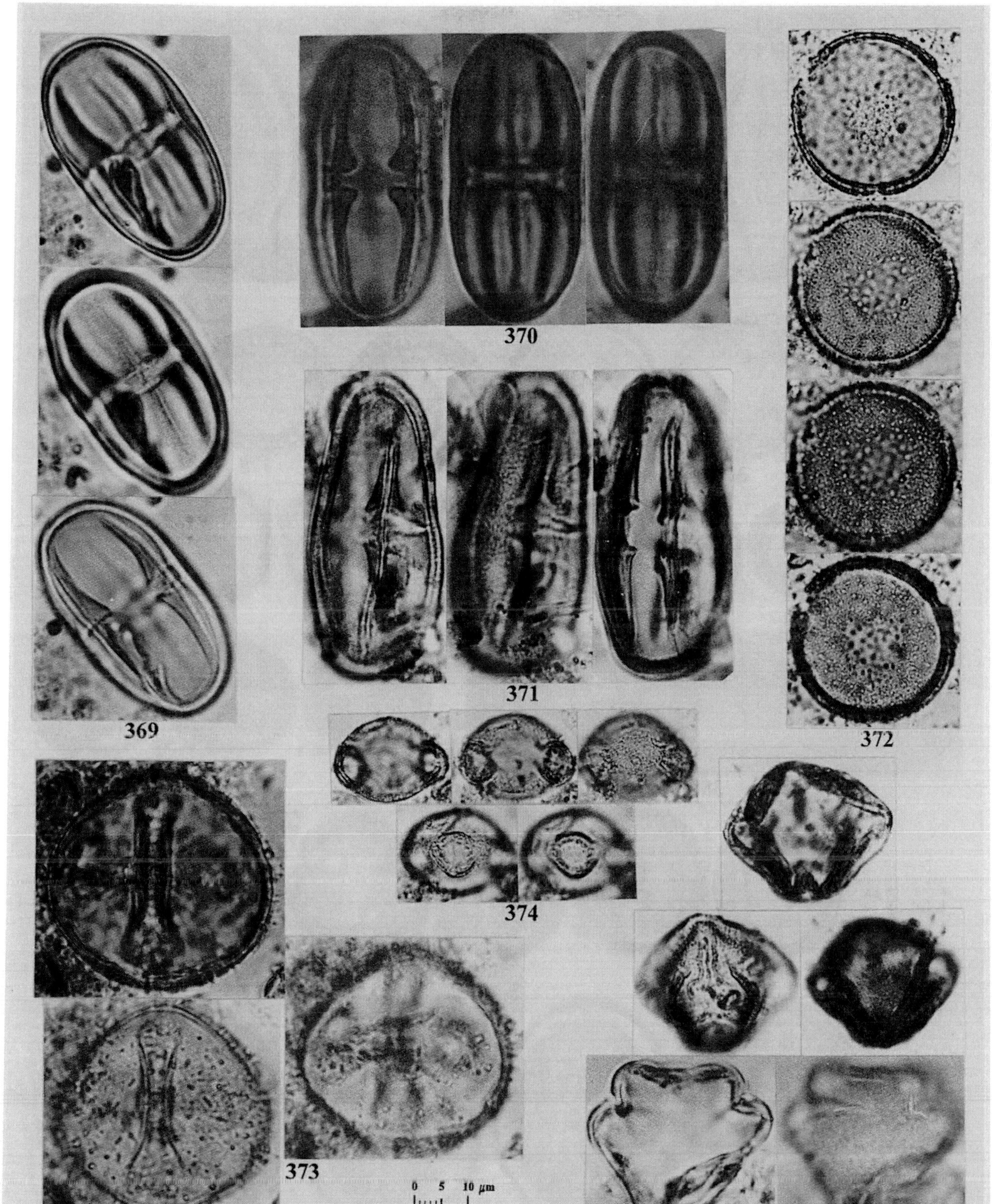

Plate/Prancha 54

Umbelliferae (Apiaceae): Umbelliferae 1 (369); Umbelliferae 2 (370); Umbelliferae 3 (371). **Verbenaceae**: *Aegiphila* sp. form 1 (372); *Aegiphila* sp. form 2 (373); *Priva* sp. (374). **Vochysiaceae**: *Vochysia* sp. (375).

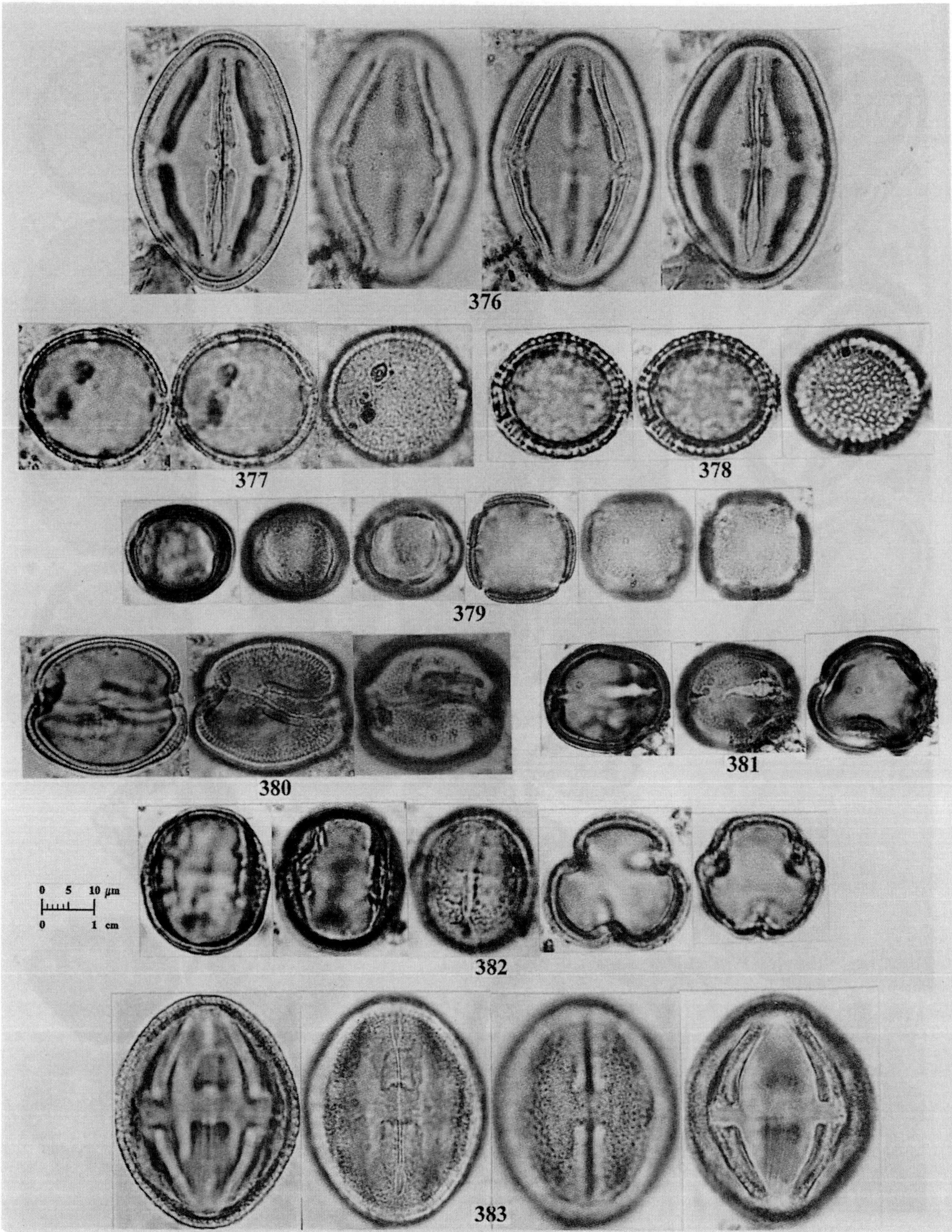

Plate/Prancha 55

Vitaceae: *Cissus* sp. (376). **Magnoliopsidae 1.** Triporate-Scabrate-28 μm (377); **Magnoliopsidae 2.** Triporate-Reticulate-27 μm (378); **Magnoliopsidae 3.** Stephanocolpate-Scabrate-21 μm (379); **Magnoliopsidae 4.** Dicolporate-Scabrate-31 μm (380); **Magnoliopsidae 5.** Tricolporate-Scabrate-24 μm (381); **Magnoliopsidae 6.** Tricolporate-Reticulate-28 μm (382); **Magnoliopsidae 7.** Tricolporate-Reticulate-44 μm (383).

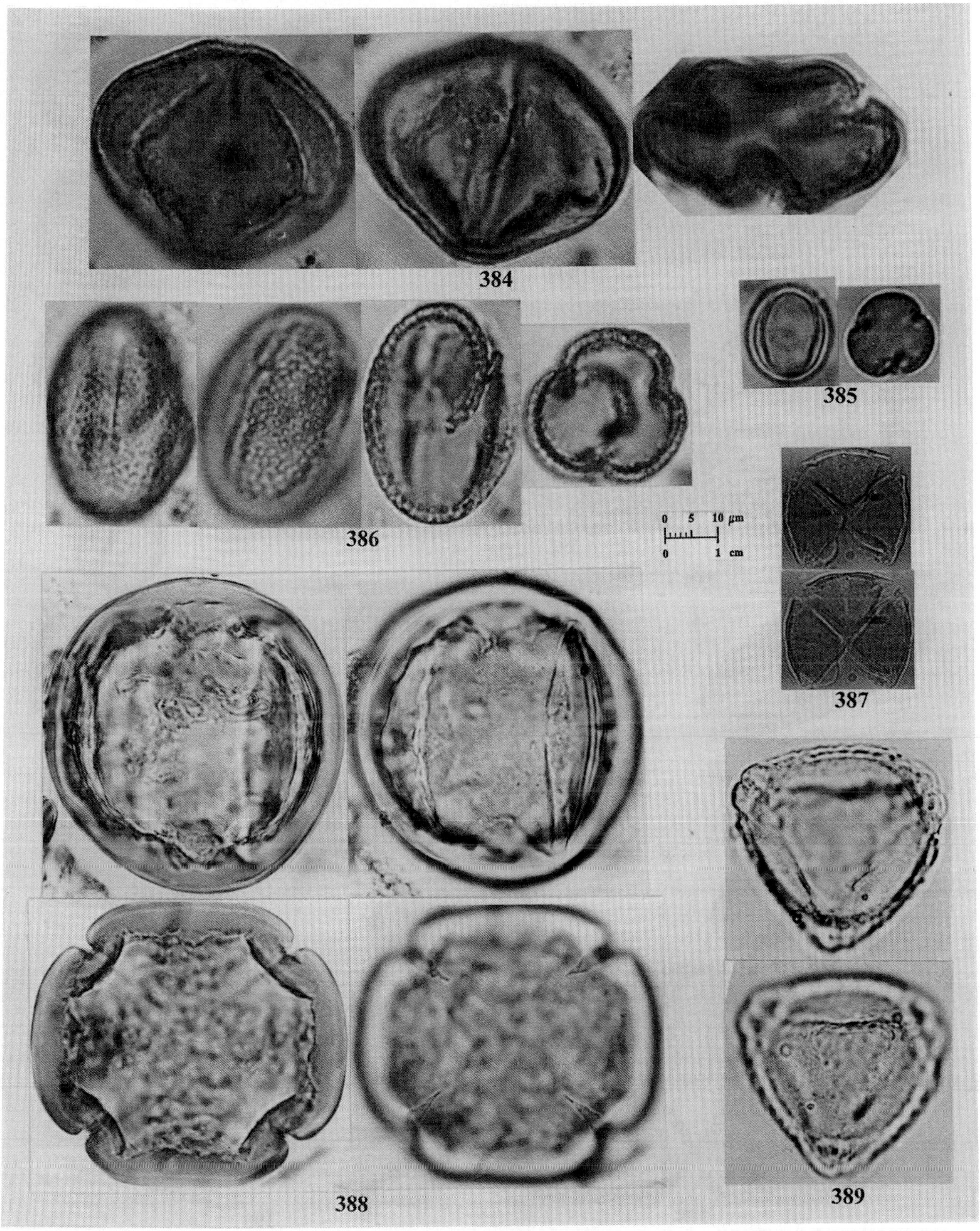

Plate/Prancha 56

Magnoliopsidae 8. Tricolporate-Psilate-52 µm (384); **Magnoliopsidae 9**. Tricolporate-Psilate-18 µm (385); **Magnoliopsidae 10**. Tricolporate-Baculate-41 µm (386); **Magnoliopsidae 11**. Heterocolpate-Psilate-24 µm (387); **Magnoliopsidae 12**. Stephanocolporate-Psilate-58 µm (388); **Magnoliopsidae 13**. Syncolporate-Verrucate-40 µm (389).

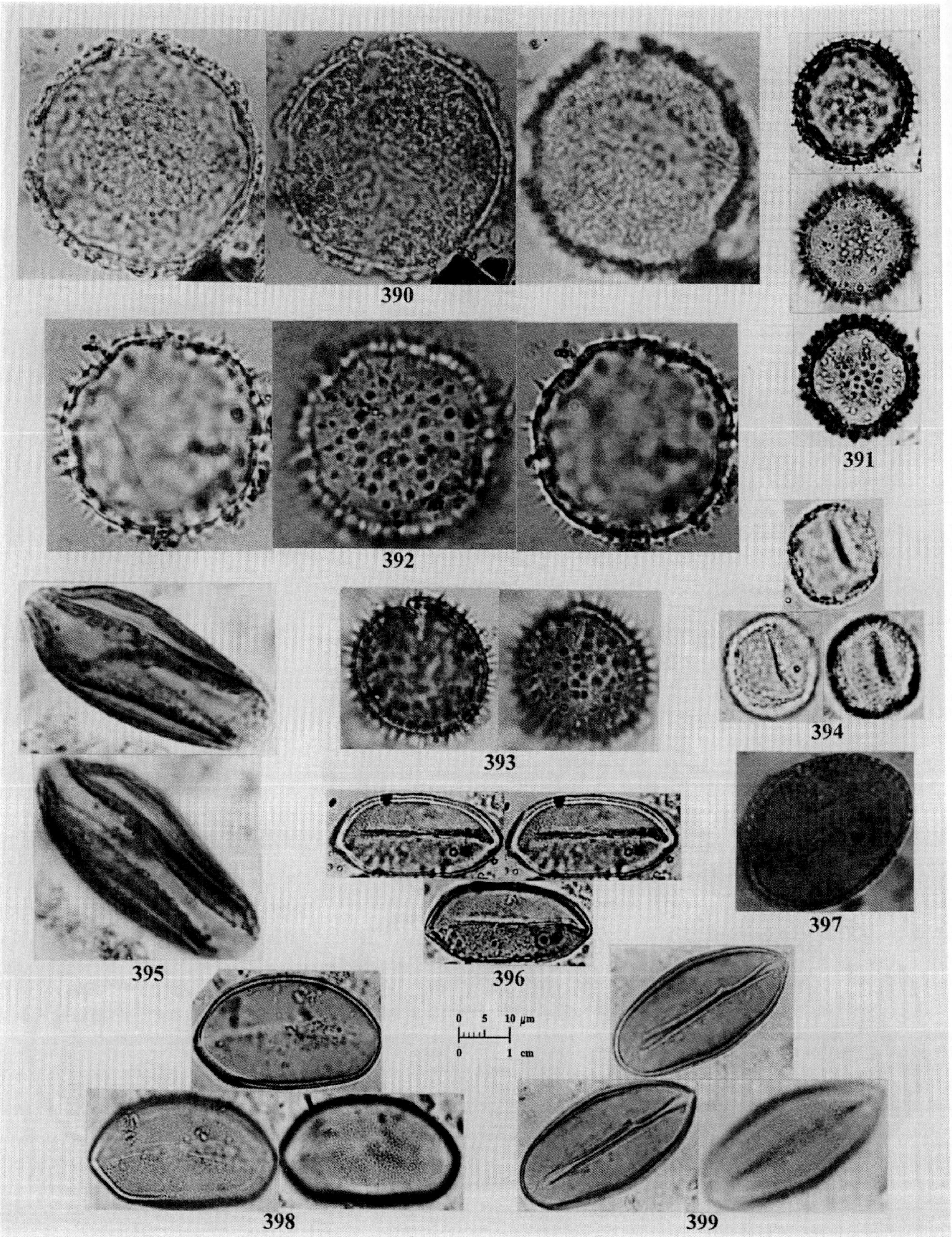

Plate/Prancha 57

ANGIOSPERMAE – LILIOPSIDAE (Monocotyledoneae). Alismataceae: *Echinodorus* sp. (390); *Sagittaria* sp. form 1 (391); *Sagittaria* sp. form 2 (392); *Sagittaria* sp. form 3 (393). **Araceae:** *Anthurium* sp. (394); *Philodendron* sp. (395). **Arecaceae (Palmae):** *Chelyocarpus* sp. (396); *Cryosophila* sp. (397); *Desmoncus* sp. (398); *Euterpe* sp. (399).

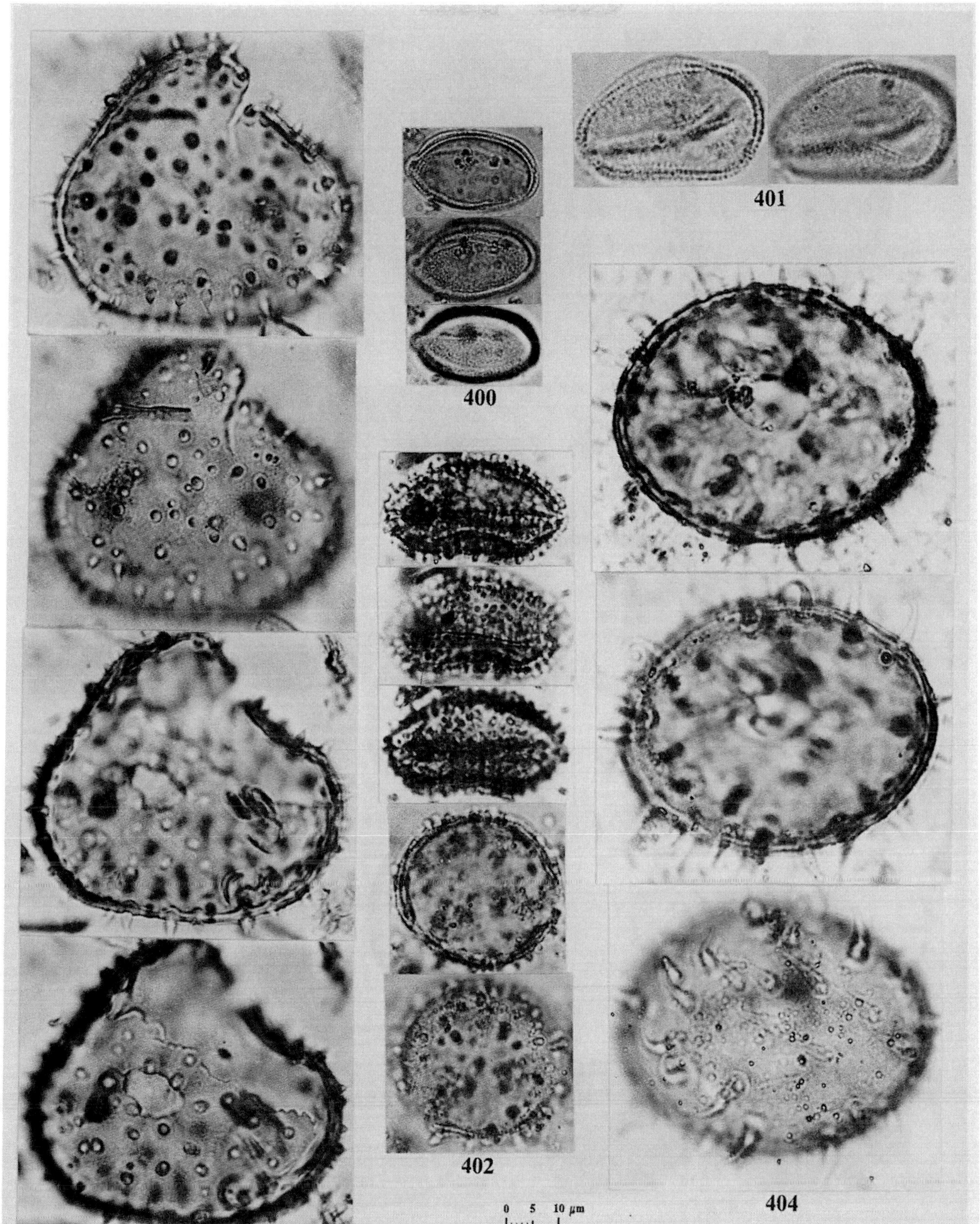

Plate/Prancha 58

Arecaceae (Palmae): *Geonoma* sp. form 1 (400); *Geonoma* sp. form 2 (401); *Iriartea aff. deltoidea* (402); *Mauritia aff. flexuosa* form 1 (403); *M. aff. flexuosa* form 2 (404).

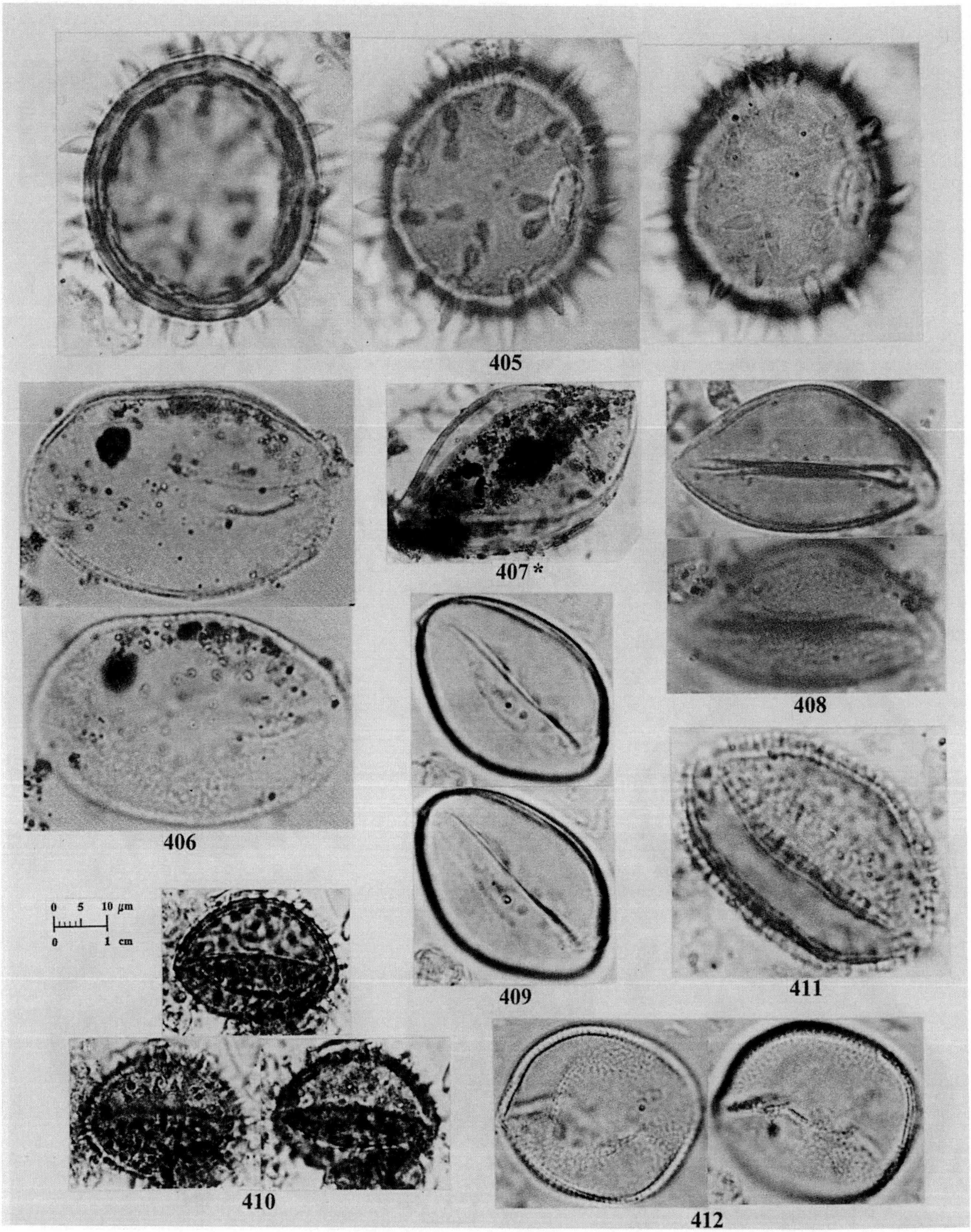

Plate/Prancha 59

Arecaceae (Palmae): *Mauritiella aff. aculeata* (405); *Oenocarpus* sp. (406); *Scheelea* sp. (407); *Syagrus* sp. form 1 (408); *Syagrus* sp. form 2 (409); Arecaceae 1 (410). **Bromeliaceae**: Bromeliaceae 1 (411). **Cyperaceae**: *Fimbrystylis* sp. (412) [*#407 is reduced to ½ scale (1 mm = 2 μm)].

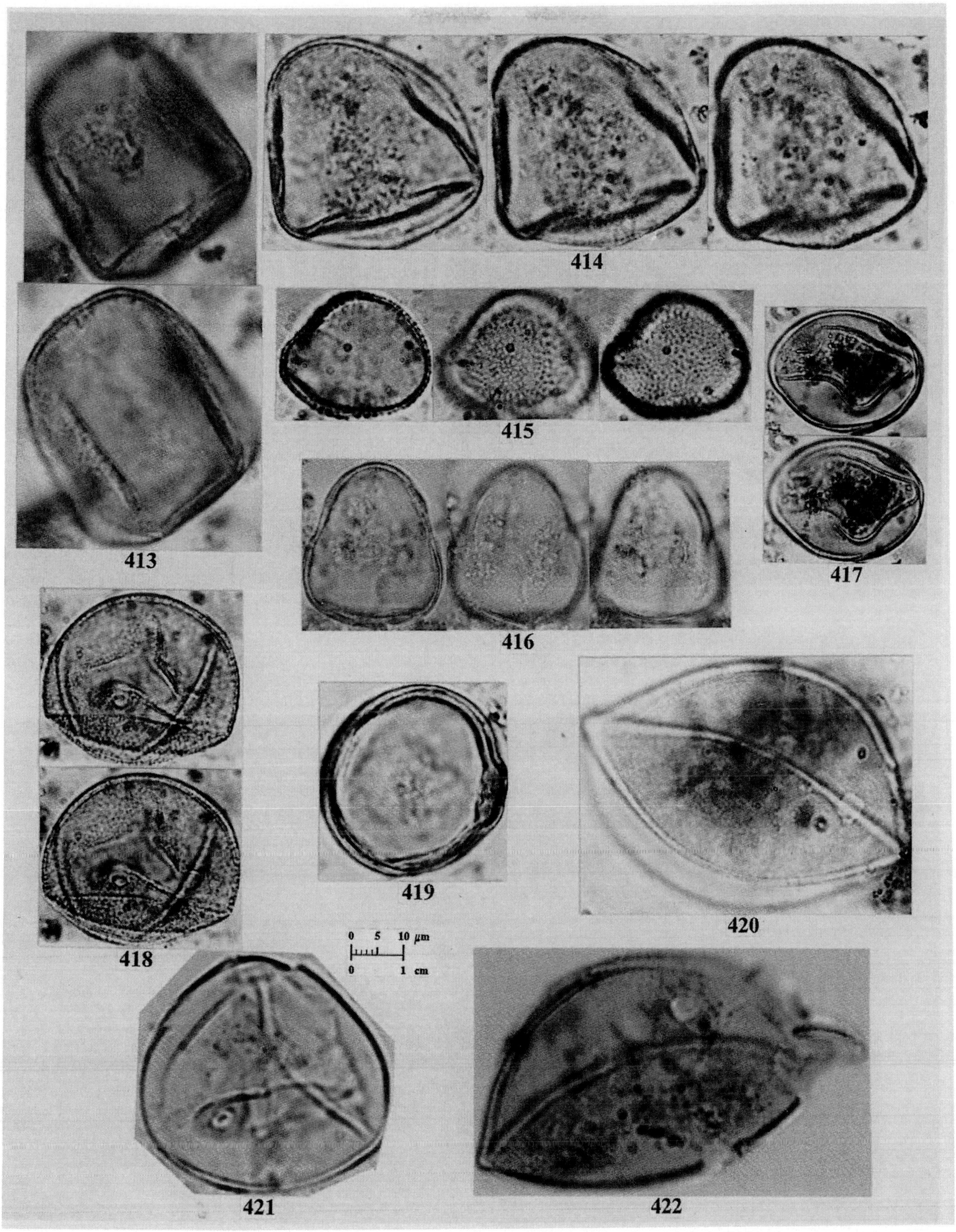

Plate/Prancha 60

Cyperaceae: *Cyperus* sp. form 1 (413); *Cyperus* sp. form 2 (414); *Rhynchospora* sp. (415); *Scleria* sp. (416). **Poaceae (Gramineae)**: Poaceae 1 (417); Poaceae 2 (418); Poaceae 3 (419); Poaceae 4 (420); Poaceae 5 (421); Poaceae 6 (422).

Glossary
Glossário

Acetolysis
Chemical technique used in the preparation of pollen, spores and other palynomorphs for palynological analyses. The acetolysis mixture, consisting of a mixture of 9 parts acetic anhydride to 1 part of sulfuric acid, removes the intine and the organic contents of the palynomorphs.
Acetólise
Técnica química usada na preparação de pólen, esporos e outros palinomorfos para análises palinológicas. A mistura de acetólise, que consiste de uma mistura de 9 partes de anidrido acético, elimina a intina e o conteúdo orgânico dos palinomorfos.

Amb
The outline of a pollen or spore grain, when seen in polar view.
Amb
Contorno de um grão de pólen ou esporo, quando visto em vista polar.

Ambitus
See **amb**.
Âmbito
Veja **amb**.

Annulus (adj. annulate)
Region of the exine around a pore possessing different thickness or ornamentation patterns as, for example, in Gramineae.
Ânulo (adj. anulado)
Região da exina, em volta do poro, que possui padrão diferente de engrossamento ou ornamentação como, por exemplo, em Gramineae.

Aperture (adj. aperturate)
Opening in the exine for growth of the pollen tube. Pores and colpi are usually the most common types of aperture in pollen grains.
Abertura (adj. aperturado)
Região da exina que permite a saída do tubo polínico, geralmente em forma de colpos, poros ou cólporos.

Apex
The tip of or corner of fossil trilete spores
Ápice
A ponta ou canto de um esporo trilete fóssil.

Apocolpial field
Described by Punt *et al.* (1994) as the region at the pole of a parasyncolpate grain, delimited by the fused colpi.
Campo apocolpial
Região de um grão parasincolpado, delimitada pela fusão dos colpos (Punt *et al.*, 1994).

Apocolpium
A region at the pole of zonocolpate or zonocolporate grain, delimited by imaginary lines conecting the apex of colpi or colpori.
Apocólpio
Diz-se da área polar de um grão zonocolpado ou zonocolporado, delimitada por linhas imaginárias que unem os ápices dos colpos ou cólporos.

Apolar
Pollen or spore grains lacking polarity
Apolar
Diz-se de grãos de pólen e esporos sem polaridade.

Arcus (pl. arcus/arci; adj. arcuate)
A thickening of the exine that extends in arch-like fashion between apertures as, for example, in *Alnus*.
Arco
Um engrossamento da exina que se estende, com forma similar a um arco, entre aberturas, como por exemplo, em *Alnus*.

Areola (pl. areolae; adj. areolate)
An ornamentation feature of the ectexine/sexine composed of circular or polygonal areas separated by grooves forming a negative reticulum as, for example, in *Phyllanthus* (Euphorbiaceae).
Aréola (pl. aréolas; adj. areolado)
Diz-se de um tipo de ornamentação na qual a ectexina/sexina é composta de áreas circulares e poligonais separadas por sulcos que formam um retículo negativo como, por exemplo, em *Phyllanthus* (Euphorbiaceae).

Aspis (pl. aspides; adj. aspidate)
Term coined by Wodehouse (1935) for a type of annulus, which is made up of a prominently protruding thickening of the exine as, for example, in *Dorstenia* (Moraceae).
Áspis (pl. áspides; adj. aspidado)
Termo proposto por Wodehouse (1935) para um tipo de ânulo, constituído de um engrossamento protuberante da exina como, por exemplo, em *Dorstenia* (Moraceae).

Atrium
Vestibulum-like cavity of a colporate grain, delimited by the separation of the inner and outer parts of the exine in the region inside the aperture.
Átrio
Cavidade similar ao vestíbulo de um grão colporado, delimitada pela separação das partes interiores e exteriores da exina, na região interior da abertura.

Baculum (pl. baculi; adj. baculate)
Sculptural elements with cylindrical shape, usually longer than 1 μm in length and width ≤1 μm.
Bácula (pl. báculas; adj. baculado)
Elementos esculturais com formato cilíndrico, usualmente com comprimento maior que 1 μm e largura ≤ 1 μm.

Bilateral (adj.)
Refers to pollen and spore grains characterized by one single plane of symmetry.
Bilateral (adj.)
Diz-se dos grãos de pólen e esporos caracterizados por um só plano principal de simetria.

Bissacate (adj.)
Pollen grain possessing two sacci or air-bladders as, for example, in *Podocarpus* (Podocarpaceae).

Bisacado (adj.)
Diz-se do grão que possue duas vesículas ou bolsas de ar como em *Podocarpus*.

Brevi (prefix)
Short
Breve (prefixo)
Curto

Bridge
Term for raised and connected margins of a colpus, in the equatorial region, forming a bridge over its exterior portion, which becomes divided into two parts (Faegri & Iversen, 1950).
Ponte
Termo usado para as margens levantadas dos colpos na região equatorial e interligadas entre si, formando uma ponte sobre a parte exterior do colpo, dividindo-o em duas partes (Faegri & Iversen, 1950).

Brochus (pl. brochi; adj. brochate)
Term defining one lumen and a half the width of the adjacent muri (Erdtman, 1952).
Brocho (adj. brochado)
Sinônimo de malha do retículo. Termo que define um lúmen e metade da largura de muros adjacentes (Erdtman, 1952). Em Português, recomenda-se o uso do termo "malha do retículo."

Cappa (pl. cappae)
In saccate or vesiculate grains, cappa refers to the thick wall found in the proximal side of the corpus.
Capa
Em grãos sacados ou vesiculados, capa refere-se à parede grossa encontrada no lado proximal do corpo.

Cappula (pl. cappulae)
Synonym of leptoma, i.e., the thin-walled of the distal side of the corpus, in vesiculate or saccate grains.
Cápula (pl. cápulas)
Sinônimo de leptoma, i.e., a parede fina do lado distal do corpus, encontrada em grãos vesiculados ou sacados.

Cicatricose (adj.)
Term used to describe the presence of scars in spores.
Cicatricose (adj.)
Diz-se dos esporos que possuem a presença de marcas que lembram cicatrizes.

Clava (pl. clavae; adj. clavate)
Column-shaped sculptural elements, in which the tip is broader than the base.
Clava (pl. clavas; adj. clavado)
Elementos esculturais com formato de colunas, onde a parte superior é mais larga que a base.

Colpus (pl. colpi; adj. colpate)
A furrow or an elongate aperture, in which the relationship of its two diameters is larger than 2:1 (Salgado-Labouriau, 1973).
Colpo (pl. colpos; adj. colpado)
Diz-se da abertura alongada onde a relação entre seus dois diâmetros é maior que 2:1 (Salgado-Labouriau, 1973).

Colporus (pl. colpori; adj. colporate)
Compound aperture made up of both a colpus and one or more pores (Punt *et al.*, 1994)
Cólporo (pl. cólporos; adj. colporado)
Abertura composta formada de um colpo e um ou mais poros (Punt *et al.*, 1994).

Colpus equatorialis (pl. colpi equatoriales)
Synonym of zonorate. The term refers to a continuous ring-like aperture around the equator of the grain.
Colpo equatorial
Sinônimo de zonorado. O termo refere-se a uma abertura contínua, em forma de anel, ao redor da região equatorial do grão.

Colpus transversalis (pl. colpi transversales)
The apertural membrane that makes up a colpus.
Colpo transversal
Diz-se da membrana apertural que constitui o colpo.

Columella (pl. columellae; adj. columellate)
Synonym of infratectum baculi. The rod-like structural elements supporting a tectum. Columellae are differentiated from baculi, by being part of the structure of the grain, whereas baculi are free-standing sculptural elements (Punt *et al.*, 1994).
Columela (pl. columelas; adj. columelado)
Sinônimo de báculo infrategilar ou báculo infratectal. Elementos estruturais em forma de bastão, que apoiam o tegilo ou tecto. Columelas são diferenciadas de báculas, por serem parte da estrutura do grão enquanto que báculas são elementos esculturais livres (Punt *et al.*, 1994).

Commissure
Term coined by Harris (1955) for the dehiscence line found in the laesurae of spores.
Comissura
Termo proposto por Harris (1955) para a linha de deiscência do esporo, encontrada na lesão do mesmo.

Contact area
Termed coined by Potonié (1934) to describe the region of the proximal face of the spore formed in contact with other tetrad members during development.
Área de contato
Termo proposto por Potonié (1934) para descrever a região da face proximal do esporo formada em contato com outros membros da tetrade, durante o seu desenvolvimento.

Corpus
In saccate or vesiculate grains, the cell corresponding to the body of the grain, i.e., the cell to which the sacci are united.
Corpo
Em grãos sacados ou vesiculados, diz se da parte correspondente ao corpo do grão, i.e., a célula onde as vesículas estão unidas.

Costa (pl. costae; adj. costate)
Term applied to a tickening of the endexine/nexine which borders an endoaperture or outlining an ectoaperture (Iversen & Troels-Smith, 1950).
Costa (adj. costado)
Termo usado para descrever um engrossamento da endexina/nexina, a qual contorna uma endo- ou ecto-abertura (Iverse & Troels-Smith, 1950).

Croton pattern
Ornamentation pattern, typically found in some genera of Euphorbiaceae, made up of five to seven raised, often triangular sexine elements, radially arranged as if forming a rosette-like structure.
Padrão Croton
Padrão de ornamentação, típicamente encontrado em alguns gêneros de Euphorbiaceae, e constituído de cinco a sete estruturas protuberantes da sexina, freqüentemente triangulares e radialmente distribuídas formando uma estrutura semelhante a uma roseta.

Decussate Tetrad
A multiplanar tetrad formed by two dyads lying across one another, and at right angles to each other (Punt *et al.*, 1994).
Tetrada decussada
Uma tétrade com vários planos de simetria, formada por duas díadas deitadas em forma de cruz, formando entre si ângulos retos (Punt *et al.*, 1994).

Demicoplus (pl. demicolpi; adj. demicolpate)
Term coined by Erdtman (1952) for an ectocolpus divided into two parts.
Demicolpo (pl. demicolpos; adj. demicolpado)
Termo proposto por Erdtman (1952) para um ectocolpo dividido em duas partes.

Dicolpate
Grain with two colpi.
Dicolpado
Diz-se do grão com dois colpos.

Dicolporate (adj.)
Grain with two compound apertures, each containing one colpus and one pore.
Dicolporado (adj.)
Diz-se do grão com duas aberturas compostas, cada uma contendo um colpo e um poro.

Diploxylonoid (adj.)
Term coined by Traverse (1988) for bisaccate pollen grains in which the outline of the sacci in polar view is discontinuous with the outline of the corpus. The diploxylonoid grain appears, therefore, to contain three distinct, more or less oval parts (Punt *et al.*, 1994). Contrast with **Haploxylonoid.**
Diploxilonóide (adj.)
Termo proposto por Traverse (1988) para grãos de pólen bisacados, nos quais o contorno das vesículas, em vista polar, é descontínua em relação ao contorno do corpo. O grão diploxilonóide parece, portanto, ser constituído de três partes mais ou menos ovais (Punt *et al.*, 1994). Contraste com **Haploxilonóide.**

Diporate (adj.)
Grain possessing two pores.
Diporado (adj.)
Diz-se do grão com dois poros.

Distal (adj.)
Term used to indicate the surface features that face outward from the center of the tetral stage of pollen or spores, even when they are no longer associated.
Distal (adj.)
Termo usado na Palinologia para indicar características da face que se encontra para fora do centro da tétrade original de grãos ou esporos, mesmo quando esses já se encontram dissociados.

Distal face
The region of the palynomorph facing outwards from the center of the tetrad, between equator and distal pole (Erdtman, 1952; Punt *et al.*, 1994).
Face distal
A região de um palinomorfo, com face oposta ao centro da tétrade, entre o equador e o pólo distal (Erdtman, 1952; Punt *et al.*, 1994).

Distal pole
The central region of the surface of the distal face.
Polo distal
A região central da superfície da face distal.

Duplibaculate (adj.)
Synonymous to duplicolumellate: presence of two rows of columellae under each murus.
Duplobaculado (adj.)
Sinônimo de duplicolumelado: presença de duas fileiras de columelas sob cada muro.

Dyads
Term for pollen and spores united in groups of two during dispersal.
Díade
Termo usado para pólen e esporos unidos em grupos de dois durante a fase de dispersão.

Echinate (adj.)
Pollen or spores possessing spines that are longer than 1 μm. Echinate should only be used in the adjectival form (Punt *et al.*, 1994). See spines.
Espinhoso (echinate) (adj.)
Diz-se dos grãos de pólen ou esporos que possuem espinhos com comprimento maior que 1 μm. Veja "spines."

Ecto (prefix)
Outer. Applied to different types of apertures, e.g., ectocolpi, ectopori. Ectoaperture is a term coined by Van Campo (1958) for an aperture found in the outer layer of the sporoderm.
Ecto (prefixo)
Exterior. Aplicado a diferentes tipos de aberturas, e.g., ectocolpo, ectoporo. Ectoabertura é um termo proposto por Van Campo (1958) para uma abertura encontrada na camada exterior da esporoderma.

Endo (prefix)
Inner. Endoaperture is a term coined by Van Campo (1958) for an aperture found in the inner layer of the sporoderm, or the inner aperture of a compound aperture (Punt *et al.*, 1994). Examples: endocolpi, endopori.
Endo (prefixo)
Interior. Endoabertura é um termo proposto por Van Campo (1958) para uma abertura encontrada na camada interior da esporoderma, ou para a abertura interior de uma abertura composta (Punt *et al.*, 1994). Exemplos: endocolpo, endoporo.

Equator
An imaginary line dividing the equatorial plane and passing through the perpendicular axis of a grain. In bilateral palynomorphs the equator is usually the longest line (Punt *et al.*, 1994).
Equador
Uma linha imaginária dividindo o plano equatorial e passando pelo eixo perpendicular de um grão, enquanto que em palinomorfos bilaterais (somente um plano de simetria), o equador é a linha mais extensa (Punt *et al.*, 1994).

Exine
The outermost layer of the cell wall of a pollen or spore grain made up of sporopollenin, an organic compound highly resistant to strong acids and bases.
Exina
A camada mais exterior da parede celular de um grão de pólen ou esporo, constituída de esporopolenina, um composto orgânico altamente resistente a ácidos e bases fortes.

Exitus
Synonym of aperture. Term coinded by Wodehouse (1935) for the exact site of exit of the pollen tube.
"Exitus"
Sinônimo de abertura. Termo proposto por Wodehouse (1935) para o exato local de saída do tubo polínico. Derivado da palavra inglêsa "exit," saída.

Exitus digitatus
A germinal aperture through which a pollen tube emerges.
"Exitus digitatus"
Abertura, através da qual o tubo polínico emerge.

Fenestrate (adj.)
Describes the ornamentation pattern of grains, in which large window-like spaces lack a tectum. The term has been used primarily to define a group of Compositae, which include lophate grains. See **lophate (adj.)**.
Fenestrado (adj.)
Termo usado para descrever um padrão de ornamentação, no qual espaços semelhantes a janelas carecem de tecto (tegilo). Este termo tem sido usado principalmente para definir um grupo de Compositae, que inclui grãos lofados. Veja: lofado (adj.).

Fossula (pl. fossulae; adj. fossulate)
Term coined by Faegri & Iversen (1950) for elongated and irregular grooves on the grain surface.
Fóssula (adj. fossulado)
Termo proposto por Faegri & Iversen (1950) para sulcos elongados e irregulares na superfície do grão.

Foveola (pl. foveolae; adj. foveolate)
A feature of ornamentation consisting of more or less rounded depressions or lumina more than 1 μm in diameter. The distance between foveolae is greater than their breadth (Erdtman, 1952).
Fovéola (pl. fovéolas; adj. foveolado)
Uma característica da ornamentação constituída de depressões mais ou menos arredondadas ou lumina maiores que 1 μm de diâmetro. A distância entre fovéolas é maior que a sua largura (Erdtman, 1952).

Gemma (pl. gemmae; adj. gemmate)
Raised and somewhat rounded ornamentation element, somewhat constricted at its base. The diameter of the base is equal or bigger than of that of its upper part (>1 μm).
Gema (adj. gemado)
Tipo de elemento de ornamentação protuberante e levemente arredondado, e mais ou menos constrito em sua base. O diâmetro da base é igual ou maior do que o da sua altura, que é maior que 1 μm.

Granulate (adj.)
Exine ornamentation comprised of small round granule-like elements (granulum, pl. granula) smaller than 1 μm.
Granulado (adj.)
Ornamentação da exina, constituída de pequenos elementos arredondados (grânulos) e menores que 1 μm de diâmetro.

"H"-shape
An endoaperture consisting of one central and two lateral, longitudinal elongations, resembling letter "H".
Forma "H"
Uma endoabertura constituída de uma elongação central, conectada a duas extensões longitudinais, formando uma estrutura parecida com a letra "H."

Haploxylonoid
Term proposed by Traverse (1988) describing the polar view of bisaccate pollen grains in which the outline of the sacci is more or less continuous with the outline of the corpus, thus giving the grain a more or less ellipsoidal form.
Haploxilonóide
Termo proposto por Traverse (1988) que descreve a vista polar de grãos de pólen bisacados, nos quais o contorno das vesículas é mais ou menos contínuo com o contorno do corpo, dando assim ao grão uma forma mais ou menos elíptica.

Heterobrochate (adj.)
Reticulate ornamentation pattern in which the brochi of the reticulum have different sizes.
Malhas heterogêneas do retículo
Padrão de ornamentação reticulada onde as malhas do retículo são de diferentes tamanhos.

Heterocolpate (adj.)
Describing pollen grains with both simple and compound colpi (colpus and pore) (Iversen & Troels-Smith, 1950) as, for example, in Melastomataceae.
Heterocolpadô (adj.)
Termo proposto por Iversen & Troels-Smith (1950) para os grãos de pólen que possuem ambos colpos simples e compostos (cólporo) como, por exemplo, em Melastomataceae.

Heteropolar
Describing grains possessing dissimilar poles, in relation to ornamentation or apertural pattern.
Heteropolar
Diz-se do grão que possui um polo diferente do outro, tanto no que se refere à ornamentação como ao sistema apertural.

Homobrochate (adj.)
Reticulate ornamentation pattern in which the brochi of the reticulum have somewhat similar sizes.
Malhas homogêneas do retículo (homobrochado)
Padrão de ornamentação reticulada onde as malhas do retículo são de tamanhos similares.

Inaperturate (adj.)
Describing a grain lacking apertures. Synonym of **ulcerate**, i.e., grains lacking well-organized pores or colpi, but possessing thinned or partially broken exine as, for example, in Cyperaceae (Traverse, 1988).
Inaperturado (adj.)
Diz-se do grão que não possui aberturas. Sinônimo de ulcerado, que é usado para grãos desprovidos de poros ou colpos organizados, mas que possuem áreas delgadas ou parcialmente quebradas da exina como, por exemplo, em Cyperaceae (Traverse, 1988).

Intectate (adj.)
Describing pollen grains lacking a tectum.
Intectado (adj.)
Diz-se do grão de pólen desprovido de tecto ou tegilo.

Isopolar (adj.)
Describing a pollen grain in which the proximal and distal faces of the exine are alike (Erdtman, 1947).
Isopolar (adj.)
Diz-se do grão de pólen, no qual as faces proximal e distal da exina são semelhantes (Erdtman, 1947).

Labrum (pl. labra; adj. labrate)
An unthickened extension of the exine at the pores.
Labrum (adj. labrado)
Uma extensão da exina, desprovida de engrossamento, nos poros.

Lacuna (pl. lacunae; adj. lacunate)
The depressed area surrounded by ridges in lophate grains.
Lacuna (adj. lacunado)
Área rebaixada e circundada por cristas, em grãos lofados.

Laesura (pl. laesurae; suffix-lete)
The arm of a proximal fissure or scar of a spore (Erdtman, 1946). In case of monolete spores, there is one laesura, though some palynologists consider the trilete condition a single triradiate laesura (Punt *et al.*, 1994).
Lesão (sufixo-lete)
O braço ou a fissura proximal ou lesão de um esporo (Erdtman, 1946). No caso de esporos monoletes, existe uma só lesão, embora alguns palinólogos considerem a condição trilete como uma única lesão tri-radiada (Punt *et al.*, 1994).

Lalongate (adj.)
Endoaperture transversally elongated.
Lalongado (adj.)
Endoabertura alongada transversalmente.

Lolongate (adj.)
Describing the shape of an endoaperture longitudinally elongated.
Lolongado (adj.)
Diz-se de uma endoabertura alongada transversalmente.

Lophate (adj.)
Term describing an ornamentation type in which the outer exine is raised forming a patttern of ridges (lophae) surrounding the depressions (lacunae).
Lofado (adj.)
Termo que descreve um tipo de ornamentação no qual a camada mais exterior da exina está erguida e formando um padrão de cristas (lofas) circundando depressões (lacunas).

Lumen (pl. lumina)
The space between the muri of a reticulum.
Lúmen
O espaço entre os muros de um retículo.

Margo (pl. margines; adj. marginate)
Area of exine around an ectocolpus differentiated from the rest of the exine, either in ornamentation or by difference in thickness (Punt *et al.*, 1994).
Margem
Área da exina em volta de um ectocolpo, diferenciada do resto da exina, tanto pela ornamentação como pela diferença em espessura.

Membrane
The exine (membrane) that forms the floor of an ectoaperture (Erdtman, 1952) as, for example, the colpus membranes or the pore membrane. The membrane may be partially or completely covered by an operculum (Punt *et al.*, 1994).
Membrana
A exina que forma o "chão" de uma ectoabertura (Erdtman, 1952) como, por exemplo, as membranas de colpos e poros. A membrana pode estar parcialmente ou completamente coberta por um opérculo (Punt *et al.*, 1994).

Mesocolpium (pl. mesocolpia)
The area of a pollen grain surface delimited by two adjacent colpi and by perdendicular lines, uniting the apex of these colpi.
Mesocólpio (mesocolpo)
A área, da superfície de um grão de pólen, delimitada por dois colpos adjacentes e por linhas perpendiculares imaginárias, unindo o ápice desses colpos.

Mesoporium
The area of a pollen grain surface delimited by lines between the margins of adjacent pores.
Mesopório
A área da superfície de um grão de pólen delimitada por linhas entre as margens de poros adjacentes.

Micron (pl. microns, micra)
A unit of length equal to one millionth (10^{-6}) of a meter and represented by "μm."
Micron (pl. micra)
Uma unidade igual a um milionésimo (10^{-6}) de um metro e representado por "μm."

Monad
A pollen or a spore grain dispersed as single unit as opposed to dyads, tetrads and polyads.
Mônade
Um grão de pólen ou esporo disperso como uma única unidade, em oposição a díades, tétrades e políades.

Monolete
Spore with only one laesura located on the proximal pole.
Monolete
Esporo contendo somente uma lesão, situada no pólo proximal.

Murus (pl. muri)
A ridge separating the lumina in a reticulate ornamentation.
Muro (adj. murado)
Uma crista separando o lúmen em uma ornamentação reticulada.

Nexine
The inner, non-sculptured part of the exine which lies below the sexine (Erdtman, 1952).
Nexina
A parte interior da exina, desprovida de elementos esculturais, que está sob a sexina (Erdtman, 1952).

Oblate (adj.)
Describing the shape of a pollen grain or spore in which the polar axis is shorter than the equatorial diameter (Erdtman, 1943). The ratio between the polar axis (P) and the equatorial diameter (E) is 0.50–0.75 (Erdtman, 1952).
Oblato (adj.)
Diz-se do grão de pólen ou esporo no qual o eixo polar é mais curto que o diâmetro equatorial (Erdtman, 1943). A relação entre o eixo polar (P) e o diâmetro equatorial (E) é de 0.50–0.75 (Erdtman, 1952).

Oblate-spheroidal (adj.)
Describing the shape of a pollen grain or spore in which the ratio between the polar axis and the equatorial diameter is 0.88–1.00 (Erdtman, 1952).
Oblato-esferoidal (adj.)
Diz-se do formato do grão de pólen ou esporo, no qual a relação entre o eixo polar e o diâmetro equatorial está entre 0.88 e 1.00 (Erdtman, 1952).

Operculum (pl. opercula; adj. operculate)
A thickening of the apertural membrane (ectexine/sexine), having a somewhat circular shape when covering pores and elongate when covering colpi.
Opérculo (adj. operculate)
Um espessamento da membrana apertural (ectexina/sexina), tendo um formato mais ou menos circular quando cobrindo poros e elongada quando cobrindo colpos.

Ornamentation
Term used for the description of the organisation of features on the surface of pollen and spores (Potonié, 1934).
Ornamentação
Termo usado para a descrição da organização dos elementos esculturais da superfície dos grãos de pólen e esporos (Potonié, 1934).

Parasyncolpate (adj.)
Describing syncolp(or)ate grains in which the apices of colpi divide into branches and anastomose towards the poles.
Parasincolpado (adj.)
Diz-se dos grãos sincolpados ou sincolporados nos quais os ápices dos colpos dividem-se em ramificações anastomosadas em direção aos pólos.

Per- (prefix)
In Palynology this prefix is used to indicate "complete" or "very" (Punt *et al.*, 1994).
Per- (prefixo)
Em Palinologia este prefixo é usado para indicar "completo" ou "muito" (Punt *et al.*, 1994).

Perforate (adj.)
Describing the presence of holes in the tectum, usually less than 1 µm in diameter.
Perforado (adj.)
Diz-se do grão cujo tecto é provido de pequenos orifícios, geralmente menores que 1 µm de diâmetro.

Peri- (prefix)
Synonym of "panto" (examples: periporate, pericolporado), indicating a global distribution (Punt *et al.*, 1994).
Peri- (prefixo)
Sinônimo do prefixo "panto" (exemplos: periporado, pericolporado) indicando uma distribuição global (Punt *et al.*, 1994).

Perine
Term coined by Erdtman (1943) describing the sporoderm layer that is not always resistant to acetolysis and is found around the exine of many spores.
Perina
Termo proposto por Erdtman (1943) que descreve a camada de esporoderma que não é sempre resistente à acetólise e que é encontrada em volta da exina de muitos esporos.

Peroblate (adj.)
Describing the shape of a pollen or spore grain when the ratio between the polar axis and the equatorial diameter is less than 0.50 (Erdtman, 1943).
Peroblato (adj.)
Diz-se do formato de um grão de pólen ou esporo, quando a relação entre o eixo polar e o diâmetro equatorial é menor que 0.50 (Erdtman, 1943).

Perprolate (adj.)
Describing the shape of a pollen or spore grain when the ratio between the polar axis and the equatorial diameter is more than 2 (Erdtman, 1943).
Perprolato (adj.)
Diz-se do formato de um grão de pólen ou esporo, quando a relação entre o eixo polar e o diâmetro equatorial é maior que 2.0 (Erdtman, 1943).

Pilum (pl. pili, adj. pilate)
Small rod-like processes of the sexine, made up of a somewhat thickened top.
Pilo (adj. pilado)
Pequenos processos cilíndricos da sexina, constituídos de uma cabeça mais espessa que o corpo.

Pole
The two extremities of the polar axis (Wodehouse, 1935). The area of the grain, still in the tetrad stage, towards its center and the area diametrically oposed and outwards. The poles can be recognized in zonoaperturate grains of the dicots as the opposed areas lacking apertures. In the monocots, the distal pole is where the aperture is located, and the proximal pole, is the area diametrically opposed to it (Salgado-Labouriau, 1973).
Pólo
As duas extremidades de um eixo polar (Wodehouse, 1935). A área do grão, ainda no estágio de tétrade, voltada para o centro e a área diametralmente oposta, voltada para fora. Os pólos podem ser reconhecidos em grãos zonoaperturados das Dicotiledôneas como sendo as áreas opostas desprovidas de aberturas. Nas Monocotiledôneas o pólo distal é onde se localiza a abertura, e o pólo proximal é a área diametralmente oposta (Salgado-Labouriau, 1973).

Polyad
A dispersal unit of polen grains united in groups of more than four grains.
Políade
Uma unidade de dispersão, de grãos de pólen unidos em grupos de mais de quatro grãos.

Pore (adj. porate)
General term for a circular or elliptic aperture, where the length/breadth ratio is less than 2:1.
Poro (adj. porado)
Termo geral para uma abertura circular ou elíptica, onde a relação comprimento/largura é menor que 2:1.

Prolate (adj.)
Describing the shape of a pollen grain or spore in which the polar axis is larger than the equatorial diameter (Erdtman, 1943). The ratio between the polar axis and the equatorial diameter is between 1.33 and 2.00 (Erdtman, 1952).
Prolado (adj.)
Diz-se do formato de um grão de pólen ou esporo, no qual o eixo polar é maior que o diâmetro equatorial (Erdtman, 1943). A relação entre o eixo polar e o diâmetro equatorial está entre 1.33 e 2.00 (Erdtman, 1952).

Prolate-spheroidal (adj.)
Describing the shape of a pollen grain or spore in which the ratio between the polar axis and the equatorial diameter is 1.00–1:14 (Erdtman, 1952).
Prolado-esferoidal (adj.)
Diz-se do formato do grão de pólen ou esporo no qual a relação entre o eixo polar e o diâmetro equatorial é 1.00–1:14 (Erdtman, 1952).

Proximal (adj.)
Term applied to the features on the surface that faces toward the center of the tetrad during its development (Wodehouse, 1935).
Proximal (adj.)
Termo empregado para as características da superfície voltadas para o centro da tétrade, durante o seu desenvolvimento (Wodehouse, 1935).

Pseudoaperture
A thinning of the exine resembling an aperture. Examples: pseudocolpi, pseudopori, etc.
Pseudoabertura
Um afinamento da exina que lembra uma abertura. Exemplos: pseudocolpos, pseudoporos, etc.

Psilate (adj.)
Describing a sexine lacking ornamentation.
Psilado (adj.)
Diz-se da sexina desprovida de ornamentação.

Radially symmetric
Pollen grains or spores with two or more vertical planes of symmetry.
Radialmente simétrico
Grãos de pólen e esporos que possuem um ou mais planos verticais de simetria.

Reticulum (pl. reticula; adj. reticulate)
Group of lumina or other spaces wider than 1 μm bordered by elements narrower than the lumina (Praglowski & Punt, 1973).
Retículo (adj. reticulado)
Conjunto de lúmens ou outros espaços, mais largos que 1 μm, rodeados por elementos mais estreitos que os lúmens (Praglowski & Punt, 1973).

Rhomboidal tetrad
Type of tetrad in which the proximal faces of two individual members are in direct contact and the remaining two are separated (Erdtman, 1945).
Tétrade rombóide
Tipo de tétrade na qual as faces proximais de dois membros individuais estão em contato direto e as outras duas estão separadas (Erdtman, 1945)

Rugulate (adj.)
A type of ornamentation made up of elongated sexine elements more than 1 μm long. The rugulate pattern is intermediary between the striate and reticulate conditions (Iversen & Troels-Smith, 1950).

Rugulado(adj.)
Um tipo de ornamentação constituída de elementos elongados da sexina com comprimento maior que 1 μm. O padrão rugulado é intermediário entre a ornamentação estriada e reticulada (Iversen & Troels-Smith, 1950).

Saccus (pl. sacci; adj. saccate)
An expanded air-bladder formed by a local separation within the exine of a pollen grain (Erdtman, 1952). Found in some gymnospermous pollen grains as, for example, *Podocarpus*.
Saco (adj. sacado)
Uma vesícula expandida e cheia de ar, formada por uma separação local dentro da exina de um grão de pólen (Erdtman, 1952). Encontrado em alguns grãos de pólen de gimnospermas como, por exemplo, *Podocarpus*.

Scabrate (adj.)
Describing an ornamentation pattern composed of granules of any shape but with diameters smaller than 1 μm.
Escabrado (adj.)
Diz-se do padrão de ornamentação constituído de grânulos de qualquer tamanho mas com diâmetros inferiores a 1 μm. Os grânulos recebem o nome de escabras.

Sclerine
Sporoderm lacking intine.
Esclerina
Esporoderma desprovida de intina.

Sculpture
The surface relief, or topography, of a pollen grain or spore (Potonié, 1934).
Escultura
O relevo ou a topografia da superfície de um grão de pólen ou esporo (Potonié, 1934).

Semitectate (adj.)
Describing a pollen grain with a discontinous tectum.
Semitectado (adj.)
Diz-se do grão de pólen cujo teto é discontínuo.

Sexine
The outermost and the sculptured layer of the exine, just above the nexine.
Sexina
A camada mais exterior da exina e provida de esculturas, logo acima da nexina.

Simplibaculate (adj.)
Synonym of simplicolumellate (Reitsma, 1970). Describing a pollen grain possessing a single row of columellae under the muri.
Simplibaculado (adj.)
Sinônimo de simplicolumelado (Reitsma, 1970). Diz-se do grão de pólen que possui uma simples fileira de columelas sob cada muro.

Spheroidal (adj.)
Describing the shape of a pollen grain or spore in which the polar axis (P) and the equatorial diameter (E) are approximatelly equal and their ratio is 0.88–1.14 (Erdtman, 1952).
Esferoidal (adj.)
Diz-se do formato do grão de pólen ou esporo, no qual o eixo polar (P) e o diâmetro equatorial (E) são aproximadamente iguais e a relação P/E varia de 0.88 a 1.14.

Spine (adj. spiny/spinose)
Long and tapering pointed sculptural elements, which are larger than 1 μm. Erdtman (1952) proposed the term spines when larger than 3 μm, **spinule** for those smaller than 3 μm and **microechinate** for spines that are not larger than 1μm.
Espinhos (adj. espinhoso)
Elementos esculturais longos e pontiagudos, maiores que 1 μm. Erdtman (1952) propôs o uso do termo "**spine**" (**espinho**) quando estes são maiores que 3 μm e "**spinule**" (**espículo**) para espinhos menores que 3 μm, e o termo "**microechinate**" (**microespiculado**) quando os espinhos não forem maiores que 1 μm.

Spinule (adj. spinulose)
See **Spine**.
Espículo (adj. espiculoso)
Veja **Espinhos**.

Spore
The microscopic unicellular and haploid reproductive unit of lower plants. In palynological preparations algal spores (resting spores) and fungal spores are also commonly found.
Esporo
Uma unidade reprodutiva haplóide, unicelular e microscópica das plantas inferiores. Em preparações palinológicas esporos (estágios de dormência) de algas e esporos de fungos são também normalmente encontrados.

Stephano (prefix)
Synonym of zono- (Erdtman & Straka, 1961). Stephano- (Faegri & Iversen, 1950) means apertures or any other features located on the equator of the grain.
Estéfano (prefixo)
Sinônimo de zono- (Erdtman & Straka, 1961). Estéfano- (Faegri & Iversen, 1950) é usado para aberturas ou qualquer outra característica localizada na região equatorial do grão.

Stephanocolpate, stephanocolporate (adj.)
Synonyms of **zonocolpate** and **zonocolporate** describing a pollen grain in which the colpus, or the unit colpus and pores, is distributed around its equatorial region.
Estefanocolpado, estefanocolporado (adj.)
Sinônimo de **zonocolpado** e **zonocolporado**. Diz-se do grão cujo colpo, ou cólporo, está distribuído em volta da sua região equatorial.

Stria (pl. striae; adj. striate)
Term coined by Faegri & Iversen (1950) for the grooves between elongated sculpturing elements.
Estria (adj. estriado)
Termo proposto pro Faegri & Iversen (1950) para os sulcos entre elementos esculturais alongados.

Striato-reticulate (adj.)
A sculptural pattern that is intermediary between the striate and the reticulate conditions.
Estriado-reticulado (adj.)
Um padrão ornamental intermediário entre os padrões estriado e reticulado.

Suboblate
A term used exclusively in the description of radially symmetric, isopolar grains where the polar axis (P): equatorial diameter (E) ratio is 0.75–0.80 (Erdtman, 1952).
Suboblato
Termo usado exclusivamente na descrição de grãos com simetria radial, isopolares, que possuem a relação entre o eixo polar (P) e o diâmetro equatorial (E) variando entre 0.75 e 0.80.

Subprolate
A term used exclusively in the description of radially symmetric, isopolar grains where the polar axis (P): equatorial diameter (E) ratio is 1.14–1.33 (Erdtman, 1952).
Subprolato
Termo usado exclusivamente na descrição de grãos com simetria radial, isopolares, que possuem a relação entre o eixo polar (P) e o diâmetro equatorial (E) variando entre 1.14 e 1.33.

Sulcus (pl. sulci; adj. sulcate)
Term used to describe the aperture of pollen grain or spore which is elongate and located at either the proximal or distal poles. The term is now used interchangeably with colpi, because they have similar shapes. According to Punt *et al.* (1994), sulci are essentially latitudinal apertures, whereas colpi are essentially longitudinally distributed.
Sulco (adj. sulcado)
Termo usado para decrever a abertura de um grão de pólen ou esporo, que é alongado e localizado, tanto no pólo distal como no proximal. De acordo com Punt *et al.* (1994) os sulcos são encontrados essencialmente como aberturas latitudinais, enquanto que os colpos são essencialmente longitudinais.

Syn- (prefix)
Prefix denoting fusion or anastomosing of any feature.
Sin- (prefixo)
Prefixo denotando a fusão ou anastomose de qualquer característica.

Syncolpate, syncolporate (adj.)
Pollen grain which has the ends of the colpus or compound colpi fusing which each other at the poles.
Sincolpado, sincolporado (adj.)
Grão de pólen possuindo a extremidades dos colpos ou cólporos fundindo-se ou anastomosando-se nas áreas polares.

Tectum (pl. tecti; adj. tectate)
The outermost layer of exine, forming a roof-like structure supported by columellae. The tectum also covers granules or other infratectal elements (Faegri & Troels-Smith, 1950; Punt *et al.*, 1994).
Teto (adj. tectado)
A camada mais externa da exina que forma uma estrutura similar a um teto apoiado por columelas. O teto também cobre grânulos e outros elementos infratectais (Faegri & Troels-Smith, 1950; Punt *et al.*, 1994).

Tetrad
Group of four pollen grains forming a dispersal unit. A tetrad stage is formed during development of the grains by meiotic divisions.
Tétrade
Grupo de quatro grãos de pólen formando uma unidade de dispersão. O estágio é formado durante o desenvolvimento dos grãos pela Meióse.

Tetragonal tetrad
Uniplanar tetrad with the four members in contact at its center.
Tétrade tetragonal
Tétrade onde os quatro membros estão situados no mesmo plano e em contato com os outros membros no seu centro.

Tetrahedral tetrad
A type of multiplanar tetrad, where three of the members are located in the same plane and the fourth member in a different plane. Each unit is in contact with three other members.
Tétrade tetraédrica
Tipo de tétrade multiplanar, onde três membros estão situados num mesmo plano e o quarto elemento em plano diferente. Cada unidade está em contato com três outros membros da tétrade.

Triaperturate (adj.)
Describing the grain with three apertures. Examples: tricolpate, tricolporate, triporate, etc.
Triaperturado (adj.)
Diz-se do grão com três aberturas. Exemplos: tricolpado, tricolporado, triporado.

Tricolpate, tricolporate (adj.)
See **triaperturate.**
Tricolpado, tricolpado (adj.)
Veja **triaperturado.**

Trilete
Spore containing three laesurae, forming a trilete mark. See **laesura.**
Trilete
Esporo com três lesões, que formam a marca trilete. Veja **lesão.**

Uniplanar tetrad
Tetrad with the four members lying in a single plane.
Tétrade uniplana
Tétrade onde os membros estão distribuídos num mesmo plano.

Verruca (pl. verrucae; adj. verrucate)
A sexine sculptural element resembling a wart, usually larger than 1 μm in diameter, base width frequently larger than height.
Verruga (adj. verrugoso)
Um elemento escultural da sexina em forma de uma verruga, geralmente maior que 1 μm em diâmetro, e com a largura da base freqüentemente maior que a sua altura.

Vesicula (pl. vesiculae; adj. vesiculate)
Synonym of saccus.
Vesícula (adj. vesiculado)
Sinônimo de saco.

Vestibulum (pl. vestibuli)
A small chamber under a protuberant pore, bordered by the exine on one side and the sexine on the other.
Vestíbulo
Uma pequena câmara abaixo de um poro aspídoto, limitado de um lado pela exina e do outro pela sexina.

Y-split mark
Synonym of trilete.
Marca em forma de "Y"
Sinônimo de trilete.

Zono- (prefix)
See **stephano-.**
Zono- (prefix)
Veja **estéfano-.**

Bibliography

Absy, M.L. 1975. Polem e esporos do Quaternário de Santos (Brasil). *Hoehnea*, **5**: 1–26.

Absy, M.L.; Cleef, A.; Fornier, M.; Servant, M.; Siffedine, A.; Da Silva, M.F.; Soubies, F.; Suguio, K.; Turcq, B.; and van der Hammen, T. 1991. Mise en evidence de quatre phases d'ouverture dela foret dense dans le sud-est de l'Amazonie au cours des 60 000 dernieres annees. Premiere comparaison avec d'autres regions tropicales. *C. R. Acad. Sci. Paris*, **313**: 673–678.

Barberi-Ribeiro, M. 1994. *Paleovegetacão e Paleoclima no Quaternário tardio da vereda de aguas emendadas* – DF. Dissertacão de Mestrado No. 93. Universidade de Brasília, Instituto de Geociências.

Barth, O.M., and Melhem, T.S. 1988. *Glossário Ilustrado de Palinologia*. Editora da Universidade de Campinas. Campinas, S.P.

Berglund, B.E., ed. 1986. *Handbook of Holocene Paleoecology and Paleohydrology*. Wiley, New York.

Birks, H.J.B., and Birks, H.H. 1980. *Quaternary Paleoecology*. University Park Press, Baltimore.

Brown, C.A. 1960. *Palynological Techniques*. Baton Rouge, LA.

Buchholz, J.T., and Gray, N.E. 1948. A taxonomic revision of Podocarpus. *J. Arnold Arboretum*. Cambridge.

Bush, M.B. 1991. Modern pollen rain data from South and Central America: a test of the feasibility of fine-resolution lowland tropical palynology. *The Holocene*, **1**: 162–167.

Bush, M. 1992. A simple yet efficient pollen trap for use in vegetation studies. *J. Vegetation Sci.*, **3**: 275–276.

Bush, M., and Colinvaux, P.A. 1988. A 7000-year pollen record from the Amazon lowlands, Ecuador. *Vegetatio*, **76**: 141–154.

Bush, M.B., and Colinvaux, P.A. 1990. A pollen record of a complete glacial cycle from lowland Panama. *J. Vegetation Sci.*, **1**: 105–118.

Bush, M.B., and Colinvaux, P.A. 1994. Tropical forest disturbance: records from Darien, Panama. *Ecology*, **75**: 1761–1768.

Bush, M.B.; Piperno, D.R.; and Colinvaux, P.A. 1989. A 6000 year history of Amazonian maize cultivation. *Nature*, **340**: 302–303.

Bush, M.B.; Colinvaux, P.A.; Weimann, M.C.; Piperno, D.R.; and Liu, K.-B. 1990. Late Pleistocene temperature depression and vegetation change in Ecuadorian Amazonia. *Quaternary Res.*, **34**: 330–345.

Bush, M.B; Piperno, D.R; Colinvaux, P.A; De Oliveira, P.E; Krissek, L.A; Miller, M.C.; and Rowe, W.E. 1992. A 14 300-yr paleoecological profile of a lowland tropical lake in Panama. *Ecological Monographs*, **62**: 251–275.

Chaloner, W.G. 1970. The rise of the first land plants. *Biol. Ver.*, **45**: 353–377.

Chavéz, R.P; Ludlow-Wiechers, B.; and Villanueva, R. 1991. *Flora palinológica de la reserva de la biosfera de Sian Ka'an, Quintana Roo, Mexico*. Centro de Investigaciones de Quintana Roo (CIQRO). Mexico.

Clapperton, C.M. 1993. *Quaternary Geology and Geomorphology of South America*. Elsevier, Amsterdam.

CLIMAP. 1976. The surface of the ice-age earth. *Science*, **191**: 1131–1144.

Colinvaux, P.A. 1993a. *Ecology* 2. Wiley, New York.

Colinvaux, P.A. 1993b. Pleistocene biogeography and diversity in tropical forests of South America. In: P. Goldblatt, ed., *Biological Relationships Between Africa and South America*. Yale University Press, New Haven, pp. 473–499.

Colinvaux, P.A. 1998. A new vicariance model for Amazonian endemics. *Global Ecology and Biogeography Letters*, **7**: 95–96.

Colinvaux, P.A., and Schofield, E.K. 1976. Historical ecology in the Galapagos Islands, II: A Holocene spore record from El Junco Lake, Isla San Cristobal. *J. Ecol.*, **64**: 1013–1026.

Colinvaux, P.A.; Olson, K.; and Liu, K.-B. 1988. Late glacial and Holocene pollen diagrams from two endorheic lakes of the Inter-Andean plateau of Ecuador. *Rev. Paleobot. Palynol.*, **55**: 83–100.

Colinvaux, P.A.; De Oliveira, P.E.; Moreno, J.E.; Miller, M.C.; and Bush, M.B. 1996a. A long pollen record from lowland Amazonia: forest and cooling in glacial times. *Science*, **247**: 85–88.

Colinvaux, P.A.; Liu, K.-B.; De Oliveira, P.E.; Bush, M.B.; Miller, M.C.; and Steinitz-Kannan, M. 1996b. Temperature depression in the lowland tropics in glacial times. *Climatic Change*, **32**: 19–33.

Colinvaux, P.A.; Bush, M.; Steinitz-Kannan, M.; and Miller, M.C. 1997. Glacial and post-glacial pollen records from the Ecuadorian Andes and Amazon. *Quaternary Res.*, **48**: 69–78.

Colinvaux, P.A.; De Oliveira, P.E.; and Bush, M.B. In Press. Amazonian and neotropical plant communities on glacial time-scales. *Quatern. Sci. Rev.*

Cronquist, A. 1988. *The Evolution and Classification of Flowering Plants*. The New York Botanical Garden. Bronx, New York.

Cwynar, L.C.; Burden, E.; and McAndrews, J.H. 1979. An inexpensive sieving method for concentrating pollen and spores from fine-grained sediments. *Can. J. Earth Sci.*, **16**: 1115–1120.

Davis, M.B. 1986. Vegetation climate equilibrium. *Vegetatio*, **67**: 1–141.

Davis, M.B.; Moeller, R.E.; and Ford, J. 1984. Sediment focusing and pollen influx. In: E.Y. Haworth and J.W.G. Lund, eds., *Lake Sediments and Environmental History*. University of Leicester Press, United Kingdom, pp. 261–293.

De Oliveira, P.E. 1992. *A Palynological Record of Late Quaternary Vegetation and Climatic Change in Southeastern Brazil*. Ph.D. Dissertation. Ohio State University, Columbus..

De Oliveira, P.E.; Barreto, A.M.F.; and Suguio, K. 1999. In Press. Late Pleistocene/Holocene climatic and vegetational history of the Brazilian caatinga: the fossil dunes of the middle São Francisco River. *Palaeogeog. Palaeoclimatol. Palaeoecol.*

Deevey, E.S; Brenner, M.; and Binford, M.W. 1983. Paleolimnology of the Peten lake district, Guatemala, III: Late Pleistocene and Gamblian environments of the Maya area. *Hydrobiologia*, **103**: 211–216.

Erdtman, G. 1943. An introduction to pollen analysis. New. Er. Pl. Sci. Books, F. *Verdoorn*, Vol. 2. Waltham, MA.

Erdtman, G. 1945. Pollen morphology and plant taxonomy, V: on the occurrence of tetrads and dyads. *Svensk Botanisk Tidskrift*, **39**: 286–297.

Erdtman, G. 1947. Suggestions for the classification of fossil and recent pollen grains and spores. *Svensk Botanisk Tidskrift*, **41**: 104–114.

Erdtman, G. 1952. *Pollen Morphology and Plant Taxonomy. An Introduction to Palynology*, Vol. 1. Almqvist & Wiksell, Stockholm.

Erdtman, G. 1969. *Handbook of Palynology*. Munksgaard, Copenhagen.

Erdtman, G., and Straka, H. 1961. Cormophyte spore classification. *Geologiska Föreningens Förenhandl.*, **83**(H1): 65–78.

Faegri, K., and Iversen, J. 1950. *Textbook of Modern Pollen Analysis*. Munksgaard, Copenhagen.

Faegri, K., and Iversen, J. 1989. *Textbook of Pollen Analyses*. 4th ed. Wiley, New York.

Faegri, K., and van der Pilj, L. 1979. *The Principles of Pollination Ecology*. 3rd ed. Munksgaard, Copenhagen.

Ferraz-Vicentini, K.R., and Salgado-Labouriau, M.L. 1996. Palynological analysis of a palm swamp in central Brazil. *J. S. Am. Earth Sci.*, **9**: 207–219.

Fisher, M.M.; Brener, M.; and Reddy, K.R. 1992. A simple, inexpensive piston corer for collecting undisturbed sediment/water interfase profiles. *J. Paleolimnol.*, **7**: 157–161.

Forster, M., and Flenley, J.R. 1993. Pollen purification and fractionation by equilibrium density gradient centrifugation. *Palynology*, **17**: 137–155.

Ganopolski, A.; Rahmstorf, S.; Petoukhov, V.; and Claussen, M. 1998. Simulation of modern and glacial climates with a coupled global model of intermediate complexity. *Nature*, **391**: 351–356.

Gentry, A.H. 1988. Tree species richness of upper Amazonian forests. *Proc. Natl. Acad. Sci. U.S.A.*, **85**: 156–159.

Gentry, A.H. 1993. *A Field Guide to the Families and Genera of Woody Plants of Northwest South America (Colombia, Ecuador, Peru) with Supplementary Notes on Herbaceous Taxa*. Conservation International. Washington, DC.

Guilderson, T.P; Fairbanks, R.G; and Rubenstone, J.L. 1994. Tropical temperature variations since 20,000 years ago: modulating interhemispheric climate change. *Science*, **263**: 663–665.

Haberle, S.G. 1997. Upper Quaternary vegetation and climate history of the Amazon basin: correlating marine and terrestrial pollen records. In: R.D. Flood, D.J.W. Piper, A. Klaus, and L.C. Peterson, eds., *Proceedings of the Ocean Drilling Program, Scientific Results*, **155**: 381–396.

Haberle, S.G., and Maslin, M. 1998. Late Quaternary vegetation and climate change in the Amazon Basin based on a 50,000-year pollen record from the Amazon Fan, ODP site 932. *Quat. Res.*

Haffer, J. 1969. Speciation in Amazonian forest birds. *Science*, **165**: 131–137.

Hansen, B.C.S.; Wright, H.E.; and Bradbury, J.P. 1984. Pollen studies in the Junin area, central Peruvian Andes. *Geolog. Soc. Am. Bull.*, **95**: 1454–1465.

Harris, W.F. 1955. A manual of the spores of New Zealand. Pteridophyta. A discussion of spore morphologyand dispersal with reference to the identification of the spores in surface samples and as microfossils. *New Zealand, Deptm. Sci. Ind. Res. Bull.*, **116**: 1–186.

Heine, K. 1994. The Mera site revisited: ice-age Amazon in the light of new evidence. *Quaternary Int.*, **21**: 113–119.

Herrera, L.F., and Urrego, L.E. 1996. *Studies on the Colombian Amazonia. Pollen atlas of useful and cultivated plants in the Colombian Amazon region*. Tropenbos Colombia. XI.

Heusser, C.J. 1971. *Pollen and Spores of Chile: Modern Pollen Types of the Pteridophyta, Gymnospermae and Angiospermae*. University Arizona Press, Tucson.

Hoffmeister, W.S. 1959. Lower Silurian plant spores from Libya. *Micropaleontology*, **5**: 331–334.

Hooghiemstra, H. 1984. Vegetational and climatic history of the high plain of Bogota, Colombia: a continuous record of the last 3.5 million years. *Dissertationes Botanicae*, Vol. 79. J. Cramer Verlag.

Hooghiemstra, H. 1989. Quaternary and Upper Pliocene glaciations and forest development in the tropical Andes: evidence from a long high-resolution pollen record from the sedimentary basin of Bogota. *Palaeogeog. Palaeoclimatol. Palaeoecol.*, **72**: 11–26.

Hueck, K. 1966. *Die Wälder Südamerikas Ökologie, Zusammensetzung und wirtschaftliche Bedeutung*. Stuttgart, G. Fischer.

Huntley, B., and Birks, H.J.B. 1983. *An Atlas of Past and Present Pollen Maps for Europe 0–13,000 Years Ago*. Cambridge University Press, Cambridge.

Huntley, B., and Webb III, T., eds. 1988. *Vegetation History*. Kluwer, Dordrecht.

Imbrie, J.D.; Hays, J.; Martison, D.G.; McIntyre, A.; Mix, A.; Morley, J.J.; Pisias, N.G.; Prell, W.L.; and Shackleton, N.J. 1984. The orbital theory of Pleistocene climate: support from revised chronology of the marine ^{18}O record. In: A.L. Berger, J. Imbrie, J. Hays, G. Kukla, and B. Saltzman, eds., *Milankovitch and Climate*. Reidel, Dordrecht, pp. 269–305.

Iversen, J., and Troels-Smith, J. 1950. Polennmorphologische Definitionen und Typen. Danm. *Geol. Unders.*, Ser. 4.3: 1–54.

Jackson, B.D. 1928. *A Glossary of Botanic Terms with Their Derivation and Accent*. Duckworth and Co., London.

Jolly, D., and Haxeltine, A. 1997. Effect of low glacial atmospheric CO_2 on tropical African montane vegetation. *Science*, **276**: 786–788.

Kremp, G.O.W. 1968. *Morphologic Encyclopedia of Palynology*. 2nd ed. University of Arizona Press, Tucson.

Lawesson, J.E., et al., eds. 1990. *Botanical Research and Management in Galapagos*. Proceedings of the Workshop on Botanical Research an Management in Galapagos held 11–18 April 1987 at the Charles Darwin Research Station, Santa Cruz, Galapagos, Ecuador.

Ledru, M. 1993. Late Quaternary environmental and climatic changes in central Brazil. *Quaternary Res.*, **39**: 90–98.

Liu, K.-B., and Colinvaux, P.A. 1985. Forest changes in the Amazon basin during the last glacial maximum. *Nature*, **318**: 556–557.

Liu, K.-B., and Colinvaux, P.A. 1988. A 5200-year history of Amazon rain forest. *J. Biogeog.*, **15**: 231–248.

Livingstone, D.A. 1955. A lightweight piston sampler for lake sediments. *Ecology*, **36**: 137–139.

Livingstone, D.A. 1962. Age of deglaciation in the Ruwenzori Range, Uganda. *Nature*, **194**: 859–860.

Lorenzi, H.; de Souza, H.M.; de Cerqueira, L.S.C.; Medeiros-Costa, J.T.; and von Behr, N. 1996. *Palmeiras no Brasil, nativas e exoticas*. Edit. Plantarum. Nova Odessa, São Paulo.

Maas, P.J.M., and Westra, L.Y.Th. 1993. *Neotropical Plant Families. A Concise Guide to Families of Vascular Plants in the Neotropics*. Koeltz Scientific Books, Berlin, Germany.

Mabberley, D.J. 1997. *The Plant Book*. 2nd ed. Cambridge University Press, Cambridge.

Markgraf, V., and D'Antoni, L. 1978. *Pollen Flora of Argentina: Modern Spore and Pollen Types of Pteridophyta, Gymnospermae and Angiospermae*. University of Arizona Press, Tucson.

Moore, P.D.; Webb, J.A.; Collinson, M.E. 1991. *Pollen Analysis*. 2nd ed. Blackwell Scientific Publications. Oxford.

Mosimann, J.E. 1965. Statistical methods for the pollen analyst: multinomial and negative multinomial techniques. In: B. Kummel and D.M. Raup, eds., *Handbook of Paleontological Techniques*. Freeman, San Francisco, pp. 636–673.

Murillo, M.T., and Bless, M.J.M. 1974. Spores of recent colombian Pteridophyta, I: Trilete spores. *Rev. Paleobot. Palynol.*, **18**: 223–269

Murillo, M.T., and Bless, M.J.M. 1978. Spores of recent Colombian Pteridophyta, II: monolete spores. *Rev. Paleobot. Palynol.*, **25**: 319–365

Potonié, R. 1934a. Zur Morphologie der fossilen Pollen und Sporen. *Arbeiten Inst. Paläobotanik Petrographie Brennsteine*, **4**: 5–24.

Potonié, R. 1934b. Zur Mikrobotanik der Kohlen und ihrer Verwandten. 2. Zur ikrobotanik des Eocänen Humodils des Geisetals. *Arbeiten Inst. Paläeobot. Petrographie rennsteine*, **4**: 25–125. Preuss. Geol. L.A., Berlin.

Praglowski, J., and Punt, W. 1973. An elucidation of the micro-reticulate structure of the exine. *Grana*, **13**: 45–50.

Prance, G.T., ed. 1982. *Biological Diversification in the Tropics*. Columbia University Press, New York.

Prentice, I.C.; Bartlein, P.J.; and Webb, T. 1991. Vegetation and climate change in eastern North America since the last glacial maximum. *Ecology*, **72**: 2038–2056.

Punt, W.; De Leeuw, J.S.; and Van Oostrum, W.A.P. 1974. *The Northwest European Pollen Flora*, Vol. 6. Elsevier, Amsterdam.

Punt, W.; Blackmore, S.; Nilsson S.; and Le Thomas, A. 1994. *Glossary of Pollen and Spore Terminology*. LPP Foundation. Utrecht. Series No. 1.

Radambrasil. 1976. *Levantamento de Recursos Naturais*. 11. Folhas Na. 19. Pico da Neblina. DNPM, Rio de Janeiro.

Reitsma, T. 1970. Suuggestions towards unification of descriptive terminology of angiosperm pollen grains. *Rev. Paleobot. Palynol.*, **10**: 39–60.

Richardson, J.B. 1969. Devonian Spores. In: R.H. Tschudy and R.A. Scott, eds., *Aspects of Palynology*, Chapter 11, pp. 193–222.

Roubik, D.W., and Moreno, J.E. 1991. Pollen and spores of Barro colorado Island. *Missouri Botanical Garden, Monographs in Systematic Botany*, 36.

Salgado-Labouriau, M.L. 1973. *Contribuição à Palinologia dos Cerrados*. Acad. Bras. Ciências, Rio de Janeiro.

Salgado-Labouriau, M.L., and Rinaldi, M. 1990. Measurements of Gramineae pollen of the Venezuelan mountains. *Rev. Brasil. Biol.* **50**: 115–122. [Also, *Grana*, **29**: 119–128.]

Salgado-Labouriau, M.L., and Rull, V. 1986. A method of introducing exotic pollen for palaeoecological analysis of sediments. *Rev. Paleobot. Palynol.*, **47**(1/2): 97–103.

Street-Perrott, F.A.; Huang, Y.; Perrot, R.A.; Eglinton, G.; Barker, P.; Khelifa, L.B.; Harkness, D.D.; and Olago, D.O. 1997. Impact of lower atmospheric carbon dioxide on tropical mountain ecosystems. *Science*, **278**: 1422–1426.

Stockmarr, J. 1971. Tablets with spores used in absolute pollen analysis. *Pollen et Spores*, **13**: 615–621.

Stute, M.; Forster, M.; Frischkorn, H.; Serejo, A.; Clark, J.F.; Schlosser, P.; Broecker, W.S.; and Bonani, G. 1995. Cooling of tropical Brazil (5°C) during the last glacial maximum. *Science*, **269**: 379–383.

Taylor, T.N., and Taylor, E. 1993. *The Biology of Fossil Plants*. Prentice-Hall, Englewood Cliffs, NJ.

Thompson, L.G.; Mosley-Thompson, E.; Davis, M.E.; Lin, P.-N., Henderson, K.A.; Cole-Dai, J.; Bolzan, J.F.; and Liu, K.-B. 1995. Late Glacial stage and Holocene tropical ice core records from Huascarán, Peru. *Science*, **269**: 46–50.

Tranquilini, W. 1979. *Physiological Ecology of the Alpine Timberline*. Springer-Verlag, New York.

Traverse, A. 1988. *Paleopalynology*. Unwin Hyman, Boston.

Vallentyne, J.R. 1955. Sedimentary chlorophyll determination as a paleobotanical method. *Can. J. Bot.*, **35**: 304–313.

Van Campo, M. 1958. Palynologie africaine, 2. *Bull. I.F.A.N. (A)*, **20**(3): 753–759.

Van der Hammen, T., and Absy, M.L. 1994. Amazonia during the last glacial. *Palaeogeog. Palaeoclimatol. Palaeoecol.*, **109**: 247–261.

Webb, R.S.; Rind, D.H.; Lehman, S.J.; Healy, R.J.; and Sigman, D. 1997. Influence of ocean heat transport on the climate of the last glacial maximum. *Nature*, 385: 695–699.

Whitmore, T.C., and Prance, G.T. 1987. *Biogeography and Quaternary history in tropical America.* Oxford University Press, Oxford.

Wodehouse, R.P. 1935. *Pollen Grains: Their Structure, Identification and Significance in Science And medicine.* Hafner Publishing Co., New York.

Wodehouse, R.P. 1959. *Pollen Grains: Their Structure, Identification, and Significance in Science and Medicine.* Hafner Publishing Co., New York.

Wright Jr., H.E. 1967. A square-rod piston sampler for lake sediments. *J. Sed. Petrol.*, **37**: 975–976.

Index of Botanical Families

ANGIOSPERMAE – LILIOPSIDAE (Monocotyledoneae)

INDEX OF BOTANICAL NAMES

About the Authors

Paul Colinvaux is the author of three ecology texts, the latest being *Ecology 2* (Wiley, 1993); a book of essays, *Why Big Fierce Animals Are Rare* (Princeton University Press, 1978); and *The Fates of Nations: A Biological Theory of History* (Simon & Schuster, 1980). These books have been translated into several languages. Colinvaux's research, in arctic regions as well as the tropics, has been to discover how the climate of Ice Ages controlled the distribution and abundance of life of the contemporary Earth. He has led numerous field parties to core lakes for environmental histories in an equatorial "slice" of the New World tropics, from the Galapagos Islands eastward to mainland Ecuador, across the high Andes to Oriente Province, and thence across Amazonian Brazil. Colinvaux's equatorial research was begun when he was a professor at Ohio State University, but was transferred in 1991 to the Smithsonian Tropical Research Institute in the Republic of Panama, where this pollen manual was written. In 1998 the research base was transferred to the Marine Biological Laboratory in Woods Hole, Massachusetts.

Paulo Eduardo De Oliveira is a research associate at the Field Museum of Natural History in Chicago and a visiting professor at the University of São Paulo. As a paleoecologist, he uses both diatoms and pollen from lake sediments in his reconstruction of the Ice-Age climates and vegetation of Brazil. He and Colinvaux have worked together since the late 1980s, first in Ecuador, then in Brazil. De Oliveira earned a doctorate from Ohio State University in 1992 for a pioneering thesis on the climatic history of southeastern Brazil.

Jorge Enrique Moreno Patiño is a palynologist at the Smithsonian Tropical Research Institute in Panama. He is co-author (with D. Roubick) of a pollen atlas for Barro Colorado Island, the Panamanian reserve for research in tropical forests.

Anyone wishing to contact Paul Colinvaux may do so at: Marine Biological Laboratory, 7 MBL Street, Woods Hole, MA 02543, USA; *e-mail*: pcolinva@mbl.edu.

Paulo Eduardo De Oliveira may be reached at: Field Museum of Natural History, Roosevelt Road at Lake Shore Drive, Chicago, IL 60605, USA; or Universidade de São Paulo, Caixa Postal 11340, São Paulo, SP 05422, Brasil; *e-mail*: pedeoliv@usp.br.

Sobre os Autores

Paul Colinvaux é o autor de três livros textos de Ecologia, sendo *Ecology 2* (Wiley, 1993) o mais recente; e *Why Big Fierce Animals Are Rare* (Princeton University Press, 1978); e *The Fates of Nations: A Biological Theory of History* (Simon & Schuster, 1980). Estes livros têm sido traduzidos em muitas línguas. A pesquisa de Colinvaux nas regiões árticas e nas tropicais tem por objetivo descobrir como o clima da Idade do Gelo controlou a distribuição e a abundância da vida na Terra. Colinvaux organizou várias expedições com o objetivo de coletar sedimentos lacustres para estudos paleoambientais na região equatorial dos trópicos do Novo Mundo, desde o arquipélago de Galápagos na costa do Equador, seguindo para leste através da região das altas elevações andinas e da província de Oriente no Equador, cruzando a Amazônia do Brasil. A pesquisa equatorial de Colinvaux teve início quando era ainda professor na Ohio State University, da qual transferiu-se para o Smithsonian Tropical Research Institute em 1991, na República do Panamá, onde este livro foi escrito. Em 1998, a base de suas pesquisas foi transferida para o Marine Biological Research Laboratory, em Woods Hole-Massachusetts.

Paulo Eduardo De Oliveira é pesquisador associado do Field Museum of Natural History, em Chicago e professor visitante da Universidade de São Paulo-USP. Como paleoecólogo usa o pólen e as algas diatomáceas encontradas em sedimentos lacustres, com o objetivo de reconstruir a vegetação e o clima da Idade do Gelo, no Brasil. Tem trabalhado com Colinvaux desde o final dos anos 80, primeiramente no Equador e depois no Brasil. De Oliveira recebeu seu título de doutor na Ohio State University, em 1992, por uma tese pioneira sobre a história climática do sudeste do Brasil.

Jorge Enrique Moreno Patiño é palinólogo do Smithsonian Tropical Research Institute, no Panamá. É co-autor, com D. Roubik, de um atlas palinológico da Ilha de Barro Colorado, reserva panamenha para pesquisas sobre a floresta tropical.